Lydia Außenhofer, Dikran Dikranjan, and Anna Giordano Bruno
Topological Groups and the Pontryagin-van Kampen Duality

De Gruyter Studies in Mathematics

Volume 83

Lydia Außenhofer, Dikran Dikranjan, and
Anna Giordano Bruno

Topological Groups and the Pontryagin-van Kampen Duality

An Introduction

DE GRUYTER

Mathematics Subject Classification 2020
22A05, 22B05, 22C05, 22D05, 22D10, 22D35, 20K45, 28C10, 43A05, 43A40, 46A20, 54B30, 18F60, 54C45, 54H11, 54H13

Authors

Prof. Lydia Außenhofer
University of Passau
Faculty of Computer Science and Mathematics
Innstraße 33
Passau 94032
Germany
lydia.aussenhofer@uni-passau.de

Prof. Dr. Dikran Dikranjan
Udine University
Department of Mathematics, Computer Science and Physics
Via della Scienze 206
33100 Udine
Italy
dikran.dikranjan@uniud.it

Prof. Anna Giordano Bruno
Udine University
Department of Mathematics, Computer Science and Physics
Via della Scienze 206
33100 Udine
Italy
anna.giordanobruno@uniud.it

ISBN 978-3-11-065349-6
e-ISBN (PDF) 978-3-11-065493-6
e-ISBN (EPUB) 978-3-11-065355-7
ISSN 0179-0986

Library of Congress Control Number: 2021943683

Bibliographic information published by the Deutsche Nationalbibliothek
The Deutsche Nationalbibliothek lists this publication in the Deutsche Nationalbibliografie; detailed bibliographic data are available on the Internet at http://dnb.dnb.de.

Typesetting: VTeX UAB, Lithuania
Printing and binding: CPI books GmbH, Leck

www.degruyter.com

Preface

The idea to write this book was born long ago, when the joint monograph [99] of Prodanov, Stoyanov and the second named author was out of print, yet it cannot be considered as a second edition of [99] as originally planned. Indeed, the overlap with [99] is contained only in Chapters 11–13 and part of Chapter 14, whereas the intersection of the rest of [99] (namely, minimal groups and their properties) with this book is limited only to the final page of §8.4 and a few exercises in §8.8. As a matter of fact, [99] was largely used as a base for various courses on topological groups held by the second named author: “Topologia 2” at Udine University (started in 1998/99), as well as four courses in the framework of the PhD programs at the Department of Mathematics at Naples University and Milan University, the Department of Geometry and Topology at the Complutense University of Madrid, and the Department of Mathematics at Nanjing Normal University in 1994, 2006, 2007, and 2016, respectively. As a result, the lecture notes of these courses gradually arose, including necessarily an introductory part (Chapters 2–10 in this book) that was not present in [99], and grew to the necessary critical level to become an independent source available online over the years. In 2019 came the generous offer to publish these notes in the series *Studies in Mathematics* of de Gruyter Editors that we gladly accepted.

Our sincere thanks go in the first place to de Gruyter who made this publication possible and to Dr. Nadja Schedensack and Dr. Apostolos Damialis for their valuable advice and assistance. It is a pleasure to thank Luchezar Stoyanov for his kind concession to use some results from [99] in this book. We thank also our colleagues F. De Giovanni, E. Martín Peinador, M. J. Chasco, M. G. Bianchi, Wei He, X. Domíngues, M. Bruguera, S. Ardanza Trevijano, and E. Pacifici, who made the above mentioned courses possible. We thank also G. Bergman, P. Spiga, and G. Lukács for their useful advice.

The first and third named authors wish to thank their respective husbands Horst Förster and Alberto Urli for their versatile support, patience, and permanent assistance during the genesis of this book.

Lydia Außenhofer
Dikran Dikranjan
Anna Giordano Bruno
Udine, Passau, July 4, 2021

https://doi.org/10.1515/9783110654936-201

This monograph is dedicated to the memory of
William Wistar Comfort, Adalberto Orsatti, and Ivan Prodanov
whose original contributions to Topological Groups and
the Pontryagin-van Kampen Duality and its applications
opened new horizons in this fascinating field.

Contents

1 Introduction

Topological groups provide a natural environment where algebra (more specifically, group theory) and topology interact in a very fruitful way. Topological groups arise in several ways: as transformation groups (in a typical nonabelian context), or as function spaces (in the framework of functional analysis), or as objects related to p-adic analysis (in number theory), etc. Given the many faces of the main object of study (a pair of a group structure and a topology on the same set, with a simple compatibility condition), there are two opposite tendencies in this field. One is mainly focused on the topological properties of the underlying topological space, by using the favorable circumstance to have homogeneity and other merely superficial ingredients due to the algebraic structure; this can be roughly called the *topological theory of topological groups*. The alternative approach, that is usually referred to as *topological algebra*, studies mainly algebraic properties of groups equipped with certain (group) topologies that frequently have a strong impact on the algebraic structure of the group.

This book is somewhat closer to the second tendency: although the topological aspect has a relevant place in the exposition (so, for example, it is more topology oriented than [169, 198]), it has a stronger focus on topological abelian groups (e. g., more than [270]). This explains our choice to enhance the role of the functorial topologies on abelian groups, and of the Pontryagin-van Kampen duality and its applications. Of those, the latter notably prevails and can be pointed out as the main topic of the book. We provide a completely self-contained exposition of this remarkable duality, following the line first adopted by Iv. Prodanov in [235, 237] and in [99] (see also the recent paper [109] and the much earlier monograph [214]).

Let $\mathcal{L}$ denote the category of locally compact abelian groups and continuous homomorphisms, and let $\mathbb{T} = \mathbb{R}/\mathbb{Z}$ be the unit circle group written additively. For $G \in \mathcal{L}$, denote by $\widehat{G}$ the group of continuous homomorphisms $G \to \mathbb{T}$ (shortly, continuous *characters* of G) equipped with the compact-open topology. The assignment $G \mapsto \widehat{G}$ induces a contravariant endofunctor $\widehat{\ }: \mathcal{L} \to \mathcal{L}$, and the celebrated Pontryagin-van Kampen duality theorem (see [228] and Theorem 13.4.17) says that this functor is an *involutive duality*: if G is a locally compact abelian group, then $\widehat{\widehat{G}}$ is topologically isomorphic to G in a canonical way. Moreover, the Pontryagin-van Kampen duality functor $\widehat{\ }: \mathcal{L} \to \mathcal{L}$ sends compact abelian groups to discrete abelian groups and vice versa, that is, it defines a *duality* between the subcategory $\mathcal{C}$ of $\mathcal{L}$ of compact abelian groups and the subcategory $\mathcal{D}$ of $\mathcal{L}$ of discrete abelian groups. This is a very efficient and fruitful tool for the study of compact abelian groups, reducing many problems related to topological properties in $\mathcal{C}$ to the corresponding problems concerning algebraic properties in $\mathcal{D}$.

The reader is advised to give a look at Mackey's beautiful survey [205] for the connection of characters and the Pontryagin-van Kampen duality with number theory, physics, and elsewhere. This duality inspired a huge amount of related research also

https://doi.org/10.1515/9783110654936-001

in category theory; a brief comment on a specific categorical aspect (uniqueness and representability) can be found in §16.2.1.

As mentioned above, we provide a self-contained proof of the Pontryagin-van Kampen duality theorem, with all necessary steps, including the basic background on topological groups and the structure theory of locally compact abelian groups. The abelian case of the Peter–Weyl theorem, asserting that the continuous characters of compact abelian groups separate the points (see Corollary 11.5.1), is certainly the most important nonelementary tool in proving the Pontryagin-van Kampen duality theorem. The Peter–Weyl theorem is valid for arbitrary compact groups, but then continuous characters must be replaced by finite-dimensional irreducible unitary representations, and the usual proof of the theorem in this general case involves Haar integration. Since in the abelian case the finite-dimensional irreducible unitary representations turn out to be one-dimensional, that is, continuous characters, we prefer here a different approach. Namely, in the abelian case the Peter–Weyl theorem can be obtained as an immediate corollary of the Følner theorem (see Theorem 11.3.5), whose (relatively) elementary proof uses nothing beyond elementary properties of finite abelian groups, a local version of the Stone–Weierstraß theorem (see Corollary B.5.22), and the Čech–Stone compactification of discrete spaces. As another application of the Følner theorem, one can characterize precompact abelian groups (i. e., the subgroups of compact abelian groups) as those having a topology generated by continuous characters. As a third application of the Følner theorem, we obtain the existence of the Haar integral on locally compact abelian groups for free (see Theorem 12.2.9, whose proof follows [99, § 2.4, Theorem 2.4.5]).

There are several topics we regretfully could not include in the book for the lack of space, the most important one being free topological groups (the reader can find an excellent presentation in [7]).

Another one is the existence of convergent sequences in topological groups; the interest in this phenomenon is triggered by the fact that compact groups have lots of nontrivial convergent sequences, yet many classes of compact-like nonmetrizable groups fail to have them (see the survey [102] for more details). Only very recently Hrušák, van Mill, Ramos-García, and Shelah [179] resolved a long standing problem by producing the first example in ZFC of a countably compact group without nontrivial convergent sequences.

Another topic, namely, topological generators of topological groups, was only superficially touched by monothetic groups and compactly generated groups in Theorem 8.3.5, Theorem 8.3.7, and Corollary 8.3.5. We refer the reader to the last 2021 edition of the monograph [177] for a recent progress in the field of suitable sets.

For dualities of non-locally compact abelian groups and the Mackey topology problem (see §13.8) the reader may consult the survey [14].

Although we pay some attention to the categorical aspects of topological groups (via pointing out a wealth of functorial topologies and functorial dualities, reflections and coreflections in the category **TopGrp** of topological groups, etc.), there is at least one important topic that remained uncovered by the book, namely, categorically compact groups introduced in [115]. The early stages of the development of this topic are nicely covered in Lukács' monograph [200]. The reader may be interested to find a remarkable Tichonov theorem obtained by Clementino and Tholen for products of categorically compact objects in a quite general context in [52]. Its value, as far as the category **TopGrp** is concerned, was recently increased by the impressive examples of noncompact categorically compact groups obtained by Klyachko, Ol′shankii, and Osin in [185] (these examples are strongly related also to nontopologizable groups, see §4.3).

The book is organized as follows. Chapter 2 contains background on topological groups, starting from scratch. Various ways of introducing a group topology are considered (see §2.2), of which the most prominent one is by means of functorial topologies, and in particular by means of characters (see §2.2.2).

Chapter 3 contains general properties of topological groups. In particular, in §3.1 we discuss subgroups and separation axioms. In §3.1.3 we briefly introduce the principle of the extension of identities in Hausdorff groups, which, roughly speaking, says that a Hausdorff group inherits nice algebraic properties (commutativity, nilpotency, etc.) from its dense subgroups. Quotients are dealt with in §3.2. Both §§3.3 and 3.4 have a more categorical flavor; more specifically, §3.3 is focused on initial and final topologies, while §3.4 explains how the general case can be reduced to that of Hausdorff groups.

In Chapter 4 Markov's problems on the existence of nondiscrete Hausdorff group topologies are discussed. In §4.1 we introduce the Markov and Zariski topologies that allow for an easier understanding of Markov's problems. These two topologies coincide on abelian and permutation groups; for the latter case, see §4.2. The first two examples of nontopologizable groups, given by Shelah and Ol′shanskii, respectively, are discussed in §4.3. The problems arising in the extension of group topologies are the topic of §4.4.

Chapter 5 is dedicated to cardinal invariants and metrizability of topological groups. In §5.1 we discuss several cardinal invariants (weight, character, pseudocharacter, density character) and their interrelations. The next section provides a proof of the Birkhoff–Kakutani metrization theorem and some corollaries. In §5.3 we recall the construction of Protasov and Zelenyuk from [241] of group topologies making a given sequence in an abelian group convergent to 0.

Connectedness and related properties in topological groups are discussed in Chapter 6. To this end, four components measuring the degree of connectedness are introduced and discussed. Special attention is paid to hereditarily disconnected groups.

Chapter 7 is dedicated to completeness. We use the Cauchy nets to define completeness for topological groups, and then to build the Raĭkov completion of a Hausdorff group. We discuss also Weil completeness and the Weil completion. Finally, we describe the completion equivalently using filters.

Chapter 8 contains specific properties of (locally) compact groups used essentially in this book. Of those, we mention here only the most relevant ones. The open mapping theorem is discussed in §8.4, where we very briefly mention also minimal and totally minimal groups, which need not be locally compact, yet the open mapping theorem holds. Section 8.5 is focused on the impact of (local) compactness on connectedness in topological groups. In §8.6 we give the first (external) construction of the Bohr compactification bG of a topological group G as the compact group that "best approximates" G.

In Chapter 9 we describe the structure of all closed subgroups of $\mathbb{R}^n$ (giving two different proofs), as well as the closure of an arbitrary subgroup of $\mathbb{R}^n$. These groups play an important role in the whole theory of locally compact abelian groups. As an application, we show that the group $\mathbb{T}^{\mathfrak{c}}$ is monothetic.

Chapter 10 starts with §10.1, in which big and small subsets of abstract groups are introduced. Section 10.2 is dedicated to precompact groups. In particular, in §10.2.1 we give an internal description of precompact groups using the notion of big subsets, and we show that the precompact groups are precisely the subgroups of compact groups. The precompact group G^+ that "best approximates" a topological group G is introduced in §10.2.2. This allows us to obtain a second, internal, construction of the Bohr compactification bG of G, since the completion of G^+ coincides with bG. In §10.2.3 we establish the precompactness of the topologies generated by characters. In §10.3 a third characterization of the Bohr compactification is given, using unitary representations.

Chapter 11 deals with the Følner theorem and its applications, following the line of [99]. Sections 11.1 and 11.2 prepare some of the ingredients for the proof of this theorem. An important feature of the proof is the crucial idea, due to Prodanov, to eliminate all discontinuous characters in the uniform approximation of continuous functions via linear combinations of characters obtained by means of the Stone–Weierstraß approximation theorem. This step is ensured in §11.3 by the Prodanov lemma, which has many other relevant applications towards independence of characters and the construction of the Haar integral for locally compact abelian groups. Then, we give various applications of the Følner theorem. The first one is a description of the precompact group topologies on abelian groups in §11.4. The main application of the Følner theorem is an immediate proof of the abelian case of the Peter–Weyl theorem in §11.5. In §11.6 compactly generated locally compact abelian groups are described explicitly and useful information on the dual group of a locally compact abelian group is gathered.

Chapter 12 is dedicated to almost periodic functions and the Haar integral. Here we introduce almost periodic functions and briefly comment their connection to the Bohr compactification. As another application of the Følner theorem, we give a proof of the Bohr–von Neumann theorem describing almost periodic functions on abelian groups as uniform limits of linear combinations of characters. Among other things, in §12.2, as a by-product of Prodanov's approach, we obtain an easy construction of the Haar integral for almost periodic functions on abelian groups, in particular for all continuous functions on a compact abelian group. In §12.2.2 we build a Haar integral on arbitrary locally compact abelian groups, using the construction from §12.2.1 in the compact case.

Chapter 13 is dedicated to the Pontryagin-van Kampen duality. In §§13.1, 13.2, and 13.3 we construct all tools for proving the Pontryagin-van Kampen duality theorem. More specifically, §§13.1 and 13.2 contain various properties of the dual group that allow for an easier computation in many cases. Using further the properties of the dual group, we see in §13.3 that many specific locally compact abelian groups G satisfy $G \cong \widehat{\widehat{G}}$. In §13.4 we stress the fact that the topological isomorphism $G \cong \widehat{\widehat{G}}$ is *natural* by studying in detail the natural transformation ω between the functors $1_{\mathcal{L}}$ and $\widehat{\widehat{\ }}$, induced by the map $\omega_G: G \to \widehat{\widehat{G}}$ connecting the locally compact abelian group G with its bidual $\widehat{\widehat{G}}$. It is shown in several steps that ω_G is a topological isomorphism, considering larger and larger classes of locally compact abelian groups where the Pontryagin-van Kampen duality theorem holds (compact abelian groups, discrete abelian groups, elementary locally compact abelian groups, compactly generated locally compact abelian groups). The last step uses the fact that the Pontryagin-van Kampen duality functor is exact, and this permits us to use all previous steps in the general case.

In Chapter 14 we give various applications of the Pontryagin-van Kampen duality theorem. As an immediate application, we obtain the main structure theorem for locally compact abelian groups, a complete description of the monothetic compact groups, the torsion compact abelian groups, and the connected compact abelian groups with a dense torsion subgroup. In §14.3.1 we focus on a topic of a more topological flavor, namely, the dimension of locally compact groups. Section 14.3.2 is dedicated to a characterization of the algebraic structure of abelian groups admitting a compact group topology (known as Halmos problem), and the following §14.3.3 to the Bohr topology of a discrete abelian group. In §14.4 we consider a precompact version of the construction from §5.3 of group topologies making a fixed sequence converge in an abelian group to 0.

In Chapter 15 we describe some relevant facts concerning countably compact and pseudocompact groups. Among them we mention the criterion for pseudocompactness due to Comfort and Ross (see Theorem 15.2.1).

Chapter 16 is dedicated to topological rings and fields (see §16.1), and to topological modules and vector spaces (see §16.2). While §16.2.1 concerns uniqueness of the Pontryagin-van Kampen duality, §16.2.4 is focused on locally linearly compact vector spaces and to the Lefschetz duality, a duality from [195] inspired by the Pontryagin-van Kampen duality. There we discuss also the more general duality for linearly compact modules due to Kaplansky, MacDonald, and Áhn.

Exercises are given at the end of each chapter to ease the understanding of the arguments treated in that chapter. At the end of most of the chapters, a brief section of notes and comments is provided.

The reader who is interested mainly in abelian groups can skip §§4.2–4.4 and take all groups abelian in Chapters 3 and 5–8. Conversely, the reader interested in the non-abelian context may dedicate more time to Chapters 4–10. Minimal groups appear in §§8.4, 8.7, 8.8, 10.5, 11.8 and minimal rings in §16.4.

For those interested in getting as soon as possible to the proof of the Pontryagin-van Kampen duality theorem, and having sufficient knowledge of topological groups, a possible route can be to read Chapters 10–13, and then Chapter 14 for the applications.

The reader of this book is not supposed to have a solid background either in group theory or in topology. In the Appendix we provide the necessary background in three directions, recalling in Appendix A, B, and C basic results and notions on group theory, general topology, and category theory, respectively, used in the book. For general notation and terminology, see the initial part of Appendix A or the Index and the Index of Symbols. Further general information on topological groups can be found in the monographs or surveys [8, 81, 82, 99, 177, 200, 223, 228].

2 Definition and examples

2.1 Basic definitions and properties

2.1.1 Definition

We start with the fundamental concept of this book.

Definition 2.1.1. A topology τ on a group $(G, \cdot)$ is a *group topology* if the map $G \times G \to G$, $(x, y) \mapsto xy^{-1}$, is continuous. A *topological group* is a pair (G, τ) of a group G and a group topology τ on G.

We simply write that G is a topological group, when there is no need to explicitly write the group topology on G.

Remark 2.1.2. A topology τ on a group G is a group topology if and only if

$$\mu: G \times G \to G,\ (x, y) \mapsto xy, \quad \text{and} \quad \iota: G \to G,\ x \mapsto x^{-1},$$

are continuous when $G \times G$ carries the product topology. Clearly, $\iota: G \to G$ is a homeomorphism. For a subset M of G, we denote $\iota(M)$ by M^{-1}.

If τ is a group topology on a group G and τ is Hausdorff (respectively, compact, locally compact, connected, etc.), then the topological group (G, τ) is called Hausdorff (respectively, compact, locally compact, connected, etc.). Analogously, if G is cyclic (respectively, abelian, nilpotent, etc.), the topological group (G, τ) is called cyclic (respectively, abelian, nilpotent, etc.).

Here we propose some examples, starting with two trivial ones.

Example 2.1.3. For every group G, the discrete topology δ_G and the indiscrete topology ι_G on G are group topologies.

A nontrivial example of a topological group is the additive group $\mathbb{R}$ of the reals, equipped with its usual Euclidean topology. Clearly, $\mathbb{R}$ is a noncompact locally compact abelian group. This extends to all finite powers $\mathbb{R}^n$, endowed with the product topology.

Example 2.1.4. For every $n \in \mathbb{N}_+$, the general linear group $\mathrm{GL}_n(\mathbb{R})$ equipped with the topology induced by $\mathbb{R}^{n^2}$ is a locally compact group and for $n \geq 2$ it is not abelian. Analogously, $\mathrm{GL}_n(\mathbb{C})$ equipped with the topology induced from $\mathbb{C}^{n^2}$ is a locally compact group.

Example 2.1.5. For every prime p, the group $\mathbb{J}_p$ of p-adic integers (see §A.4.2), considered as the ring of all endomorphisms of the Prüfer group $\mathbb{Z}(p^\infty)$, embeds into the product $\mathbb{Z}(p^\infty)^{\mathbb{Z}(p^\infty)}$. Then $\mathbb{J}_p$ is a topological group with the topology induced by the product topology of the discrete topology of $\mathbb{Z}(p^\infty)$.

https://doi.org/10.1515/9783110654936-002

The same topology on $\mathbb{J}_p$ is induced by the product topology of $\prod_{n\in\mathbb{N}} \mathbb{Z}/p^n\mathbb{Z}$, when we consider $\mathbb{J}_p$ as the inverse limit $\varprojlim_{n\in\mathbb{N}} \mathbb{Z}/p^n\mathbb{Z}$ (for more details, see Example 3.3.8).

Other examples of group topologies are given in §2.2.

Lemma 2.1.6. *Let G be a topological group. For every $a \in G$, the* left translation ${}_at: G \to G$, $x \mapsto ax$, *the* right translation $t_a: G \to G$, $x \mapsto xa$, *and the* inner automorphism $\phi_a: G \to G$, $x \mapsto axa^{-1}$, *are homeomorphisms.*

Proof. Since $\mu: G \times G \to G$, $(x, y) \mapsto xy$ is continuous by Remark 2.1.2, both ${}_at = \mu(a, -)$ and its inverse ${}_{a^{-1}}t$, are continuous, for every $a \in G$. Hence, ${}_at$ is a homeomorphism for every $a \in G$. A similar proof works for t_a. Since, for every $a \in G$, $\phi_a = {}_at \circ t_{a^{-1}}$, it follows that ϕ_a is a homeomorphism. □

Making use of Lemma 2.1.6, one can prove that a topological group is homogeneous as a topological space (see Exercise 2.4.1). Consequently, a topological group G is discrete if and only if e_G is isolated, namely, the singleton $\{e_G\}$ is open.

For example, this permits showing the following property.

Lemma 2.1.7. *Let G be a countable topological group. If G is of second category, then G is discrete. Consequently, if G is a Baire space then G is discrete.*

Proof. The second assertion follows from the first, since every Baire space is of second category, by Lemma B.5.19. Assume that G is of second category. As the union $G = \bigcup_{g\in G}\{g\}$ is countable, there exists $g \in G$ with $\operatorname{Int}\{g\} \neq \emptyset$, so $\{g\}$ is open. By the homogeneity of G as a topological space, we deduce that G is discrete. □

Notation 2.1.8. For a group G, a subset M of G, and $a \in G$, we denote by

$$aM := {}_at(M) = \{am : m \in M\}$$

the image of M under the left translation ${}_at$. This is extended also to families of subsets of G, in particular, for every filter $\mathcal{F}$ on G, we denote $a\mathcal{F} := \{aF : F \in \mathcal{F}\}$. Analogously we define Ma and $\mathcal{F}a$.

2.1.2 The neighborhood filter of the neutral element

For a topological group (G, τ) and $g \in G$, the filter $\mathcal{V}_{G,\tau}(g)$ of all neighborhoods of g in G is also denoted by $\mathcal{V}_G(g)$, $\mathcal{V}_\tau(g)$, or $\mathcal{V}(g)$, when no confusion is possible. As an immediate consequence of Lemma 2.1.6, we have the following property.

Lemma 2.1.9. *For a topological group G and $a \in G$, $\mathcal{V}(a) = a\mathcal{V}(e_G) = \mathcal{V}(e_G)a$.*

This property determines the central role played by the filter $\mathcal{V}(e_G)$. The next theorem collects its properties, which completely describe the situation. Of those (gt3) is vacuously satisfied for abelian groups.

Theorem 2.1.10. *Let G be a group. If τ is a group topology on G, then:*
(gt1) *for every $U \in \mathcal{V}_\tau(e_G)$, there exists $V \in \mathcal{V}_\tau(e_G)$ with $VV \subseteq U$;*
(gt2) *for every $U \in \mathcal{V}_\tau(e_G)$, there exists $V \in \mathcal{V}_\tau(e_G)$ with $V^{-1} \subseteq U$;*
(gt3) *for every $U \in \mathcal{V}_\tau(e_G)$ and $a \in G$, there exists $V \in \mathcal{V}_\tau(e_G)$ with $aVa^{-1} \subseteq U$.*

Conversely, if $\mathcal{V}$ is a filter on G satisfying (gt1), (gt2), and (gt3), then there exists a unique group topology τ on G such that $\mathcal{V} = \mathcal{V}_\tau(e_G)$.

Proof. Let τ be a group topology on G. Then (gt1) and (gt2) hold by the continuity of $\mu: G \times G \to G$, $(x,y) \mapsto xy$, at (e_G, e_G) and the continuity of $\iota: G \to G$, $x \mapsto x^{-1}$, at e_G (see Remark 2.1.2), while (gt3) follows from the continuity of the inner automorphism ϕ_a at e_G for every $a \in G$ proved in Lemma 2.1.6.

Now let $\mathcal{V}$ be a filter on G satisfying the conditions (gt1) and (gt2).

First, we verify that every $U \in \mathcal{V}$ contains e_G. Indeed, by (gt1) there exists $W \in \mathcal{V}$ with $WW \subseteq U$. By (gt2), there exists $O \in \mathcal{V}$ with $O^{-1} \subseteq W$. Then, for $V = O \cap O^{-1} \subseteq W$, $e_G \in VV^{-1} \subseteq U$. This shows that

$$\text{for every } U \in \mathcal{V} \text{ there exists } V \in \mathcal{V} \text{ such that } VV^{-1} \subseteq U. \tag{2.1}$$

Let $\tau = \{O \subseteq G: \forall x \in O,\ \exists V \in \mathcal{V},\ xV \subseteq O\}$. Obviously, $\emptyset, G \in \tau$ and τ is stable under taking arbitrary unions. If $O_1, O_2 \in \tau$ and $O_1 \cap O_2 \neq \emptyset$, let $x \in O_1 \cap O_2$; by definition of τ, there exist $V_1, V_2 \in \mathcal{V}$ such that $xV_1 \subseteq O_1$ and $xV_2 \subseteq O_2$; then $V = V_1 \cap V_2 \in \mathcal{V}$ and $xV \subseteq O_1 \cap O_2$, therefore $O_1 \cap O_2 \in \tau$. We conclude that τ is a topology on G.

Now we verify that

$$g\mathcal{V} = \mathcal{V}_\tau(g) \quad \text{for every } g \in G. \tag{2.2}$$

The inclusion $\mathcal{V}_\tau(g) \subseteq g\mathcal{V}$ follows from the definition of τ. To prove the converse inclusion, let $U \in \mathcal{V}$ and let $O = \{h \in gU: \exists W \in \mathcal{V}, hW \subseteq gU\}$. Obviously, $g \in O \subseteq gU$. It remains to check that $O \in \tau$. In fact, for every $x \in O$, by definition, there exists $W \in \mathcal{V}$ with $xW \subseteq gU$, and by (gt1) there exists $V \in \mathcal{V}$ with $VV \subseteq W$; then $xV \subseteq O$, since for every $v \in V$, $xv \in xV \subseteq xW \subseteq gU$ and $xvV \subseteq xVV \subseteq xW \subseteq gU$.

To verify that τ is a group topology, we assume that $\mathcal{V}$ satisfies also (gt3). We have to check the continuity of $f: G \times G \to G$, $(x,y) \mapsto xy^{-1}$, at a fixed pair $(x,y) \in G \times G$. By (2.2), a τ-neighborhood of xy^{-1} has the form $xy^{-1}U$, where $U \in \mathcal{V}$. By (gt3), there exists $W \in \mathcal{V}$ with $Wy^{-1} \subseteq y^{-1}U$, and by (2.1) there exists $V \in \mathcal{V}$ with $VV^{-1} \subseteq W$. Then $O = xV \times yV$ is a neighborhood of (x,y) in $G \times G$ and $f(O) \subseteq xVV^{-1}y^{-1} \subseteq xWy^{-1} \subseteq xy^{-1}U$. Hence, f is continuous at (x,y). □

Definition 2.1.11. For a topological group G, a base $\mathcal{B}$ of the filter $\mathcal{V}(e_G)$ is called *base of the neighborhoods of e_G* (or briefly, *local base at e_G*) in G and the elements of $\mathcal{B}$ are called *basic neighborhoods*. A neighborhood U of e_G in G is *symmetric* if $U = U^{-1}$.

For every $U \in \mathcal{V}(e_G)$ in a topological group G, $U \cap U^{-1} \in \mathcal{V}(e_G)$ is symmetric, hence the symmetric neighborhoods of e_G form a base of $\mathcal{V}(e_G)$.

Remark 2.1.12. Similarly to the reduction to symmetric neighborhoods, in a topological group G one may achieve further nice properties of the elements of a base $\mathcal{B}$ of $\mathcal{V}(e_G)$, including global properties of $\mathcal{B}$, as having small size $|\mathcal{B}|$. Of those the best instances are the topological groups G admitting a countable base $\mathcal{B} = \{U_n : n \in \mathbb{N}\}$ of $\mathcal{V}(e_G)$. Indeed, under this assumption, by further applying (gt1) and (gt2), one can assume without loss of generality that

$$U_{n+1}U_{n+1} \subseteq U_n \quad \text{and} \quad U_n = U_n^{-1} \quad \text{for every } n \in \mathbb{N}. \tag{2.3}$$

Example 2.1.13. Consider the group $\mathbb{R}$ with the Euclidean topology. Then

$$\mathcal{V}(0) = \{U \subseteq \mathbb{R} : \exists \varepsilon \in \mathbb{R}_{>0}, (-\varepsilon, \varepsilon) \subseteq U\}.$$

The base $\mathcal{B} = \{(-\varepsilon, \varepsilon) : \varepsilon \in \mathbb{R}_{>0}\}$ of $\mathcal{V}(0)$ consisting of symmetric open neighborhoods of 0 has size $\mathfrak{c}$. One may choose also the countable base $\mathcal{B}_1 = \{(-1/n, 1/n) : n \in \mathbb{N}_+\}$ of $\mathcal{V}(0)$, or even the smaller one $\mathcal{B}_2 = \{U_n : n \in \mathbb{N}\}$, where $U_n = (2^{-n}, 2^n)$ for $n \in \mathbb{N}_+$; note that $\mathcal{B}_2$ has the additional property (2.3).

The following observation is of fundamental relevance for our treatment of topological groups in the sequel, when we use mainly filter bases instead of filters.

Remark 2.1.14. In Theorem 2.1.10, for a topological group (G, τ), instead of the filter $\mathcal{V}_\tau(e_G)$ one can characterize a base of $\mathcal{V}_\tau(e_G)$. To this end, take a *filter base* $\mathcal{B}$ on G satisfying (gt1), (gt2), and (gt3). In this case, for the unique group topology τ on G ensured by the conclusion of Theorem 2.1.10, the filter $\mathcal{V}_\tau(e_G)$ coincides with the filter generated by $\mathcal{B}$.

To characterize the local base $\mathcal{B}$ at e_G formed by the *open* neighborhoods of e_G in τ, one has to add to (gt1), (gt2), and (gt3) also the following property:

(gt4) for every $U \in \mathcal{B}$ and every $x \in U$, there exists $V \in \mathcal{B}$ such that $Vx \subseteq U$.

It guarantees that $\mathcal{B}$ is a filter base of open sets of $\mathcal{V}_\tau(e_G)$, where τ is again the group topology on G given by the conclusion of Theorem 2.1.10.

Definition 2.1.15. The *core* of a topological group (G, τ) is $\operatorname{core}(G, \tau) := \bigcap \mathcal{V}_\tau(e_G)$; we denote it simply by $\operatorname{core}(G)$ when the topology is clear.

Lemma 2.1.16. *For a topological group G, $\operatorname{core}(G)$ is a normal subgroup of G.*

Proof. Let $N = \operatorname{core}(G)$. Clearly, $e_G \in N$. If $x, y \in N$, then $xy \in N$ by (gt1), while $x^{-1} \in N$ for $x \in N$ can be deduced from (gt2). Finally, for $g \in G$, the inclusion $gNg^{-1} \subseteq N$ follows from (gt3). □

Remark 2.1.17. Let (G, τ) be a topological group and let $N = \operatorname{core}(G, \tau)$.

(a) Then $N = \overline{\{e_G\}}^\tau$. In fact, for $x \in G$, $x \notin \overline{\{e_G\}}^\tau$ if and only if there exists a symmetric $U \in \mathcal{V}_\tau(e_G)$ with $e_G \notin xU$ (i. e., $x \notin U$).

(b) In case $N \in \mathcal{V}_\tau(e_G)$, we have $N \in \tau$ (if $U \in \tau$ with $e_G \in U \subseteq N$, then for every $y \in N$, $y \in yU \subseteq N$, as N is a subgroup of G; so $N \in \tau$). In this case $\mathcal{V}_\tau(e_G)$ has the smallest possible base, namely, $\mathcal{B} = \{N\}$. The τ-open sets are unions of cosets of N, hence τ is an Alexandrov topology (see Example B.1.16).

Definition 2.1.18. An *Alexandrov group topology* τ on a group G is a group topology that is an Alexandrov topology, and (G, τ) is called *Alexandrov group*.

The Alexandrov group topologies are the simplest examples of group topologies that are neither discrete nor indiscrete.

The morphisms in the category **TopGrp** of topological groups are the continuous group homomorphisms.

Definition 2.1.19. Let G, H be topological groups and let $f: G \to H$ be a homomorphism. If f is simultaneously an isomorphism and a homeomorphism, then f is called *topological isomorphism*; in this case, we say that G and H are *topologically isomorphic* and write $G \cong H$. In case $G = H$, a topological isomorphism $G \to G$ is called *topological automorphism*.

The topological isomorphisms are the isomorphisms in the category **TopGrp**.

The continuity of a group homomorphism can be characterized in terms of the neighborhood filters of the neutral element, or filter bases generating them, as follows. We often use this characterization without giving reference to it.

Lemma 2.1.20. *Let G, H be topological groups and $f: G \to H$ a homomorphism. Then the following conditions are equivalent:*
(a) *f is continuous;*
(b) *f is continuous at e_G;*
(c) *for every $U \in \mathcal{V}_H(e_H)$ there exists $V \in \mathcal{V}_G(e_G)$ such that $f(V) \subseteq U$;*
(d) *for $\mathcal{B}_G, \mathcal{B}_H$ local bases at e_G, e_H in G, H, respectively, for every $U \in \mathcal{B}_H$ there exists $V \in \mathcal{B}_G$ such that $f(V) \subseteq U$.*

Moreover, f is open if and only if $f(U) \in \mathcal{V}_H(e_H)$ for every $U \in \mathcal{V}_G(e_G)$.

Proof. (a)⇔(b) immediately follows from the homogeneity of topological groups (see Exercise 2.4.1), while (c) and (d) are clearly equivalent forms of (b). □

Example 2.1.21. If G, H are Alexandrov groups, a homomorphism $f: G \to H$ is continuous precisely when $f(\mathrm{core}(G)) \subseteq \mathrm{core}(H)$. Therefore, the category of Alexandrov groups is isomorphic to the category of pairs (G, N) of a group G and a normal subgroup N of G, where the morphisms $(G, N) \to (H, L)$ are group homomorphisms $f: G \to H$ with $f(N) \subseteq L$.

2.1.3 Comparing group topologies

For a group G, denote by $\mathfrak{L}(G)$ the set of all group topologies on G. Clearly, $\mathfrak{L}(G) \subseteq \mathcal{T}(G)$, where $\mathcal{T}(G)$ is the complete lattice of topologies on G with the partial order given by inclusion (see Definition B.1.13 and Remark B.1.14).

Consider $\mathfrak{L}(G)$ with the partial order inherited from $\mathcal{T}(G)$, and note that if τ, τ' are group topologies on G, then the identity map $(G, \tau) \to (G, \tau')$ is continuous if and only if $\tau' \subseteq \tau$. We say that τ' is *coarser* than τ or that τ is *finer* than τ', and we denote this partial order also by $\tau' \leq \tau$.

Both posets $(\mathcal{T}(G), \leq)$ and $(\mathfrak{L}(G), \leq)$ share the same top and bottom element (i. e., the discrete and the indiscrete topology, respectively) and $\mathfrak{L}(G)$ is likewise a complete lattice, but there is a subtle difference between these two cases: the lattice operations in $\mathfrak{L}(G)$ *are not the same* as in $\mathcal{T}(G)$, as pointed out below. We consider separately both underlying semilattice structures.

Remark 2.1.22. For a group G, $\mathfrak{L}(G)$ is a complete join-semilattice considered as a subposet of $\mathcal{T}(G)$. Take a family $\{\tau_i : i \in I\} \subseteq \mathfrak{L}(G)$. To check that the join $\sup_{i\in I} \tau_i$ taken in the larger lattice $\mathcal{T}(G)$ is also a group topology on G, it suffices to consider the filter base $\mathcal{B}$ formed by the family of all finite intersections $U_1 \cap \cdots \cap U_n$, where $U_k \in \mathcal{V}_{\tau_{i_k}}(e_G)$ for $k \in \{1, \dots, n\}$, $\{i_1, \dots, i_n\}$ runs over all finite subsets of I and $n \in \mathbb{N}_+$. Since $\mathcal{B}$ satisfies (gt1), (gt2), and (gt3) from Theorem 2.1.10, it gives a group topology τ on G having $\mathcal{B}$ as a local base at e_G. Since all topologies τ_i are group topologies, $\tau = \sup_{i\in I} \tau_i$. This completely describes the semilattice structure of $(\mathfrak{L}(G), \sup)$.

Since the poset $\mathfrak{L}(G)$ has a bottom element, this suffices to conclude that $\mathfrak{L}(G)$ is a complete lattice (see [30]):

Proposition 2.1.23. *If G is a group, then $\mathfrak{L}(G)$ is a complete lattice.*

The simple argument leading to Proposition 2.1.23 gives no idea about how the meets look like in this complete lattice. In fact, the meet in $\mathfrak{L}(G)$ fails to coincide with the meet in the lattice $\mathcal{T}(G)$, since *the intersection of group topologies need not be a group topology* (see Exercise 2.4.5). In case G is abelian, the meet of two topologies in $\mathfrak{L}(G)$ is described as follows (see Exercise 2.4.2 for the general case).

Example 2.1.24. If G is an abelian group and $\tau_1, \tau_2 \in \mathfrak{L}(G)$, then it is easy to check that the family $\{U_1 + U_2 : U_i \in \mathcal{V}_{\tau_i}(0),\ i = 1, 2\}$ is a base of the filter $\mathcal{V}_{\inf\{\tau_1,\tau_2\}}(0)$ ((gt1) and (gt2) from Theorem 2.1.10 are satisfied).

Remark 2.1.25. Due to Theorem 2.1.10, for a group G, one can conveniently describe the poset $\mathfrak{L}(G)$ by the poset $\mathfrak{F}(G)$ of all filters $\mathcal{F}$ on G satisfying conditions (gt1), (gt2), and (gt3). The poset $\mathfrak{F}(G)$ is ordered again by inclusion (of filters).

2.2 Examples of group topologies

Here we give several series of examples of group topologies, introducing them by means of the filter of neighborhoods of the neutral element, as explained above. Furthermore, in all cases we avoid to treat the whole filter and we prefer to deal with a conveniently described filter base (see Remark 2.1.14).

2.2.1 Linear topologies and functorial topologies

Let G be a group and let $\mathcal{B} = \{N_i : i \in I\}$ be a filter base consisting of *normal subgroups* of G. Then $\mathcal{B}$ satisfies (gt1), (gt2), (gt3), and also (gt4), hence by Remark 2.1.14, $\mathcal{B}$ generates a group topology τ on G with $\mathcal{B} \subseteq \tau$ a base of the filter $\mathcal{V}_\tau(e_G)$. For every $g \in G$, the family of cosets $\{gN_i : i \in I\}$ is a base of the filter $\mathcal{V}_\tau(g)$. The group topologies of this type are called *linear topologies*, and a group endowed with a linear topology is called *linearly topologized.*

The simplest example of linear topologies are the Alexandrov group topologies. It is easy to see that every linear group topology on a group G is a supremum, in $\mathfrak{L}(G)$, of Alexandrov group topologies on G.

Example 2.2.1. Let G be a group and let p be a prime. We list examples of filter bases $\mathcal{B} \ni G$ consisting of normal subgroups of G giving rise to group topologies:

(a) the *profinite* topology ϖ_G, with $\mathcal{B}$ the family of all normal subgroups of G with finite index;
(b) the *pro-p-finite* topology ϖ_G^p, with $\mathcal{B}$ the family of all normal subgroups of G with finite index that is a power of p;
(c) the *pro-countable* topology ϱ_G, with $\mathcal{B}$ the family all normal subgroups of G with at most countable index;
(d) the *p-adic* topology ν_G^p, with $\mathcal{B}_p = \{N_n : n \in \mathbb{N}\}$ and, for $n \in \mathbb{N}$, N_n the subgroup (necessarily normal) of G generated by $\{g^{p^n} : g \in G\}$;
(e) the *natural* topology (or *$\mathbb{Z}$-topology*) ν_G, with $\mathcal{B}_\nu = \{M_n : n \in \mathbb{N}_+\}$ and, for $n \in \mathbb{N}_+$, M_n the subgroup (necessarily normal) of G generated by $\{g^n : g \in G\}$.

Clearly, $\varpi_G^p \leq \varpi_G$ and $\nu_G^p \leq \nu_G$ for every prime p.

When G is an abelian group, for every $n \in \mathbb{N}$ the basic subgroup N_n defining the p-adic topology of G has the form $N_n = p^n G$, and analogously for every $n \in \mathbb{N}_+$ the basic subgroup M_n defining the natural topology of G has the form $M_n = nG$.

Example 2.2.2. Let p be a prime. The basic open neighborhoods of 0 in the topology of $\mathbb{J}_p$ described in Example 2.1.5 are the subgroups $p^n\mathbb{J}_p$ of $(\mathbb{J}_p, +)$ for $n \in \mathbb{N}$, that is, the topology of $\mathbb{J}_p$ is its p-adic topology $\nu_{\mathbb{J}_p}^p$. Actually, each $p^n\mathbb{J}_p$ is an ideal of the ring $\mathbb{J}_p$ (see Claim A.4.16).

Moreover, the p-adic topology of $\mathbb{J}_p$ coincides with its natural topology, that is, $\nu^p_{\mathbb{J}_p} = \nu_{\mathbb{J}_p}$; in fact, for $m \in \mathbb{N}_+$, $m\mathbb{J}_p = p^s\mathbb{J}_p$, where $s \in \mathbb{N}$ is such that $m = m_1 p^s$ for some $m_1 \in \mathbb{N}_+$ with $(m_1, p) = 1$. Moreover, for $\mathbb{J}_p$ the p-adic topology coincides also with the profinite topology, as well as with the pro-p-finite topology (see Exercise 2.4.11), but it differs from the pro-countable topology (see Exercise 3.5.15).

The usual topology τ on the field $\mathbb{Q}_p$ of p-adic numbers is given by declaring $\mathbb{J}_p$ open in $\mathbb{Q}_p$, that is, $\{p^n\mathbb{J}_p : n \in \mathbb{N}\}$ is a filter base of $\mathcal{V}_\tau(0)$ in $\mathbb{Q}_p$.

Remark 2.2.3. Furstenberg [144] used the natural topology $\nu_{\mathbb{Z}}$ of $\mathbb{Z}$ to find a new proof of the infinitude of prime numbers (see Exercise 2.4.12).

In the next lemma one can appreciate the special behavior of the above topologies with respect to the continuity of group homomorphisms.

Lemma 2.2.4. *Let G, H be groups and let $f: G \to H$ be a homomorphism. Then f is continuous when both groups G, H are equipped with their profinite (respectively, pro-p-finite, pro-countable, p-adic, natural) topology.*

Proof. If N is a subgroup of H of finite index, then $f^{-1}(N)$ is a subgroup of finite index of G. The other cases are similar. □

This lemma suggests that the topologies from Example 2.2.1 have a "natural" origin, that is made more precise by the following notion.

Definition 2.2.5. Assume that every abelian group G is equipped with a group topology τ_G such that for every group homomorphism $f: G \to H$ between abelian groups, $f: (G, \tau_G) \to (H, \tau_H)$ is continuous. Then we say that the class of group topologies $\tau = \{\tau_G : G \text{ abelian group}\}$ is a *functorial topology*.

Clearly, every functorial topology gives rise to a functor **AbGrp** $\to$ **TopGrp** through the assignments $G \mapsto (G, \tau_G)$ and $f \mapsto f$ for $G \in$ **AbGrp** and $f \in \mathrm{Hom}(\mathbf{AbGrp})$.

All five topologies in Example 2.2.1 are functorial, other examples of functorial topologies are given below.

Remark 2.2.6. In view of Lemma 2.2.4, one could extend the above definition to non-abelian groups. Nevertheless, we prefer to limit our field of interest in the abelian case, where these topologies were originally introduced (see [40, 43, 206, 207] and also [138, p. 33]).

2.2.2 Topologies generated by characters

Here we introduce a functorial topology that is not linear, and which is of vital importance for our exposition. The argument $\mathrm{Arg}(z)$ of a complex number z is taken in $(-\pi, \pi]$.

Definition 2.2.7. A *character* of an abelian group G is a homomorphism $\chi: G \to \mathbb{S}$, where $\mathbb{S}$ is the unitary circle in $\mathbb{C}$. Let $G^* := \operatorname{Hom}(G, \mathbb{S})$.

For an abelian group G, $\chi \in G^*$, and $\delta > 0$, let

$$U_G(\chi; \delta) := \{x \in G: |\operatorname{Arg}(\chi(x))| < \delta\}.$$

For characters $\chi_1, \dots, \chi_n$ of G and $\delta > 0$, let

$$U_G(\chi_1, \dots, \chi_n; \delta) := \{x \in G: |\operatorname{Arg}(\chi_i(x))| < \delta, \ \forall i \in \{1, \dots, n\}\}, \tag{2.4}$$

namely,

$$U_G(\chi_1, \dots, \chi_n; \delta) = \bigcap_{i=1}^{n} U_G(\chi_i; \delta). \tag{2.5}$$

Remark 2.2.8. One can describe (2.4) alternatively, using as target group the abelian group $\mathbb{T} = \mathbb{R}/\mathbb{Z}$ instead of $\mathbb{S}$ (clearly, $\mathbb{T} \cong \mathbb{S}$). For $z = r + \mathbb{Z} \in \mathbb{T}$, let $\|z\| = \|r + \mathbb{Z}\| := d(r, \mathbb{Z}) = \min\{(d(r, m)): m \in \mathbb{Z}\}$ where d denotes the Euclidean metric on $\mathbb{R}$. In such a case, characters $\chi: G \to \mathbb{T}$ of an abelian group G must be used and, for $x \in G$ and $\delta > 0$, the inequality $|\operatorname{Arg}(\chi(x))| < \delta$ must be replaced by $\|\chi(x)\| < \delta/2\pi$. We use the multiplicative version with $\mathbb{S}$ mainly when some more subtle computations are involved.

For an abelian group G and a subset H of G^*, the family

$$\mathcal{B}_H := \{U_G(\chi_1, \dots, \chi_n; \delta): \delta > 0, \ n \in \mathbb{N}_+, \ \chi_i \in H, \ i \in \{1, \dots, n\}\}$$

is a filter base satisfying the conditions (gt1), (gt2), and (gt3). Moreover, $\mathcal{B}_H$ satisfies also (gt4): consider $U_G(\chi_1, \dots, \chi_n; \delta) \in \mathcal{B}_H$, $x \in U_G(\chi_1, \dots, \chi_n; \delta)$ and let $\eta > 0$ be such that $|\operatorname{Arg}(\chi_i(x))| + \eta < \delta$ for every $i \in \{1, \dots, n\}$; then $x + U_G(\chi_1, \dots, \chi_n; \eta) \subseteq U_G(\chi_1, \dots, \chi_n; \delta)$. Therefore, $\mathcal{B}_H$ is a filter base of open neighborhoods of e_G in a group topology $\mathcal{T}_H$ on G by Remark 2.1.14.

Definition 2.2.9. Let G be an abelian group. For a subset H of G^*, the group topology $\mathcal{T}_H$ is the *topology generated by the characters* of H. The topology $\mathfrak{B}_G := \mathcal{T}_{G^*}$ generated by all characters of G is the *Bohr topology* of G and we use the abbreviation $G^\# := (G, \mathcal{T}_{G^*})$.

Example 2.2.10. Let G be an abelian group, fix $\chi \in G^*$, and denote $\mathcal{T}_\chi = \mathcal{T}_{\{\chi\}}$.

(a) Then $\chi: (G, \mathcal{T}_\chi) \to \mathbb{S}$ is continuous, so $\ker \chi$ is a closed subgroup of $(G, \mathcal{T}_\chi)$ contained in $U_G(\chi; \delta)$ for every $\delta > 0$. On the other hand, every subgroup of G contained in $U_G(\chi; \pi/2)$ is contained in $\ker \chi$ as well, since $\mathbb{S}_+ = \{z \in \mathbb{S}: \operatorname{Re}(z) \geq 0\}$ contains no nontrivial subgroups.

For a subset H of G^*, one has $\mathcal{T}_H = \sup\{\mathcal{T}_\chi : \chi \in H\}$, by (2.5).

(b) For $n \in \mathbb{Z} \setminus \{0\}$, consider the character $\chi^n\colon G \to \mathbb{S}$ defined by $(\chi^n)(x) = (\chi(x))^n$ for every $x \in G$. Then $\mathcal{T}_{\chi^n} \subseteq \mathcal{T}_\chi$ as $U_G(\chi; \delta/|n|) \subseteq U_G(\chi^n; \delta)$ for every $\delta > 0$.
(c) Obviously, $\mathcal{T}_{\chi^{-1}} = \mathcal{T}_\chi$. Moreover, for $G = \mathbb{Z}$ and $\chi, \xi \in \mathbb{Z}^*$ with $\ker\chi = \ker\xi = 0$ the equality $\mathcal{T}_\chi = \mathcal{T}_\xi$ holds true if and only if $\xi \in \{\chi, \chi^{-1}\}$ (see Exercise 13.7.1).

Lemma 2.2.11. *Let G be an abelian group and H a subset of G^*. The assignment $\mathcal{P}(G^*) \to \mathfrak{L}(G)$, $H \mapsto \mathcal{T}_H$, is monotone increasing and $\mathcal{T}_{\langle H\rangle} = \mathcal{T}_H$.*

Proof. If $H \subseteq H' \subseteq G^*$, then $\mathcal{T}_H \leq \mathcal{T}_{H'}$; this proves the first assertion and that $\mathcal{T}_H \leq \mathcal{T}_{\langle H\rangle}$. The converse inclusion $\mathcal{T}_{\langle H\rangle} \leq \mathcal{T}_H$ follows from Example 2.2.10(b) and the fact that, for $\chi_1, \chi_2 \in H$ and $\delta > 0$, $U_G(\chi_1\chi_2; \delta) \supseteq U_G(\chi_1, \chi_2; \delta/2)$, and thus $\mathcal{T}_{\chi_1\chi_2} \leq \mathcal{T}_H$. □

Due to Lemma 2.2.11, it is worth studying the topologies $\mathcal{T}_H$ only when H is a subgroup of G^*.

Next we verify that the Bohr topology is functorial.

Lemma 2.2.12. *Let G, H be abelian groups and let $f\colon G \to H$ be a homomorphism. Then $f\colon G^\# \to H^\#$ is continuous.*

Proof. Let $\chi_1, \dots, \chi_n \in H^*$ and $\delta > 0$. Then

$$f^{-1}(U_H(\chi_1, \dots, \chi_n; \delta)) = U_G(\chi_1 \circ f, \dots, \chi_n \circ f; \delta).$$

To conclude apply Lemma 2.1.20. □

This lemma shows that for an abelian group G, among all topologies of the form $\mathcal{T}_H$, with H a subgroup of G^*, the Bohr topology $\mathcal{T}_{G^*}$ plays a special role, being a functorial topology. This is why we introduced the special notation $\mathfrak{B}_G$ for $\mathcal{T}_{G^*}$.

As we shall see below, for an abelian group some of the linear topologies introduced above are also generated by appropriate families of characters.

2.2.3 Interrelations among functorial topologies

In the next proof we make use of the simple observation:

Remark 2.2.13. If τ, τ' are two linear topologies on a group G and $\mathcal{B}_\tau, \mathcal{B}_{\tau'}$ are local bases at e_G of τ, τ', respectively, formed by normal subgroups, then a local base at e_G of $\inf\{\tau, \tau'\}$ is given by the family of normal subgroups $\{NN'\colon N \in \mathcal{B}_\tau, N' \in \mathcal{B}_{\tau'}\}$.

Proposition 2.2.14. *For every group G and every prime p,*

$$\varpi_G^p \leq \varpi_G \leq \nu_G \geq \nu_G^p \quad \text{and} \quad \varpi_G^p = \inf\{\varpi_G, \nu_G^p\}.$$

Proof. The first and the last inequality are obvious. To prove the inequality $\varpi_G \leq \nu_G$, it suffices to note that if N is a finite-index normal subgroup of G, with $[G : N] = m$, then N contains the subgroup $M_m = \langle g^m\colon g \in G\rangle$.

To prove the equality $\varpi_G^p = \inf\{\varpi_G, \nu_G^p\}$, take into account that $\varpi_G^p \leq \nu_G^p$, by an argument similar to that above. So, it only remains to check the inequality $\varpi_G^p \geq \inf\{\varpi_G, \nu_G^p\}$. In view of Remark 2.2.13, a basic neighborhood of e_G in $\inf\{\varpi_G, \nu_G^p\}$ has the form NN_n, where N is a finite-index normal subgroup of G, while N_n is the normal subgroup $\langle g^{p^n} : g \in G\rangle$ of G for some fixed $n \in \mathbb{N}$. Then G/NN_n is a finite p-group, hence $NN_n \in \varpi_G^p$. □

The profinite topology of an abelian group is contained in its Bohr topology:

Proposition 2.2.15 ([86]). *For every abelian group G, $\varpi_G \leq \inf\{\mathfrak{P}_G, \nu_G\}$.*

Proof. In view of the inequality $\varpi_G \leq \nu_G$ established in Proposition 2.2.14, it is enough to prove that $\varpi_G \leq \mathfrak{P}_G$.

Let H be a subgroup of G of finite index; we show that H is open in $G^\#$. Being a finite abelian group, G/H has the form $C_1 \times \cdots \times C_n$, where each C_i is a finite cyclic group for $i \in \{1, \ldots, n\}$. Let $q: G \to G/H$ be the canonical projection. For every $i \in \{1, \ldots, n\}$, let $p_i: C_1 \times \cdots \times C_n \to C_i$ be the ith projection and $q_i = p_i \circ q: G \to C_i$; then $G/H_i \cong C_i$ with $H_i = \ker q_i$. Identifying C_i with $\mathbb{Z}(m_i) \leq \mathbb{T}$, we can consider $q_i: G \to C_i \hookrightarrow \mathbb{T}$ as a character of G. Then, $H_i = U_G(q_i; 1/2m_i) \in \mathfrak{P}_G$. Hence, $H = \bigcap_{i=1}^n H_i \in \mathfrak{P}_G$. □

For an abelian group G, a character $\chi: G \to \mathbb{T}$ is *torsion* if there exists $n \in \mathbb{N}_+$ such that $n\chi$ is trivial; equivalently, the character χ is a torsion element of the group G^*. In other words, χ vanishes on the subgroup $nG = \{nx : x \in G\}$ of G, and this means that the subgroup $\chi(G)$ of $\mathbb{T}$ is finite cyclic. Therefore, G^* is torsion-free when G is divisible.

Lemma 2.2.16. *Let G be an abelian group and H a subset of G^*. Then $\mathcal{T}_H \leq \varpi_G$ if and only if every $\chi \in H$ is torsion (i. e., $H \subseteq t(G^*)$).*

Proof. Let $\chi \in H$ be torsion and $\delta > 0$. Then $U_G(\chi; \delta)$ contains $\ker \chi$. Since $\chi(G) \cong G/\ker \chi$ is finite, $\ker \chi$ is an open neighborhood of e_G in ϖ_G, and so $U_G(\chi; \delta) \in \mathcal{V}_{\varpi_G}(e_G)$. Hence, $\mathcal{T}_H \leq \varpi_G$. Now assume that $\mathcal{T}_H \leq \varpi_G$ and let $\chi \in H$. Then $U_G(\chi; \pi/2)$ must contain a finite-index subgroup N of G, and so $N \subseteq \ker \chi$ by Example 2.2.10(a). Thus, $\ker \chi$ has finite index in G, and consequently χ is torsion. □

The inequality $\varpi_G \leq \inf\{\mathfrak{P}_G, \nu_G\}$ proved in Proposition 2.2.15 is actually an equality that we prove in the sequel. Now we establish it when the abelian group G is bounded (ν_G is discrete for such a group G by Exercise 2.4.8).

Theorem 2.2.17. *For an abelian group G, $\mathfrak{P}_G = \varpi_G$ if and only if G is bounded.*

Proof. By Proposition 2.2.15, $\varpi_G \leq \mathfrak{P}_G$. If G is bounded, then every character of G is torsion, so Lemma 2.2.16 gives $\mathfrak{P}_G \leq \varpi_G$.

Now assume that $\mathfrak{B}_G = \varpi_G$. According to Lemma 2.2.16, the group G^* is torsion. This immediately implies that G is torsion, since otherwise G has an infinite cyclic subgroup C and by Theorem A.2.4 any embedding $j: C \to \mathbb{S}$ can be extended to a character of G, that results to be nontorsion. It remains to prove that G is bounded.

Assume that G is not bounded. If $r_{p_n}(G) \neq 0$ for infinitely many pairwise distinct primes p_n, $n \in \mathbb{N}_+$, then we find a subgroup G_1 of G isomorphic to $\bigoplus_{n\in\mathbb{N}_+} \mathbb{Z}(p_n)$. Take an embedding $j: G_1 \to \mathbb{S}$ and extend j to a character of the whole group G (using Theorem A.2.4). It cannot be torsion, a contradiction. Then $r_p(G) \neq 0$ for finitely many primes p. Since G is infinite, this yields that at least one of the primary components $t_p(G)$ is not bounded. If $r_p(G) < \infty$, then $t_p(G)$ contains a copy of the group $\mathbb{Z}(p^\infty)$ by Example A.4.12(b). Now take an embedding $j: \mathbb{Z}(p^\infty) \to \mathbb{S}$ and extend j to a character of the whole group G (using Theorem A.2.4). It cannot be torsion, a contradiction. Hence, $r_p(G)$ is infinite and $t_p(G)$ contains no copies of the group $\mathbb{Z}(p^\infty)$; then there exists a subgroup of $t_p(G)$ isomorphic to $L = \bigoplus_{n\in\mathbb{N}} \mathbb{Z}(p^n)$. In the latter case it is easy to build a surjective homomorphism $h: L \to \mathbb{Z}(p^\infty) \subseteq \mathbb{S}$. Now extend h to a character of G, that cannot be torsion, a contradiction. □

We introduce a partial order between functorial topologies by letting $\mathcal{T} \leq \mathcal{S}$ whenever $\mathcal{T}_G \leq \mathcal{S}_G$ for every abelian group G. This makes the class $\mathfrak{Ft}$ of all functorial topologies a large complete lattice with top element δ and bottom element ι.

We enrich here our supply of examples of functorial topologies.

Example 2.2.18. Let G be an abelian group and p a prime. The *p-Bohr topology* is $\mathfrak{B}^p_G := \mathcal{T}_{\mathrm{Hom}(G,\mathbb{Z}(p^\infty))}$ where we consider $\mathbb{Z}(p^\infty) \leq \mathbb{T}$, and so $\mathrm{Hom}(G, \mathbb{Z}(p^\infty)) \leq G^*$.

Remark 2.2.19. It can be proved that:
(a) $\varpi = \inf\{\mathfrak{B}, \nu\}$ (see Exercise 3.5.17);
(b) analogously, $\varpi^p = \inf\{\nu^p, \mathfrak{B}^p\} = \inf\{\nu^p, \varpi\} = \inf\{\mathfrak{B}^p, \varpi\}$.

In the large complete lattice $\mathfrak{Ft}$ of functorial topologies one has the following diagram:

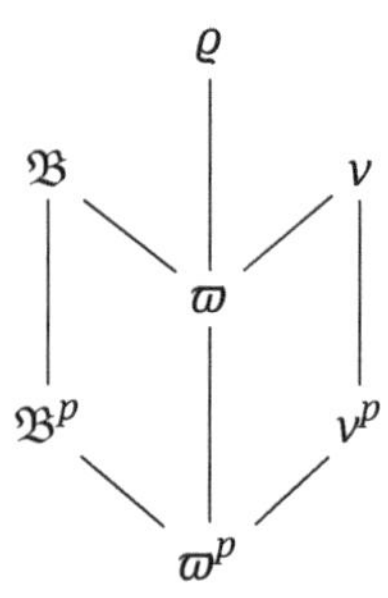

2.2.4 The pointwise convergence topology

The topology that we consider in this section mainly appears in transformation groups and in function spaces. The function spaces were the first instances of topological spaces.

Definition 2.2.20. Let X be a nonempty set and let Y be a topological space. The *pointwise convergence topology* on the set Y^X of all maps $X \to Y$ has as a base the family $\{f \in Y^X : \forall x \in F, f(x) \in U_x\}$, where F is a finite subset of X and all U_x, $x \in F$, are nonempty open sets of Y.

Remark 2.2.21. Let X be a nonempty set and let Y be a topological space. If we identify the function space Y^X with the Cartesian power, then the pointwise convergence topology coincides with the product topology of Y^X.

The name pointwise convergence topology comes from the fact that $\{f_\alpha\}_{\alpha\in A}$ is a net in Y^X that converges to $f \in Y^X$ with respect to the pointwise convergence topology precisely when for every $x \in X$ the net $\{f_\alpha(x)\}_{\alpha\in A}$ converges to $f(x)$ in Y. In particular, if Y is discrete this means that for every $x \in X$ there exists $\alpha_0 \in A$ such that, for all $\alpha \geq \alpha_0$ in A, $f_\alpha(x) = f(x)$.

Since in Y^X the target Y of the functions (e. g., the reals, the complex number field, etc.) usually has a topological group structure itself, the function spaces Y^X have a very rich structure from both points of view – topological and algebraic. Indeed, when Y is a topological group, the pointwise convergence topology on Y^X is a *group topology*.

Example 2.2.22. For a nonempty set X, let $S(X)$ denote the group of all permutations of X. The *stabilizer* $S_x = \{f \in S(X) : f(x) = x\}$ of $x \in X$ in $S(X)$ is a subgroup of $S(X)$. Consider on $S(X)$ the filter base

$$\mathcal{B}_X := \{S_F : F \subseteq X \text{ finite}\}, \quad \text{where } S_F := \bigcap_{x\in F} S_x = \{f \in S(X) : \forall x \in F,\ f(x) = x\}.$$

Since the subgroups S_x are pairwise conjugated, one can easily check that $\mathcal{B}_X$ satisfies all conditions (gt1), (gt2), (gt3), and (gt4), so by Remark 2.1.14, $\mathcal{B}_X$ induces a group topology $\mathcal{T}_X$ on $S(X)$ for which all subgroups S_F are open.

Remark 2.2.23. The topology-minded reader has already observed that for an infinite set X, the topology $\mathcal{T}_X$ introduced in Example 2.2.22 on $S(X)$ can be described also as the topology induced by the natural embedding of $S(X)$ into the Cartesian power X^X equipped with the product topology, where X has the discrete topology. In other words, $\mathcal{T}_X$ coincides with the pointwise convergence topology on $S(X)$ inherited from X^X (see Remark 2.2.21).

Here comes a further specialization of the pointwise convergence topology from linear algebra and module theory.

Example 2.2.24. Fix U, V vector spaces over a field K and consider on the space $\mathrm{Hom}(V, U)$ of all linear maps $V \to U$ the so-called *finite topology* τ_{fin}. This is the topology generated (in the sense of Theorem 2.1.10 and Remark 2.1.14) by the filter base $\mathcal{B} = \{W(F) : F \subseteq V, F \text{ finite}\}$, where $W(F) = \{f \in \mathrm{Hom}(V, U) : \forall x \in F, f(x) = 0\}$. It is easy to see that each $W(F)$ is a linear subspace of $\mathrm{Hom}(V, U)$, and so $\mathcal{B}$ satisfies conditions (gt1), (gt2), (gt3), and (gt4).

The finite topology τ_{fin} on $\mathrm{Hom}(V, U)$ coincides with the pointwise convergence topology when $\mathrm{Hom}(V, U)$ is considered as a subset of U^V and U carries the discrete topology (see Exercise 2.4.19(a)).

Remark 2.2.25. The finite topology τ_{fin} is especially useful when imposed on the dual $V^* := \mathrm{Hom}(V, K)$ of the vector space V over the discrete field K. In this case $(V^*, \tau_{\mathrm{fin}})$ is a linearly compact vector space (see [195] and §16.2.3).

The finite topology can be introduced in the same way for modules over an arbitrary ring, so in particular for abelian groups.

2.3 Semitopological, paratopological, and quasitopological groups

The group topology τ built in Theorem 2.1.10 starting from a filter $\mathcal{V}$ on a group G with the properties (gt1), (gt2), and (gt3) was defined by letting the neighborhood filter at $g \in G$ to be the filter $g\mathcal{V} = \mathcal{V}g$. The coincidence of these two filters is ensured by (actually, equivalent to) property (gt3). In case (gt3) fails, one obtains two topologies τ_r and τ_l on G, having as local bases at $g \in G$ the filters $\mathcal{V}g$ and $g\mathcal{V}$, respectively. For the pair (G, τ_r), all right translations $x \mapsto xg$ are continuous (actually, homeomorphisms); respectively, all left translations $x \mapsto gx$ are continuous for the pair (G, τ_l). Pairs of a group and a topology on it with this property are called *right topological groups*, respectively *left topological groups*.

Definition 2.3.1. A pair (G, τ), of a group G and a topology τ on G, is a *semitopological group* if it is simultaneously a left and a right topological group.

The filter $\mathcal{V}$ of neighborhoods of the neutral element of a semitopological group satisfies (gt3).

The right topological groups were introduced by Namioka [218] and largely used since then. Ellis, in his two fundamental papers [131, 132], proved that locally compact semitopological groups are topological. Later this was generalized by Bouziad [39] to Čech-complete groups.

Along with right topological groups, left topological groups and semitopological groups, one can find in the literature also the following weak versions of the notion of topological group.

Definition 2.3.2. A pair (G, τ), of a group G and a topology τ on G, is called:
(i) a *quasitopological group* when the multiplication map is separately continuous in both variables and the inverse map is continuous (i. e., it is a semitopological group such that the inverse is continuous);
(ii) a *paratopological group* when the multiplication map is jointly continuous in both variables.

The filter of neighborhoods of the neutral element of a quasitopological group has the properties (gt2) and (gt3), while that of a paratopological group has the properties (gt1) and (gt3).

For paratopological groups, having some nice compact-like properties (as pseudocompactness, etc.), one has analogues of Ellis' result. More details on this issue can be found in the survey [274] and the monograph [7].

Now we provide an example of a paratopological group that is not topological:

Example 2.3.3. Consider the filter base $\mathcal{B} = \{[x, x + 1/n) : n \in \mathbb{N}_+, x \in \mathbb{R}\}$ on $\mathbb{R}$. Since $\mathcal{V}(0) = \{[0, 1/n) : n \in \mathbb{N}_+\}$ satisfies (gt1) and (gt3), it gives rise to a topology τ_ℓ on $\mathbb{R}$, known as the *Sorgenfrey topology*, while $\mathbb{R}_\ell = (\mathbb{R}, \tau_\ell)$ is known as *Sorgenfrey line*. Then $\mathbb{R}_\ell$ is a paratopological group, but it is not a topological group since (gt2) fails.

In Chapter 4 we introduce two topologies $\mathfrak{Z}_G$ and $\mathfrak{M}_G$ (called Zariski topology and Markov topology, respectively) for every group G. Then $(G, \mathfrak{Z}_G)$ and $(G, \mathfrak{M}_G)$ are always quasitopological groups, but quite rarely topological groups.

2.4 Exercises

Exercise 2.4.1. Prove that every topological group is a homogeneous topological space.

Exercise 2.4.2. If G is an abelian group, describe $\inf_{i \in I} \tau_i$ for a family $\{\tau_i : i \in I\}$ in $\mathfrak{L}(G)$.

Exercise 2.4.3. Let G be a group and $\{\tau_i : i \in I\}$ a family in $\mathfrak{L}(G)$. Prove that if $\{a_n\}_{n \in \mathbb{N}}$ is a sequence in G with $a_n \to e_G$ in τ_i for every $i \in I$, then $a_n \to e_G$ also in $\sup_{i \in I} \tau_i$.

Exercise 2.4.4. Prove that the join (supremum) of all p-adic topologies on $\mathbb{Z}$, when p runs over the set of all primes, coincides with the natural topology on $\mathbb{Z}$, that is, $\sup_{p \in \mathbb{P}} \nu_{\mathbb{Z}}^p = \nu_{\mathbb{Z}}$.

Exercise 2.4.5. Let $p \neq q$ be prime numbers. Show that the intersection $\nu_{\mathbb{Z}}^p \cap \nu_{\mathbb{Z}}^q$ of the p-adic topology $\nu_{\mathbb{Z}}^p$ and the q-adic topology $\nu_{\mathbb{Z}}^q$ is not a group topology on $\mathbb{Z}$.

Hint. Using the fact that $p^n\mathbb{Z} + q^m\mathbb{Z} = \mathbb{Z}$ for all $n, m \in \mathbb{N}$, deduce that $\inf\{\nu_{\mathbb{Z}}^p, \nu_{\mathbb{Z}}^q\}$ is the indiscrete topology. On the other hand, the set $\mathbb{Z} \setminus \{0\}$ is open in both topologies, so the intersection $\nu_{\mathbb{Z}}^p \cap \nu_{\mathbb{Z}}^q$ of these topologies is not the indiscrete topology, hence cannot be a group topology.

Exercise 2.4.6. Let G be a group. An *atom* (respectively, *coatom*) in the lattice $\mathfrak{L}(G)$ is an element $\tau \in \mathfrak{L}(G) \setminus \{\iota_G\}$ (respectively, $\tau \in \mathfrak{L}(G) \setminus \{\delta_G\}$) such that there exists no element $\eta \in \mathfrak{L}(G)$ with $\iota_G < \eta < \tau$ (respectively, $\tau < \eta < \delta_G$).

Let N be a normal subgroup of G and denote by τ_N the Alexandrov topology on G with $\mathrm{core}(G, \tau_N) = N$.

(a) Prove that if the index $[G : N]$ is a prime number, then τ_N is an atom in $\mathfrak{L}(G)$; if N is a finite minimal nontrivial normal subgroup of G, then τ_N is a coatom of $\mathfrak{L}(G)$.
(b) Prove that if $\tau \in \mathfrak{L}(G)$ is an atom, then $G/\mathrm{core}(G, \tau)$ is a simple group.
(c) Prove that all atoms in $\mathfrak{L}(G)$ have the form described in (a), when G is abelian.
(d) Describe the supremum of all atoms in $\mathfrak{L}(\mathbb{Z})$.
(e) For which abelian groups G the lattice $\mathfrak{L}(G)$ has no atoms?

Exercise 2.4.7. For a prime p, prove that $\sum_{n=1}^{\infty} p^n$ converges in $\mathbb{Q}_p$ (see Example 2.2.2) and compute the sum.

Exercise 2.4.8. Let G be a group. Prove that:

(a) the profinite topology of G is discrete (respectively, indiscrete) if and only if G is finite (respectively, G has no subgroups of finite index); in case G is abelian, the profinite topology of G is indiscrete if and only if G is divisible;
(b) the pro-p-finite topology of G is discrete (respectively, indiscrete) if and only if G is a finite p-group (respectively, G has no normal subgroups of index power of p);
(c) the p-adic topology of G is discrete if and only if G is a p-group of finite exponent; moreover, if G is abelian, then the p-adic topology of G is indiscrete if and only if G is p-divisible;
(d) the natural topology of G is discrete if and only if G is a group of finite exponent; moreover, if G is abelian, the natural topology of G is indiscrete if and only if G is divisible;
(e) the pro-countable topology of G is discrete (respectively, indiscrete) if and only if G is countable (respectively, G has no normal subgroups of countable index); in case G is abelian, the pro-countable topology of G is indiscrete if and only if G is trivial; give an example of a nonabelian group where the pro-countable topology is indiscrete;
(f) if G is abelian and $m, k \in \mathbb{N}_+$ are coprime, then $mG \cap kG = mkG$; hence $\nu_G = \sup\{\nu_G^p : p \in \mathbb{P}\}$.

Hint. (a) Note that every finite-index subgroup H of G contains a finite-index normal subgroup (namely, its normal core $H_G := \bigcap_{x\in G} x^{-1}Hx$).

(e) Prove that in every abelian group G the intersection of all subgroups of G with countable index is trivial. Deduce from this fact that every uncountable abelian group has (plenty of) countable-index subgroups. On the other hand, under the assumption $\mathfrak{c} = \omega_2 = 2^{\omega_1}$, for a set X with $|X| = \omega_1$ every proper subgroup G of $S(X)$ with $[S(X) : G] < 2^{\omega_1}$ must have index precisely ω_1 in view of [125, Observation 3], so $S(X)$ has no proper subgroups of countable index.

Exercise 2.4.9. Prove that the Euclidean topology τ on $\mathbb{R}$ is not functorial, and that the same holds for $\mathbb{T}$.

Hint. If τ were functorial, then every automorphism $\mathbb{R} \to \mathbb{R}$ would be τ-continuous. But while the continuous automorphisms of $\mathbb{R}$ are of the form $\phi(x) = ax$ for some $a \in \mathbb{R} \setminus \{0\}$, i. e., $\mathfrak{c}$-many, the automorphisms of $\mathbb{R}$ are $2^{\mathfrak{c}}$-many. The same argument works for $\mathbb{T}$ (the only continuous automorphisms of $\mathbb{T}$ are $id_{\mathbb{T}}$ and $-id_{\mathbb{T}}$).

Exercise 2.4.10. Let p be a prime, $n \in \mathbb{N}_+$, and $G = \mathbb{Z}^n$.
(a) Show that $\varpi_G = \nu_G$.
(b) Show that $\varpi_G^p = \nu_G^p$.
(c) Show that (a) and (b) remain true for every subgroup G of $\mathbb{Q}^n$.
(d) Describe the abelian groups G for which (a) holds true.

Exercise 2.4.11. Let p be a prime. Show that the p-adic topology of $\mathbb{J}_p$ coincides with its profinite topology as well as with the pro-p-finite topology.

Hint. It suffices to note that if H is a subgroup of $\mathbb{J}_p$ containing $m\mathbb{J}_p$ for some $m \neq 0$, then H has finite index.

Exercise 2.4.12. Prove that there are infinitely many primes in $\mathbb{Z}$ using the natural topology $\nu_{\mathbb{Z}}$ of $\mathbb{Z}$.

Hint. If $p_1, p_2, \dots, p_n$ were the only primes, then consider the union of the open (hence, closed) subgroups $p_1\mathbb{Z}, \dots, p_n\mathbb{Z}$ of $(\mathbb{Z}, \nu_{\mathbb{Z}})$ and use the fact that every $n \in \mathbb{Z} \setminus \{0, \pm 1\}$ has a prime divisor, so belongs to the closed set $F = \bigcup_{i=1}^n p_i\mathbb{Z}$. Therefore, the set $\{0, \pm 1\} = \mathbb{Z} \setminus F$ is an open neighborhood of 0, so must contain a nonzero subgroup $m\mathbb{Z}$, a contradiction.

Exercise 2.4.13. (a) Give an example of a group G where the profinite topology of G and the Bohr topology of G differ.
(b) Let G be an abelian group and $H = t(G^*)$. Prove that $\mathcal{T}_H = \varpi_G$.
(c) Let H be the family of all characters χ of an abelian group G such that $\chi(G)$ is finite and contained in the subgroup $\mathbb{Z}(p^\infty)$ of $\mathbb{T}$. Prove that $\mathcal{T}_H = \varpi_G^p$.

Hint. (b) Lemma 2.2.16 implies that $\mathcal{T}_H \leq \varpi_G$. For the proof of the other inclusion it remains to argue as in the proof of Proposition 2.2.15 and to observe that the characters appearing there are torsion.

Exercise 2.4.14. Prove that, for an abelian group G and a prime p, $\varpi_G^p = \mathfrak{B}_G$ if and only if G is a bounded p-group.

Exercise 2.4.15. For an abelian group G, let $\mathcal{S}_G$ be the group topology on G with the family $\{nG + \mathrm{Soc}(G) : n \in \mathbb{N}\}$ as a local base at 0. Prove that:
(a) $\mathcal{S}_G$ is a functorial topology and $\mathcal{S}_G = \iota_G$ when G is divisible;
(b) ([234]) every nondiscrete group topology on G is contained in a coatom of $\mathfrak{L}(G)$;
(c) * ([234]) the infimum $\mathcal{M}_G$ of all Hausdorff coatoms of $\mathfrak{L}(G)$ coincides with $\sup\{\mathcal{S}_G, \mathfrak{B}_G\}$.

Exercise 2.4.16. Let G be a group. Prove that:
(a) (Taĭmanov [271]) the family of centralizers $\{c_G(F): F \subseteq G \text{ finite}\}$ is a local base at e_G in a group topology $\mathfrak{T}_G$, called *Taĭmanov topology*, that is not necessarily linear;
(b) $\mathfrak{T}_G$ is discrete if and only if there exists a finite subset of G with trivial centralizer;
(c) $\mathfrak{T}_G$ is indiscrete if and only if G is abelian;
(d) if $f: G \to H$ is a surjective homomorphism, then $f: (G, \mathfrak{T}_G) \to (H, \mathfrak{T}_H)$ is continuous;
(e) is (d) true if f is not surjective?

Exercise 2.4.17. Let X be an infinite set. Prove that $S_\omega(X)$ is a dense normal subgroup of $(S(X), \mathrm{T}_X)$.

Exercise 2.4.18. Let X be a nonempty set. Prove that $(S(X), \mathrm{T}_X)$ has a local base at id_X formed by open subgroups, yet its topology is not linear.

Exercise 2.4.19. Let U, V be vector spaces over a field K. Prove that:
(a) the finite topology τ_{fin} of $\mathrm{Hom}(V, U)$ coincides with the pointwise convergence topology when $\mathrm{Hom}(V, U)$ is considered as a subset of U^V and U carries the discrete topology;
(b) the finite topology of $\mathrm{Hom}(V, U)$ is discrete if and only if $\dim V$ is finite;
(c) if $\dim U$ is finite, then for every finite subset F of V, the linear subspace $W(F)$ has finite codimension in $\mathrm{Hom}(V, U)$ (i. e., $\dim \mathrm{Hom}(V, U)/W(F) < \infty$).

Exercise 2.4.20. Equip a group G with the cofinite topology γ_G (see Example B.1.5). Then (G, γ_G) is a quasitopological group. Prove that (G, γ_G) is a topological group if and only if G is finite; in this case γ_G is discrete.

2.5 Further readings, notes, and comments

For an infinite abelian group G, $\mathfrak{L}(G)$ has the maximum possible cardinality $|\mathfrak{L}(G)| = 2^{2^{|G|}}$, but this size may collapse in the nonabelian case. The group G built in Example 4.3.5 has size $|G| = \mathfrak{c}$, while G is simple (see Remark 4.3.6(a)) and so $|\mathfrak{L}(G)| = 2$, the minimum possible size when G is a nontrivial group. Atoms and coatoms (i. e., maximal nondiscrete topologies) in the lattice of group topologies were studied by many authors (see [27, 50, 243]); these authors describe, among other things, also the atoms in $\mathfrak{L}(\mathbb{Z})$, although this was done much earlier by Mutylin [215].

The functorial topology $\mathcal{M}_G$ from Exercise 2.4.15 was introduced by Prodanov [234] under the name *submaximal topology*.

The lattice $\mathcal{T}(X)$, for a topological space X, is complemented. The situation completely changes in the lattice $\mathfrak{L}(G)$ for a group G. Therefore, one can split the notion of complementation into two natural components, of those we consider only one. Following [112], call a pair $\tau, \sigma \in \mathfrak{L}(G) \setminus \{\delta_G\}$ *transversal* if $\tau \vee \sigma = \delta_G$. The submaximal topology is related to complementation in $\mathfrak{L}(G)$ as follows: *a Hausdorff* $\tau \in \mathfrak{L}(G)$ *has a transversal topology if and only if* $\tau \not\leq \mathcal{M}_G$ (see [112]). This implies that every infinite abelian group admits a pair of transversal group topologies, since $\mathcal{M}_G$ is never maximal for an infinite abelian group G (i. e., G admits at least two distinct maximal topologies which are, of course, transversal).

Separation axioms for topological groups will be discussed in Chapters 3 and 5. More precisely, we show that for topological groups all separations axioms T_0–$T_{3.5}$ are equivalent (Proposition 3.1.15 and Theorem 5.2.14).

For paratopological groups the situation changes dramatically: all T_0–T_3 are distinct [274]. The question of whether T_3 and $T_{3.5}$ coincide as well remained open for about 60 years and was affirmatively resolved only recently by Banakh and Ravsky [17]. The Ellis theorem mentioned in §2.3 requires the Hausdorff axiom, for paratopological groups it was improved in Ravsky's PhD thesis in 2003 (see also the survey [274]) as follows: every locally compact (not necessarily Hausdorff) paratopological group is a topological group. There is a wealth of results in the spirit of the Ellis-Ravsky theorem ensuring that paratopological groups with some compactness type condition are topological groups. The reader can find more in this direction in [274, § 3], in particular [274, § 3.4] is focused on the case of semitopological and quasitopological groups (e. g., a pseudocompact quasitopological group need not be a topological group, in contrast with the case of paratopological groups mentioned in §2.3).

Finally, we briefly mention a useful tool in the study of paratopological groups. Namely, for a paratopological group (G,τ) with neighborhood filter $\mathcal{V}_\tau(e)$ at e, one can easily observe that the family $\{U^{-1} : U \in \mathcal{V}_\tau(e)\}$ is a neighborhood base of a paratopological group topology, named the *conjugate topology* of τ and denoted by τ^{-1}. The join $\tau^* = \tau \vee \tau^{-1}$ is a group topology, with neighborhood base $\{U \cap U^{-1} : U \in \mathcal{V}_\tau(e)\}$. It is easy to see that this is the coarsest group topology on G finer than τ. Therefore, the assignment $(G,\tau) \mapsto (G,\tau^*)$ is a coreflection from the category of paratopological groups to its subcategory of topological groups. For the nice properties of this coreflection, the reader may consult [274, § 4.1].

3 General properties of topological groups

3.1 Subgroups and separation

We start computing the closure $\overline{H}$ of a subset H of a topological group G.

Lemma 3.1.1. *Let G be a topological group. Then:*
(a) *for a subset A of G,*

$$\overline{A} = \bigcap_{U\in\mathcal{V}(e_G)} UAU = \bigcap_{U\in\mathcal{V}(e_G)} UA = \bigcap_{U\in\mathcal{V}(e_G)} AU = \bigcap_{U,V\in\mathcal{V}(e_G)} UAV;$$

(b) *if H is a (normal) subgroup of G, then $\overline{H}$ is a (normal) subgroup of G.*

Proof. Let $\mathcal{V} = \mathcal{V}(e_G)$.

(a) For $x \in G$, one has $x \notin \overline{A}$ if and only if there exists a neighborhood W of x such that $W \cap A = \emptyset$. Since $G \times G \to G$, $(a, b) \mapsto axb$, is continuous, we can find a symmetric open $U \in \mathcal{V}$ such that $UxU \subseteq W$. The latter implies $UxU \cap A = \emptyset$, and hence $x \notin UAU$. The other inclusions can be shown similarly.

(b) Let $x, y \in \overline{H}$. According to (a), to verify that $xy \in \overline{H}$, it suffices to see that $xy \in UHU$ for every $U \in \mathcal{V}$, and this follows from $x \in UH$ and $y \in HU$ for every $U \in \mathcal{V}$. Since the inversion is a homeomorphism, we obtain $\imath(\overline{H}) = \bigcap_{U\in\mathcal{V}} H^{-1}U^{-1} = \overline{H}$, hence $\overline{H}$ is a subgroup.

Assume H is normal. To prove that its closure $\overline{H}$ is normal, take an inner automorphism f of G. Then $f(H) = H$. Since f is a homeomorphism by Lemma 2.1.6, f takes $\overline{H}$ to itself, and this gives that $\overline{H}$ is normal. □

From Lemma 3.1.1(b) we deduce that, in every topological group G, $\overline{\{e_G\}}$ is a closed normal subgroup of G, as observed in Remark 2.1.17(a) and Lemma 2.1.16. Moreover, we get the next useful description of the closure of singletons in a topological group.

Corollary 3.1.2. *Let G be a topological group and $N = \overline{\{e_G\}}$. For every $x \in G$, $\overline{\{x\}} = xN = Nx$.*

Corollary 3.1.3. *If A, B are nonempty subsets of a topological group G, then $\overline{A}\,\overline{B} \subseteq \overline{AB}$. If one of the sets A, B is a singleton, then $\overline{A}\,\overline{B} = \overline{AB}$.*

Proof. The inclusion follows from Lemma 3.1.1(a), as $\overline{A}\,\overline{B} \subseteq UABU$ for every $U \in \mathcal{V}(e_G)$. In case $B = \{b\}$ is a singleton, $AB = Ab = t_b(A)$. Since t_b is a homeomorphism by Lemma 2.1.6, $\overline{AB} = \overline{Ab} = \overline{t_b(A)} = t_b(\overline{A}) = \overline{A}b \subseteq \overline{A}\,\overline{B}$. □

Clearly, $\overline{A}\,\overline{B}$ is dense in $\overline{AB}$, as it contains the dense subset AB of $\overline{AB}$. Therefore, the equality $\overline{A}\,\overline{B} = \overline{AB}$ holds true precisely when $\overline{A}\,\overline{B}$ is closed. Later we show that this often fails even in the group $\mathbb{R}$ (see Example 3.1.12). On the other hand, we show in Lemma 8.2.1(a) that equality holds when B is compact.

https://doi.org/10.1515/9783110654936-003

Remark 3.1.4. For a subset A of a topological group G and $x \in G$, $x \in \overline{A}$ precisely when there is a net $\{x_\alpha\}_{\alpha \in A}$ in A converging to x. We can assume without loss of generality that all converging nets $\{x_\alpha\}_{\alpha \in A}$ in G have the same directed set A as domain, namely, the filter $\mathcal{V}(e_G)$ ordered by inverse inclusion, or a fixed base $\mathcal{B}$ of $\mathcal{V}(e_G)$. Indeed, if $\{x_\alpha\}_{\alpha \in A}$ is a net in G that converges to $x \in G$, then for every $U \in \mathcal{B}$ there exists $\alpha_U \in A$ with $x_\beta \in U$ for all $\beta \geq \alpha_U$ in A. This gives a net $\{x_{\alpha_U}\}_{U \in \mathcal{B}}$ that converges to x. If necessary, we specify this fixed $\mathcal{B}$ and call $\{x_{\alpha_U}\}_{U \in \mathcal{B}}$ a $\mathcal{B}$*-net*, otherwise we write generic nets of G as $\{x_\alpha\}_{\alpha \in A}$, $\{y_\alpha\}_{\alpha \in A}$, etc., knowing that we can choose the same domain A.

3.1.1 Closed and dense subgroups

A subgroup H of a topological group G becomes a topological group when endowed with the subspace topology $\tau\restriction_H$ induced by the topology τ of G. Sometimes we refer to this situation by saying that H is a *topological subgroup* of G.

Example 3.1.5. The circle group $\mathbb{S}$ is a closed subgroup of the multiplicative group $\mathbb{C} \setminus \{0\}$, equipped with its usual topology. So, $\mathbb{S}$ itself is a topological group. Moreover, $\mathbb{S}$ is a compact and connected abelian group.

For a subgroup H of a topological group (G, τ), the filter

$$\mathcal{V}\restriction_H := \{H \cap V : V \in \mathcal{V}_\tau(e_G)\} = \mathcal{V}_{\tau\restriction_H}(e_H)$$

on H satisfies (gt1), (gt2), and (gt3).

Definition 3.1.6. Let G be a topological group and H a subgroup of G. If $f: H \to f(H) \subseteq G$ is a topological isomorphism, where $f(H)$ is a topological subgroup of G, then f is called a *topological group embedding*, or shortly *embedding*.

We start with properties of the open subgroups.

Proposition 3.1.7. *Let G be a topological group and H a subgroup of G. Then:*
(a) *H is open in G if and only if H has nonempty interior;*
(b) *if H is open, then H is also closed;*
(c) *if $[G : H] < \infty$, then H is closed if and only if H is open.*

Proof. (a) Let $\emptyset \neq V \subseteq H$ be an open set of G, and let $h_0 \in V$. Then $e_G \in h_0^{-1}V \subseteq H = h_0^{-1}H$. Now $U := h_0^{-1}V \subseteq H$ is open, contains e_G, and $h \in hU \subseteq H$ for every $h \in H$. Therefore, H is open.

(b) If H is open, then every coset gH, with $g \in G$, is open and consequently $G \setminus H = \bigcup_{g \in G \setminus H} gH$ is open. So, H is closed.

(c) In view of (b), it remains to add that when H is closed, $G \setminus H$ is closed as well, being a finite union of cosets of H. □

Definition 3.1.8. A subgroup H of a topological abelian group G is called:
(i) *dually embedded* if every continuous character of H can be extended to a continuous character of G;
(ii) *dually closed* if every point $x \in G \setminus H$ can be separated from H by a continuous character of G; more precisely, there exists a continuous character χ of G such that $\chi(x) \neq 0$ and $\chi \restriction_H = 0$.

If H is a dually closed subgroup of a topological abelian group G, then for every $x \in G \setminus H$ there exists a continuous character χ_x of G with $\chi_x \in A_{\widehat{G}}(H)$ such that $\chi_x(x) \neq 0$, so $H = \bigcap\{\ker \chi_x : x \in G \setminus H\}$. Since each $\ker \chi_x$ is closed, we conclude that H is closed. Hence, dually closed subgroups are closed.

Next we see that every subgroup is dually embedded and dually closed, hence closed, in the Bohr topology.

Proposition 3.1.9. *Let G be an abelian group and H a subgroup of G. Then H is (dually) closed in $(G, \mathfrak{B}_G)$ and $\mathfrak{B}_G \restriction_H = \mathfrak{B}_H$.*

Proof. To see that H is dually closed, pick $x \in G \setminus H$. Now it suffices to apply Corollary A.2.6 to the quotient G/H and deduce that for the nonzero element $x + H$ of G/H there exists a character $\xi: G/H \to \mathbb{T}$ with $\xi(x+H) \neq 0$. Now ξ composed with the canonical projection $G \to G/H$ gives a character χ of G that vanishes on H and $\chi(x) \neq 0$.

The inclusion $j: H^{\#} \hookrightarrow G^{\#}$ is continuous, by Lemma 2.2.12. To see that $j: H^{\#} \to j(H)$ is open, take a basic neighborhood $U_H(\chi_1, \dots, \chi_n; \delta)$ of 0 in $H^{\#}$, where $\chi_1, \dots, \chi_n \in H^*$. By Theorem A.2.4, each χ_i can be extended to some character $\xi_i \in G^*$, hence $U_H(\chi_1, \dots, \chi_n; \delta) = H \cap U_G(\xi_1, \dots, \xi_n; \delta)$ is open in $j(H)$. This proves that the subgroup topology $\mathfrak{B}_G \restriction_H$ of H coincides with $\mathfrak{B}_H$. □

Remark 3.1.10. In Lemma 2.2.4 we have seen that if $f: G \to H$ is a homomorphism of abelian groups, then f is continuous when both groups are equipped with their profinite (respectively, pro-p-finite, p-adic, natural, pro-countable) topology. Proposition 3.1.9 shows that f is actually a topological embedding if f is simply the embedding of a subgroup and one takes the Bohr topology on both groups. One can show that this fails for the profinite, pro-p-finite, p-adic or the natural topology (take, for example, $G = \mathbb{Z}$ and $H = \mathbb{Q}$).

The next example shows that the closed subgroups of $\mathbb{R}$ have a very simple description. The closed subgroups of $\mathbb{R}^n$ are completely described in Chapter 9.

Proposition 3.1.11. *For a proper subgroup H of $\mathbb{R}$, the following are equivalent:*
(a) *H is cyclic;*
(b) *H is discrete;*
(c) *H is closed;*
(d) *H is not dense in $\mathbb{R}$.*

Proof. (a)⇒(b) It is easy to see that a cyclic subgroup of $\mathbb{R}$ is discrete.

(b)⇒(c) Let $\varepsilon > 0$ and let the neighborhood $U = (-\varepsilon, \varepsilon)$ of 0 witness the discreteness of H, that is, $U \cap H = \{0\}$. Then obviously $|h - h'| \geq \varepsilon$ whenever h, h' are distinct elements of H. To see that H is closed, pick $x \in \mathbb{R} \setminus H$ and note that the neighborhood $V = (x - \varepsilon/2, x + \varepsilon/2)$ meets at most one element of H in view of the above observation. Hence, we can pick $0 < \eta < \varepsilon/2$ such that the smaller neighborhood $V' = (x - \eta, x + \eta)$ of x misses H.

(c)⇒(d) This is obvious.

(d)⇒(a) Since subgroups of cyclic groups are cyclic, and since the closure of a nondense subgroup is still nondense, we can replace H by $\overline{H}$ and assume without loss of generality that H is a closed proper subgroup of $\mathbb{R}$.

Let $h_0 = \inf\{h \in H : h > 0\}$. If $h_0 = 0$, then for every $\varepsilon > 0$ there exists $h \in (0, \varepsilon) \cap H$, and this yields that H hits every open interval of $\mathbb{R}$ of length ε. This implies that H is dense in $\mathbb{R}$, a contradiction. Therefore, $h_0 > 0$ and $h_0 \in H$ as H is closed. We prove now that $H = \langle h_0 \rangle$. Indeed, for $h \in H$, pick the greatest $m \in \mathbb{Z}$ such that $mh_0 \leq h < (m+1)h_0$; then $0 \leq h - mh_0 < h_0$ and $h - mh_0 \in H$, hence $h - mh_0 = 0$, that is, $h \in \langle h_0 \rangle$. □

It is not hard to see that every subgroup G of $\mathbb{R}$ has the same property, namely, for every subgroup H of G all properties (a)–(d) of Proposition 3.1.11 are equivalent (with (d) replaced by "H is not dense in G"). Obviously, the equivalence between (c) and (d) holds for subgroups H of finite prime index of every topological group.

Example 3.1.12. According to Proposition 3.1.11, a proper subgroup of $\mathbb{R}$ is dense if and only if it is not cyclic. In other words, the topological property of being dense is completely described by the purely algebraic one of being not cyclic.

This gives easy examples of closed subgroups H_1, H_2 of $\mathbb{R}$ such that $H_1 + H_2$ is not closed in $\mathbb{R}$. Since such H_1, H_2 are necessarily cyclic, we can take $H_1 = \mathbb{Z}$ and H_2 any cyclic subgroup generated by an irrational number. Then $H_1 + H_2$ is a proper noncyclic subgroup of $\mathbb{R}$, so by Proposition 3.1.11 it is dense in $\mathbb{R}$.

It is natural to expect that a cyclic subgroup need not be closed in general.

Definition 3.1.13. For a topological group G, let $M_G := \{x \in G : \overline{\langle x \rangle} = G\}$. We say that G is:
(i) *monothetic* if $M_G \neq \emptyset$;
(ii) *strongly monothetic* if every nontrivial subgroup of G is dense.

As shown in the next example, $\mathbb{T}$ is monothetic; moreover, the groups $\mathbb{T}^n$, with $n \in \mathbb{N}_+$, as well as $\mathbb{T}^{\mathbb{N}}$, are monothetic (see Exercise 3.5.19). Actually, even $\mathbb{T}^{\mathfrak{c}}$ is monothetic and contains a dense homomorphic image of $\mathbb{R}$, as we shall see later (see Theorem 9.4.8).

Example 3.1.14. Let $q_0 : \mathbb{R} \to \mathbb{T}$ be the canonical projection.

(a) A proper subgroup H of $\mathbb{T}$ is either closed (precisely when H is finite and cyclic), or dense (when H is infinite). Indeed, $L = q_0^{-1}(H)$ is a proper subgroup of $\mathbb{R}$ containing $\mathbb{Z}$. So, if H is closed then L is closed, hence cyclic by Proposition 3.1.11 and generated by a rational since $\mathbb{Z} \subseteq L$; therefore, H is a finite cyclic subgroup of $\mathbb{T}$. If H is infinite, then L is dense in $\mathbb{R}$ by Proposition 3.1.11, and so $H = q_0(L)$ is dense in $\mathbb{T}$ as well.

(b) The group $\mathbb{T}$ is monothetic. Indeed, pick an irrational $a \in \mathbb{R}$. Then the subgroup $N = \mathbb{Z}+\langle a\rangle$ of $\mathbb{R}$ is noncyclic, hence it is dense by Proposition 3.1.11. So, $\overline{\langle q_0(a)\rangle} = \mathbb{T}$ and $q_0(a) \in M_{\mathbb{T}}$. This proves that $M_{\mathbb{T}} = \mathbb{T} \setminus t(\mathbb{T})$.

We see in Corollary 3.1.23 that a Hausdorff monothetic group is necessarily abelian.

Clearly, strongly monothetic groups are monothetic, and Example 3.1.14(b) shows that this implication cannot be inverted.

Easier examples are offered by the cyclic group $\mathbb{Z}$, which is monothetic even when equipped with the discrete topology, whereas a Hausdorff group topology τ on $\mathbb{Z}$ makes it a strongly monothetic group precisely when $(\mathbb{Z}, \tau)$ has no proper open subgroups. For more properties of these groups see Corollary 3.1.23, Exercises 3.5.21 and 3.5.22 (see also Corollary 9.4.7 as well as Theorems 9.4.8 and 10.2.9).

3.1.2 Separation axioms

Here we discuss separation axioms in topological groups. Making use of Lemma 3.1.1, now we show that for a topological group all separation axioms T_0–T_3 are equivalent. Here, T_3 stands for "regular and T_1" (see Appendix B).

Proposition 3.1.15. *Let G be a topological group. Then G is a regular topological space and the following conditions are equivalent:*
(a) *G is T_0;*
(b) *G is T_1 (i. e., $\overline{\{e_G\}} = \{e_G\}$);*
(c) *G is Hausdorff;*
(d) *G is T_3.*

Proof. Since G is a homogeneous topological space, to prove regularity of G it suffices to check that for every $U \in \mathcal{V}(e_G)$ there exists $V \in \mathcal{V}(e_G)$ such that $\overline{V} \subseteq U$. According to Lemma 3.1.1, it suffices to pick a $V \in \mathcal{V}(e_G)$ such that $VV \subseteq U$. Then $\overline{V} \subseteq VV \subseteq U$. This property proves also the implication (b)⇒(d). Indeed, beyond being regular, G is a T_1 space. The latter property follows from the fact that all singletons $\{g\}$ of G are closed, as $\{e_G\}$ is closed.

On the other hand, clearly (d)⇒(c)⇒(b). Therefore, the properties (b), (c), and (d) are equivalent, and obviously imply (a). It remains to prove the implication (a)⇒(b).

Let $N = \overline{\{e_G\}}$ and assume for a contradiction that there exists $x \in N \setminus \{e_G\}$. Then $\overline{\{x\}} = xN = N$ according to Corollary 3.1.2, and hence $e_G \in \overline{\{x\}}$. This contradicts our assumption that G is T_0. □

The following is a consequence of Proposition 3.1.15 and Remark 2.1.17(a).

Corollary 3.1.16. *A topological group G is Hausdorff if and only if $\bigcap \mathcal{V}(e_G) = \{e_G\}$.*

It is not hard to check that a topological subgroup of a Hausdorff group is Hausdorff. Next we see that a discrete subgroup of a Hausdorff group is always closed.

Proposition 3.1.17. *Let G be a topological group and H a subgroup of G. If H is discrete and G is Hausdorff, then H is closed.*

Proof. Since H is discrete there exists $U \in \mathcal{V}(e_G)$ with $U \cap H = \{e_G\}$. Choose $V \in \mathcal{V}(e_G)$ with $V^{-1}V \subseteq U$. Then $|xV \cap H| \leq 1$ for every $x \in G$, as $h_1 = xv_1 \in xV \cap H$ and $h_2 = xv_2 \in xV \cap H$ give $h_1^{-1}h_2 \in (V^{-1}V) \cap H = \{e_G\}$, hence $h_1 = h_2$. Therefore, if $x \in G \setminus H$, one can find a neighborhood W of x such that $W \subseteq xV$ and $W \cap H = \emptyset$ (i. e., $x \notin \overline{H}$). Indeed, if $xV \cap H = \emptyset$, just take $W = xV$. In case $xV \cap H = \{h\}$ for some $h \in H$, one has $h \neq x$ as $x \notin H$. Since G is Hausdorff, $W = xV \setminus \{h\}$ is open, so the desired neighborhood of x. □

One cannot relax T_2 in Proposition 3.1.17, as shown in Exercise 3.5.4(a).

We discuss below stronger separation axioms and we shall see that every Hausdorff group is also a Tichonov space (see Theorem 5.2.14).

Remark 3.1.18. (a) Countable Hausdorff groups are normal, since they are regular by Proposition 3.1.15 and they are obviously Lindelöff: moreover, every regular Lindelöff space is normal (see Theorem B.5.10(b)).
(b) As far as normality is concerned, the situation is not so clear for uncountable topological groups. In fact, uncountable Hausdorff groups need not be normal as topological spaces (see Exercise 3.5.8). A nice "uniform" counterexample to this was given by Trigos [275]: for every uncountable abelian group G, the topological group $G^\#$ is not normal as a topological space.

Here we briefly discuss the question of when some functorial topologies are Hausdorff, other instances can be found in §3.5.

Definition 3.1.19. Let G be an abelian group and H a subset of G^*. We say that the characters of H *separate the points of* G (or that H *separates the points of* G) if for every $x \in G \setminus \{0\}$ there exists $\chi \in H$ with $\chi(x) \neq 1$.

According to Corollary A.2.6, G^* separates the points of G.

Proposition 3.1.20. *For an infinite abelian group G and a subgroup H of G^*:*
(a) *$\mathcal{T}_H$ is Hausdorff if and only if the characters of H separate the points of G;*
(b) *$\mathcal{T}_H$ is nondiscrete.*

Proof. (a) Assume that H separates the points of G and pick a nonzero $a \in G$. Then there exists $\chi \in H$ such that $\chi(a) \neq 1$. Let $\delta = \frac{1}{2}|\mathrm{Arg}(\chi(a))|$. Then $U(\chi;\delta)$ is a neighborhood of 0 in G missing a. This shows that $\mathcal{T}_H$ is Hausdorff.

Now assume that $\mathcal{T}_H$ is Hausdorff. To show that H separates the points of G pick $a \in G \setminus \{0\}$. Then there exists a basic $\mathcal{T}_H$-neighborhood W of 0 that misses a. We can take $W = U_G(\chi_1,\dots,\chi_n;\delta)$, with $n \in \mathbb{N}_+$, $\chi_1,\dots,\chi_n \in H$ and $\delta > 0$. Then $a \notin W$ gives $\mathrm{Arg}(\chi_i(a)) \geq \delta$ for some $i \in \{1,\dots,n\}$. Hence, $\chi_i(a) \neq 1$.

(b) Suppose, for a contradiction, that $\mathcal{T}_H$ is discrete. Then there exist $\chi_1,\dots,\chi_n \in H$ and $\delta > 0$ such that $U(\chi_1,\dots,\chi_n;\delta) = \{0\}$. In particular, $\bigcap_{i=1}^n \ker\chi_i = \{0\}$, hence the diagonal homomorphism $f = \chi_1 \times \cdots \times \chi_n \colon G \to \mathbb{S}^n$, $g \mapsto (\chi_1(g),\dots,\chi_n(g))$, is injective and $f(G) \cong G$ is an infinite discrete subgroup of $\mathbb{S}^n$, by our hypothesis. According to Proposition 3.1.7, $f(G)$ is closed in $\mathbb{S}^n$, and consequently compact. This is a contradiction, since compact discrete spaces are finite. □

Proposition 3.1.21. *For a group G the following conditions are equivalent:*
(a) *the profinite topology ϖ_G of G is Hausdorff;*
(b) *G is residually finite.*

If G is abelian, then they are equivalent also to:
(c) *the natural topology ν_G of G is Hausdorff;*
(d) $G^1 = \{0\}$.

Proof. (a)⇔(b) and (c)⇔(d) are obvious, in view of Corollary 3.1.16. Since $\varpi_G \leq \nu_G$, the first pair of conditions imply those of the second one. (d)⇒(a) follows from the fact that G^1 coincides with the intersection of all subgroups of G of finite index (see Proposition A.4.6(d)). □

By the above proposition, one obtains a purely topological proof of the fact that a reduced abelian group G has no nontrivial divisible subgroups: if D is a divisible subgroup of G, equip G and D with their profinite topologies; this makes the inclusion $j\colon D \hookrightarrow G$ continuous. Since D is indiscrete, $j(D)$ is an indiscrete subgroup of the Hausdorff group G. Therefore, $D = j(D)$ is trivial.

3.1.3 Extension of identities in Hausdorff groups

Hausdorff groups have a remarkable property that we shall briefly refer to as *extension of identities*, where identity is meant in the sense of Remark C.1.9. The following theorem treats the specific identity $xy = yx$ (written also in the form $[x,y] = 1$) that defines the variety of all abelian groups:

Theorem 3.1.22. *If G is a Hausdorff group containing a dense abelian subgroup H, then G is abelian.*

Proof. Take $x, y \in G$. Then by Remark 3.1.4, there exist nets $\{h_\alpha\}_{\alpha\in A}$, $\{g_\alpha\}_{\alpha\in A}$ in H such that $x = \lim_{\alpha\in A} h_\alpha$ and $y = \lim_{\alpha\in A} g_\alpha$, where it is easy to see that $[x,y] = \lim_{\alpha\in A}[h_\alpha, g_\alpha] = e_G$ as H is abelian. Then $[x,y] = e_G$ by the uniqueness of the limit in Hausdorff groups (see Lemma B.2.6(d)). □

An immediate application of Theorem 3.1.22 gives:

Corollary 3.1.23. *A Hausdorff monothetic group is necessarily abelian.*

Theorem 3.1.22 has the following counterpart for nilpotent groups.

Theorem 3.1.24. *If G is a Hausdorff group containing a dense nilpotent subgroup H of class s, then G is nilpotent of class s.*

Proof. For $g_0, g_1, \dots, g_s \in G$, by the density of H in G and Remark 3.1.4, we can write, for every $n \in \{0, 1, \dots, s\}$, $g_n = \lim_{\alpha\in A} h_{n,\alpha}$ where $\{h_{n,\alpha}\}_{\alpha\in A}$ is a net in H. Then, as H is nilpotent of class $\le s$,

$$[[\dots[[g_0, g_1], g_2], \dots], g_s] = \lim_{\alpha\in A}[[\dots[[h_{0,\alpha}, h_{1,\alpha}], h_{2,\alpha}]\dots], h_{s,\alpha}] = e_G.$$

Therefore, $[[\dots[[g_0, g_1], g_2], \dots], g_s] = e_G$ by the uniqueness of the limit in Hausdorff groups (see Lemma B.2.6(b)). □

One can prove similarly that if G is a Hausdorff group containing a dense solvable group, then G is solvable (see Exercise 3.5.3). More generally, if $\mathfrak{V}$ is a variety of groups (see Definition C.1.8 and Remark C.1.9), then one can prove that, if G is a Hausdorff group containing a dense subgroup H that belongs to $\mathfrak{V}$, then $G \in \mathfrak{V}$.

Example 3.1.25. For a Hausdorff group G, the centralizer $c_G(g)$ of each $g \in G$ and so the center $Z(G)$ are closed subgroups of G. Indeed, for every $g \in G$, by Exercise B.7.12 we conclude that $c_G(g)$ is closed in G. Then also $Z(G) = \bigcap_{g\in G} c_G(g)$ is closed in G.

3.2 Quotients of topological groups

Let G be a topological group, H a normal subgroup of G, and $q\colon G \to G/H$ the canonical projection. The quotient G/H can be equipped with a topology, named *quotient topology*, in two equivalent ways. First, as $\{q(U)\colon U \in \mathcal{V}_G(e_G)\}$ is a filter on G/H satisfying (gt1), (gt2), and (gt3), it gives rise to a group topology on G (by Theorem 2.1.10) that makes q continuous and open (see Lemma 3.2.1). The second alternative is to just take the finest topology on G/H that makes the canonical projection $q\colon G \to G/H$ continuous, that is, the quotient topology defined in the category of *topological spaces*. Since we have a group topology on G, the quotient topology consists of all sets $q(U)$, where U runs over the family of all open sets of G (as $q^{-1}(q(U))$ is open in G in such a case). So, both approaches give *the same topology* on G/H, that we refer to as the quotient topology of G/H.

Lemma 3.2.1. *Let G be a topological group, H a normal subgroup of G, and let G/H be equipped with the quotient topology. Then:*
(a) *the canonical projection $q: G \to G/H$ is open;*
(b) *if $f: G/H \to G_1$ is a homomorphism to a topological group G_1, then f is continuous if and only if $f \circ q$ is continuous.*

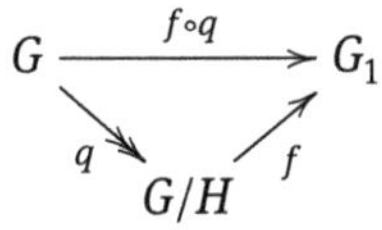

Proof. (a) Let $U \neq \emptyset$ be an open set of G. Then $q^{-1}(q(U)) = HU = \bigcup_{h\in H} hU$ is open, since each hU is open. Therefore, $q(U)$ is open in G/H.

(b) If f is continuous, then the composition $f \circ q$ is obviously continuous. If $f \circ q$ is continuous and W in an open set of G_1, then $(f \circ q)^{-1}(W) = q^{-1}(f^{-1}(W))$ is open in G, so $f^{-1}(W)$ is open in G/H. This proves that f is continuous. □

Remark 3.2.2. If G is a topological group and H simply a subgroup of G, the quotient set G/H consisting of left cosets of G need not be a group, yet it carries the quotient topology that makes it a homogeneous topological space, and the continuous map $q: G \to G/H$ is open.

Theorem 3.2.3 (Frobenius theorem). *Let G, H be topological groups, $f: G \to H$ a continuous surjective homomorphism, and $q: G \to G/\ker f$ the canonical projection, and let $f_1: G/\ker f \to H$ be the unique homomorphism with $f = f_1 \circ q$:*

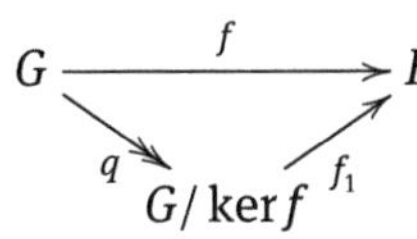

Then f_1 is a continuous isomorphism. Moreover, f_1 is a topological isomorphism if and only if f is open.

Proof. It follows immediately from the definitions of quotient topology and open map, and from Lemma 3.2.1. □

As a first application of Frobenius theorem 3.2.3, we show that the quotient topology is invariant under topological isomorphism in the following sense.

Corollary 3.2.4. *Let G, H be topological groups and $f: G \to H$ a topological isomorphism. Then for every normal subgroup N of G the quotient $H/f(N)$ is topologically isomorphic to G/N.*

Proof. Obviously, $f(N)$ is a normal subgroup of H and the canonical projection $q: H \to H/f(N)$ is continuous and open by Lemma 3.2.1. Therefore, $h = q \circ f: G \to H/f(N)$ is an open continuous surjective homomorphism with $\ker h = N$.

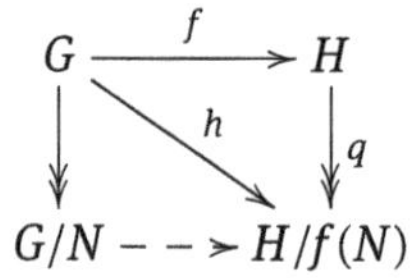

We conclude that $H/f(N)$ is topologically isomorphic to G/N by Frobenius theorem 3.2.3. □

One can order the continuous surjective homomorphisms with common domain the topological group G calling a continuous surjective homomorphism $f\colon G \to H$ *projectively larger* than another continuous surjective homomorphism $f'\colon G \to H'$ when there exists a continuous homomorphism $\eta\colon H \to H'$ such that $f' = \eta \circ f$:

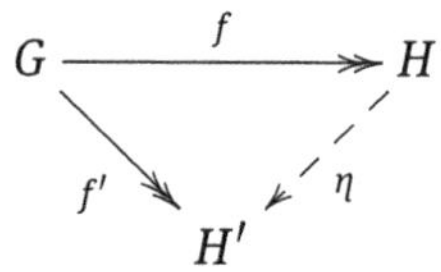

In the next proposition we show that, roughly speaking, the projective order between open continuous surjective homomorphisms with the same domain corresponds to the order by inclusion of their kernels.

Proposition 3.2.5. *Let G, H_1, H_2 be topological abelian groups and $f_i\colon G \to H_i$, $i = 1, 2$, open continuous surjective homomorphisms. Then there exists a continuous homomorphism $\eta\colon H_1 \to H_2$ such that $f_2 = \eta \circ f_1$ if and only if $\ker f_1 \subseteq \ker f_2$. Moreover, η is a topological isomorphism if and only if $\ker f_1 = \ker f_2$.*

Proof. The necessity is obvious. So, assume that $\ker f_1 \subseteq \ker f_2$. By Frobenius theorem 3.2.3 applied to f_i, for $i = 1, 2$, there exists a topological isomorphism $j_i\colon G/\ker f_i \to H_i$ such that $f_i = j_i \circ q_i$, where $q_i\colon G \to G/\ker f_i$ is the canonical projection. As $\ker f_1 \subseteq \ker f_2$, we get a homomorphism $t\colon G/\ker f_1 \to G/\ker f_2$ that makes the diagram

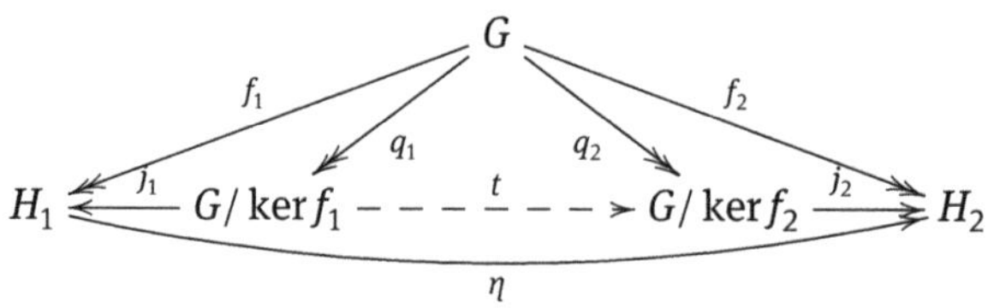

commute. Moreover, t is continuous by Lemma 3.2.1(b). Obviously, $\eta = j_2 \circ t \circ j_1^{-1}$ works.

If $\ker f_1 = \ker f_2$, then t is a topological isomorphism, hence η is a topological isomorphism as well. □

Regardless of its simplicity, Frobenius theorem 3.2.3 is very useful since it produces topological isomorphisms as in the above proof. The openness of f is its main ingredient, so from now on we are interested in providing conditions that ensure openness.

Lemma 3.2.6. *Let X, Y be topological spaces and let $f: X \to Y$ be an open continuous map. Then for every subspace P of Y with $P \cap f(X) \neq \emptyset$, the restriction $f' = f\restriction_{H_1}: H_1 \to P$ to the subspace $H_1 = f^{-1}(P)$ is open.*

Proof. To see that f' is open, choose a point $x \in H_1$ and a neighborhood U of x in H_1. Then there exists a neighborhood W of x in X such that $U = H_1 \cap W$. To see that $f'(U)$ is a neighborhood of $f'(x)$ in P, note that if $f(w) \in P$ for $w \in W$, then $w \in H_1$, hence $w \in H_1 \cap W = U$. Thus, $f(W) \cap P \subseteq f(U) = f'(U)$. □

We shall apply Lemma 3.2.6 when $X = G$ and $Y = G_1$ are topological groups and $f = q: G \to G_1$ is an open continuous homomorphism. Then the restriction $q^{-1}(P) \to P$ of q is open for every subgroup P of G_1.

Nevertheless, even in the particular case when q is surjective, the restriction $H \to q(H)$ of q to an arbitrary closed subgroup H of G need not be open:

Example 3.2.7. For $G = \mathbb{T}$ the continuous homomorphism $\mu_2: G \to G$, $x \mapsto 2x$, is surjective and open. Let now $H = \mathbb{Z}(3^\infty) \leq \mathbb{T}$. The restriction $\mu_2': H \to 2H = H$ of μ_2 is a continuous isomorphism.

To see that μ_2' is not open, it suffices to notice that the sequence $\{x_n\}_{n\in\mathbb{N}_+}$ in H, defined by $x_n = \sum_{k=1}^n 1/3^k$ for every $n \in \mathbb{N}_+$, is not convergent in H (as it converges to the point $1/2 \in \mathbb{T} \setminus H$). On the other hand, obviously $\mu_2'(x_n) = 2x_n \to 0$ in H.

Next we see some relevant isomorphisms related to the quotient groups.

Theorem 3.2.8. *Let G be a topological group, N a normal subgroup of G, and $q: G \to G/N$ the canonical projection.*

(a) *If H is a subgroup of G, then the homomorphism $q_1: HN/N \to q(H)$, defined by $q_1(xN) = q(x)$ for every $x \in H$, is a topological isomorphism.*
(b) *If H is a normal subgroup of G with $N \subseteq H$, then $q(H) = H/N$ is a normal subgroup of G/N and the map $j: G/H \to (G/N)/(H/N)$, defined by $j(xH) = (xN)(H/N)$ for every $x \in G$, is a topological isomorphism. If H is closed, then also $q(H)$ is closed.*
(c) *If H is a subgroup of G, then the map $s: H/H\cap N \to (HN)/N$, defined by $s(x(H\cap N)) = xN$ for every $x \in H$, is a continuous isomorphism. It is a topological isomorphism if and only if the restriction $q\restriction_H: H \to (HN)/N$ is open.*

(The quotient groups are always equipped with the quotient topology.)

Proof. (a) As $HN = q^{-1}(q(H))$, we can apply Lemma 3.2.6 and conclude that the restriction $q': HN \to q(H)$ of q is open. Now Frobenius theorem 3.2.3 applies to q' and implies that the unique homomorphism $q_1: HN/N \to q(H)$ satisfying $q_1 \circ p = q'$, where $p: HN \to HN/N$ is the canonical projection, is a topological isomorphism.

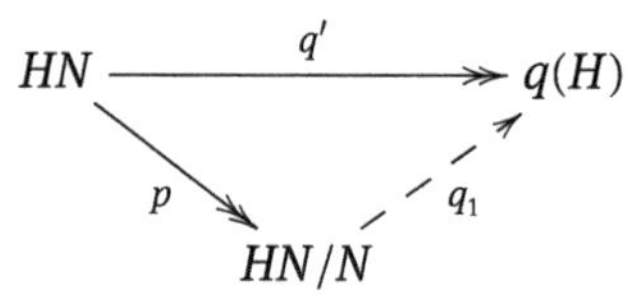

(b) Since $H = HN$, item (a) implies that the induced topology of $q(H)$ coincides with the quotient topology of H/N. Hence, we can identify H/N with the topological subgroup $q(H)$ of G/N. The composition $f: G \to (G/N)/(H/N)$ of q with the canonical projection $\pi: G/N \to (G/N)/(H/N)$ is open, the latter being open. Let $s: G \to G/H$ be the canonical projection. Applying Frobenius theorem 3.2.3 to the open homomorphism f with $\ker f = H$, we find a topological isomorphism j' making the diagram

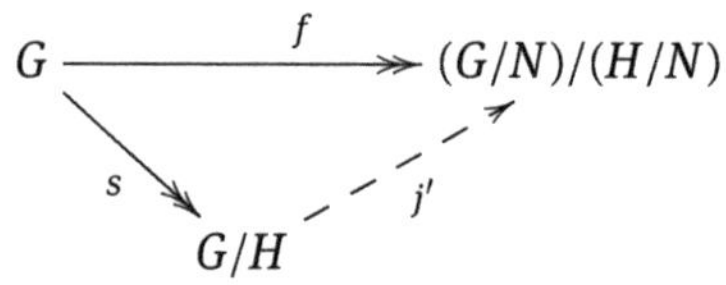

commute. Since also j makes the diagram commutative and s is surjective, we deduce that $j = j'$ is a topological isomorphism.

If $H = HN$ is closed, then $G \setminus HN$ is open, so $(G/N) \setminus q(HN) = q(G \setminus HN)$ is open as well. Therefore, $q(H)$ is closed.

(c) To the continuous surjective homomorphism $q \restriction_H: H \to (HN)/N$ we apply Frobenius theorem 3.2.3 to find a continuous isomorphism $s: H/H \cap N \to (HN)/N$, that is necessarily defined by $s(x(H \cap N)) = xN$ for every $x \in H$, as it makes the diagram

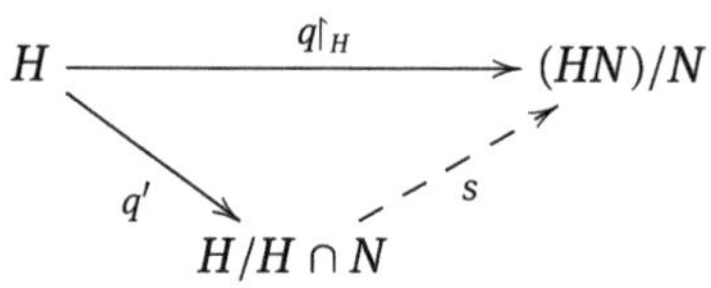

commute, where q' is surjective. By the same theorem, s is a topological isomorphism if and only if $q \restriction_H: H \to (HN)/N$ is open. □

Example 3.2.7 shows that the continuous isomorphism $s: H/H \cap N \to (HN)/N$ need not be open: take $G = \mathbb{T}$, $H = \mathbb{Z}(3^\infty)$, and $N = \langle 1/2 \rangle$, so that $H \cap N = \{0\}$ and $s: H \to (H + N)/N = q(H)$ is not open.

The next theorem gives a precise description when this occurs:

Theorem 3.2.9. *Let G be a topological group, N a closed normal subgroup of G, and $q: G \to G/N$ the canonical projection. If H is a dense subgroup of G, then $q_1 := q \restriction_H: H \to q(H)$ is open if and only if $N \cap H$ is dense in N.*

Proof. Let $N_1 = \overline{H \cap N}$.

Assume that q_1 is open and consider first the case when $N_1 = \{e_G\}$. This assumption implies that G is Hausdorff, and hence the limits of nets are unique. Assume for a contradiction that $N \neq N_1$ and pick $x \in N \setminus \{e_G\}$. By the density of H in G, there exists a net $\{g_\alpha\}_{\alpha\in A}$ in H converging to x. Then $q(g_\alpha) \to e_{G/N}$ in $q(H)$ equipped with the topology induced by G/N. Therefore, the isomorphism (in this case) $q_1: H \to q(H)$ is not open, as $g_\alpha \not\to e_G$ in G, a contradiction.

In the general case, consider the canonical projections $\pi: G \to G/N_1$ and $q': G/N_1 \to G/N$, and their restrictions $\pi_1 = \pi \restriction_H: H \to \pi(H)$ and $q_1' = q' \restriction_{\pi(H)}: \pi(H) \to q(H) = q'(\pi(H))$:

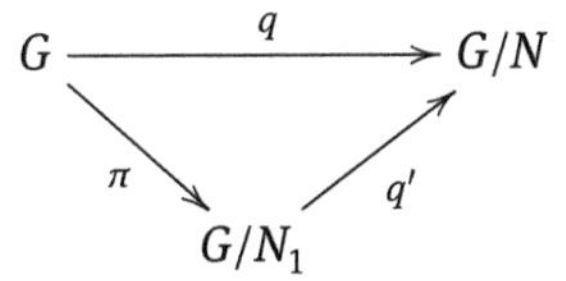

The openness of $q_1 = q_1' \circ \pi_1$ and the surjectivity of π_1 imply that q_1' is open. Hence, the above argument applied to G/N_1, its closed subgroup N/N_1, and its dense subgroup $\pi(H)$ implies $N = N_1$. In fact, $\pi(H) \cap N/N_1 = \{e_{G/N_1}\}$, and so also $\overline{\pi(H) \cap N/N_1} = \{e_{G/N_1}\}$ since G/N_1 is Hausdorff; by the above argument, $N/N_1 = \{e_{G/N_1}\}$ as well, that is, $N = N_1$.

Vice versa, assume that $N = N_1$ and pick $U, V \in \mathcal{V}_H(e_G)$ with $VV \subseteq U$. Then $\overline{V} \in \mathcal{V}_G(e_G)$, so $q(\overline{V}) \in \mathcal{V}_{G/N}(e_{G/N})$. Hence, it suffices to check that $q(\overline{V}) \cap q(H) \subseteq q(U)$. Pick $h \in H$ such that $q(h) \in q(\overline{V})$ so that $h = v'x$ with $v' \in \overline{V}$, $x \in N = \overline{H \cap N}$. Then there exist nets $\{h_\alpha\}_{\alpha\in A}$ and $\{v_\alpha\}_{\alpha\in A}$ in $H \cap N$ and V, respectively, such that $h_\alpha \to x$ and $v_\alpha \to v'$. This implies $v_\alpha h_\alpha \to v'x = h$, and so $hh_\alpha^{-1}v_\alpha^{-1} \to e_G$ in H. Therefore, there exists $\beta \in A$ such that $hh_\beta^{-1}v_\beta^{-1} \in V$, hence $h \in Vv_\beta h_\beta \subseteq VVN$. This implies that $q(h) \subseteq q(VV) \subseteq q(U)$, and we can conclude that $q(U) \in \mathcal{V}_{q(H)}(e_{G/N})$. □

Now, for a normal subgroup H of a topological group G, we relate properties of the quotient G/H to those of H.

Lemma 3.2.10. *For a topological group G and a normal subgroup H of G:*
(a) *G/H is discrete if and only if H is open;*
(b) *G/H is Hausdorff if and only if H is closed.*

Proof. Let $q: G \to G/H$ be the canonical projection.

(a) If G/H is discrete, then $H = q^{-1}(e_{G/H})$ is open since the singleton $\{e_{G/H}\}$ is open. If H is open, then $\{e_{G/H}\} = q(H)$ is open since the map q is open.

(b) If G/H is Hausdorff, then $H = q^{-1}(e_{G/H})$ is closed since $\{e_{G/H}\}$ is closed. If H is closed, then $\{e_{G/H}\} = q(H)$ is closed, by Theorem 3.2.8(b). □

Example 3.2.11. An Alexandrov group G is simply an extension of an indiscrete group $N = \overline{\{e_G\}}^\tau$ by the discrete one G/N.

Here we consider the counterpart of Remark 3.1.10 for quotient groups.

Proposition 3.2.12. *Let G, H be abelian groups and $f: G \to H$ a surjective homomorphism. Then f is (continuous and) open when both G, H are equipped with their profinite (respectively, pro-p-finite, p-adic, natural, pro-countable) topology.*

Proof. For the profinite and for the pro-countable topology use that fact that for a subgroup N of G the homomorphism $f_1: G/N \to H/f(N)$ induced by f is surjective. The remaining cases are trivial. □

We shall see in the sequel (see Corollary 11.2.8) that if G is an abelian group equipped with its Bohr topology and H is a subgroup of G, then the quotient topology of G/H coincides with the Bohr topology of G/H.

3.3 Initial and final topologies: products of topological groups

Proposition 3.3.1. *Let $\{G_i : i \in I\}$ be a family of topological groups. The direct product $G = \prod_{i \in I} G_i$, equipped with the product topology, is a topological group.*

Proof. The filter $\mathcal{V}(e_G)$ in the product topology of G has as a base the family of neighborhoods

$$\bigcap_{k=1}^{n} p_{j_k}^{-1}(U_{j_k}) = U_{j_1} \times \cdots \times U_{j_n} \times \prod_{i \in I \setminus J} G_i,$$

where $J = \{j_1, \ldots, j_n\}$ varies among all finite subsets of I and $U_j \in \mathcal{V}(e_{G_j})$ for all $j \in J$. It is easy to check that the filter $\mathcal{V}(e_G)$ satisfies the conditions (gt1), (gt2), and (gt3) from Theorem 2.1.10. Due to (gt3), for an arbitrary element $a \in G$, one can easily check that $\mathcal{V}(a) = a\mathcal{V}(e_G) = \mathcal{V}(e_G)a$. Hence, G is a topological group. □

Definition 3.3.2. For a group G, a family $\{K_i : i \in I\}$ of topological groups, and a family $\mathcal{F} = \{f_i : i \in I\}$ of group homomorphisms $f_i : G \to K_i$, the *initial topology* of $\mathcal{F}$ is the coarsest group topology on G that makes continuous all $f_i \in \mathcal{F}$.

Namely, the initial topology of $\mathcal{F}$ is obtained by taking as local base at e_G the family $\{\bigcap_{i \in J} f_i^{-1}(U_i) : J \subseteq I \text{ finite}, \forall i \in J, U_i \in \mathcal{V}_{K_i}(e_{K_i})\}$.

The initial topology is introduced in the same way for topological spaces and continuous maps (see Exercise B.7.6). For a group G, a family $\{K_i : i \in I\}$ of topological groups, and a family $\mathcal{F} = \{f_i : i \in I\}$ of group homomorphisms $f_i : G \to K_i$, the initial topology of $\mathcal{F}$ introduced above coincides with the initial topology for topological spaces.

Example 3.3.3. (a) For a topological group G and a subgroup H of G, the subgroup topology of H is the initial topology of the inclusion map $H \hookrightarrow G$.

(b) For a family $\{G_i : i \in I\}$ of topological groups, the product topology of $G = \prod_{i \in I} G_i$ is the initial topology of the family $\{p_i : i \in I\}$ of the projections $p_i : G \to G_i$.
(c) Let G be a group, let $\{K_i : i \in I\}$ be the family of all finite quotient groups $K_i = G/N_i$ of G equipped with the discrete topology and, for every $i \in I$, let $f_i : G \to K_i$ be the canonical projection. Then the profinite topology ϖ_G of G coincides with the initial topology of the family $\{f_i : i \in I\}$.
(d) For a fixed prime p, the pro-p-topology of a group G can be described in a similar manner as the profinite topology, using the finite quotients G/N_i of G that are p-groups. The p-adic topology of G is obtained if one takes all quotients of G of finite exponent that is a power of p.
(e) To obtain the natural topology of a group G as an initial topology in the above sense, one has to make recourse to all quotients of G of finite exponent.
(f) The cocountable topology of a group G can be obtained as the initial topology in the above sense, if one takes all countable quotients of G.
(g) For G an abelian group and a set $H = \{f_i : i \in I\}$ of characters $f_i : G \to \mathbb{T}$ of G, the initial topology of H coincides with the topology $\mathcal{T}_H$ defined in §2.2.2. In particular, the Bohr topology $\mathfrak{B}_G$ of G is the initial topology of G^*.

Remark 3.3.4. Let G be a group, $\{K_i : i \in I\}$ a family of topological groups, and $\mathcal{F} = \{f_i : i \in I\}$ a family of group homomorphisms $f_i : G \to K_i$. The initial topology of $\mathcal{F}$ on G coincides with the initial topology of the single diagonal map

$$\Delta_{\mathcal{F}} : G \to \prod_{i \in I} K_i,\ x \mapsto (f_i(x)).$$

The map $\Delta_{\mathcal{F}} : G \to \prod_{i \in I} K_i$ is injective if and only if $\mathcal{F}$ separates the points of G. In this case the initial topology of $\mathcal{F}$ coincides with the subgroup topology of G induced by the product topology of $\prod_{i \in I} K_i$, when G is identified with $\Delta_{\mathcal{F}}(G)$.

Now we define an inverse system of topological groups and its inverse limit.

Definition 3.3.5. Let $(I, \leq)$ be a directed set.
(i) An *inverse system* of topological groups, indexed by a directed poset $(I, \leq)$, is a family $\{G_i : i \in I\}$ of topological groups and continuous homomorphisms $\nu_{ij} : G_j \to G_i$ for every pair $i \leq j$ in I, such that $\nu_{ii} = id_{G_i}$ for every $i \in I$ and, for every triple $i \leq j \leq k$ in I, $\nu_{ij} \circ \nu_{jk} = \nu_{ik}$:

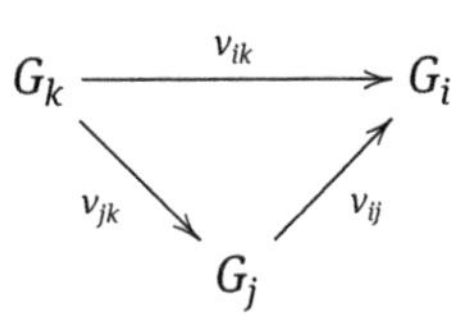

We briefly denote by $[G_i, (\nu_{jk}), I]$ this inverse system.

(ii) An *inverse limit* of an inverse system $[G_i, (v_{jk}), I]$ is a topological group G and a family $\{p_i : i \in I\}$ of continuous homomorphisms $p_i : G \to G_i$ satisfying $p_i = v_{ij} \circ p_j$ for every pair $i \leq j$ in I,

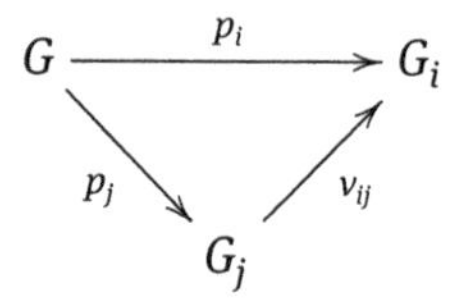

such that, for every topological group H and every family of continuous homomorphisms $q_i : H \to G_i$ satisfying $q_i = v_{ij} \circ q_j$ for every pair $i \leq j$ in I, there exists a unique continuous homomorphism $t : H \to G$ such that $q_i = p_i \circ t$ for every $i \in I$.

As the inverse limit of the inverse system $[G_i, (v_{jk}), I]$ determined in item (b) exists and it is unique up to isomorphism, we denote it by $\varprojlim_{i \in I} G_i$:

Proposition 3.3.6. *Let $[G_i, (v_{ij}), I]$ be an inverse system of topological groups.*
(a) *In the product $H = \prod_{i \in I} G_i$, consider the subgroup*

$$G = \{x = (x_i)_{i \in I} \in H : v_{ij}(x_j) = x_i \text{ whenever } i \leq j\}$$

and, for every $i \in I$, denote by $p_i : G \to G_i$ the restriction to G of the canonical projection $H \to G_i$. Then:
(a_1) *G along with the family $\{p_i : i \in I\}$ is an inverse limit of $[G_i, (v_{ij}), I]$;*
(a_2) *G has the initial topology of $\{p_i : i \in I\}$.*
(b) *The inverse limit $\varprojlim_{i \in I} G_i$ is unique up to isomorphism.*

Proof. (a) This is straightforward, since $p_i : G \to G_i$ is continuous for every $i \in I$.

(b) Suppose that G with $\{p_i : i \in I\}$ and L with $\{q_i : i \in I\}$ are inverse limits of $[G_i, (v_{ij}), I]$. Then $N = \bigcap_{i \in I} \ker q_i = \{0\}$, since there must be a unique $t : N \to L$ such that $q_i \circ t = 0$ for all $i \in I$ and both the inclusion map $j : N \hookrightarrow L$ and the zero morphism $z : N \to L$ satisfy this condition, so $j = z$ yields $N = \{0\}$.

There exists a unique continuous homomorphism $f : L \to G$ such that $q_i = p_i \circ f$ for every $i \in I$, and there exists a unique continuous homomorphism $f' : G \to L$ such that $p_i = q_i \circ f'$ for every $i \in I$. For every $i \in I$, $p_i \circ (f \circ f') = p_i$:

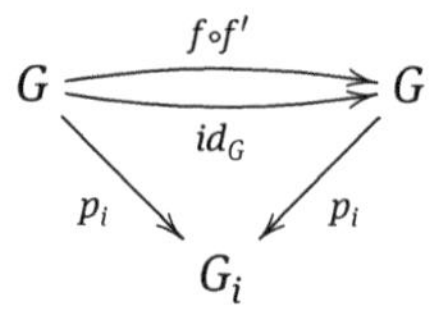

hence $f \circ f' = id_G$. Analogously, for every $i \in I$, $q_i \circ (f' \circ f) = q_i$, hence $f' \circ f = id_L$ (using, in both cases, the equalities $\bigcap_{i \in I} \ker p_i = \{0\}$ and $\bigcap_{i \in I} \ker q_i = \{0\}$). We conclude that f is a topological isomorphism. □

Remark 3.3.7. Proposition 3.3.6 yields that, for an inverse system $[G_i, (\nu_{ij}), I]$ of topological groups, $\varprojlim_{i\in I} G_i$ (identified with G) is a closed subgroup of $\prod_{i\in I} G_i$.

Very often the inverse systems have a more simple form as follows.

Example 3.3.8. Let $\{G_n : n \in \mathbb{N}\}$ be a family of topological groups and let $\{\phi_n : n \in \mathbb{N}\}$ be a family of continuous homomorphisms $\phi_n : G_{n+1} \to G_n$. Putting, for every pair $m, n \in \mathbb{N}$ with $m > n$, $\varphi_{nm} = \phi_n \circ \cdots \circ \phi_{m-1} : G_m \to G_n$,

$$G_m \xrightarrow{\phi_{m-1}} G_{m-1} \xrightarrow{\phi_{m-2}} \cdots \xrightarrow{\phi_{n+1}} G_{n+1} \xrightarrow{\phi_n} G_n f$$

we obtain an inverse system $[G, (\varphi_{nm}), \mathbb{N}]$ that we simply write as $[G_n, (\phi_n), \mathbb{N}]$.

In these terms, if p is a prime and $\phi_n : \mathbb{Z}(p^{n+1}) \to \mathbb{Z}(p^n)$ is the canonical projection for $n \in \mathbb{N}$, the inverse limit of the inverse system $[\mathbb{Z}(p^n), (\phi_n), \mathbb{N}]$ is $\mathbb{J}_p$.

Definition 3.3.9. Let G be a group and $\{K_i : i \in I\}$ a family of topological groups. For a given family $\mathcal{F} = \{f_i : i \in I\}$ of group homomorphisms $f_i : K_i \to G$, the *final topology* of the family $\mathcal{F}$ is the finest group topology on G that makes continuous all homomorphisms $f_i \in \mathcal{F}$.

Analogously to the initial topology, the final topology is introduced in the same way for topological spaces and continuous maps (see Exercise B.7.7).

The main example in this direction is the quotient topology of a quotient group $G = K/N$ of a topological group K. It is the final topology of the canonical projection $q : K \to G$. This is a simple example, since the map q is surjective.

When surjectivity is missing, the final topology is less transparent even in the case of a *single injective homomorphism*, say a subgroup embedding $i : K \hookrightarrow G$. In this specific case, denote by τ the topology on K; the final topology $\bar{\tau}$ of $\{i\}$ on G exists, it is simply the supremum of all group topologies σ on G with $\sigma\restriction_K \leq \tau$. This yields, of course, $\bar{\tau}\restriction_K \leq \tau$. Nevertheless, although somewhat surprising, this final topology $\bar{\tau}$ need not satisfy $\bar{\tau}\restriction_K = \tau$, i. e., need not be an “extension” of τ. More details on this subtle issue are given in Chapter 4.

3.4 The Hausdorff reflection of a topological group

This subsection is motivated by the simple fact reported in Lemma 3.4.1. According to Lemma 2.1.16 and Remark 2.1.17(a), the core of a topological group G is a closed normal subgroup of G and coincides with the closure of $\{e_G\}$.

Lemma 3.4.1. *Let (G, τ) be a topological group, $N := \mathrm{core}(G) = \overline{\{e_G\}}$, and $q : G \to G/N$ the canonical projection. Then:*
(a) *N is an indiscrete closed normal subgroup of G and G/N is Hausdorff;*
(b) *τ coincides with the initial topology of q;*

(c) *every continuous homomorphism $f: G \to H$, where H is a Hausdorff group, factorizes through q:*

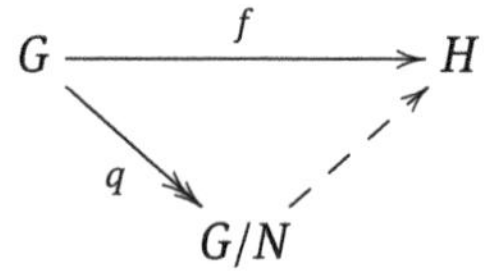

Proof. (a) Since $N \subseteq U$ for every $U \in \mathcal{V}_\tau(e_G)$, N is indiscrete. The last assertion follows from Lemma 3.2.10(b).

(b) Let $V \in \mathcal{V}_\tau(e_G)$ be open. Then $N \subseteq V$. Fix arbitrarily $x \in V$. Then there exists $U \in \mathcal{V}_\tau(e_G)$ such that $xU \subseteq V$. Since $N \subseteq U \subseteq V$, we conclude that $xN \subseteq V$. This proves that $VN \subseteq V$. On the other hand, $V \subseteq VN$, hence $V = VN = q^{-1}(q(N))$. Therefore, τ coincides with the initial topology of $q: G \to G/N$.

(c) Let $L = \ker f$. Then L is a closed normal subgroup of G, so $N \subseteq L$. By Frobenius theorem 3.2.3, there exists a continuous injective homomorphism $f_1: G/L \to H$ such that $f = f_1 \circ \pi$, where $\pi: G \to G/L$ is the canonical projection. Since $N \subseteq L$, there exists a homomorphism $h: G/N \to G/L$ such that $\pi = h \circ q$. Moreover, h is continuous by the continuity of $\pi = h \circ q$. Now the composition $\eta = f_1 \circ h: G/N \to H$ provides the desired factorization $f = \eta \circ q$:

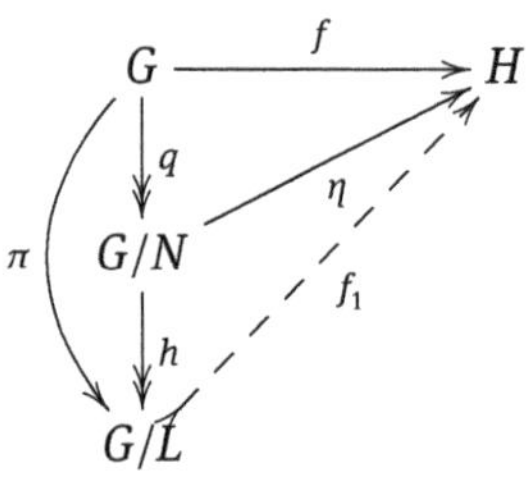

□

This lemma shows that the properties of a topological group G are easily determined from the corresponding properties of the Hausdorff quotient $G/\mathrm{core}(G)$. This is why it is not restrictive to work mainly with Hausdorff groups. Therefore, most often the topological groups in the sequel are assumed to be Hausdorff.

Now we put this observation in a more categorical framework.

Proposition 3.4.2. *Let (G, τ) be a topological group.*

(a) *The quotient topology of the group $\mathfrak{h}G := G/\mathrm{core}(G, \tau)$ is Hausdorff.*

(b) *If $f: (G, \tau) \to H$ is a continuous homomorphism to a Hausdorff group H, then there exists a unique continuous homomorphism $f_1: \mathfrak{h}G \to H$ such that $f_1 \circ q = f$, where $q: G \to \mathfrak{h}G$ is the canonical projection:*

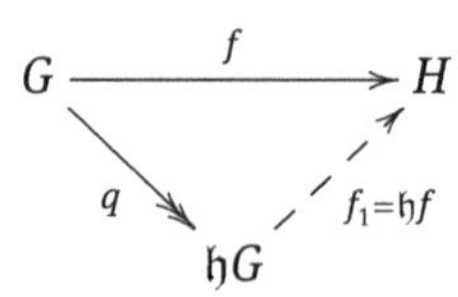

Proposition 3.4.2(b) shows that the Hausdorff quotient group $\mathfrak{h}G$ associated to a topological group G is its best approximation by Hausdorff groups.

Let us see that the assignment $G \mapsto \mathfrak{h}G$ induces a functor from the category of all topological groups to its subcategory $\mathbf{TopGrp_2}$ of Hausdorff groups, so that $f_1 = \mathfrak{h}f$ in the above proposition (while $H = \mathfrak{h}H$, as H is Hausdorff).

Proposition 3.4.3. *Let G, H be topological groups and $f: G \to H$ a continuous homomorphism. Then $\mathfrak{h}f: \mathfrak{h}G \to \mathfrak{h}H$, defined by $\mathfrak{h}f(x\,\mathrm{core}(G)) = f(x)\,\mathrm{core}(H)$ for every $x \in G$, is a continuous homomorphism commuting with the canonical projections $q_G: G \to \mathfrak{h}G$ and $q_H: H \to \mathfrak{h}H$. If f is an embedding, then so is $\mathfrak{h}f$.*

Proof. Since $f(e_G) = e_H$, Proposition 3.4.2(a) implies that $f(\mathrm{core}(G)) \subseteq \mathrm{core}(H)$. This proves the correctness of the definition of $\mathfrak{h}f$ and the commutativity of the following diagram:

$$\begin{array}{ccc} G & \xrightarrow{\ f\ } & H \\ {\scriptstyle q_G}\downarrow & & \downarrow{\scriptstyle q_H} \\ \mathfrak{h}G & \xrightarrow[\mathfrak{h}f]{} & \mathfrak{h}H \end{array}$$

The continuity of $\mathfrak{h}f$ easily follows from the continuity of f and q_H, and the properties of the quotient topology of $\mathfrak{h}G$ and q_G.

Now assume that f is an embedding. For simplicity we assume that G is a topological subgroup of H and $f: G \hookrightarrow H$ is the inclusion. Furthermore, we let $q = q_H$ for the sake of brevity.

Obviously, $\mathrm{core}(G) = \mathrm{core}(H) \cap G$, so that $\mathfrak{h}f$ is injective, as $\ker q_G = \mathrm{core}(G)$ and $\ker q = \mathrm{core}(H)$. It remains to prove that $\mathfrak{h}f: \mathfrak{h}G \to q(G) = \mathfrak{h}f(\mathfrak{h}G)$ is open, when $q(G) = \mathfrak{h}G$ carries the topology induced by $\mathfrak{h}H$. According to Theorem 3.2.8(c), it suffices to see that $q\restriction_G: G \to q(G)$ is open. To this end, take $U \in \mathcal{V}_G(e_G)$; we have to prove that $q(U) \in \mathcal{V}_{q(G)}(e_{\mathfrak{h}H})$. Pick $W, W_1 \in \mathcal{V}_H(e_H)$ such that $U = G \cap W$ and W_1 is symmetric with $W_1 W_1 \subseteq W$. We prove that

$$q(W_1) \cap q(G) \subseteq q(W \cap G). \tag{3.1}$$

This equality implies that $q(W \cap G)$ is a neighborhood of $e_{\mathfrak{h}H}$ in $q(G)$ equipped with the topology induced by $\mathfrak{h}H$.

To prove (3.1), pick $w \in W_1$ such that $q(w) \in q(W_1) \cap q(G)$, namely, $q(w) \in q(G)$. Then there exists $g \in G$ such that $w = gy$, where $y \in \ker q = \mathrm{core}(H)$. As $\mathrm{core}(H) \subseteq W_1$, this implies that $g = wy^{-1} \in W_1 W_1 \subseteq W$, so $g \in W \cap G$. Therefore, $q(w) = q(g) \in q(W \cap G)$ and (3.1) is proved. □

This shows that $\mathfrak{h}: \mathbf{TopGrp} \to \mathbf{TopGrp_2}$ is a reflection, since $\mathfrak{h}G = G$ in case G is a Hausdorff group (in fact, $\mathrm{core}(G) = \{e_G\}$). We collect further properties of the reflection functor $\mathfrak{h}$ and the map q_G:

Remark 3.4.4. We shall often use in the sequel the following facts making use of the notation above:
(a) If H is a topological group and G is a closed (respectively, open) subgroup of H, then G contains core(H), and $q_H(G)$ is a closed (respectively, open) subgroup of $\mathfrak{h}H$.
(b) If a continuous homomorphism $f: G \to H$ of topological groups is closed (respectively, open), then also $\mathfrak{h}f: \mathfrak{h}G \to \mathfrak{h}H$ is closed (respectively, open). This follows from Theorem 3.2.8(b) and the fact that G contains core(H).
(c) If $\{G_i : i \in I\}$ is a family of topological groups, then $\mathfrak{h}(\prod_{i\in I} G_i) \cong \prod_{i\in I} \mathfrak{h}G_i$.
(d) $\mathfrak{h}G$ is discrete for a topological group G precisely when G is an Alexandrov group.

3.5 Exercises

Exercises on separation axioms and products

Exercise 3.5.1. Let H be a discrete subgroup of a topological group G. Prove that $\overline{H}$ is topologically isomorphic to the semidirect product of H and $\overline{\{e_G\}}$, carrying the product topology, where H is discrete and $\overline{\{e_G\}}$ is indiscrete.

Exercise 3.5.2. Consider a group G endowed with the Taĭmanov topology $\mathfrak{T}_G$ (see Exercise 2.4.16).
(a) Prove that $\mathfrak{T}_G$ is Hausdorff if and only if the center $Z(G)$ of G is trivial.
(b) If $Z(G)$ is trivial, consider G as a subgroup of Aut(G) (making use of the inner automorphisms). Identify Aut(G) with a subgroup of the power G^G and show that $\mathfrak{T}_G$ coincides with the topology induced by Aut(G) equipped with the pointwise convergence topology τ (when G carries the discrete topology).
(c) Under the standing hypothesis that $Z(G)$ is trivial, consider Aut(G) as a subgroup of $S(G)$ in the natural way and show that τ coincides with the topology induced on Aut(G) by T_G.

Exercise 3.5.3. Prove that a Hausdorff group containing a dense solvable group is solvable.

Exercise 3.5.4. (a) Let H be a nontrivial discrete group and let $G = H \times N$, where N is an indiscrete nontrivial group. Prove that $H \times \{e_G\}$ is a discrete dense (hence, nonclosed) subgroup of G.
(b) If H is a normal Hausdorff subgroup of a topological group G and $N = \overline{\{e_G\}}$, prove that $\overline{H}$ contains HN and the subgroup HN is topologically isomorphic to the direct product $H \times N$ equipped with the product topology.

Hint. (a) follows from the more general item (b). The first assertion of (b) is obvious. Since $\{e_G\}$ is dense in the Hausdorff subgroup $H \cap N$, it follows that $H \cap N = \{e_G\}$. The set-wise product HN is a subgroup of G (as N is normal subgroup of G) and the map $f: H \times N \to HN$, $(h, n) \mapsto hn$ is an isomorphism, since $H \cap N = \{e_G\}$. Since $f \restriction_{H\times\{e\}}$ and $f \restriction_{\{e\}\times N}$ are continuous, f is continuous as well. On the other hand, the compositions of f^{-1} with the canonical projections of the product $H \times N$ are continuous, so f^{-1} is continuous.

Exercise 3.5.5. Prove that $\mathbb{T}/\mathbb{Z}(m) \cong \mathbb{T}$ for every $m \in \mathbb{N}_+$.

Hint. The subgroup $H = \langle 1/m \rangle$ of $\mathbb{R}$ containing $\mathbb{Z}$ satisfies $H/\mathbb{Z} \cong \mathbb{Z}(m)$, so $\mathbb{T}/\mathbb{Z}(m) = (\mathbb{R}/\mathbb{Z})/(H/\mathbb{Z}) \cong \mathbb{R}/H$. It remains to note that $\mathbb{R}/H \cong \mathbb{R}/\mathbb{Z} = \mathbb{T}$, as the automorphism $\phi: \mathbb{R} \to \mathbb{R}$ defined by $\phi(x) = x/m$ takes $\mathbb{Z}$ to H.

Exercise 3.5.6. Let H be a group, $G = H \times H$ and $\tau, \tau' \in \mathfrak{L}(H)$. Endow G with the product topology $\tau \times \tau'$ and consider the diagonal map $\Delta_H: H \to G$, defined by $\Delta_H(h) = (h, h)$ for $h \in H$. Show that Δ_H becomes an embedding when H is equipped with $\sup\{\tau, \tau'\}$. Formulate and prove the corresponding property with respect to the diagonal map $\Delta_H: H \to H^I$ in the case of arbitrary Cartesian powers.

Exercise 3.5.7. Let G_1, G_2 be groups and $G = G_1 \times G_2$. Identify G_1 and G_2 with the subgroups $G_1 \times \{e_{G_2}\}$ and $\{e_{G_1}\} \times G_2$ of G, respectively. For a group topology τ on G, denote by τ_i the topology induced on G_i by τ, with $i = 1, 2$.
(a) Prove that τ is coarser than the product topology $\tau_1 \times \tau_2$ of G.
(b) In case G_1 and G_2 are abelian, conclude that $(G, \tau_1 \times \tau_2)$ is the coproduct of (G_1, τ_1) and (G_2, τ_2) in the category of topological abelian groups.

Hint. Let W be a neighborhood of e_G in τ and V be a τ-open neighborhood of e_G with $VV \subseteq W$. Now $V \cap G_i$ is a τ_i-open neighborhood of e_{G_i} in (G_i, τ_i) for $i = 1, 2$, and

$$\tau_1 \times \tau_2 \ni U = (V \cap G_1) \times (V \cap G_2) = ((V \cap G_1) \times \{e_2\})(\{e_1\} \times (V \cap G_2)) \subseteq VV \subseteq W,$$

therefore, W is a neighborhood of e_G also in $\tau_1 \times \tau_2$.

Exercise 3.5.8 ([174]).* Prove that the group $\mathbb{Z}^{\aleph_1}$ equipped with the product topology of the discrete topology of $\mathbb{Z}$ is not a normal space.

Exercises on initial and final topologies

Exercise 3.5.9. For a group G, a family $\{K_i : i \in I\}$ of topological groups, and a family $\mathcal{F} = \{f_i : i \in I\}$ of group homomorphisms $f_i: G \to K_i$, prove that:
(a) the initial topology of the family $\mathcal{F}$ coincides with $\sup_{i\in I} \tau_i$, where τ_i denotes the initial topology of the single homomorphism $f_i \in \mathcal{F}$;
(b) a homomorphism $h: H \to G$ is continuous with respect to the initial topology of the family $\mathcal{F}$ on G if and only if all compositions $f_i \circ h: H \to K_i$ are continuous.

Exercise 3.5.10. Let G be a group and $\mathcal{T} = \{\tau_i : i \in I\}$ a family of group topologies on G. Prove that $\sup_{i\in I} \tau_i$ coincides with the initial topology of the family $\mathcal{F}$ of all maps $id_G : G \to (G, \tau_i)$ and also with the topology induced on G by the diagonal map $\Delta_G : G \to \prod_{i\in I} G = G^I$, in other words, $(G, \sup_{i\in I} \tau_i)$ is topologically isomorphic to the diagonal subgroup $\Delta = \{x = (x_i)_{i\in I} \in G^I : x_i = x_j \text{ for all } i, j \in I\}$ of $\prod_{i\in I}(G, \tau_i)$.

Exercise 3.5.11. Let G be a group, $\{K_i : i \in I\}$ a family of topological groups, and $\mathcal{F} = \{f_i : i \in I\}$ a family of group homomorphisms $f_i : K_i \to G$. Prove that a homomorphism $h : G \to H$ is continuous with respect to the final topology of the family $\mathcal{F}$ on G if and only if all compositions $h \circ f_i : K_i \to H$ are continuous.

Exercise 3.5.12. Prove that:
(a) if V, U are vector spaces over a field K, then the finite topology of $\mathrm{Hom}(V, U)$ is the initial topology of the family $\{f_v : v \in V\}$ of all linear transformations $f_v : \mathrm{Hom}(V, U) \to U$, of the form $\phi \mapsto \phi(v)$ for $\phi \in \mathrm{Hom}(V, U)$, when U is equipped with the discrete topology;
(b) if $U = K = \mathbb{Z}(p)$ is a prime finite field, then the finite topology of V^* coincides with the profinite topology of V^*.

Exercises on functorial topologies

Exercise 3.5.13. Let $\mathcal{T}$ be a functorial topology. Prove that:
(a) $\mathcal{T}_{G_1\times G_2} = \mathcal{T}_{G_1} \times \mathcal{T}_{G_2}$ for every pair G_1, G_2 of abelian groups;
(b) $\mathcal{T}_G \geq \prod_{i\in I} \mathcal{T}_{G_i}$ for every family $\{G_i : i \in I\}$ of abelian groups with $G = \prod_{i\in I} G_i$;
(c) $\mathcal{T}_H \geq \mathcal{T}_G \restriction_H$ for every abelian group G and every subgroup H of G;
(d) $(\mathcal{T}_G)_q \geq \mathcal{T}_{G/H}$ for every abelian group G and every subgroup H of G where $(\mathcal{T}_G)_q$ denotes the quotient topology of $(G, \mathcal{T}_G)$.

Hint. (a) Consider the projections $p_j : (G, \mathcal{T}_{G_1\times G_2}) \to (G_j, \mathcal{T}_j)$ for $j = 1, 2$, which are continuous by the definition of functorial topology. Then, for every neighborhood $U_1 \times U_2$ of 0 in $(G_1 \times G_2, \mathcal{T}_1 \times \mathcal{T}_2)$, there exists a neighborhood W of 0 in $(G_1 \times G_2, \mathcal{T}_{G_1\times G_2})$ such that $p_j(W) \subseteq U_j$ for $j = 1, 2$, that is, $W \subseteq U_1 \times U_2$. Hence, $\mathcal{T}_{G_1\times G_2} \geq \mathcal{T}_{G_1} \times \mathcal{T}_{G_2}$. To prove the converse inequality, note that the inclusions $i_j : (G_i, \mathcal{T}_{G_i}) \to (G_1\times G_2, \mathcal{T}_{G_1\times G_2})$, $j = 1, 2$, are continuous by the definition of functorial topology. Then $\mathcal{T}_{G_1\times G_2} \leq \inf\{\mathcal{T}_{G_1} \times \delta_{G_2}, \delta_{G_1} \times \mathcal{T}_{G_2}\} = \mathcal{T}_{G_1} \times \mathcal{T}_{G_2}$.

To prove (b), proceed as in the first part of the proof of item (a). For (c) and (d), it suffices to note that by definition the inclusion $(H, \mathcal{T}_H) \hookrightarrow (G, \mathcal{T}_G)$ and the projection $(G, \mathcal{T}_G) \to (G/H, \mathcal{T}_{G/H})$ are continuous.

Exercise 3.5.14. Let G be an abelian group. Show that every countable subgroup is discrete in the pro-countable topology of G.

Hint. Let A be a countable subgroup of G. It is enough to find an open subgroup H in the pro-countable topology of G with $H \cap A = \{0\}$. To this end find (using the Zorn lemma) a subgroup H of G with $H \cap A = \{0\}$ and maximal with this property. Then the quotient projection $q : G \to G/H$ has the property to be injective when restricted to A. Let us see that $q(A)$ is an essential subgroup of G/H. Indeed, if

$C \cap q(A) = \{\bar{0}\}$ in G/H, then $q^{-1}(C) \cap A = H \cap A = \{0\}$. Since $H \subseteq q^{-1}(C)$, the maximality of H implies $H = q^{-1}(C)$ and hence $C = q(q^{-1}(C)) = \{\bar{0}\}$. Now the essentiality of $q(A)$ in G/H implies that G/H is countable, so H is open in the pro-countable topology of G.

Exercise 3.5.15. Let p be a prime. Show that the p-adic topology of $\mathbb{J}_p$ differs from the pro-countable topology.

Hint. While $\mathbb{Z}$ is not discrete in the p-adic topology, it is discrete in the pro-countable topology, due to Exercise 3.5.14.

Exercise 3.5.16. Let G be a group, N a normal subgroup of G, p a prime, and $q: G \to G/N$ the canonical projection. Prove that:

(a) q is open whenever both groups carry the profinite, pro-p-finite, natural, p-adic, or pro-countable topology;

(b) if G is abelian, then q is open whenever both groups carry their Bohr topology.

Hint. (a) The openness of $q: (G, \varpi_G) \to (G, \varpi_{G/N})$ follows from the fact that whenever H is a finite-index subgroup of G, then $q(H)$ is a finite-index subgroup of G/N. A similar proof goes for the other four functorial topologies.

(b) The fact that the Bohr topology of G/N coincides with the quotient of the Bohr topology of G is proved in Corollary 11.2.8.

Exercise 3.5.17. Prove that $\varpi_G = \inf\{\mathcal{T}_{G^*}, \nu_G\}$ for an abelian group G.

Hint. The inequality $\varpi_G \le \inf\{\mathcal{T}_{G^*}, \nu_G\}$ was already proved in Proposition 2.2.15 and the equality was proved in case G is bounded in Theorem 2.2.17.

To prove the inequality $\varpi_G \ge \inf\{\mathcal{T}_{G^*}, \nu_G\}$, suppose that G is not bounded and take a basic neighborhood of 0 in the latter topology of the form $mG + U$, where $U \in \mathcal{V}_{\mathcal{T}_{G^*}}(0)$. It is enough to check that $mG + U \in \varpi_G$. Let $q: G \to G/mG$ be the canonical projection. If we use item (b) of the previous exercise, then it suffices to use the fact that $q(U)$ is a neighborhood of 0 in the Bohr topology of G/mG, so $q(U)$ contains a finite-index subgroup H of G/mG in view of the equality $\varpi_{G/mG} = \inf\{\mathcal{T}_{(G/mG)^*}, \nu_{G/mG}\}$. Hence, $U + mG = q^{-1}(q(U))$ contains a finite-index subgroup of G and we are done.

Here we give a proof, without any recourse to Corollary 11.2.6, of the weaker inequality

$$\varpi_G \ge \sup_{\chi \in G^*}\{\inf\{\mathcal{T}_\chi, \nu_G\}\}.$$

To this end, fix $\chi \in G^*$ and $H = \chi(G) \cong G/\ker\chi$. Pick $U \in \mathcal{V}_{\mathcal{T}_\chi}(0)$. So, $U = \chi^{-1}(V)$, where V is a neighborhood of 1 in $\mathbb{S}$. Working with $U + U$ (still a basic neighborhood of $\mathcal{V}_{\mathcal{T}_\chi}(0)$) in place of U, we have to prove that $U + U + mG \in \varpi_G$. If H is finite, then $\ker\chi$ is a finite-index subgroup of G, so $U \supseteq \ker\chi$ contains a finite-index subgroup and we are done. Suppose that H is infinite, then H is dense in $\mathbb{S}$. Now consider the subgroup $mH = \chi(mG)$ of H. Then $mH + \chi(U) = mH + (\Delta_\delta \cap H)$. Since H is an infinite subgroup of $\mathbb{S}$, it is not bounded, so mH is infinite as well, and mH is dense in $\mathbb{S}$. In particular, mH is dense in H, hence $mH + (V \cap H) = H$. This proves that

$$mG + U + U \supseteq mG + U + \ker\chi \supseteq \chi^{-1}(mH + \chi(U)) = \chi^{-1}(mH + V \cap H) = \chi^{-1}(H) = G.$$

Clearly, this is a neighborhood of 0 in any topology, in particular $U + mG \in \varpi_G$.

Miscellanea

Exercise 3.5.18. Prove that if a compact abelian group G has a dense divisible subgroup, then G is divisible as well.

Exercise 3.5.19. Is $\mathbb{T}^2$ monothetic? What about $\mathbb{T}^{\mathbb{N}}$?

Hint. The questions have positive answer, see Chapter 9.

Exercise 3.5.20. For a continuous surjective homomorphism $f: G \to H$ of topological groups, prove that $f(M_G) \subseteq M_H$. In particular, M_H is dense in H whenever M_G is dense in G.

Exercise 3.5.21. Prove that:
(a) every infinite strongly monothetic Hausdorff group is torsion-free;
(b) every torsion-free subgroup of $\mathbb{T}$ is strongly monothetic.

Exercise 3.5.22. Let τ be a Hausdorff group topology on $\mathbb{Z}$. Prove that either τ admits a coarser Hausdorff linear topology or there exists a τ-open subgroup H of $\mathbb{Z}$ that is strongly monothetic.

Exercise 3.5.23. Let G be a linearly topologized abelian group and let H be a topological subgroup of G. Prove that H is linearly topologized and for every open subgroup U of H there exists an open subgroup V of G with $V \cap H = U$.

Exercise 3.5.24. Prove that a Hausdorff group topology on $\mathbb{Z}$ which is coarser than a nondiscrete linear group topology is linear itself.

Hint. Fix $U \in \mathcal{V}_{\mathbb{Z}}(0)$. By hypothesis there exists $n \in \mathbb{N}_+$ such that $n\mathbb{Z} \subseteq U$ and hence $\overline{n\mathbb{Z}} \subseteq \overline{U}$, the closure taken in τ. By Lemma 3.1.1, $\overline{n\mathbb{Z}}$ is a τ-closed subgroup of $\mathbb{Z}$ containing $n\mathbb{Z}$, so $\overline{n\mathbb{Z}}$ is also τ-open by Proposition 3.1.7(c). This, combined with the regularity of $(\mathbb{Z}, \tau)$, shows that τ is a linear group topology.

Exercise 3.5.25. Let τ be a Hausdorff group topology on $\mathbb{Z}$ coarser than $\nu_{\mathbb{Z}}^p$ for a prime p. Prove that $\tau = \nu_{\mathbb{Z}}^p$.

Hint. By Exercise 3.5.24, τ is linear. If $m\mathbb{Z}$ is a τ-open subgroup of $\mathbb{Z}$, then it must be also $\nu_{\mathbb{Z}}^p$-open. So, $m\mathbb{Z}$ contains $p^n\mathbb{Z}$ for some $n \in \mathbb{N}_+$, i. e., $m \mid p^n$. Hence, $m = p^k$ for some $k \in \mathbb{N}$. Since all τ-open subgroups have this form and since τ is Hausdorff, $\tau = \nu_{\mathbb{Z}}^p$.

4 Markov's problems

4.1 The Zariski and Markov topologies

Let G be a Hausdorff group, $a \in G$, and $n \in \mathbb{N}$. Then the set $\{x \in G: x^n = a\}$ is obviously closed in G. This simple fact motivates the following:

Definition 4.1.1. A subset S of a group G is called:
(i) *elementary algebraic* if there exist $n \in \mathbb{N}_+$, $a_1, \dots, a_n \in G$ and $\varepsilon_1, \dots, \varepsilon_n \in \{-1, 1\}$ such that $S = \{x \in G : x^{\varepsilon_1} a_1 x^{\varepsilon_2} a_2 \cdots a_{n-1} x^{\varepsilon_n} = a_n\}$;
(ii) *algebraic* if S is an intersection of finite unions of elementary algebraic subsets;
(iii) *unconditionally closed* if S is closed in *every* Hausdorff group topology on G.

We denote by $\mathfrak{E}_G$ the family of all elementary algebraic subsets of G, by $\mathfrak{E}_G^{\cup}$ the family of all finite unions of elementary algebraic sets of G, and by $\mathfrak{A}_G$ the family of all algebraic subsets of G. Clearly,

$$\mathfrak{E}_G \subseteq \mathfrak{E}_G^{\cup} \subseteq \mathfrak{A}_G.$$

In the sequel we assume that G is nontrivial, this allows us to obtain $\emptyset \in \mathfrak{E}_G$. Since $\mathfrak{A}_G$ is closed under intersections and finite unions and contains all finite subsets of G, it is the family of all closed sets of some T_1 topology $\mathfrak{Z}_G$ on G, called *Zariski topology*. In other words, the $\mathfrak{Z}_G$-closed sets in G are precisely the algebraic subsets of G.

Example 4.1.2. (a) For $G = \mathbb{Z}$, $\mathfrak{E}_{\mathbb{Z}} = \{\mathbb{Z}, \emptyset\} \cup \{\{n\}: n \in \mathbb{Z}\}$, and consequently $\mathfrak{A}_{\mathbb{Z}} = \mathfrak{E}_{\mathbb{Z}}^{\cup} = \{\mathbb{Z}\} \cup [\mathbb{Z}]^{<\omega}$. Hence, $\mathfrak{Z}_{\mathbb{Z}}$ is the cofinite topology $\gamma_{\mathbb{Z}}$ of $\mathbb{Z}$.
(b) Analogously, if G is a torsion-free abelian group and $g \in G$, then the set $S = \{x \in G: nx + g = 0\} \in \mathfrak{E}_G$ either coincides with G or $|S| \leq 1$, so $\mathfrak{Z}_G$ is the cofinite topology γ_G of G.

More generally, $\mathfrak{Z}_G$ is non-Hausdorff for all infinite abelian groups G (see Exercise 4.5.9).

Example 4.1.3. (a) Let G be a group. Clearly, each centralizer $c_G(a)$ of some $a \in G$ is an elementary algebraic subset of G. Consequently, $Z(G)$ is an algebraic set.
(b) We saw in Example 4.1.2(b) that for a torsion-free abelian group G, $\mathfrak{Z}_G = \gamma_G$. This property fails if we replace "abelian" with its closest approximation "nilpotent of class 2". (Indeed, if G is the Heisenberg group $\mathcal{H}_{\mathbb{Z}}$, then $Z(G) \cong \mathbb{Z}$ is a proper infinite $\mathfrak{Z}_G$-closed subgroup of G.)

The family of all unconditionally closed sets of G is closed under arbitrary intersections and finite unions and contains all finite subsets of G, hence it coincides with the family of all closed sets of a T_1 topology $\mathfrak{M}_G$ on G, namely, the infimum in $\mathcal{T}(G)$ of all Hausdorff group topologies on G. We call $\mathfrak{M}_G$ the *Markov topology* of G.

https://doi.org/10.1515/9783110654936-004

Since an elementary algebraic subset of a group G must be closed in every Hausdorff group topology on G, one always has

$$\mathfrak{Z}_G \leq \mathfrak{M}_G.$$

Remark 4.1.4. In 1944 Markov [209] asked if the equality $\mathfrak{Z}_G = \mathfrak{M}_G$ holds for every group G. He showed that the answer is positive in case G is countable. Moreover, in the same manuscript Markov attributes to Perel′man the fact that $\mathfrak{Z}_G = \mathfrak{M}_G$ for every abelian group G (a proof has never appeared in print until [103]). An example of a group G with $\mathfrak{Z}_G \neq \mathfrak{M}_G$ was given by Hesse [173].

Remark 4.1.5. For any group G, $(G, \mathfrak{Z}_G)$ and $(G, \mathfrak{M}_G)$ are quasitopological groups, i. e., the inversion and the translations are continuous (see Exercise 4.5.2). On the other hand, when G is infinite abelian, $(G, \mathfrak{Z}_G)$ and $(G, \mathfrak{M}_G)$ are not group topologies since they are T_1, but non-Hausdorff (for the fact that $\mathfrak{Z}_G$ is not T_2 for every infinite abelian group see Exercise 4.5.9, for infinite torsion-free abelian groups see also Example 4.1.3(b)).

4.2 The Markov topology of the symmetric group

Let X be an infinite set. In the sequel we consider the pointwise convergence topology $\mathcal{T}_X$ of the infinite symmetric group $S(X)$ introduced in Example 2.2.22. It turns out that the Markov topology of $S(X)$ coincides with $\mathcal{T}_X$.

Theorem 4.2.1. *Let X be an infinite set. Then $\mathfrak{M}_{S(X)} = \mathcal{T}_X$.*

This theorem follows immediately from an old result due to Gaughan:

Theorem 4.2.2 (Gaughan theorem). *Let X be an infinite set. Every Hausdorff group topology on $S(X)$ is finer than $\mathcal{T}_X$.*

The proof of this theorem that we give below follows more or less the line of the proof exposed in [99, § 7.1] with several simplifications. The final stage of the proof is preceded by a number of claims (and their corollaries) and two facts about *purely algebraic* properties of the group $S(X)$ (namely, Lemmas 4.2.4 and 4.2.7).

Definition 4.2.3. Let X be an infinite set. A subset A of $S(X)$ is *m-transitive* for some $m \in \mathbb{N}_+$ if for every subset Y of X of size at most m and every injection $f\colon Y \to X$, there exists $a \in A$ that extends f.

We briefly say *transitive* in place of 1-transitive. A countable subset H of $S(X)$ cannot be transitive unless X itself is countable.

The leading idea is that a transitive subset A of $S(X)$ is placed "generically" in $S(X)$, whereas a nontransitive one is a subset of some subgroup of $S(X)$ that is a direct product $S(Y) \times S(X \setminus Y)$ where $\emptyset \neq Y \subset X$. (Here and in the sequel, for a nonempty subset Y of X, we tacitly identify the group $S(Y)$ with the subgroup of $S(X)$ consisting of all permutations of $S(X)$ that are identical on $X \setminus Y$.)

The first fact concerns the stabilizers $S_x = \{f \in S(X): f(x) = x\}$ of points $x \in X$, which constitute a prebase of the filter of $\mathfrak{T}_X$-neighborhoods of id_X.

Lemma 4.2.4. *For an infinite set X and $x \in X$, the subgroup S_x of $S(X)$ is maximal.*

Proof. Assume that H is a subgroup of $S(X)$ properly containing S_x. To show that $H = S(X)$, take any $f \in S(X)$. If $f(x) = x$, then $f \in S_x \subseteq H$, and we are done. Assume that $y := f(x) \neq x$ and let $h \in H \setminus S_x$. Then $z = h(x) \neq x$, so $x \notin \{y, z\}$. There exists $g \in S(X)$ such that $g(x) = x$, $g(y) = z$ and $g(z) = y$. Then $g \in S_x \subseteq H$ and $f(x) = g(h(x)) = y$, that is, $h^{-1}(g^{-1}(f(x))) = x$. Therefore, $h^{-1}g^{-1}f \in S_x \subseteq H$, and so $f \in ghH = H$. □

Claim 4.2.5. *Let X be an infinite set, T a Hausdorff group topology on $S(X)$, and $x \in X$. If S_x is T-closed, then S_x is also T-open.*

Proof. As S_x is T-closed, for every fixed $y \in X$, $y \neq x$, letting $\sigma_{x,y} = (xy)$ be the transposition with support $\{x, y\}$, the set

$$V_y := \{f \in S(X): f(x) \neq y\} = S(X) \setminus \sigma_{x,y} S_x$$

is T-open and contains id_X. So, there exists a symmetric neighborhood W of id_X in T such that $WW \subseteq V_y$. Define $Wx = \{f(x): f \in W\}$ and similarly Wy. Then $Wx \cap Wy = \emptyset$, by the definition of V_y. Since $X = (X \setminus Wx) \cup (X \setminus Wy)$, either $|X \setminus Wx| = |X|$ or $|X \setminus Wy| = |X|$. Suppose that $|X \setminus Wx| = |X|$. Then one can find $f \in S(X)$ with $f(Wx \setminus \{x\}) \subseteq X \setminus Wx$ and $f(x) = x$. Such an f satisfies $f^{-1}Wf \cap W \subseteq S_x$, as $f^{-1}(Wf(x)) \cap Wx = \{x\}$ by the choice of f. In view of Proposition 3.1.7(a), this proves that S_x is T-open.

An analogous argument works for S_y when $|X \setminus Wy| = |X|$. Since S_x and S_y are conjugated, this will prove that S_x is T-open. □

Corollary 4.2.6. *If X is an infinite set and T is a Hausdorff group topology on $S(X)$ such that S_x is T-closed in $S(X)$ for some $x \in X$, then $\mathfrak{T}_X \leq T$.*

Proof. Since all S_x are conjugated, the hypothesis implies that S_x is T-closed for every $x \in X$. By Claim 4.2.5, S_x is T-open for every $x \in X$. As the subgroups S_x of $S(X)$ form a prebase of the filter of neighborhoods of id_X in $(S(X), \mathfrak{T}_X)$, this implies that $\mathfrak{T}_X \leq T$. □

This was the first step in the proof. The next step consists in establishing that all $S_{x,y} := S_{\{x,y\}}$, with distinct $x, y \in X$, are never dense in any Hausdorff group topology on $S(X)$ (see Corollary 4.2.10). We need the subgroup

$$\widetilde{S}_{x,y} := \{f \in S(X): f(\{x, y\}) = \{x, y\}\} = S_{x,y}\langle (xy) \rangle \subseteq S(X)$$

that contains $S_{x,y}$ as a subgroup of index 2. Note that $\widetilde{S}_{x,y}$ is precisely the subgroup of all permutations in $S(X)$ that leave the doubleton $\{x, y\}$ setwise invariant.

Lemma 4.2.7. *For an infinite set X and any doubleton $\{x, y\}$ in X:*

(a) *the subgroup* $\widetilde{S}_{x,y}$ *of* $S(X)$ *is maximal;*

(b) *every proper subgroup of* $S(X)$ *properly containing* $S_{x,y}$ *coincides with one of the subgroups* S_x, S_y*, or* $\widetilde{S}_{x,y}$*.*

Proof. (a) Let H be a subgroup of $S(X)$ that properly contains $\widetilde{S}_{x,y}$. Our goal is to prove that $S_x \subseteq H$. Then, analogously, $S_y \subseteq H$, and an application of Lemma 4.2.4 yields $H = S(X)$.

Fix $g \in S_x$. If $g(y) = y$, then $g \in S_{x,y} \subseteq \widetilde{S}_{x,y} \subseteq H$. We may assume that $z := g(y) \notin \{x, y\}$. By assumption, $\widetilde{S}_{x,y}$ is a proper subgroup of H. So, we can choose $h_0 \in H \setminus \widetilde{S}_{x,y}$. Hence, either $h_0(x) \notin \{x, y\}$ or $h_0(y) \notin \{x, y\}$. Since $(xy) \in \widetilde{S}_{x,y} \subseteq H$, we may assume that $z' := h_0(y) \notin \{x, y\}$. If $z' = z$, let $h = h_0$, otherwise $h = (zz')h_0$; so $h \in H$ in both cases. Choose $t \in X \setminus \{x, y, h^{-1}(x), h^{-1}(y)\}$; then let $v = h(t) \in X \setminus \{x, y, z\}$, and hence $(zt), (zv) \in S_{x,y} \subseteq H$. Then $(yt) = h^{-1}(zv)h \in H$, and it is straightforward to check that $h_1 := (yt)(zt)g \in S_{x,y} \subseteq H$. This implies that $g \in H$, as desired.

(b) Assume that H is a proper subgroup of $S(X)$ properly containing $S_{x,y}$ such that $H \neq S_x$ and $H \neq S_y$. We aim to show that $H = \widetilde{S}_{x,y}$, i. e., $(xy) \in H$.

By Lemma 4.2.4 applied to $S_x = S(X \setminus \{x\})$ and its subgroup $S_{x,y}$ (the stabilizer of y in S_x), we conclude that $S_{x,y}$ is a maximal subgroup of S_x. It follows that $H \cap S_x = S_{x,y}$. Analogously, $H \cap S_y = S_{x,y}$. Take $f \in H \setminus S_{x,y}$. Then $f \notin S_x$ and $f \notin S_y$. Let $z = f(x)$ and $t = f(y)$; then $z \neq x$, $t \neq y$ and $z \neq t$. Consider the following cases.

Case 1: $\{z, t\} = \{x, y\}$. This is possible precisely when $z = y$ and $t = x$. Then $(xy)f \in S_{x,y} \subseteq H$, and thus $(xy) \in H$.

Case 2: $\{z, t\} \cap \{x, y\} = \emptyset$. Then $(zt) \in S_{x,y} \subseteq H$, so $(xy) = f^{-1}(zt)f \in H$.

Case 3: $\{z, t\} \cap \{x, y\} = \{z\} = \{y\}$. So, $x \neq t$. Since $(tyx)f \in S_{x,y} \subseteq H$, we deduce that $(xyt) = (tyx)^{-1} \in H$. Choose $v \in X \setminus \{x, y, t\}$; then $(tv) \in S_{x,y}$, and so $f_1 = (xt)(yv) = (xyt)(tv)(xyt)(tv) \in H$. Since $f_1(x) = t \notin \{x, y\}$ and $f_1(y) = v \notin \{x, y\}$, we can apply the argument from Case 2 with $f_1 \in H$ to get $(xy) \in H$. □

Claim 4.2.8. *Let X be an infinite set and T a Hausdorff group topology on $S(X)$. Then there exists a T-neighborhood of id_X that is not 2-transitive.*

Proof. Assume for a contradiction that all T-neighborhoods of id_X are 2-transitive. Fix arbitrarily distinct $u, v, w \in X$. We show that the 3-cycle $(uvw) \in V$ for every arbitrarily fixed T-neighborhood V of id_X; this contradicts the Hausdorffness of T. Indeed, choose a symmetric T-neighborhood W of id_X such that $WW \subseteq V$. For f the transposition (uv), $U := fWf \cap W$ is a symmetric T-neighborhood of id_X and $fUf = U$. Since U is 2-transitive, there exists $g \in U$ such that $g(u) = u$ and $g(v) = w$. Then $(uvw) = (uw)(uv) = gfg^{-1}f \in W(fUf) \subseteq WW \subseteq V$. □

Claim 4.2.9. *Let X be an infinite set, $m \in \mathbb{N}_+$, and T a group topology on $S(X)$. Then every T-neighborhood V of id_X in $S(X)$ is m-transitive if and only if every stabilizer S_F with $|F| = m$ is T-dense.*

Proof. Assume that S_F is T-dense for every finite subset F of X with $|F| = m$. Let V be a T-neighborhood of id_X. Let $F = \{x_1, \dots, x_m\}$ and $\{y_1, \dots, y_m\}$ be two subsets of X of size m. We have to show that there exists $f \in V$ such that $f(x_i) = y_i$ for all $i \in \{1, \dots, m\}$. Take an arbitrary $g \in S(X)$ such that $g(x_i) = y_i$ for all $i \in \{1, \dots, m\}$. Since S_F is dense in $S(X)$ by assumption, $S(X) = VS_F$ holds. So, there exist $f \in V$ and $h \in S_F$ such that $g = fh$. Then $y_i = g(x_i) = f(h(x_i)) = f(x_i)$ for every $i \in \{1, \dots, m\}$. This shows that V is m-transitive.

Conversely, assume that all T-neighborhoods V of id_X are m-transitive. Let F be a finite subset of X of cardinality m. We have to check that S_F is T-dense in $S(X)$. So, fix a T-neighborhood V of id_X and $g \in S(X)$. By assumption, V is m-transitive, hence there exists $f \in V$ such that $f(x) = g(x)$ for all $x \in F$. This means that $f^{-1}g \in S_F$, and therefore $g \in fS_F \subseteq VS_F$, that is, $S(X) = VS_F$. □

What we really need further on is that the density of the stabilizers $S_{x,y}$ implies that every T-neighborhood V of id_X in $S(X)$ is 2-transitive.

Corollary 4.2.10. *Let X be an infinite set and T a Hausdorff group topology on $S(X)$. Then for every pair $x, y \in X$ with $x \neq y$, the stabilizer $S_{x,y}$ is not T-dense in $S(X)$.*

Proof. By Claims 4.2.8 and 4.2.9, there exist distinct $x', y' \in X$ such that $S_{x',y'}$ is not T-dense. The assertion follows from the fact that all stabilizers of the form $S_{x,y}$ with distinct x, y are conjugated. □

Proof of Theorem 4.2.2. Assume for a contradiction that T is a Hausdorff group topology on $S(X)$ that does not contain $\mathcal{T}_X$. By Corollaries 4.2.6 and 4.2.10, all subgroups of the form S_x are T-dense and no subgroup of the form $S_{x,y}$ is T-dense.

Fix a pair $x, y \in X$ with $x \neq y$, and let $G_{x,y}$ denote the T-closure of $S_{x,y}$. Then $G_{x,y}$ is a proper subgroup of $S(X)$ containing $S_{x,y}$. Since S_x is T-dense, $G_{x,y}$ cannot contain S_x, so $S_x \cap G_{x,y}$ is a proper subgroup of S_x containing $S_{x,y}$. By Lemma 4.2.4, $S_{x,y}$ is a maximal subgroup of S_x. Therefore, $S_x \cap G_{x,y} = S_{x,y}$. This shows that $S_{x,y}$ is a T-closed subgroup of S_x. By Claim 4.2.5 applied to $S_x = S(X \setminus \{x\})$ and its subgroup $S_{x,y}$, we conclude that $S_{x,y}$ is a T-open subgroup of S_x. Since S_x is T-dense in $S(X)$, the closure $G_{x,y}$ of $S_{x,y}$ is T-open in $S(X)$. Since S_x is a proper T-dense subgroup of $S(X)$, S_x cannot contain $G_{x,y}$. Analogously, S_y cannot contain $G_{x,y}$ either. So, $G_{x,y} \neq S_{x,y}$, that is, $G_{x,y}$ is a proper subgroup of $S(X)$ containing $S_{x,y}$ and $S_x \neq G_{x,y} \neq S_y$. Therefore, $G_{x,y} = \widetilde{S}_{x,y}$ by Lemma 4.2.7(b). This proves that $\widetilde{S}_{x,y}$ is T-open. Since all subgroups of the form $\widetilde{S}_{x,y}$ are pairwise conjugated, all subgroups $\widetilde{S}_{x,y}$ are T-open.

Now we can see that the stabilizers S_F with $|F| > 2$ are T-open, as

$$S_F = \bigcap \{\widetilde{S}_{x,y} : x, y \in F, x \neq y\}.$$

This proves that all basic neighborhoods S_F of id_X in $\mathcal{T}_X$ are T-open. In particular, also the subgroups S_x are T-open, contrary to our hypothesis. □

A generalization of Theorem 4.2.2 was obtained by Banakh, Guran, and Protasov [16]. Indeed, for an infinite set X, denote by $S_\omega(X)$ the subset of all permutations of finite support in $S(X)$. In [69] Dierolf and Schwanengel proved that for each infinite set X and each subgroup G of $S(X)$ containing $S_\omega(X)$, the topology $\mathcal{T}_X \restriction_G$ is a minimal Hausdorff group topology on G (see Definition 8.4.3), and it was conjectured in [99, p. 220] (see also [201]) that for an infinite set X and a subgroup G of $S(X)$ containing $S_\omega(X)$, the topology $\mathcal{T}_X \restriction_G$ is the coarsest Hausdorff group topology on G. In [16] this conjecture was proved:

Theorem 4.2.11. *For an infinite set X and a subgroup G of $S(X)$ containing $S_\omega(X)$, the topology $\mathcal{T}_X \restriction_G$ is the coarsest Hausdorff group topology on G, i. e., $\mathfrak{M}_G = \mathcal{T}_X \restriction_G$.*

This result is based on another fact, which is interesting on its own:

Theorem 4.2.12 ([16]). *Let X be an infinite set and G a subgroup of $S(X)$ containing $S_\omega(X)$. Then $\mathfrak{Z}_G$ coincides with the topology induced by $\mathcal{T}_X$ on G.*

The following consequence of the above theorem answered [102, Question 38] about the coincidence of $\mathfrak{Z}_{S(X)}$ and $\mathfrak{M}_{S(X)}$ (see Remark 4.1.4).

Corollary 4.2.13. *Let X be an infinite set and G a subgroup of $S(X)$ containing $S_\omega(X)$. Then $\mathfrak{Z}_G = \mathfrak{M}_G$.*

We conclude with a relevant property of the groups $(S(X), \mathcal{T}_X)$.

Call a topological group G *topologically simple* if G has no proper closed normal subgroups.

Corollary 4.2.14. *If G is a nontrivial Hausdorff group, X is an infinite set, and $f: (S(X), \mathcal{T}_X) \to G$ is a continuous surjective homomorphism, then f is a topological isomorphism.*

Proof. Since $(S(X), \mathcal{T}_X)$ is topologically simple (see Exercise 8.7.7) and f is surjective and nontrivial, we deduce that f is a continuous isomorphism. Theorem 4.2.2 implies that f is open. □

4.3 Existence of Hausdorff group topologies

According to Proposition 3.1.20, every infinite abelian group admits a nondiscrete Hausdorff group topology, for example, the Bohr topology. This gives:

Proposition 4.3.1. *Every group with infinite center admits a nondiscrete Hausdorff group topology.*

Proof. The center $H = Z(G)$ of the group G has a nondiscrete Hausdorff group topology τ, by Proposition 3.1.20. Obviously, $\mathcal{V}_{(H,\tau)}(e_H)$ is a filter base satisfying conditions (gt1), (gt2), and (gt3), so it forms a local base at e_G of a nondiscrete Hausdorff group topology on G. □

In 1946 Markov set the problem of the existence of a (countably) infinite group G that admits no Hausdorff group topology beyond the discrete one. Let us call such a group a *Markov group*, or *nontopologizable*. Obviously, G is a Markov group precisely when $\mathfrak{M}_G$ is discrete, and a Markov group must have finite center, by Proposition 4.3.1.

Remark 4.3.2. According to Remark 2.1.17(a) and Lemma 2.1.16, the closure of the neutral element of every topological group G is always a normal subgroup of G. Therefore, a simple topological group is either Hausdorff or indiscrete. Therefore, a simple Markov group G admits only two group topologies, the discrete and the indiscrete.

The equality $\mathfrak{Z}_G = \mathfrak{M}_G$ established by Markov for countable groups G was intended to help in finding a countably infinite Markov group G. Indeed, a countable group G is Markov precisely when $\mathfrak{Z}_G$ is discrete. Nevertheless, Markov failed in building a countable group G with discrete Zariski topology. This was done much later, in 1980, by Ol′shanskii [222] making use of the so-called Adian groups $A = \mathcal{A}(m,n)$ (constructed by Adian to negatively resolve the famous 1902 Burnside problem on finitely generated groups of finite exponent).

In order to sketch here Ol′shanskii's elegant short proof, we need to give first a description of $\mathcal{A}(m,n)$.

Example 4.3.3 ([222]). Let m and n be odd integers ≥ 665. The *Adian group* $A = \mathcal{A}(m,n)$ has the following properties:
(a) A is generated by n elements;
(b) A is torsion-free;
(c) the center C of A is infinite cyclic;
(d) the quotient A/C is infinite, of exponent m (i. e., $y^m \in C$ for every $y \in A$).

The finitely generated infinite quotient A/C of exponent m negatively resolves the Burnside problem.

Ol′shanskii used the group $A = \mathcal{A}(m,n)$, which is countable by (a), and its subgroup $C_m := \{c^m : c \in C\}$ as follows:

Theorem 4.3.4. *Let m and n be odd integers ≥ 665 and $A = \mathcal{A}(m,n)$. The group $G = A/C_m$ has discrete Zariski topology.*

Proof. Let us see that (b), (c), and (d) jointly imply that the Zariski topology of the infinite quotient $G = A/C_m$ is discrete (so G is a countable Markov group). Let d be a generator of C. Then $x^m \in C \setminus C_m$ for every $x \in A \setminus C$. Indeed, if $x^m = d^{ms}$ for $s \in \mathbb{Z}$, then $(xd^{-s})^m = e_A$, as d is central; by (b), $xd^{-s} = e_A$, so $x \in C$, a contradiction. Hence,

$$\forall u \in G \setminus \{e_G\}, \exists a \in C/C_m \setminus \{e_G\}, \quad \text{such that either } u = a \text{ or } u^m = a. \tag{4.1}$$

As $|C/C_m| = m$, this means that every $u \in G \setminus \{e_G\}$ is a solution of some of the $2(m-1)$ equations in (4.1). Thus, $G \setminus \{e_G\}$ is closed in the Zariski topology $\mathfrak{Z}_G$ of G. Therefore, $\mathfrak{Z}_G$ is discrete. □

Now we recall an example, due to Shelah, of an uncountable nontopologizable group. It appeared somewhat earlier than the above ZFC-example of Ol′shanskii.

Example 4.3.5 ([260]). Under CH, there exists a group G of size ω_1 satisfying the following conditions (a) (with $m = 10000$) and (b) (with $n = 2$):

(a) there exists $m \in \mathbb{N}$ such that $A^m = G$ for every subset A of G with $|A| = |G|$;

(b) for every subgroup H of G with $|H| < |G|$, there exist $n \in \mathbb{N}_+$ and $x_1, \dots, x_n \in G$ such that the intersection $\bigcap_{i=1}^n x_i^{-1} H x_i$ is finite.

To see that G is a Markov group (i. e., $\mathfrak{M}_G$ is discrete), assume that T is a Hausdorff group topology on G. There exists a T-neighborhood V of e_G with $V \neq G$. Choose a $W \in \mathcal{V}_T(e_G)$ with $\underbrace{W \cdots W}_{m} \subseteq V$. Now $V \neq G$ and (a) yield $|W| < |G|$. Let $H = \langle W \rangle$. Then $|H| \leq |W| \cdot \omega < |G|$. By (b), the intersection $O = \bigcap_{i=1}^n x_i^{-1} H x_i$ is finite for some $n \in \mathbb{N}_+$ and $x_1, \dots, x_n \in G$. Since each $x_i^{-1} H x_i$ is a T-neighborhood of e_G, this proves that $e_G \in O \in T$. Since T is Hausdorff and O is finite, it follows that $\{e_G\}$ is T-open, and therefore T is discrete.

Hesse showed in [173] that the use of CH in Shelah's construction of a Markov group of size ω_1 can be avoided.

One can see that even the weaker form of (a) with m depending on a subset A of G with $|A| = |G|$ yields that every proper subgroup of G has size $< |G|$. In the case $|G| = \omega_1$, the groups with this property are known as *Kurosh groups*. By the time when Shelah's paper appeared, the existence of Kurosh groups was an open problem. The group G built by Shelah (see Example 4.3.5) is a Kurosh group. Actually, it is a *Jónsson semigroup* of size ω_1, i. e., an uncountable semigroup whose proper subsemigroups are countable (the existence of Jónsson semigroups was an open problem as well). The group G by Shelah furnished also the first consistent example to a third open problem that we discuss in item (c) of the next remark.

Remark 4.3.6. We list some properties of Shelah's group G described in Example 4.3.5.

(a) The group G is simple. Indeed, assume that N is a proper normal subgroup of G. Since G is a Kurosh group, $|N| < |G|$. Then Example 4.3.5(b) implies that N is finite. Hence, every proper normal subgroup of G is finite. Since G is torsion-free (see [260]), we deduce that N is trivial.

(b) Clearly, G has no maximal subgroups, as for every proper subgroup H of G one has $|H| < |G|$, so any $x \in G \setminus H$ gives rise to a larger subgroup $H_1 = \langle H, x \rangle$ of size at most $\max\{|H|, \omega\}$, hence $|H_1| < |G|$. Therefore, $\mathrm{Fratt}(G) = G$.

(c) Since the diagonal subgroup Δ_G of $G \times G$ is a maximal subgroup (see Exercise 4.5.15), this shows that taking the Frattini subgroup "does not commute" with taking finite direct products, in the sense that $\mathrm{Fratt}(G \times G) \neq \mathrm{Fratt}(G) \times \mathrm{Fratt}(G) = G \times G$.

In conclusion, it is good to mention the following nice result of Zelenyuk [291].

Theorem 4.3.7. *Every infinite group admits a nondiscrete Hausdorff topology τ such that (G, τ) is a quasitopological group.*

The Markov and Zariski topologies cannot provide an alternative solution to this issue, since they need not be Hausdorff (in fact, they are never Hausdorff if G is infinite and abelian).

Sipacheva showed in [261] that an appropriate version of Shelah's example can produce, under the assumption of CH, a group G with $\mathfrak{M}_G \neq \mathfrak{Z}_G$.

4.4 Extension of group topologies

The problem of the *existence* of Hausdorff nondiscrete group topologies can be considered also as a problem of *extension* of Hausdorff nondiscrete group topologies.

Remark 4.4.1. The theory of extensions of topological spaces is well understood. If a subset Y of a set X carries a topology τ, then it is easy to extend τ to a topology τ^* on X such that (Y, τ) is a subspace of (X, τ^*).

The easiest way to do it is to consider $X = Y \cup (X \setminus Y)$ as a partition of the new space (X, τ^*) into clopen sets and define the topology of $X \setminus Y$ arbitrarily. Usually, one prefers to define the extension topology τ^* on X in such a way to have Y *dense* in X. In such a case the extensions of a given space (Y, τ) can be described by means of appropriate families of *open filters* of Y (i. e., filters on Y having a base of τ-open sets).

The counterpart of this problem for groups and group topologies is more complicated because of the presence of the group structure. Indeed, let H be a subgroup of a group G and assume that τ is a group topology of H. Now one has to build a group topology τ^* on G such that (H, τ) is a topological subgroup of (G, τ^*).

The first idea to extend τ is to imitate the first case of extension considered above by declaring the subgroup H a τ^*-open topological subgroup of the new topological group (G, τ^*). Let us note that this would immediately determine the topology τ^* in a unique way. Indeed, every coset gH of H in G must carry the topology transported from H to gH by the translation $_gt: G \to G$, $x \mapsto gx$; this means that the τ^*-open sets of gH must have the form gU, where U is an open set of (H, τ) and $g \in G$. In other words, the family $\{gU : \emptyset \neq U \in \tau, g \in G\}$ is a base of τ^*.

This idea has worked in the proof of Corollary 4.3.1 where H is the center of G. Indeed, this idea works in the following more general case.

Lemma 4.4.2 ([103]). *Let H be a subgroup of a group G such that $G = Hc_G(H)$. Then for every group topology τ on H, the above described topology τ^* is a group topology on G such that (H, τ) is an open topological subgroup of (G, τ^*).*

Proof. The first two axioms, (gt1) and (gt2), on the neighborhood base are easy to check. For (gt3), pick a basic τ^*-neighborhood U of e_G in G. Since H is τ^*-open, we can assume without loss of generality that $U \subseteq H$, so U is a τ-neighborhood of e_G. Let $x \in G$. We have to produce a τ^*-neighborhood V of e_G in G such that $x^{-1}Vx \subseteq U$. By our hypothesis, there exist $h \in H$ and $z \in c_G(H)$ such that $x = hz$. Since τ is a group topology on H, there exists $V \in \mathcal{V}_{H,\tau}(e_G)$ such that $h^{-1}Vh \subseteq U$. Then, as $z \in c_G(H)$, $x^{-1}Vx = z^{-1}h^{-1}Vhz \subseteq z^{-1}Uz = U$. This proves that τ^* is a group topology on G. □

Clearly, the condition $G = Hc_G(H)$ is satisfied when H is a central subgroup of G. It is satisfied also when H is a direct summand of G. On the other hand, subgroups H satisfying $G = Hc_G(H)$ are normal. The condition $G = Hc_G(H)$ imposed in the above lemma for the extension problem is only sufficient, it need not be necessary. In Example 4.4.8 we show that the extension problem may have negative outcome even for subgroups H of index 2.

Our next theorem shows that the difficulties of the extension problem are not hidden in the special features of the extension τ^*.

Theorem 4.4.3 ([103]). *Let H be a normal subgroup of the group G and let τ be a group topology on H. Then the following conditions are equivalent:*
(a) *the extension τ^* is a group topology on G;*
(b) *τ can be extended to a group topology of G;*
(c) *for every $x \in G$, the automorphism $\phi_x \restriction_H$ of H induced by the inner automorphism ϕ_x of G is τ-continuous.*

Proof. (a)⇒(b) is obvious, while (b)⇒(c) follows from the fact that the conjugations are continuous in any topological group (see Lemma 2.1.6).

(c)⇒(a) Assume that all automorphisms of H induced by the conjugation by elements of G are τ-continuous. Take the filter of all neighborhoods of e_G in (H, τ) as a base of neighborhoods of e_G in the group topology τ^* of G. This works since we only have to check the axiom (gt3), i. e., to find for every $x \in G$ and every τ^*-neighborhood U of e_G a τ^*-neighborhood V of e_G such that $x^{-1}Vx \subseteq U$. Since we can choose U, V contained in H, this immediately follows from our assumption of τ-continuity of the restrictions to H of the conjugations in G and property (gt3) for the topological group H. □

Now we see that, under the hypotheses of the above theorem, one can always extend τ if it makes all automorphisms of the group H continuous:

Corollary 4.4.4. *If H is a normal subgroup of a group G and τ is a group topology on H such that every automorphism of H is τ-continuous, then τ can be extended to a group topology on G.*

Corollary 4.4.5. *For a Hausdorff group (H,τ), the following conditions are equivalent:*
(a) *every automorphism of H is τ-continuous;*
(b) *for every group G containing H as a normal subgroup, τ can be extended to a group topology on G;*
(c) *τ can be extended to a group topology on $G = H \rtimes \mathrm{Aut}(H)$.*

Proof. (a)⇒(b) follows from Corollary 4.4.4 and (b)⇒(c) is trivial.

(c)⇒(a) Extend τ to a group topology τ' on G and note that the automorphisms of H act as restrictions of inner automorphisms of G on H. As the inner automorphisms of G are τ'-continuous in G, their restrictions to H are obviously τ-continuous. □

Corollary 4.4.4 gives a series of examples when the extension problem has always a positive solution.

Example 4.4.6. Let p be a prime number. If the group of p-adic integers $N = \mathbb{J}_p$ is a normal subgroup of some group G, then the p-adic topology of N can be extended to a group topology on G. Indeed, it suffices to note that if $\xi: N \to N$ is an automorphism of N, then obviously $\xi(p^n N) = p^n N$. Since the subgroups $p^n N$ define the topology of N, this proves that every automorphism of N is continuous. Now Theorem 4.4.3 applies.

Clearly, the p-adic integers can be replaced by any topological group N such that every automorphism of N is continuous (e. g., products of the form $\prod_{p\in\mathbb{P}} \mathbb{J}_p^{k_p} \times F_p$, where $k_p < \omega$ and F_p is a finite abelian p-group).

It remains to see that the extension problem cannot be resolved for certain triples G, H, τ of a group G, its normal subgroup H, and a group topology τ on H. By Corollary 4.4.4, to produce an example when the extension is not possible, we need to produce a triple G, H, τ such that at least some conjugation by an element of G is not τ-continuous when considered as an automorphism of H. Inspired by Corollary 4.4.5, we are going to use semidirect products. So, to produce the required example of such a triple G, H, τ, it suffices to find a group K and a group homomorphism $\theta: K \to \mathrm{Aut}(H)$ such that at least one of the automorphisms $\theta(k)$ of H is τ-discontinuous. One can simplify the construction by taking the cyclic group $K_1 = \langle k\rangle$ instead of the whole group K, by choosing $k \in K$ such that the automorphism $\theta(k)$ of H is not τ-continuous. A further simplification can be possibly arranged by taking k in such a way that the automorphism $f = \theta(k)$ of H is also an involution, i. e., $f \circ f = id_H$. Then H is an index 2 subgroup of G.

The following lemma is needed to build an example as above. Its proof exploits the fact that the arcs are the only connected subsets of $\mathbb{T}$. Hence, $\chi \in \widehat{\mathbb{T}}$ sends any arc to an arc, and end points to end points.

Lemma 4.4.7. *The only topological automorphisms* $\chi\colon \mathbb{T} \to \mathbb{T}$ *are* $\pm id_{\mathbb{T}}$.

Proof. For $n \in \mathbb{N}_+$, let $c_n = q_0(1/2^n)$ be the generators of $\mathbb{Z}(2^\infty) \leq \mathbb{T}$. Then $c_1 = q_0(1/2)$ is the only element of $\mathbb{T}$ of order 2, hence $\chi(c_1) = c_1$. Therefore, the arc $A_1 = q_0([0,1/2])$ either goes onto itself, or goes onto its symmetric image $-A_1$. Assume that $\chi(A_1) = A_1$. Then $\chi(c_2) = c_2$ as $o(g(c_2)) = 4$ and $\pm c_2$ are the only elements of order 4 of $\mathbb{T}$. Now the arc $A_2 = [0, c_2]$ is sent to itself by χ, hence for c_3 we must have $\chi(c_3) = c_3$ as c_3 is the only element of order 8 on the arc A_2. We see in the same way that $\chi(c_n) = c_n$ for every $n \in \mathbb{N}_+$, hence χ is identical on the whole subgroup $\mathbb{Z}(2^\infty)$. As this subgroup is dense in $\mathbb{T}$, we conclude that $\chi = id_{\mathbb{T}}$. In the case $\chi(A_1) = -A_1$, we replace χ by $-\chi$ and the previous proof gives $-\chi = id_{\mathbb{T}}$, that is, $\chi = -id_{\mathbb{T}}$. □

Example 4.4.8 ([103]). Here is an example of a topological abelian group (H,τ) admitting an involution f that is not τ-continuous. Then the triple G, H, τ such that τ cannot be extended to G is obtained by simply taking $G = H \rtimes \langle f\rangle$, where the involution f acts on H. Take as (H,τ) the circle group $\mathbb{T}$ with the usual topology. Then $\mathbb{T}$ is algebraically isomorphic to $(\mathbb{Q}/\mathbb{Z}) \oplus \mathbb{Q}^{(\mathfrak{c})}$, so $\mathbb{T}$ has $2^{\mathfrak{c}}$ involutions. Of these, only the involutions $\pm id_{\mathbb{T}}$ of $\mathbb{T}$ are continuous, by Lemma 4.4.7. A more general example will be given in Chapter 13.

4.5 Exercises

Exercise 4.5.1. Prove that for every infinite set X and every group topology on $S(X)$ the stabilizers S_x with $x \in X$ are either closed or dense.

Hint. Use the fact that S_x is a maximal subgroup of $S(X)$ by Lemma 4.2.4.

Exercise 4.5.2. Show that $(G, \mathfrak{Z}_G)$ and $(G, \mathfrak{M}_G)$ are quasitopological groups.

Exercise 4.5.3. Show that if $(G,\cdot)$ is an abelian group, then every elementary algebraic set of G has the form $\{x \in G\colon x^n = a\}$ for $a \in G$ and $n \in \mathbb{N}$.

Exercise 4.5.4. Show that in an abelian group G every nonempty set in $\mathfrak{E}_G$ has the form $a + G[n]$, for some $a \in G$ and $n \in \mathbb{N}_+$. Deduce from this that every descending chain in $\mathfrak{E}_G$ stabilizes, and that $\mathfrak{E}_G$ is stable under finite, and consequently arbitrary, intersections.

Hint. For the first assertion, use Exercise 4.5.3. To prove the second assertion, it suffices to note that every descending chain of subgroups of G of the form $G[n]$ stabilizes. Furthermore, to prove that $\mathfrak{E}_G$ is stable under finite intersections, show that if $I = (a + G[n]) \cap (b + G[m]) \neq \emptyset$, then $I = c + G[d]$ for an appropriate $c \in G$ and $d = (n, m)$, the greatest common divisor of n and m.

Exercise 4.5.5 ([103]).* Show that $\mathfrak{Z}_G = \mathfrak{M}_G$ for an abelian group G.

Exercise 4.5.6. Let G be an abelian group.

(a) Prove that $\mathfrak{Z}_G$ coincides with the cofinite topology γ_G in case G is torsion-free.

(b) Show that the conclusion of item (a) remains true also when $r_p(G)$ is finite for every prime p or when the group G has exponent p for some prime p.
(c) Prove that if $\mathfrak{Z}_G = \gamma_G$, then $r_p(G)$ is finite for every prime p or the group G has exponent p for some prime p.

Hint. (b) In case G has prime exponent p, deduce with Exercise 4.5.4 that G and the singletons are the only elementary algebraic sets of G. In the second case show that for every prime p the subgroup $G[p^n]$ is finite for every $n \in \mathbb{N}$ and deduce that every subgroup of the form $G[m]$ is finite.

(c) Assume that $\mathfrak{Z}_G = \gamma_G$ and that G is not of prime exponent. Then for every prime p, the subgroup $G[p]$ must be finite, so $r_p(G)$ is finite.

Exercise 4.5.7. Show that in an abelian group G every descending chain of $\mathfrak{Z}_G$-closed sets of G stabilizes and deduce that $\mathfrak{A}_G = \mathfrak{E}_G^{\cup}$.

Hint. Obviously, $\mathfrak{E}_G^{\cup}$ is stable under taking finite intersections. We need to show that every descending chain in $\mathfrak{E}_G^{\cup}$ stabilizes. Assume for a contradiction that there exists a descending chain

$$F_1 \supset F_2 \supset \cdots \supset F_n \supset \cdots \tag{4.2}$$

in $\mathfrak{E}_G^{\cup}$. We can assume without loss of generality that $F_1 = E \in \mathfrak{E}_G$. (If $F_1 = E_1 \cup \cdots \cup E_k$ with $E_i \in \mathfrak{E}_G$, then at least one of the descending chains $\{E_i \cap F_n : n \in \mathbb{N}_+\}$ does not stabilize, and so we can replace (4.2) by the chain $E := E_i \supset E_i \cap F_2 \supset \cdots \supset E_i \cap F_n \supset \cdots$.) Since $\mathfrak{E}_G$ satisfies the descending chain condition, we can assume additionally that $E = F_1$ is minimal with this property. Now write $F_2 = E_1' \cup \cdots \cup E_l'$ with $E_i' \in \mathfrak{E}_G$ and consider for each $i \in \{1, \ldots, l\}$ the chain $E \supset E_i' \supset E_i' \cap F_3 \supset \cdots \supset \cdots$. Since the inclusion $E \supset E_i'$ is proper, the chain $E_i' \supset E_i' \cap F_3 \supset \cdots \supset \cdots$ stabilizes by our assumption of minimality of E. Since all these l-many chains stabilize, also (4.2) stabilizes.

Exercise 4.5.8. Show that for an abelian group G the space $(G, \mathfrak{Z}_G)$ is Noetherian (i. e., every ascending chain of $\mathfrak{Z}_G$-open sets of G stabilizes) and *hereditarily compact* (i. e., every subspace of $(G, \mathfrak{Z}_G)$ is compact).

Hint. According to Exercise 4.5.7, descending chains of $\mathfrak{Z}_G$-closed sets of G stabilize. This means that the space $(G, \mathfrak{Z}_G)$ is Noetherian. Since this property is stable under taking arbitrary subspaces, this proves that $(G, \mathfrak{Z}_G)$ is hereditarily compact.

Exercise 4.5.9. Prove that $\mathfrak{Z}_G$ is Hausdorff for an abelian group G if and only if G is finite.

Hint. Assume that $(G, \mathfrak{Z}_G)$ is Hausdorff. Then, in view of Exercise 4.5.8, every subset of G is compact, so closed. This means that $(G, \mathfrak{Z}_G)$ is discrete. Since $(G, \mathfrak{Z}_G)$ is also compact, we deduce that G is finite.

Exercise 4.5.10. Prove that the Zariski topology is not functorial, in the sense that group homomorphisms need not be continuous when both the domain and the codomain carry their Zariski topology.

Hint. Note first that if $f: X \to Y$ is a map between two topological spaces carrying their cofinite topology, then f is continuous if and only if it is finitely-many-to-one.

Consider the canonical projection $q: \mathbb{Q} \to \mathbb{Q}/\mathbb{Z}$. By Exercise 4.5.6, the Zariski topology of both $\mathbb{Q}$ and $\mathbb{Q}/\mathbb{Z}$ is the cofinite topology. According to the general fact we mentioned above, $\ker q$ must be finite. But $\ker q = \mathbb{Z}$ is infinite.

Another easy example can be obtained by taking any infinite group G of prime exponent and an infinite proper subgroup H of G. Then the canonical projection $q: G \to G/H$ is not continuous by the above argument.

Exercise 4.5.11. Let G be an abelian group and $m \in \mathbb{N}_+$. Prove that the endomorphism $\mu_m: G \to G, x \mapsto mx$, is $\mathfrak{Z}_G$-continuous.

Hint. By Exercise 4.5.3, every $E \in \mathfrak{E}_G$ has the form $E = \{x \in G: nx = a\}$ for appropriate $a \in G$ and $n \in \mathbb{N}$. Deduce that $\mu_m^{-1}(E) \in \mathfrak{E}_G$.

Exercise 4.5.12. For an abelian group G and a subgroup H of G, prove that $\mathfrak{Z}_H$ coincides with $\mathfrak{Z}_G \restriction_H$.

Hint. Check that $E \cap H \in \mathfrak{E}_H$ for all $E \in \mathfrak{E}_G$, by applying Exercise 4.5.3.

Exercise 4.5.13.* Prove that for an abelian group G, $\mathfrak{Z}_G$ is completely determined by its subsets that carry the cofinite topology in the following sense: an infinite subset A of G is a $\mathfrak{Z}_G$-*atom* if its induced topology is the cofinite one (see [104]).

(a) Prove that, for an infinite subset M of G, $g \in \overline{M}^{\mathfrak{Z}_G}$ if and only if either $g \in M$ or there exists a $\mathfrak{Z}_G$-atom $A \subseteq M$ such that $\{g\} \cup A$ is still a $\mathfrak{Z}_G$-atom.

(b) Deduce from (a) that every infinite abelian group G contains a $\mathfrak{Z}_G$-atom.

Hint. A proof of (a) can be found in [104].

Exercise 4.5.14. Let $G = \prod_{p\in\mathbb{P}} \mathbb{J}_p^{k_p} \times F_p$, where $k_p \in \mathbb{N}$ and F_p is a finite abelian p-group. Endow G with the product topology and show that all endomorphisms of G are continuous.

Hint. Use the fact that the product topology of G coincides with ν_G.

Exercise 4.5.15. Show that for the group G from Example 4.3.5 the diagonal subgroup Δ_G of $G \times G$ is a maximal subgroup.

Hint. Assume that H is a subgroup of $G \times G$ properly containing Δ_G. Then $H_0 = \{h \in G: (h, e_G) \in H\}$ is a nontrivial subgroup of G. Let us see that H_0 is a normal subgroup of G, so $H_0 = G$, since G is simple. This would imply $G \times \{e_G\} \subseteq H$ which, along with $\Delta_G \subseteq H$, would entail $H = G \times G$.

Pick $g \in G$ and $h \in H_0$, so that $(h, e_G) \in H$. Then $(g, g), (g^{-1}, g^{-1}) \in H$. Hence,

$$(g,g)(h,e_G)(g^{-1},g^{-1}) = (ghg^{-1}, e_G) \in H,$$

so $ghg^{-1} \in H_0$.

Exercise 4.5.16.* Let X be an infinite set. Prove that:

(a) if H is a subgroup of $S(X)$ containing S_F for some finite F of X and F is not H-invariant, then there is a proper subset F' of F such that $H \supseteq S_{F'}$;

(b) $A_\omega(X) := \bigcup\{A(F): F \in [X]^{<\omega}\}$ is dense in $(S(X), \mathcal{T}_X)$.

Hint. (a) Argue by induction on $n = |F|$ (for $n = 1$ use Lemma 4.2.4). See [99, Proposition 7.1.1] for more detail.

(b) We prove that $S_F A_\omega(X) = S(X)$ for every $F \in [X]^{<\omega}$. Since $A_\omega(X)$ is a normal subgroup of $S(X)$, each $S_F A_\omega(X)$ is a subgroup of $S(X)$.

We proceed by induction on $n = |F|$. The case $n = 1$ follows from Lemma 4.2.4. Assume that $|F| = n > 1$ and let $H = S_F A_\omega(X)$. Clearly, F is not $A_\omega(X)$-invariant and so not H-invariant. By item (a), there exists a proper subset F' of F such that $H \supseteq S_{F'}$. So, by inductive hypothesis, $H = S_{F'} A_\omega(X) = S(X)$.

4.6 Further readings, notes, and comments

Definition 4.1.1 is due to Markov [209], but the Zariski topology was explicitly defined only in [42] under the name *verbal topology*. Our preference goes for the term *Zariski topology* coined by most authors as in [26]. The Markov topology was introduced in [102, 103]. Various applications of these topologies are given in [105, 107], more details can be found in the survey papers [113, 114].

The property from Exercise 4.5.13 shows the analogy between the Zariski topology of an abelian group and metric (or, more generally, Fréchet–Urysohn) spaces, where the converging sequences completely describe closures, hence the topology (see [104] for further details).

Following Markov [209], call a subset A of a group G *potentially dense in* G if there exists a Hausdorff group topology τ on G such that A is τ-dense in G. Markov proved that every infinite subset of $\mathbb{Z}$ is potentially dense in $\mathbb{Z}$ and asked which subsets of a group G are potentially dense in G. The Zariski topology was applied in [105] for a complete answer in the case of countable subsets as follows. First, one can easily see that a potentially dense subset A of a group G is also $\mathfrak{Z}_G$-dense; moreover, if A is countable, then necessarily $|G| \leq 2^{\mathfrak{c}}$, in view of Lemma 5.1.5. It was proved in [105] that these necessary conditions turn out to be also sufficient in case G is abelian: a countable subset A of an abelian group G is potentially dense if and only if A is $\mathfrak{Z}_G$-dense and $|G| \leq 2^{\mathfrak{c}}$.

Further applications of the Zariski topology will be given in Chapter 10.

The following questions, inspired by Theorem 4.3.7, seem to be open.

Question 4.6.1. Does every infinite group G admit a nondiscrete Hausdorff topology τ such that (G, τ) is a paratopological group? In particular, do the groups from Theorem 4.3.4 and Example 4.3.5 admit such topologies?

More information on extension of group topologies can be found in [48, 49].

Note added in proofs September 20, 2021: Recently Shakhmatov and Yañez proved that $\mathfrak{Z}_F = \mathfrak{M}_F$ for every free non-abelian group F, obtaining in this way a positive answer, for free non-abelian groups, to the Markov problem on coincidence of these two topologies (see Remark 4.1.4). The proof is heavily based on the reflection principle characterizing the groups G with $\mathfrak{Z}_G = \mathfrak{M}_G$ in [103].

5 Cardinal invariants and metrizability

5.1 Cardinal invariants of topological groups

The *cardinal invariants* of topological groups are cardinal numbers, say $\rho(G)$, associated to every topological group G such that $\rho(G) = \rho(H)$ for topologically isomorphic groups G and H. For example, the cardinality $|G|$ is the simplest cardinal invariant. The cardinal invariants that we study here for topological groups G are the *weight* $w(G)$, the *character* $\chi(G)$, the *pseudocharacter* $\psi(G)$, and the *density character* $d(G)$ (see §B.3). Due to the homogeneity of topological groups, the character $\chi(G)$ coincides with the character $\chi(G, e_G)$ at e_G.

Remark 5.1.1. Let G be a topological group. Here we pay special attention to the case when the cardinal invariants $w(G)$, $\chi(G)$, and $d(G)$ are finite.

(a) Assume that $d(G)$ is finite. If X is a finite dense subset of G and G is Hausdorff (i. e., $\mathfrak{h}G = G$), then $X = G$ (as X is closed), so G is finite discrete. Otherwise, if G is not Hausdorff, G need not have a finite dense subgroup (of size $d(G)$). Indeed, take $G = \mathbb{Z}$ with the topology given by taking $2\mathbb{Z}$ with the indiscrete topology and declaring $2\mathbb{Z}$ open in $\mathbb{Z}$; in particular, $\mathfrak{h}G = \mathbb{Z}/2\mathbb{Z}$ is discrete. Now $d(G) = 2$ and G has no finite dense subgroups.

(b) In particular, $d(G) = 1$ if and only if G is indiscrete, if and only if $w(G) = 1$. Moreover, both $w(G)$ and $d(G)$ may attain all possible finite values $m \in \mathbb{N}_+$: take, for example, the discrete group $\mathbb{Z}(m)$ with $w(\mathbb{Z}(m)) = d(\mathbb{Z}(m)) = m$.

(c) On the other hand, either $\chi(G) = 1$ or $\chi(G)$ is infinite, and the same applies to $\psi(G)$. Moreover, $\chi(G) = 1$ if and only if $\mathfrak{h}G$ is discrete (i. e., there exists a local base at e_G in G consisting of a single element N, and so $N = \overline{\{e_G\}}$). This occurs, in particular, when G is discrete or when G is indiscrete.

(d) More generally, $\chi(G) = \chi(\mathfrak{h}G)$, $d(G) = d(\mathfrak{h}G)$, and $w(G) = w(\mathfrak{h}G)$. Combining with (a), we conclude that $\mathfrak{h}G$ is finite if and only if $d(G)$ is finite, in such case $d(G) = w(G) = |\mathfrak{h}G| \geq \chi(G) = 1$. If $\mathfrak{h}G$ is infinite, then $d(G)$ is infinite as well, and so G has a dense subgroup of size $d(G)$.

We start relating the local bases at e_G in G to those of a subgroup H of G:

Lemma 5.1.2. *If H is a subgroup of a topological group G and $\mathcal{B}$ is a base (respectively, local base at e_G) of G, then $\{U \cap H : U \in \mathcal{B}\}$ is a base (respectively, local base at e_G) of H.*

Now we consider the case when H is a dense subgroup of G.

Lemma 5.1.3. *If H is a dense subgroup of a Hausdorff group G and $\mathcal{B}$ is an open local base at e_G in H, then $\{\overline{U}^G : U \in \mathcal{B}\}$ is a local base at e_G in G.*

Proof. Since G is a regular space by Proposition 3.1.15, the closed neighborhoods of e_G in G form a local base at e_G in G. Hence, for $V \in \mathcal{V}_G(e_G)$, one can find $V_0 \in \mathcal{V}_G(e_G)$ such

https://doi.org/10.1515/9783110654936-005

that $\overline{V_0} \subseteq V$. Since $V_0 \cap H$ is a neighborhood of e_G in H, there exists $U \in \mathcal{B}$ such that $U \subseteq V_0 \cap H$. By Lemma 5.1.2, and since U is open in H, there exists an open $W \in \mathcal{V}_G(e_G)$ such that $U = W \cap H$ and $W \subseteq V_0$. Hence, by Lemma B.1.19, $\overline{U}^G = \overline{W}^G$ as H is dense in G and W is open in G. Thus, $\overline{U}^G = \overline{W}^G \subseteq \overline{V_0}^G \subseteq V \in \mathcal{V}_G(e_G)$. □

Proposition 5.1.4. *Let H be a dense subset of a topological group G and let $\mathcal{B}$ be a local base at e_G in G consisting of symmetric neighborhoods. Then $\{hU : U \in \mathcal{B}, h \in H\}$ is a base of the topology of G.*

Proof. Let $x \in G$ and let O be an open set of G containing x. Then there exists a symmetric $U \in \mathcal{B}$ with $xUU \subseteq O$. Pick $h \in H \cap xU$. Then $x^{-1}h \in U$, so $h^{-1}x \in U^{-1} = U$. Therefore, $x \in hU = xx^{-1}hU \subseteq xUU \subseteq O$. □

Lemma 5.1.5. *Let G be a topological group. Then:*
(a) $d(G) \leq w(G) \leq 2^{d(G)}$;
(b) *in case G is Hausdorff,* $\psi(G) \leq |G| \leq 2^{w(G)}$.

Proof. (a) To see that $d(G) \leq w(G)$, choose a base $\mathcal{B}$ of the topology on G with $|\mathcal{B}| = w(G)$, and for every $\emptyset \neq U \in \mathcal{B}$ pick a point $d_U \in U$. Then the set $D = \{d_U : U \in \mathcal{B}\}$ is dense in G and $|D| \leq w(G)$.

To prove that $w(G) \leq 2^{d(G)}$, note that in case $\mathfrak{h}G$ is finite this obviously follows from Remark 5.1.1(d), hence we can assume in the sequel that $\mathfrak{h}G$ is infinite. Consequently, G has a dense subgroup D of size $d(G)$. According to Proposition 3.1.15, G is a regular space, hence every base $\mathcal{B}$ of the topology on G consisting of open subsets of G contains a base $\mathcal{B}_r$ of the same size consisting of regular open subsets of G. Fix such a base $\mathcal{B}_r$ of size $w(G)$. If $U, V \in \mathcal{B}_r$ and $U \cap D = V \cap D$, then $\overline{U} = \overline{U \cap D} = \overline{V \cap D} = \overline{V}$. With U and V being regular open, the equality $\overline{U} = \overline{V}$ implies $U = V$. Hence, the map $\mathcal{B}_r \to \mathcal{P}(D)$, $U \mapsto U \cap D$, is injective, so $w(G) \leq 2^{d(G)}$.

(b) Since G is T_1, the complement of every singleton $\{x\}$, where $x \in G \setminus \{e_G\}$, is an open neighborhood of e_G in G. The intersection of these neighborhoods is $\{e_G\}$, and this witnesses the inequality $\psi(G) \leq |G|$.

Let $\mathcal{B}$ a base of the topology on G with $|\mathcal{B}| = w(G)$. To every point $x \in G$ assign the set $O_x = \{U \in \mathcal{B} : x \in U\}$. Then the axiom T_2 guarantees that the map $G \to \mathcal{P}(\mathcal{B})$, $x \mapsto O_x$, is injective. Therefore, $|G| \leq 2^{w(G)}$. □

Remark 5.1.6. Three observations related to Lemma 5.1.5(b) are in order here.
(a) Let G be a dense cyclic subgroup of $\mathbb{T}^{\mathfrak{c}}$ (see Theorem 9.4.8). Then $w(G) = w(\mathbb{T}^{\mathfrak{c}}) = \mathfrak{c} = 2^{|G|}$, in view of Lemma 5.1.8(b) and Theorem 5.1.15.
(b) Although $\chi(G)$ does not appear in Lemma 5.1.5(b), one can say at least that $\psi(G) \leq \chi(G) \leq w(G)$ for a Hausdorff group G, but one cannot prove that $\chi(G) \leq |G|$. Actually, we shall see that $\chi(G^{\#}) = 2^{|G|} > |G|$ for every infinite abelian group G (see Exercise 13.7.2).

(c) Without the assumption that G is Hausdorff, one can only prove that $|\mathfrak{h}G| \le 2^{w(G)}$ in view of Remark 5.1.1(d), while $|G| = |\mathrm{core}(G)| \cdot |\mathfrak{h}G| \le |\mathrm{core}(G)| \cdot 2^{w(G)}$. So, the inequality $|G| \le 2^{w(G)}$ fails for every indiscrete G with $|G| > 2$.

Now we see another precise relation among the weight, character, and density character.

Lemma 5.1.7. *Let G be a topological group. Then $w(G) = \chi(G) \cdot d(G)$.*

Proof. The inequality $w(G) \ge \chi(G)$ is obvious. The inequality $w(G) \ge d(G)$ was proved in Lemma 5.1.5(a). This gives $w(G) \ge \chi(G) \cdot d(G)$.

To prove the inequality $w(G) \le \chi(G) \cdot d(G)$, pick a dense subset D of G of size $d(G)$ and a base $\mathcal{B}$ of symmetric open sets of $\mathcal{V}(e_G)$ with $|\mathcal{B}| = \chi(G)$, and apply Proposition 5.1.4. □

The next lemma concerns the behavior of the cardinal invariants with respect to taking subgroups.

Lemma 5.1.8. *Let H be a subgroup of a topological group G. Then:*
(a) *$w(H) \le w(G)$ and $\chi(H) \le \chi(G)$; moreover, $\psi(H) \le \psi(G)$ if G is T_2;*
(b) *if H is dense in G, then $\chi(G) = \chi(H)$, $w(G) = w(H)$ and $d(G) \le d(H)$.*

Proof. (a) This follows from Lemma 5.1.2, as $|\{U \cap G : U \in \mathcal{B}\}| \le |\mathcal{B}|$ for every base (respectively, local base at e_G) $\mathcal{B}$ of the topology on G. A similar argument applies for ψ.

(b) We prove first $\chi(G) = \chi(H)$. The inequality $\chi(G) \ge \chi(H)$ follows from item (a). To prove the opposite inequality, fix an open local base $\mathcal{B}$ of $\mathcal{V}_H(e_G)$ with $|\mathcal{B}| = \chi(H)$. By Lemma 5.1.3, $\mathcal{B}^* = \{\overline{U}^H : U \in \mathcal{B}\}$ is a base of $\mathcal{V}_G(e_G)$. Since $|\mathcal{B}^*| \le |\mathcal{B}| = \chi(H)$, this proves $\chi(G) \le \chi(H)$.

The inequality $d(G) \le d(H)$ follows from the fact that the dense subsets of H are dense in G as well.

The inequality $w(G) \ge w(H)$ follows from item (a). By the above argument, $\chi(G) = \chi(H)$. Now $w(G) = \chi(G) \cdot d(G) \le \chi(H) \cdot d(H) = w(H)$, where the first and the last equality follow from Lemma 5.1.7 and the inequality from $d(G) \le d(H)$. □

The strict inequality $d(G) < d(H)$ in item (b) of Lemma 5.1.8 is possible. For example, $G = \mathbb{T}^{\mathfrak{c}}$ has $d(\mathbb{T}^{\mathfrak{c}}) = \omega$, by Remark 5.1.6(a). On the other hand, the Σ-product $H = \Sigma\mathbb{T}^{\mathfrak{c}}$ is a dense subgroup of G with $d(H) > \omega$.

Concerning the pseudocharacter, the equality $\psi(H) = \psi(G)$ may fail in general (see Exercise 5.4.5).

Next we see that these cardinal invariants behave well with respect to continuous images.

Lemma 5.1.9. *Let G, H be topological groups. If $f : G \to H$ is a continuous surjective homomorphism, then $d(H) \le d(G)$. If f is open, then also $w(H) \le w(G)$ and $\chi(H) \le \chi(G)$.*

Proof. If D is a dense subset of G, then $f(D)$ is a dense subset of H with $|f(D)| \leq |D|$. This proves the first assertion. The second assertion follows from the fact that if $\mathcal{B}$ is a base (respectively, a local base at e_G) of the topology on G, then $\mathcal{B}_0 = \{f(B): B \in \mathcal{B}\}$ is a base (respectively, local base at e_H) of the topology on H, if f is open, and $|\mathcal{B}_0| \leq |\mathcal{B}|$. □

The inequality $w(H) \leq w(G)$ may fail if f is not open:

Example 5.1.10. Since in Lemma 5.1.9 H is a continuous image of G by means of the continuous surjective homomorphism $f: G \to H$, the canonical projection $G \to G/\ker f$ is open, and there exists a continuous isomorphism $G/\ker f \to H$ (which is open if and only if f is open), we have to find a group G with two group topologies $\tau_1 < \tau_2$ and $w(G, \tau_1) < w(G, \tau_2)$.

(a) Let G be an infinite abelian group and $\tau_2 = \delta_G$, so that $w(G, \tau_2) = |G|$. Moreover, let $\tau_1 = \mathfrak{B}_G$. Then $w(G, \tau_1) = w(G^\#) = 2^{|G|} > |G| = w(G, \tau_2)$, by Corollary 13.3.12.

(b) Let G be a countably infinite abelian group and consider $\tau_2 = \delta_G$, so $w(G, \tau_2) = \omega$. In view of Remark 5.3.10(a), there exists a T-sequence $A = \{a_n\}_{n\in\mathbb{N}}$ of G (see Definition 5.3.1). Then the topology $\tau_1 = \tau_{\{a_n\}}$ (defined right after Definition 5.3.1) is not metrizable in view of Remark 5.3.10(b), and so $\chi(G, \tau_2) = 1 < \omega = w(G, \tau_2) < \chi(G, \tau_1) \leq w(G, \tau_1)$ by Birkhoff–Kakutani theorem 5.2.17.

Now we discuss the behavior of the cardinal invariants $d(-)$, $w(-)$, and $\chi(-)$ with respect to taking direct products of topological groups.

Theorem 5.1.11. *If $\{G_i : i \in I\}$ is a family of topological groups and $G = \prod_{i\in I} G_i$, then $\sup\{d(G_i): i \in I\} \leq d(G) \leq |I| \cdot \sup\{d(G_i): i \in I\}$.*

Proof. Applying Lemma 5.1.9 to the projection $p_i: G \to G_i$, we deduce $d(G_i) \leq d(G)$. Hence, $\sup\{d(G_i): i \in I\} \leq d(G)$.

For each $i \in I$, let D_i be a dense subset of G_i containing e_{G_i} and such that $d(G_i) = |D_i|$. Then $D = \bigoplus_{i\in I} D_i$ is a dense subset of G with

$$|D| \leq |I| \cdot \sup\{|D_i|: i \in I\} = |I| \cdot \sup\{d(G_i): i \in I\}.$$

So, $d(G) \leq |I| \cdot \sup\{d(G_i): i \in I\}$. □

Example 5.1.12. The topological groups G with $d(G) \leq \omega$ are precisely the separable groups. If $\{G_i : i \in I\}$ is a family of separable groups, then $d(\prod_{i\in I} G_i) \leq \max\{\omega, |I|\}$, by Theorem 5.1.11.

A stronger, yet nontrivial, inequality holds for this type of products in view of Hewitt–Marczewski–Pondiczery theorem B.3.15: for an infinite cardinal κ, the inequality $d(\prod_{i\in I} G_i) \leq \kappa$ holds, whenever $|I| \leq 2^\kappa$ and all groups G_i are separable; in particular, $\prod_{i\in I} G_i$ is separable whenever $|I| \leq \mathfrak{c}$. We give a proof of this result for compact abelian groups in Exercise 14.5.2 by means of the Pontryagin-van Kampen duality.

Remark 5.1.13. (a) If $\{G_i : i \in I\}$ is a family of topological groups, then $G = \prod_{i\in I} G_i$ is indiscrete if and only if each G_i is indiscrete; in view of Remark 5.1.1(b), this can be equivalently stated as $w(G) = 1$ if and only if $w(G_i) = 1$ for every $i \in I$.

(b) If G_1, G_2 are topological groups, $G = G_1 \times G_2$ and G_2 is indiscrete, then $\mathfrak{h}G \cong \mathfrak{h}G_1$, and hence $d(G) = d(G_1), \chi(G) = \chi(G_1)$, and $w(G) = w(G_1)$.

(c) More generally, consider a family $\{G_i : i \in I\}$ of topological groups and $G = \prod_{i \in I} G_i$. Let $J = \{j \in I : G_j \text{ nonindiscrete}\}$, let $G_J = \prod_{i \in J} G_i$ and $G_0 = \prod_{i \in I \setminus J} G_i$ (we put $G_0 = \{e_G\}$ if $J = I$, and similarly $G_J = \{e_G\}$ if $J = \emptyset$). Then $G = G_J \times G_0$ and G_0 is indiscrete according to item (a). Therefore, $\chi(G) = \chi(G_J)$, $d(G) = d(G_J)$ and $w(G) = w(G_J)$ by item (b).

By Remark 5.1.13(c), it is safe to assume in the sequel that all topological groups appearing in the following products are nonindiscrete.

Lemma 5.1.14. *Let $n \in \mathbb{N}_+$, let $G_1, \dots, G_n$ be topological groups and $G = G_1 \times \cdots \times G_n$. Then*

$$\chi(G) = \chi(G_1) \cdots \chi(G_n) \quad \text{and} \quad w(G) = w(G_1) \cdots w(G_n).$$

If the groups G_i are Hausdorff, then $\psi(G) = \psi(G_1) \cdots \psi(G_n)$.

Proof. We prove the assertions for $n = 2$. Then one can proceed by induction. Let $\mathcal{B}_1, \mathcal{B}_2$ be local bases at e_{G_1}, e_{G_2}, respectively in G_1, G_2. Then

$$\mathcal{B} = \{U \times V : U \in \mathcal{B}_1, V \in \mathcal{B}_2\}$$

is a local base at e_G in G with $|\mathcal{B}| = |\mathcal{B}_1| \cdot |\mathcal{B}_2|$. This shows that $\chi(G) \leq \chi(G_1) \cdot \chi(G_2)$. The missing inequality $\chi(G_1) \cdot \chi(G_2) \leq \chi(G)$ follows from Lemma 5.1.8(a).

According to Remark 5.1.1, $w(G) = w(\mathfrak{h}G)$, $w(G_1) = w(\mathfrak{h}G_1)$, and $w(G_2) = w(\mathfrak{h}G_2)$. Since $\mathfrak{h}G = \mathfrak{h}G_1 \times \mathfrak{h}G_2$, this allows us to assume from this point till the very end of the proof that G_1 and G_2 are Hausdorff. If both groups are finite, then $w(G_i) = |G_i|$, so the desired equality obviously holds. Assume now that at least one of the groups, say G_i, is infinite. Then both $w(G_i)$ and $d(G_i)$ (hence also $w(G)$ and $d(G)$) are infinite as well, so Theorem 5.1.11 yields $d(G) = d(G_1) \cdot d(G_2)$, while $\chi(G) = \chi(G_1) \cdot \chi(G_2)$, by the first part of the proof.

By Lemma 5.1.7, $w(G) = d(G) \cdot \chi(G)$, so

$$\begin{aligned} w(G) &= (d(G_1) \cdot d(G_2)) \cdot (\chi(G_1) \cdot \chi(G_2)) = \\ &= (d(G_1) \cdot \chi(G_1)) \cdot (d(G_2) \cdot \chi(G_2)) = w(G_1) \cdot w(G_2). \end{aligned}$$

A proof, similar to the case of $\chi(G)$, applies for $\psi(G)$. □

An alternative proof of the last assertion of the above lemma can be obtained also from Exercise 5.4.6.

In view of Lemma 5.1.14, in the sequel we can consider infinite products.

Theorem 5.1.15. *Let $\{G_i : i \in I\}$ be an infinite family of nonindiscrete topological groups and $G = \prod_{i \in I} G_i$. Then*

$$\chi(G) = |I| \cdot \sup\{\chi(G_i) : i \in I\} \quad \text{and} \quad w(G) = |I| \cdot \sup\{w(G_i) : i \in I\}. \tag{5.1}$$

Moreover, if all G_i are Hausdorff, then $\psi(G) = |I| \cdot \sup\{\psi(G_i) : i \in I\}$.

Proof. For every $i \in I$, let $\mathcal{B}_i$ be a base of $\mathcal{V}_{G_i}(e_{G_i})$ with $|\mathcal{B}_i| = \chi(G_i)$. For any finite subset $J \subseteq I$ and for $U_i \in \mathcal{B}_i$ when $i \in J$, let

$$W_{J,(U_i)_{i\in J}} = \prod_{i\in J} U_i \times \prod_{i\in I\setminus J} G_i,$$

which is a basic neighborhood of e_G in the product topology of G. Then the family $\mathcal{B} = \{W_{J,(U_i)_{i\in J}} : J \subseteq I \text{ finite}, U_i \in \mathcal{B}_i \text{ for } i \in J\}$ is a base of $\mathcal{V}_G(e_G)$ and

$$|\mathcal{B}| \le |I| \cdot \sup\{|\mathcal{B}_i| : i \in I\} = |I| \cdot \sup\{\chi(G_i) : i \in I\}.$$

Hence, $\chi(G) \le |\mathcal{B}| \le |I| \cdot \sup\{\chi(G_i) : i \in I\}$.

Clearly, $\chi(G_i) \le \chi(G)$ for every $i \in I$ by Lemma 5.1.8(a). Hence, to prove the first equality in (5.1), it remains to show that $|I| \le \chi(G)$. Let $\mathcal{B}$ be a base of $\mathcal{V}_G(e_G)$ with $|\mathcal{B}| = \chi(G)$. By assumption, G_i is not indiscrete for every $i \in I$. Hence, for every $i \in I$, there exists $B_i \in \mathcal{B}$ such that $p_i(B_i) \ne G_i$, where p_i denotes the canonical projection. To see that the mapping $\phi : I \to \mathcal{B}$, defined by $\phi(i) = B_i$, is finitely-many-to-one (so, $|I| \le |\mathcal{B}|$), note that by the definition of open sets in the product topology, for $B \in \mathcal{B}$ the set $F_B = \{i \in I : p_i(B) \ne G_i\}$ is finite. Since $\phi^{-1}(B) \subseteq F_B$ for every $B \in \mathcal{B}$, we are done.

To prove the remaining inequality in (5.1), put $\kappa = \sup\{w(G_i) : i \in I\}$. Clearly, $w(G) \ge w(G_i)$ for every $i \in I$, by Lemma 5.1.8(a). Hence, $w(G) \ge \kappa$. Moreover, $w(G) \ge \chi(G) \ge |I|$, by the above proof. This proves the "$\ge$" part of the second equality in (5.1). It remains to check that $w(G) \le |I| \cdot \kappa$. By Lemma 5.1.7, $w(G) = d(G) \cdot \chi(G)$. Since $d(G_i) \le w(G_i) \le \kappa$ for every $i \in I$, Theorem 5.1.11 implies $d(G) \le |I| \cdot \kappa$. Similarly, $\chi(G_i) \le w(G_i) \le \kappa$ for every $i \in I$, so the first equality in (5.1) implies $\chi(G) \le |I| \cdot \kappa$. Therefore, $w(G) = d(G) \cdot \chi(G) \le |I| \cdot \kappa$.

When all G_i are Hausdorff, a proof, similar to the case of $\chi(G)$, applies for $\psi(G)$. □

The equalities in (5.1) may fail when $1 < |I| < \infty$.

5.2 Metrizability of topological groups

5.2.1 Pseudonorms and invariant pseudometrics in a group

Following Markov [211], we recall the following definition.

Definition 5.2.1. A *pseudonorm* on a group G is a function $\nu : G \to \mathbb{R}$ such that:
(i) $\nu(e_G) = 0$;
(ii) $\nu(x^{-1}) = \nu(x)$ for every $x \in G$;
(iii) $\nu(xy) \le \nu(x) + \nu(y)$ for every $x, y \in G$.

A pseudonorm ν on G is a *norm* if it has the additional property:
(i*) $\nu(x) = 0$ if and only if $x = e_G$.

If (iii) is replaced by the stronger property:
(iii*) $v(xy) \leq \max\{v(x), v(y)\}$ for every $x, y \in G$,

the (pseudo)norm is called *non-Archimedean.*

Remark 5.2.2. The values of a pseudonorm on a group G are necessarily *nonnegative* reals, since for every $x \in G$,

$$0 = v(e_G) = v(x^{-1}x) \leq v(x^{-1}) + v(x) = v(x) + v(x) = 2v(x).$$

Example 5.2.3. The simplest example of a norm is the usual norm on $\mathbb{R}$ defined by the absolute value $|a|$ for every $a \in \mathbb{R}$.

Remark 5.2.4. The norms defined on a real or complex vector space V are obviously norms of the underlying abelian group, although they have a property stronger than (ii), namely, $v(\lambda x) = |\lambda| v(x)$ for all $\lambda \in K \in \{\mathbb{R}, \mathbb{C}\}$ and $x \in V$.

Sometimes a norm on a real vector space V can be induced by an *inner product* (or, *scalar product*) defined as a symmetric bilinear function $\langle - \mid - \rangle : V \times V \to \mathbb{R}$ such that $\langle x \mid x \rangle > 0$ for every $x \in V \setminus \{0\}$. Then, letting $\|x\| = \sqrt{\langle x \mid x \rangle}$ for every $x \in V$, one obtains a norm on V.

In $\mathbb{R}^n$ the *standard scalar product* is defined by letting $(x \mid y) := \sum_{i=1}^n x_i y_i$ for every $x = (x_1, \dots, x_n), y = (y_1, \dots, y_n) \in \mathbb{R}^n$.

Keeping the idea from pseudometrics (see §B.3.2), for a pseudonorm v on a group G, we still denote by $B_r^v(e_G) := \{x \in G : v(x) < r\}$ the open ball with center e_G and radius $r \in \mathbb{R}_{>0}$. Every pseudonorm v on an abelian group G generates a group topology τ_v on G by way of the filter base $\mathcal{B}^v := \{B_r^v(e_G) : r \in \mathbb{R}_{>0}\}$ (see Exercise 5.4.11). It is a relevant fact that $\mathcal{V}_{\tau_v}(e_G)$ has a countable base given by $\{B_{1/n}^v(e_G) : n \in \mathbb{N}_+\}$.

Remark 5.2.5. Let G be a group. Every pseudonorm v on G generates a pseudometric d_v on G defined by letting

$$d_v(x, y) = v(x^{-1}y) \quad \text{for every } x, y \in G.$$

This pseudometric is *left invariant*, i. e., $d_v(ax, ay) = d_v(x, y)$ for every $a, x, y \in G$. In particular, for every $r \in \mathbb{R}_{>0}$, $B_r^v(e_G) = B_r^{d_v}(e_G)$.

Conversely, every left invariant pseudometric d on G gives rise to a pseudonorm on G defined by

$$v_d(x) = d(x, e_G) \quad \text{for every } x \in G.$$

The assignment $v \mapsto d_v$ defines a bijective correspondence between pseudonorms on G and left invariant pseudometrics on G, as $d = d_{v_d}$ for a left invariant pseudometric d.

Clearly, d_v is a metric if and only if v is a norm. The above bijective correspondence between norms and left invariant metrics takes non-Archimedean norms to left invariant ultrametrics.

Similarly, *right invariant pseudometrics* are defined, which are in a similar bijective correspondence with pseudonorms.

Example 5.2.6. Let ℓ_2 denote the set of all sequences $x = \{x_n\}_{n\in\mathbb{N}}$ of real numbers such that the series $\sum_{n\in\mathbb{N}} x_n^2$ converges. Then ℓ_2 has a natural structure of vector space (induced by the Cartesian product $\mathbb{R}^{\mathbb{N}} \supseteq \ell_2$) usually called *Hilbert space*. Let $\|x\| = \sqrt{\sum_{n\in\mathbb{N}} x_n^2}$ for every $x = \{x_n\}_{n\in\mathbb{N}} \in \ell_2$. This defines a norm on the abelian group $(\ell_2, +)$, that provides an invariant metric on ℓ_2 making it a metric space and a topological group.

Example 5.2.7. For $n \in \mathbb{N}_+$, $q \in [1,\infty)$ and $x = (x_1,\dots,x_n) \in \mathbb{R}^n$, let

$$\|x\|_q = \left(\sum_{i=1}^{n} |x_i|^q\right)^{\frac{1}{q}}.$$

(i) For $q = 2$, the norm $\|x\|_2 = \sqrt{\sum_{i=1}^n x_i^2}$ provides the Euclidean metric.
(ii) For $q = 1$, the norm $\|x\|_1 = \sum_{i=1}^n |x_i|$ provides the taxi driver metric.
(iii) For $q = \infty$, the norm $\|x\|_\infty = \sup_{i\in\{1,\dots,n\}} |x_i|$ (called also sup-norm) gives the chess-board metric.

See Example B.3.6.

Example 5.2.8. For a prime p, the p-adic norm $|-|_p\colon \mathbb{Q} \to \mathbb{R}_{>0}$ (see Example B.3.12) is a non-Archimedean norm on the abelian group $(\mathbb{Q}, +)$. It gives rise to the p-adic ultrametric on $\mathbb{Q}$.

Now we provide an example of a norm on a typically nonabelian group.

Example 5.2.9. For $S(\mathbb{N})$, the permutation group of $\mathbb{N}$, let $v(id_{\mathbb{N}}) = 0$, and for $g \in S(\mathbb{N}) \setminus \{id_{\mathbb{N}}\}$, let $s(g) = \min\{n \in \mathbb{N}\colon g(n) \neq n\}$ and $v(g) = 2^{-s(g)}$. Then v is a non-Archimedean norm on $S(\mathbb{N})$, which induces an ultrametric on $S(\mathbb{N})$.

The next example comes from geometric group theory.

Example 5.2.10. Let G be a finitely generated group and let $S = S^{-1}$ be a finite set of generators of G. Given $g \in G \setminus \{e_G\}$, its *word norm* $\ell_S(g)$ with respect to S is the shortest length of a word in the alphabet S whose evaluation is equal to g:

$$\ell_S(g) = \min\{n \in \mathbb{N}_+\colon g = s_1 \cdots s_n,\ s_1,\dots,s_n \in S\}.$$

Moreover, let $\ell_S(e_G) = 0$. It is easy to check that the function ℓ_S satisfies the conditions that determine a word norm:
- $\ell_S(g^{-1}) = \ell_S(g)$ for every $g \in G$;
- $\ell_S(gh) \leq \ell_S(g) + \ell_S(h)$ for every $g, h \in G$.

The *word metric* d_S on G with respect to S is the left invariant metric associated to ℓ_S.

The growth of the size of the balls $B_n^{d_S}(e_G) = \{g \in G : \ell_S(g) \leq n\}$ is one of the main topics in geometric group theory, starting from the Milnor problem (see [212]) and the Gromov theorem (see [161]). More precisely, for every $n \in \mathbb{N}$, let $\gamma_S(n) = |B_n^{d_S}(e_G)|$. The function

$$\gamma_S : \mathbb{N} \to \mathbb{N},\ n \mapsto |B_n^{d_S}(e_G)|,$$

is called the *growth function* of G with respect to S. If G is infinite, the function γ_S is strictly monotone increasing, i. e., $\gamma_S(n) < \gamma_S(n+1)$ for every $n \in \mathbb{N}$.

Depending on the type of growth of γ_S, the group $G = \langle S \rangle$ is said to be of *polynomial, exponential,* or *intermediate* growth rate. It can be shown that this type of growth does not depend on S. See [208] for a further reading on this topic.

5.2.2 Continuous pseudonorms and pseudometrics

Definition 5.2.11. Let G be a topological group. A pseudonorm ν on G is *continuous* if $\nu : G \to \mathbb{R}_{\geq 0}$ is continuous. Similarly, a pseudometric d on G is *continuous* if $d : G \times G \to \mathbb{R}_{\geq 0}$ is continuous.

If (G, τ) is a topological group and d is a pseudometric on G, then d is continuous precisely when the topology τ_d induced by d is coarser than the topology τ (i. e., every open set with respect to the metric d is τ-open).

Remark 5.2.12. Clearly, a left invariant pseudometric d is continuous precisely when the related pseudonorm ν_d is continuous.

Our next aim is to build a continuous pseudometric on a topological group G starting with a decreasing chain $U_0 \supseteq U_1 \supseteq \cdots \supseteq U_n \supseteq \cdots$ of symmetric neighborhoods of e_G in G with

$$U_{n+1}U_{n+1} \subseteq U_n \quad \text{for every } n \in \mathbb{N}. \tag{5.2}$$

To this end, we need the following lemma providing a continuous pseudonorm that we use further to produce the desired continuous pseudometric.

Lemma 5.2.13. *Let (G, τ) be a topological group and $\{U_n : n \in \mathbb{N}\}$ a decreasing chain of symmetric neighborhoods of e_G in G as in (5.2). Then there exists a continuous pseudonorm $\nu : G \to [0, 1]$ such that, for every $n \in \mathbb{N}_+$,*

$$B^{\nu}_{2^{-n}}(e_G) \subseteq U_n \subseteq B^{\nu}_{2^{-n+2}}(e_G) \tag{5.3}$$

and $H = \bigcap_{n \in \mathbb{N}} U_n$ is a closed subgroup of G with $H = \{x \in G : \nu(x) = 0\}$. So, ν is a norm if and only if $H = \{e_G\}$.

Proof. Denote by $D = \{\frac{m}{2^n} : m \in \mathbb{N}_+,\ n \in \mathbb{N}\}$ the set of positive dyadic rational numbers. We are going to construct a collection $\{V_r : r \in D\}$ of neighborhoods of e_G as follows. If

$r > 1$, we put $V_r = G$ and for $l \in \mathbb{N}$, let $V_{2^{-l}} = U_l$, in particular $V_1 = U_0$. For $r \in D$ of the form

$$r = \frac{1}{2^{l_1}} + \frac{1}{2^{l_2}} + \cdots + \frac{1}{2^{l_n}} \tag{5.4}$$

with $l_1, \ldots, l_n \in \mathbb{N}_+$, $l_1 < l_2 < \cdots < l_n$, $n \in \mathbb{N}_+$, we define V_r as follows:

$$V_r = V_{2^{-l_1}} V_{2^{-l_2}} \cdots V_{2^{-l_n}} = U_{l_1} U_{l_2} \cdots U_{l_n}. \tag{5.5}$$

Let us prove that

$$V_r \subseteq V_s \quad \text{when } r \le s \text{ in } D. \tag{5.6}$$

We may assume that $s < 1$ and write $s \in D$ as

$$s = \frac{1}{2^{m_1}} + \frac{1}{2^{m_2}} + \cdots + \frac{1}{2^{m_N}} \tag{5.7}$$

with $m_1, \ldots, m_N \in \mathbb{N}_+$, $m_1 < m_2 < \cdots < m_N$. Clearly, $l_1 \ge m_1$, otherwise $l_1 \le m_1 - 1$, and this would imply

$$s < \frac{1}{2^{m_1 - 1}} \le \frac{1}{2^{l_1}} \le r,$$

a contradiction. Assume that $m_1 = l_1, \ldots, m_{k-1} = l_{k-1}$, but $m_k < l_k$. Taking into account the definition in (5.5) of V_r and V_s, it is not restrictive to assume that $k = 1$. Now, from $m_1 < l_1$ we deduce that $m_1 \le l_1 - 1$, so

$$r < \frac{1}{2^{l_1 - 1}} \le \frac{1}{2^{m_1}} \le s,$$

as above. From (5.2), (5.5), and these inequalities, we deduce $V_r \subseteq V_{2^{l_1 - 1}} \subseteq V_{2^{m_1}} \subseteq V_s$.

Further, we have:

$$V_r V_{2^{-l}} \subseteq V_{r + \frac{1}{2^l}} \quad \text{for } r \ge 1 \quad \text{or} \quad r = \sum_{k=1}^{n} 2^{-l_k} \in D \quad \text{and} \quad l > l_n. \tag{5.8}$$

The first assertion is trivial as $V_{r + \frac{1}{2^l}} = G$ if $r \ge 1$. The second assertion follows directly from the definition of $V_{r + \frac{1}{2^l}} = V_r U_l = V_r V_{2^{-l}}$, since $l > l_n$.

Now, for $x \in G$, let

$$f(x) = \inf\{r \in D : x \in V_r\} \quad \text{and} \quad \nu(x) = \sup\{|f(zx) - f(z)| : z \in G\}.$$

Observe that $f(e_G) = 0$ and hence $\nu \ge f$, so that $x \in V_r$ implies $f(x) \le r$, and that $f(x) \le 1$ for all $x \in G$. This implies $\nu(G) \subseteq [0, 1]$. Conversely, if $f(z) < r \in D$, then $z \in V_r$. Indeed, by the definition of f, there exists $s \in D$, $s < r$, such that $x \in V_s$ and (5.6) yields $x \in V_r$. Further,

$$\nu(x^{-1}) = \sup\{|f(zx^{-1}) - f(z)| : z \in G\} = \sup\{|f(zxx^{-1}) - f(zx)| : z \in G\} = \nu(x)$$

holds for all $x \in G$. Next, for $x, y \in G$,

$$\begin{aligned} v(xy) &= \sup\{|f(zxy) - f(z)| : z \in G\} \\ &\leq \sup\{|f(zxy) - f(zx)| + |f(zx) - f(z)| : z \in G\} \\ &\leq \sup\{|f(zxy) - f(zx)| : zx \in G\} + \sup\{|f(zx) - f(z)| : z \in G\} \\ &= v(y) + v(x). \end{aligned}$$

This proves that v is a pseudonorm and implies that H is a subgroup.

Let $n \in \mathbb{N}_+$. If $u \in G \setminus U_n$, then $u \notin V_{2^{-n}} = U_n$, therefore $v(u) \geq f(u) \geq 2^{-n}$, that is, $u \notin B^v_{2^{-n}}(e_G)$. This shows that $B^v_{2^{-n}}(e_G) \subseteq U_n$, that is, the first inclusion in (5.3).

Now assume that $n \in \mathbb{N}_+$ and $u \in U_n$, so $u, u^{-1} \in V_{2^{-n}}$. Pick an arbitrary $z \in G$. There exists $k \in \mathbb{N}$ such that $\frac{k}{2^{n-1}} \leq f(z) < \frac{k+1}{2^{n-1}} =: r$. This implies $z \in V_r$. As a consequence of (5.8), we obtain $zu, zu^{-1} \in V_r V_{2^{-n}} \subseteq V_{r+\frac{1}{2^n}}$ and hence $f(zu) \leq r + \frac{1}{2^n}$ and $f(zu^{-1}) \leq r + \frac{1}{2^n}$. Since $f(z) \geq \frac{k}{2^{n-1}}$, we conclude that

$$f(zu) - f(z) \leq r + \frac{1}{2^n} - \frac{k}{2^{n-1}} = \frac{k+1}{2^{n-1}} - \frac{k}{2^{n-1}} + \frac{1}{2^n} = \frac{3}{2^n}, \tag{5.9}$$

and analogously

$$f(zu^{-1}) - f(z) < \frac{3}{2^n}. \tag{5.10}$$

These inequalities hold for all $z \in G$. So, replacing z by zu in (5.10) gives $f(z) - f(zu) \leq \frac{3}{2^n}$. Combining this with (5.9) yields $v(u) \leq \frac{3}{2^n} < \frac{1}{2^{n-2}}$. Now we have shown that $U_n \subseteq B^v_{2^{-n+2}}(e_G)$, namely, the second inclusion in (5.3).

It remains to check that $v: (G, \tau) \to \mathbb{R}_{\geq 0}$ is continuous. For $x, y \in G$, $v(x) = v(x(y^{-1}y)) = v((xy^{-1})y) \leq v(xy^{-1}) + v(y)$ and this proves that $v(x) - v(y) \leq v(xy^{-1})$. Exchanging the roles of x, y, we get $v(y) - v(x) \leq v(yx^{-1}) = v(xy^{-1})$. Therefore, the desired inequality $|v(x) - v(y)| \leq v(xy^{-1})$ follows, and so the continuity of $v: (G, \tau_v) \to \mathbb{R}_{\geq 0}$. Since $\tau_v = \tau$ in view of (5.3), the continuity of $v: (G, \tau) \to \mathbb{R}$ is proved.

Finally, the continuity of v implies that H is closed. □

The following result improves Proposition 3.1.15 as promised.

Theorem 5.2.14. *Every Hausdorff group G is a Tichonov space.*

Proof. Let $\emptyset \neq F$ be a closed set of G with $a \notin F$. By the homogeneity of G, we may assume that $a = e_G$. Then we can find a chain $\{U_n : n \in \mathbb{N}\}$ as in (5.2) of symmetric open neighborhoods of e_G with $U_{n+1}U_{n+1} \subseteq U_n$ for every $n \in \mathbb{N}$, and such that $F \cap U_0 = \emptyset$. Let v be the continuous pseudonorm $v: G \to [0, 1]$ given by Lemma 5.2.13. Then $v(F) = \{1\}$ and $v(e_G) = 0$. □

The following result can be easily deduced from Lemma 5.2.13.

Corollary 5.2.15. *Let G be a topological group and let $\{U_n : n \in \mathbb{N}\}$ be as in (5.2) a decreasing chain of symmetric neighborhoods of e_G in G with $U_{n+1}U_{n+1} \subseteq U_n$ for every $n \in \mathbb{N}$. Then there exists a continuous left invariant pseudometric d on G such that, for every $n \in \mathbb{N}_+$, $B^d_{2^{-n}}(e_G) \subseteq U_n \subseteq B^d_{2^{-n+2}}(e_G)$ and $H = \bigcap_{n\in\mathbb{N}} U_n$ is a closed subgroup of G with the property $H = \{x \in G : d(x, e_G) = 0\}$. So, d is a metric if and only if $H = \{e_G\}$.*

Remark 5.2.16. Let G be a topological group.
(a) If the chain $\{U_n : n \in \mathbb{N}\}$ as in (5.2) and in Corollary 5.2.15 has also the property $xU_{n+1}x^{-1} \subseteq U_n$ for every $x \in G$ and for every $n \in \mathbb{N}$, then the subgroup $H = \bigcap_{n\in\mathbb{N}} U_n$ is also normal. Moreover, the pseudometric d produced by Corollary 5.2.15 defines a metric $\tilde{d}$ on the quotient group G/H by letting $\tilde{d}(xH, yH) = d(x, y)$ for every $x, y \in G$. The metric $\tilde{d}$ induces the quotient topology on G/H.
(b) Assume that U_0 is a subgroup of G and that $U_n = U_0$ in (5.2) for every $n \in \mathbb{N}$. Then this stationary chain satisfies the hypothesis of Corollary 5.2.15. The pseudometric d is defined as follows: $d(x, y) = 0$ if $xU_0 = yU_0$, otherwise $d(x, y) = 1$.

5.2.3 The Birkhoff–Kakutani theorem

The metrizability problem for topological groups has a relatively simple solution:

Theorem 5.2.17 (Birkhoff–Kakutani theorem). *A Hausdorff group G is metrizable if and only if $\chi(G) \leq \omega$.*

Proof. The necessity is obvious as every metrizable space is first countable.

Suppose now that $\mathcal{V}(e_G)$ has a countable base. Then one can build a chain $\{U_n : n \in \mathbb{N}\}$ as in (5.2) of symmetric neighborhoods of e_G in G with $U_{n+1}U_{n+1} \subseteq U_n$ for every $n \in \mathbb{N}$, that is a base of $\mathcal{V}(e_G)$. In particular, $\bigcap_{n\in\mathbb{N}} U_n = \{e_G\}$ since G is Hausdorff. Then the pseudometric produced by Corollary 5.2.15 is a metric that induces the topology of the group G. □

Example 5.2.18. Let $\{G_i : i \in I\}$ be an infinite family of nontrivial metrizable Hausdorff groups. Then $G = \prod_{i\in I} G_i$ satisfies $\chi(G) = |I|$ by Birkhoff–Kakutani theorem 5.2.17 and by Theorem 5.1.15.

If G_i is separable for every $i \in I$, in view of the equality $w(G) = \chi(G) \cdot d(G)$ from Lemma 5.1.7 and the inequality $d(G) \leq |I|$ from Example 5.1.12, we conclude that $w(G) = \chi(G) = |I|$.

The following are consequences of Birkhoff–Kakutani theorem.

Corollary 5.2.19. *Every Hausdorff abelian group G embeds into a product of metrizable abelian groups.*

Proof. Denote by τ the topology of G and let $\mathcal{B} = \{U \in \mathcal{V}_G(0) : U \text{ open}\}$. For every $U \in \mathcal{B}$, build a decreasing chain $\{U_n : n \in \mathbb{N}\}$ of symmetric open neighborhoods of

0 with $U_0 \subseteq U$ and $U_{n+1} + U_{n+1} \subseteq U_n$ for every $n \in \mathbb{N}$. By Corollary 5.2.15, the countable family $\{U_n : n \in \mathbb{N}\}$ is a local base at 0 of a pseudometrizable group topology τ_U on G. Moreover, $H_U = \bigcap_{n\in\mathbb{N}} U_n$ is a τ_U-closed subgroup of G, by Corollary 5.2.15. Denote by $\pi_U : G \to G/H_U$ the canonical projection and let $\bar{\tau}_U$ be the quotient topology of G/H_U. Then $(G/H_U, \bar{\tau}_U)$ is Hausdorff, hence metrizable.

Let $P = \prod_{U\in\mathcal{B}}(G/H_U, \bar{\tau}_U)$ and let $j : (G,\tau) \to P$ be the diagonal map of the family $\{\pi_U : U \in \mathcal{B}\}$. Then $\ker j = \bigcap_{U\in\mathcal{B}} H_U$ is trivial since τ is Hausdorff, so j is a continuous injection. To check that $j : G \to j(G)$ is open, take $U \in \mathcal{B}$. To prove that $j(U) \in \mathcal{V}_{j(G)}(0)$, it suffices to show that $j(U) \supseteq j(G) \cap W$, where $W = \pi_U(U_1) \times \prod_{V\in\mathcal{B}\setminus\{U\}} G/H_V \in \mathcal{V}_P(0)$. Indeed, if $j(x) \in W$ for some $x \in G$, then $\pi_U(x) \in \pi_U(U_1)$, so $x \in U_1 + H_U \subseteq U_1 + U_1 \subseteq U$, so $j(x) \in j(U)$. This proves that j is an embedding. □

This theorem fails for nonabelian groups. Indeed, for an uncountable set X, the permutation group $S(X)$, equipped with the topology T_X described in Example 2.2.22, admits no nontrivial continuous homomorphism to a metrizable abelian group G. Indeed, such a homomorphism $S(X) \to G$ must be a topological isomorphism, by Corollary 4.2.14. This contradicts the fact that $(S(X), \mathsf{T}_X)$ is not metrizable when X is uncountable.

Call a topological group *submetrizable* if it admits a coarser metrizable group topology.

Corollary 5.2.20. *Every Hausdorff abelian group (G, τ) of countable pseudocharacter is submetrizable. In particular, every countable Hausdorff abelian group is submetrizable.*

Proof. Let $\{U_n : n \in \mathbb{N}\}$ be open neighborhoods of 0 in τ with $\bigcap_{n\in\mathbb{N}} U_n = \{0\}$. It is not restrictive to assume that they form a decreasing chain as in (5.2) and Corollary 5.2.15. Call τ_m the group topology on G having as a local base at 0 the countable family $\{U_n : n \in \mathbb{N}\}$. By Corollary 5.2.19, τ_m is metrizable and $\tau_m \leq \tau$.

By Lemma 5.1.5, G has countable pseudocharacter when G is countable. □

5.2.4 Function spaces as topological groups

We already introduced in §2.2.4 a topology on function spaces, namely, the pointwise convergence topology on the set Y^X of all maps $X \to Y$ from a nonempty set X to a topological space Y. This topology was nothing else but the product topology on Y^X.

Now we define two more topologies for function spaces (i. e., subsets of Y^X) in case Y is a metric space. We start by an example where the target space is just $\mathbb{C}$.

Example 5.2.21. Following the counterpart of Example 5.2.7(iii) for $\mathbb{C}^n$ in place of $\mathbb{R}^n$, here we extend the definition of the sup-norm from $\mathbb{C}^n$ to the $\mathbb{C}$-algebra $C^*(X)$ of all bounded complex-valued functions of an arbitrary nonempty set X, by simply letting

$$\|f\| := \sup\{|f(x)| : x \in X\} \quad \text{for every } f \in C^*(X).$$

This norm gives rise to the invariant metric $d = d_{\|-\|}$ described by

$$d(f,g) = \|f - g\| = \sup\{|f(x) - g(x)| : x \in X\} \quad \text{for every } f,g \in C^*(X).$$

The metric topology related to this metric takes the name *uniform convergence topology*. It is a group topology finer than the pointwise convergence topology.

The above example can be generalized as follows:

Definition 5.2.22. Let X be a nonempty set, (Y,d) a metric space and let $C^*(X,Y)$ be the set of all maps $f: X \to Y$ with bounded range (i. e., $\operatorname{diam}(f(X)) < \infty$). Define

$$d(f,g) = \sup\{d(f(x),g(x)) : x \in X\} \quad \text{for every } f,g \in C^*(X,Y).$$

The metric topology related to d is called the *uniform convergence topology* on $C^*(X,Y)$.

There is a third topology that is coarser than the uniform convergence topology:

Definition 5.2.23. Let X be a topological space, (Y,d) a metric space, and let $C(X,Y)$ be the set of all continuous maps $X \to Y$. The *compact-open topology* is defined on $C(X,Y)$ by taking as a base of the filter of the neighborhoods of $f \in C(X,Y)$ the family

$$W(K,\varepsilon,f) := \{g \in C(X,Y) : \forall x \in K, d(f(x),g(x)) < \varepsilon\},$$

where K is a compact subset of X and $\varepsilon > 0$.

The compact-open topology need not be metrizable in general; indeed, when X is an uncountable discrete space and Y a metric space, the compact-open topology on $C(X,Y) = Y^X$ coincides with the pointwise convergence topology, which is nonmetrizable (see Theorem 5.1.15 in case Y is a metrizable topological group).

In case X is a compact topological space and (Y,d) a metric space, $C(X,Y) \subseteq C^*(X,Y)$, so one can consider the uniform convergence topology on $C(X,Y)$, which coincides with the compact-open topology in this case.

Remark 5.2.24. (a) The compact-open topology can be introduced more generally on the set $C(X,Y)$ of all continuous maps $X \to Y$ between two topological spaces, with prebase the family

$$W(K,U) := \{f \in C(X,Y) : f(K) \subseteq U\},$$

where $K \subseteq X$ and $U \subseteq Y$ vary among all compact (respectively, open) sets of X (respectively, Y).

(b) In case X is a topological space and Y is a metrizable topological abelian group with an invariant metric d, the compact-open topology makes $C(X,Y)$ a topological group (see Exercise 5.4.12), and a local base at the constant function $f \equiv e_Y$ is given by the subsets

$$W(K,\varepsilon) := W(K,\varepsilon,f) = \{g \in C(X,Y) : \forall x \in K, d(g(x),e_Y) < \varepsilon\},$$

where $K \subseteq X$ is compact and $\varepsilon > 0$.

(c) Since finite sets are compact, the compact-open topology on $C(X, Y)$, where X, Y are topological spaces, is finer than the restriction of the pointwise convergence topology to $C(X, Y) \subseteq Y^X$. Clearly, the pointwise convergence topology may be obtained as a compact-open topology when X is equipped with the discrete topology (and so $C(X, Y) = Y^X$).

These topologies have many applications in analysis and topological algebra. In particular, for $X = G$ a topological abelian group, the compact-open topology is used to define the Pontryagin dual $\widehat{G}$ of G, taking as target the group $Y = \mathbb{T}$ (see Chapter 13).

We conclude with a further example of a group topology on a function space, by taking $Y = X$.

Example 5.2.25. Let (X, d) be a compact metric space and consider on X the metric topology induced by d. Then the group Homeo(X) of all homeomorphisms of X admits a norm ν defined by

$$\nu(f) = \sup\{d(x, f(x)) + d(x, f^{-1}(x)) : x \in X\} \quad \text{for every } f \in \text{Homeo}(X).$$

This norm induces an invariant metric d_ν on Homeo(X), and the metric topology relative to d_ν is a group topology on Homeo(X) known as *Birkhoff topology*.

Since surjective isometries $X \to X$ of a compact metric space X are homeomorphisms with respect to the metric topology on X, it makes sense to consider the subgroup Iso(X) of all surjective isometries $X \to X$. The restriction of the Birkhoff topology on Iso(X) coincides with the uniform convergence topology on Iso(X).

5.3 Topologies and subgroups determined by sequences

5.3.1 *T*-sequences

Let G be an abelian group and let $A = \{a_n\}_{n\in\mathbb{N}}$ be a sequence in G. The question of the existence of a Hausdorff group topology on G that makes A converge to 0 is not only a mere curiosity. Indeed, assume that some Hausdorff group topology τ on $\mathbb{Z}$ makes the sequence $\{p_n\}_{n\in\mathbb{N}}$ of all primes converge to zero. Then $p_n \to 0$ in τ would yield $p_n - p_{n+1} \to 0$ in τ, so this sequence cannot contain infinitely many entries equal to 2. This would provide a very easy negative solution to the celebrated problem of the infinitude of twin primes (actually this argument would show that the shortest distance between two consecutive primes converges to ∞).

Definition 5.3.1 ([240]). A one-to-one sequence $A = \{a_n\}_{n\in\mathbb{N}}$ in an abelian group G is a *T-sequence* if there exists a Hausdorff group topology τ on G such that $a_n \to 0$ in τ.

Since we consider only sequences without repetitions, the convergence to zero of the sequence $\{a_n\}_{n\in\mathbb{N}}$ depends only on the set $\{a_n : n \in \mathbb{N}\}$ and does not depend on the enumeration of the sequence.

Example 5.3.2. We shall see below that the sequence $\{p_n\}_{n\in\mathbb{N}}$ of all primes is not a T-sequence in $\mathbb{Z}$ (see Exercise 10.4.12). So, the above mentioned possibility to resolve the problem of the infinitude of twin primes does not work.

We consider a couple of examples and nonexamples in $\mathbb{Z}$.

Example 5.3.3. (a) The sequences $\{n^2\}_{n\in\mathbb{N}}$ and $\{n^3\}_{n\in\mathbb{N}}$ are not T-sequences in $\mathbb{Z}$. Indeed, suppose for a contradiction that some Hausdorff group topology τ on $\mathbb{Z}$ makes $\{n^2\}_{n\in\mathbb{N}}$ converge to 0. Then $\{(n+1)^2\}_{n\in\mathbb{N}}$ converges to 0 as well. Taking the difference of $\{(n+1)^2\}_{n\in\mathbb{N}}$ and $\{n^2\}_{n\in\mathbb{N}}$, we conclude that $\{2n+1\}_{n\in\mathbb{N}}$ converges to 0 as well, and so also its subsequence $\{2n+3\}_{n\in\mathbb{N}}$ converges to 0. After subtraction of the latter two sequences, we conclude that the constant sequence 2 converges to 0. This is a contradiction, since τ is Hausdorff.
We leave the case of $\{n^3\}_{n\in\mathbb{N}}$ as an exercise to the reader.

(b) A similar argument proves that the sequence $\{P(n)\}_{n\in\mathbb{N}}$, where $P(x) \in \mathbb{Z}[x]$ is a fixed polynomial with $\deg P = d > 0$, is not a T-sequence in $\mathbb{Z}$.

(c) The celebrated Fibonacci sequence $\{f_n\}_{n\in\mathbb{N}}$, defined by $f_0 = f_1 = 1$ and $f_{n+2} = f_{n+1}+f_n$ for all $n \in \mathbb{N}$, is a T-sequence, witnessed by the Hausdorff group topology τ on $\mathbb{Z}$ induced by the embedding $\mathbb{Z} \to \mathbb{T}$ defined by $1 \mapsto \alpha + \mathbb{Z} \in \mathbb{T}$, where $\alpha = \frac{1+\sqrt{5}}{2}$ is the golden ratio (see Exercise 5.4.18 for a more general statement).

(d) For a prime p, the sequence $\{p^n\}_{n\in\mathbb{N}}$ is a T-sequence in $\mathbb{Z}$ (witnessed by the p-adic topology); this remains true for any geometric progression $\{a^n\}_{n\in\mathbb{N}}$, where $a > 1$ is an integer (this is witnessed by the p-adic topology for any prime p dividing a).

(e) The natural topology (or the p-adic topology for any prime p) witnesses that the sequence $\{n!\}_{n\in\mathbb{N}}$ is a T-sequence.

5.3.2 Topologically torsion elements and subgroups

Braconnier [41] and Vilenkin [281] gave the following related notion.

Definition 5.3.4. Let G be a Hausdorff group and p a prime. An element $x \in G$ satisfying $x^{p^n} \to e_G$ (respectively, $x^{n!} \to e_G$) is *topologically p-torsion* (respectively, *topologically torsion*). The sets

$$G_p := \{x \in G : x \text{ topologically } p\text{-torsion}\} \quad \text{and} \quad G! := \{x \in G : x \text{ topologically torsion}\}$$

are the *topological p-component* and the *topological torsion part* of G, respectively. Moreover, G is *topologically p-torsion* (respectively, *topologically torsion*) if $G = G_p$ (respectively, $G = G!$).

When G is abelian, G_p and $G!$ are subgroups of G. If G is discrete, then G is topologically p-torsion (respectively, topologically torsion) precisely when G is p-torsion (respectively, torsion).

Remark 5.3.5. Let G be a residually finite group. Since (G, ν_G) is Hausdorff, every element of G is topologically torsion and for every non-torsion element x, $\{n!x\}_{n\in\mathbb{N}}$ is a T-sequence. If (G, ν_G^p) is Hausdorff, then for every nontorsion element x of G, the sequence $\{p^n x\}_{n\in\mathbb{N}_+}$ is a T-sequence of G.

For other equivalent forms with respect to the following notions, see Exercise 5.4.13.

Definition 5.3.6. Let G be a Hausdorff group and p be a prime. Call $x \in G$:
(i) *quasi-p-torsion* if $\langle x\rangle$ is either a finite p-group or isomorphic to $(\mathbb{Z}, \nu_{\mathbb{Z}}^p)$;
(ii) *quasitorsion* if $\langle x\rangle$ is either finite or nondiscrete and has a linear topology.

Let $td_p(G) := \{x \in G : x \text{ quasi-}p\text{-torsion}\}$ and $td(G) := \{x \in G : x \text{ quasitorsion}\}$.

Remark 5.3.7. Let G be a topological abelian group. Then $td_p(G)$ and $td(G)$ are subgroups of G (see Exercise 5.4.13(e)), and obviously $t_p(G) \subseteq td_p(G) \subseteq G_p$ and $t(G) \subseteq td(G) \subseteq G$!

We shall see in Corollary 11.6.9 that $td_p(G) = G_p$ when G is a locally compact abelian group. In this case G_p carries a natural structure of a $\mathbb{J}_p$-module, since for $x \in G_p \setminus t_p(G)$ the subgroup $\overline{\langle x\rangle}$ isomorphic to $\mathbb{J}_p$; moreover, the multiplication $\mathbb{J}_p \times G_p \to G_p$ is continuous (see [99, 3.5.8]). In particular, G_p is q-divisible for every prime $q \neq p$. In case G is p-divisible, G_p is obviously p-divisible as well, so G_p is divisible.

5.3.3 Characterization of T-sequences

The existence of a finest group topology τ_A on an abelian group G that makes an arbitrary given sequence $A = \{a_n\}_{n\in\mathbb{N}}$ in G converge to 0 is easy to prove as far as we are not interested in imposing the Hausdorff axiom. Indeed, as $a_n \to 0$ in the indiscrete topology, τ_A is simply the supremum of all group topologies τ on G such that $a_n \to 0$ in τ (see Exercise 2.4.3). Clearly, τ_A is Hausdorff if and only if A is a T-sequence.

In [241] Protasov and Zelenyuk established a number of nice properties of this topology. One can easily describe τ_A as follows.

Remark 5.3.8. For an abelian group G and a family $\{A_n : n \in \mathbb{N}_+\}$ of subsets of G, let

$$\sum_{n\in\mathbb{N}_+} A_n = A_1 + \cdots + A_n + \cdots = \bigcup_{n\in\mathbb{N}_+} (A_1 + \cdots + A_n).$$

If $A = \{a_n\}_{n\in\mathbb{N}}$ is a sequence in G, for $m \in \mathbb{N}_+$, let

$$A_m^* := \{a_m, a_{m+1}, \dots\} \quad \text{and} \quad A_m := \{0\} \cup A_m^* \cup -A_m^*,$$

calling A_m^* the "*m-tail*" of A. For $k \in \mathbb{N}_+$, let $A(k, m) := \underbrace{A_m + \cdots + A_m}_{k \text{ times}}$.

For $n \in \mathbb{N}_+$ and a sequence $\{m_k\}_{k\in\mathbb{N}_+}$ in $\mathbb{N}$, let

$$A(m_1,\dots,m_n) := A_{m_1} + \dots + A_{m_n} \quad \text{so } A(m) = A_m$$

and $A(m_1,\dots,m_n) = \{0\}$ when $n = 0$ (i. e., the sum is empty).

Furthermore,

$$A(m_1,\dots,m_n,\dots) := \sum_{n\in\mathbb{N}_+} A_{m_n} = \bigcup_{n\in\mathbb{N}_+} A(m_1,\dots,m_n),$$

Then the family

$$\mathcal{B}_A := \{A(m_1,\dots,m_n,\dots) : \{m_n\}_{n\in\mathbb{N}_+} \text{ sequence of natural numbers}\}$$

is a filter base, satisfying the axioms of group topology (gt1), (gt2), and (gt3). Let τ be the group topology given by Remark 2.1.14. Then $\tau = \tau_A$. Indeed, obviously $a_n \to 0$ in (G,τ) and τ contains any other group topology with this property.

Next, by using the topology τ_A, we characterize the T-sequences $A = \{a_n\}_{n\in\mathbb{N}}$ in an abelian group G. In the above notation, for every $k \in \mathbb{N}_+$, we have

$$A(k,m) \subseteq A(m_1,\dots,m_n,\dots), \tag{5.11}$$

where $m = \max\{m_1,\dots,m_k\}$. The sets $A(k,m)$, for $k,m \in \mathbb{N}_+$, form a filter base, but the filter they generate need not be the filter of neighborhoods of 0 in a group topology. The utility of this family becomes clear now.

Theorem 5.3.9. *A sequence $A = \{a_n\}_{n\in\mathbb{N}}$ in an abelian group G is a T-sequence if and only if*

$$\bigcap_{m\in\mathbb{N}_+} A(k,m) = \{0\} \quad \textit{for every } k \in \mathbb{N}_+. \tag{5.12}$$

Proof. Obviously, $A = \{a_n\}_{n\in\mathbb{N}}$ is a T-sequence if and only if τ_A is Hausdorff. If τ_A is Hausdorff, then (5.12) obviously holds by (5.11).

Clearly, τ_A is Hausdorff if and only if

$$\bigcap_{\{m_n\}_{n\in\mathbb{N}_+}\subseteq\mathbb{N}^{\mathbb{N}_+}} A(m_1,\dots,m_n,\dots) = \{0\}. \tag{5.13}$$

So, it remains to see that (5.12) implies (5.13).

We prove first that, for every $l \in \mathbb{N}$ and every l-tuple $(m_1,\dots,m_l)$,

$$\bigcap_{i\in\mathbb{N}_+} (A(m_1,\dots,m_l) + A_i) = A(m_1,\dots,m_l). \tag{5.14}$$

The inclusion $\bigcap_{i\in\mathbb{N}_+}(A(m_1,\dots,m_l) + A_i) \supseteq A(m_1,\dots,m_l)$ is obvious since $0 \in A_i$.

We argue by induction on $l \in \mathbb{N}$. The case $l = 0$ follows directly from (5.12) with $k = 1$.

Now assume that $l > 0$ and that (5.14) is true for $l-1$ and all $(l-1)$-tuples $m_1, \dots, m_{l-1}$. Take $g \in \bigcap_{i\in\mathbb{N}}(A(m_1, \dots, m_l) + A_i)$. Then for every $j \in \{1, \dots, l\}$ and every $i \in \mathbb{N}_+$ one can find $b_j(i) \in A_{m_j}$ and $a(i) \in A_i$ such that

$$g = b_1(i) + \cdots + b_l(i) + a(i).$$

If there exists some $j \in \{1, \dots, l\}$, such that $b_j(i) = 0$ for infinitely many $i \in \mathbb{N}_+$, then $g \in \sum_{\nu\in\{1,\dots,l\}\setminus\{j\}} A_{m_\nu} + A_i$ for infinitely many $i \in \mathbb{N}_+$. Since $\{A_i : i \in \mathbb{N}_+\}$ is a decreasing chain, the inductive hypothesis gives

$$g \in \bigcap_{i\in\mathbb{N}_+} \Big(\sum_{\nu\in\{1,\dots,l\}\setminus\{j\}} A_{m_\nu} + A_i \Big) = \sum_{\nu\in\{1,\dots,l\}\setminus\{j\}} A_{m_\nu} \subseteq A(m_1, \dots, m_l).$$

Hence, we may assume that there exists $i_0 \in \mathbb{N}_+$ such that $b_j(i) \neq 0$ for all $i > i_0$ and for all $j \in \{1, \dots, l\}$. Then for all $i > i_0$ and for all $j \in \{1, \dots, l\}$ there exists $m_j(i) \geq m_j$ so that $b_j(i) = a_{m_j(i)} \in A_{m_j}$.

If $m_j(i) \to \infty$ for all $j \in \{1, \dots, l\}$, then $g \in A(l + 1, i)$ for infinitely many $i \in \mathbb{N}_+$, so $g \in \bigcap_{i\in\mathbb{N}_+} A(l + 1, i) = \{0\}$ by (5.12). Hence, $g = 0 \in A(m_1, \dots, m_n)$.

Otherwise there exist some $j \in \{1, \dots, l\}$ and $h \in \mathbb{N}_+$ such that $m_j(i) = h \geq m_j$ for infinitely many $i \in \mathbb{N}_+$. Let $N_h = \{i \in \mathbb{N}_+ : m_j(i) = h\}$. Then $g^* = g - a_h \in \bigcap_{i\in N_h} (\sum_{\nu\in\{1,\dots,l\}\setminus\{j\}} A_{m_\nu} + A_i) = \bigcap_{i\in\mathbb{N}} (\sum_{\nu\in\{1,\dots,l\}\setminus\{j\}} A_{m_\nu} + A_i)$, since $(\sum_{\nu\in\{1,\dots,l\}\setminus\{j\}} A_{m_\nu} + A_i)$ is decreasing. By inductive hypothesis,

$$g^* \in \bigcap_{i\in\mathbb{N}_+} \Big(\sum_{\nu\in\{1,\dots,l\}\setminus\{j\}} A_{m_\nu} + A_i \Big) = \sum_{\nu\in\{1,\dots,l\}\setminus\{j\}} A_{m_\nu}.$$

Therefore, since $a_h \in A_{m_j}$,

$$g = a_h + g^* \in a_h + \sum_{\nu\in\{1,\dots,l\}\setminus\{j\}} A_{m_\nu} \subseteq A(m_1, \dots, m_n).$$

This proves (5.14).

To prove (5.13), assume that $g \in G\setminus\{0\}$. Then using our assumption (5.12) and (5.14), it is easy to build inductively a sequence $\{m_n\}_{n\in\mathbb{N}_+}$ such that $g \notin A(m_1, \dots, m_n)$ for every $n \in \mathbb{N}_+$, i. e., $g \notin A(m_1, \dots, m_n, \dots)$, namely, for $g \in G\setminus\{0\}$, there exists $m_1 \in \mathbb{N}$ such that $g \notin A(1, m_1) = A_{m_1}$. Suppose that $n \geq 2$ and $m_1, \dots, m_{n-1}$ have already been constructed such that $g \notin A(m_1, \dots, m_{n-1})$. According to (5.14), $g \notin \bigcap_{i\in\mathbb{N}_+} (A(m_1, \dots, m_{n-1}) + A_i)$. So, there exists $m_n \in \mathbb{N}_+$ such that $g \notin A(m_1, \dots, m_{n-1}) + A_{m_n} = A(m_1, \dots, m_n)$. □

Remark 5.3.10. (a) According to Exercise 5.4.9, every infinite abelian group G admits a nondiscrete metrizable group topology. This gives rise to a plenty of T-sequences $A = \{a_n\}_{n\in\mathbb{N}}$ of G.

(b) The topology $\tau_{\{a_n\}}$ is never metrizable (see Exercise 7.3.2), but it is always complete (see Example 7.1.6(b)). More precisely, $\tau_{\{a_n\}}$ is sequential, but not Fréchet–Urysohn (see [240]).

The topology $\tau_{\{a_n\}}$ can be studied essentially on countable abelian groups:

Lemma 5.3.11. *Let G be an abelian group and $\{a_n\}_{n\in\mathbb{N}}$ a T-sequence of G. Then the subgroup H of G generated by the countable set $\{a_n : n \in \mathbb{N}\}$ is $\tau_{\{a_n\}}$-open.*

Proof. Denote by τ^* the supremum of $\tau_{\{a_n\}}$ and the Alexandrov group topology on G with H as the smallest neighborhood of 0. Then $\tau^* \geq \tau_{\{a_n\}}$ and obviously $a_n \to 0$ in τ^*. Since $\tau_{\{a_n\}}$ is the finest group topology with this property, we deduce that $\tau^* = \tau_{\{a_n\}}$, and consequently H is $\tau_{\{a_n\}}$-open. □

This lemma helps us understand better the properties of $\tau_{\{a_n\}}$, since the fact that the countable subgroup $H = \langle a_n : n \in \mathbb{N}\rangle$ is $\tau_{\{a_n\}}$-open and carries its own topology generated by the T-sequence $\{a_n\}_{n\in\mathbb{N}}$ of H reduces completely the study of the topology $\tau_{\{a_n\}}$ only to countable abelian groups.

A notion similar to that of T-sequence, but defined only with respect to topologies induced by characters, is given in §14.4. From many points of view, it turns out to be easier to deal with that notion than with the notion of a T-sequence. In particular, we shall see easy sufficient conditions for a sequence of integers to be a T-sequence.

We give without proof the following technical lemma that is useful in §14.4.

Lemma 5.3.12 ([240]). *For every T-sequence $A = \{a_n\}_{n\in\mathbb{N}}$ in $\mathbb{Z}$, there exists a sequence $\{b_n\}_{n\in\mathbb{N}}$ in $\mathbb{Z}$ such that for every choice of the sequence $\{e_n\}_{n\in\mathbb{N}}$, where $e_n \in \{0,1\}$ for every $n \in \mathbb{N}$, the sequence $\{q_n\}_{n\in\mathbb{N}}$ defined by $q_{2n} = b_n + e_n$ and $q_{2n-1} = a_n$ for $n \in \mathbb{N}$ is a T-sequence.*

5.4 Exercises

Exercises on cardinal invariants

Exercise 5.4.1. Let G be a Hausdorff group. Prove that:
(a) $w(G) = w(\mathfrak{h}G)$, $\chi(G) = \chi(\mathfrak{h}G)$, and $d(G) = d(\mathfrak{h}G)$;
(b) $d(U) = d(G)$ for every nonempty open set U of G, if G is nondiscrete Lindelöff;
(c) if G is Hausdorff, then $\chi(G)$ is finite if and only if G is discrete, and in such a case $\chi(G) = 1$;
(d) if G is Hausdorff, then $w(G)$ is finite if and only if G is finite, if and only if $d(G)$ is finite.

Hint. (b) Show first that $d(U) \leq d(G)$ using the fact that U is open. There exists a countable subset X of G such that $G = XU$, since G is Lindelöff. If D is a dense subset of U with $|D| = d(U)$, then $|D| \geq \omega$ and XD is a dense subset of G with $|XD| = |D| = d(U)$. This proves the equality $d(U) = d(G)$.

Exercise 5.4.2. Let G be an abelian group and H an infinite subset of G^*. Prove that $w(G, \mathcal{T}_H) \leq |H|$.

Hint. Since $(G, \mathcal{T}_H)$ is a topological subgroup of $\mathbb{S}^H$, one has $w(G, \mathcal{T}_H) \leq w(\mathbb{S}^H) = |H|$ by Theorem 5.1.15 (see also Example 5.2.18).

Exercise 5.4.3. Show that $w(-)$, $\chi(-)$, and $d(-)$ are cardinal invariants in the sense explained above: if the topological groups G, H are topologically isomorphic, then $w(G) = w(H)$, $\chi(G) = \chi(H)$ and $d(G) = d(H)$.

Exercise 5.4.4. Making use of Lemma 5.1.5, prove that $w(G) \leq \mathfrak{c}$ for a monothetic group G.

Exercise 5.4.5. Show that $\psi(H) < \psi(G)$ may occur for a dense subgroup H of a Hausdorff group G.

Hint. Take $G = \mathbb{Z}(2)^{\mathfrak{c}}$ with $\psi(G) = \mathfrak{c} > \omega = \psi(H)$ for every dense countable subgroup H of G. By Hewitt–Marczewski–Pondiczery theorem B.3.15, G is a separable Hausdorff group with $\psi(G) > \omega$.

Exercise 5.4.6. Prove that if N is a closed normal subgroup of a Hausdorff group G, then $\psi(G) \leq \max\{\psi(N), \psi(G/N)\}$.

Hint. Let $\{U_i : i \in I\} \subseteq \mathcal{V}_G(e_G)$ be a family with $|I| \leq \max\{\psi(N), \psi(G/N)\}$ such that $N \cap \bigcap_{i\in I} U_i = \{e_G\}$ and $\bigcap_{i\in I} U_i N = \{e_{G/N}\}$ in G/N. If $x \in G \setminus \{e_G\}$, then $x \notin \bigcap_{i\in I} U_i$ in case $x \in N$. If $x \in G \setminus N$, then $xN \neq e_{G/N}$ in G/N, so there exists $i \in I$ such that $xN \notin U_i N$, hence $x \notin U_i$. This proves that $\bigcap_{i\in I} U_i = \{e_G\}$, so $\psi(G) \leq \max\{\psi(N), \psi(G/N)\}$.

Exercises on metrizable topological groups

Exercise 5.4.7. Prove that subgroups, countable products, and Hausdorff quotients of metrizable groups are metrizable. Prove that a product of metrizable groups is metrizable if and only if all but countably many of these groups are trivial.

Exercise 5.4.8. Let G be an abelian group and H a countable set of characters of G separating the points of G. Prove that $\mathcal{T}_H$ is metrizable.

Exercise 5.4.9. Show that every infinite abelian group admits a nondiscrete metrizable group topology.

Hint. Let G_1 be a countably infinite subgroup of G. Then there exists a countable set H of characters of G_1 separating the points of G_1. By Exercise 5.4.8, this gives a nondiscrete metrizable topology $\mathcal{T}_H$ on G_1. Now extend $\mathcal{T}_H$ to a group topology τ on G with $\mathcal{V}_{G_1}(0)$ as a local base at 0. Clearly, τ is metrizable and nondiscrete.

Exercise 5.4.10. Prove that the profinite topology of an infinite bounded abelian group is not metrizable.

Hint. Since metrizability is preserved under taking subgroups, it suffices to assume that G is a p-group. For the same reason, and since G is bounded, we can assume that $G = G[p]$, that is, G has exponent p. Let $\{U_n : n \in \mathbb{N}\}$ be a decreasing chain of neighborhoods of 0 in the profinite topology ϖ_G. To show that it cannot be a base, we find a finite-index subgroup of G that contains none of these neighborhoods. One can inductively choose a sequence $\{x_n\}_{n\in\mathbb{N}}$ of elements of G such that $x_0 = 0$ and $x_n \in U_n \setminus \langle x_1, \dots, x_{n-1}\rangle$ for all $n \in \mathbb{N}_+$, because U_n is a subgroup of finite index while $\langle x_1, \dots, x_{n-1}\rangle$ is a finite subgroup. Then the family $\{x_n : n \in \mathbb{N}_+\}$ is independent and we can define a homomorphism $f: G \to \mathbb{Z}(p)$ by letting $f(x_n) = a \in \mathbb{Z}(p)$ for all $n \in \mathbb{N}$, where $a \neq 0$ is a fixed element of $\mathbb{Z}(p)$; this homomorphism can be extended then to the whole group G. The finite-index subgroup $V = \ker f$ does not contain any U_n, since $x_n \in U_n$, while $x_n \notin V$ for every $n \in \mathbb{N}_+$.

Exercise 5.4.11. Let v be a pseudonorm on an abelian group G. Show that the filter base $\mathcal{B} = \{B_r^v(e_G) : r \in \mathbb{R}_{>0}\}$ satisfies (gt1) and (gt2), as well as, vacuously, (gt3). Hence, $\mathcal{B}$ generates a pseudometrizable group topology τ_v on G.

Exercise 5.4.12. Establish that the compact-open topology defined in Remark 5.2.24(b) in case X is a topological space and Y is a metrizable topological abelian group with an invariant metric d is a group topology.

Hint. Fix compact subsets K, K_1, K_2 of X and $\varepsilon, \varepsilon_1, \varepsilon_2 > 0$. Since $W(K_1, \varepsilon_1) \cap W(K_2, \varepsilon_2) \supseteq W(K_1 \cup K_2, \min\{\varepsilon_1, \varepsilon_2\})$, $-W(K, \varepsilon) = W(K, \varepsilon)$ and $W(K, \epsilon) + W(K, \epsilon) \subseteq W(K, 2\epsilon)$, the family of sets $\mathcal{B} = \{W(K, \varepsilon) : K \subseteq G \text{ compact}, \varepsilon > 0\}$ forms a base of open neighborhoods of the constant function $f \equiv e_Y$. (For $g \in W(K, \varepsilon)$, the compactness of $g(K) \subseteq B_\varepsilon(e_Y)$ implies that $B_\delta(g(K)) \subseteq B_\varepsilon(e_Y)$ for some $\delta > 0$, hence $g + W(K, \delta) \subseteq W(K, \varepsilon)$, so $W(K, \varepsilon)$ is open.) Finally, observe that $W(K, \varepsilon, f) = f + W(K, \varepsilon)$ for $f \in C(X, Y)$ and $\varepsilon > 0$, so the topology generated by $\mathcal{B}$ coincides with the compact-open one.

Exercises on T-sequences

Exercise 5.4.13. Let p be a prime. Show that:
(a) if τ is a Hausdorff group topology on $\mathbb{Z}$ such that $k_n p^n \to 0$ for every sequence $\{k_n\}_{n\in\mathbb{N}}$ in $\mathbb{Z}$, then $\tau = \nu_{\mathbb{Z}}^p$;
(b) if τ is a Hausdorff group topology on $\mathbb{Z}$ such that $k_n n! \to 0$ for every sequence $\{k_n\}_{n\in\mathbb{N}}$ in $\mathbb{Z}$, then τ is a nondiscrete linear topology on $\mathbb{Z}$;
(c) if G is a Hausdorff abelian group, then $x \in G$ is:
  (c_1) quasi-p-torsion if and only if $k_n p^n x \to 0$ for every sequence $\{k_n\}_{n\in\mathbb{N}}$ in $\mathbb{Z}$;
  (c_2) quasitorsion if and only if $k_n n! x \to 0$ for every sequence $\{k_n\}_{n\in\mathbb{N}}$ in $\mathbb{Z}$;
(d) $x \in td_p(G)$ (respectively, $x \in td(G)$) if and only if there exists a continuous homomorphism $f: (\mathbb{Z}, \nu_{\mathbb{Z}}^p) \to G$ (respectively, $f: (\mathbb{Z}, \nu_{\mathbb{Z}}) \to G$) with $f(1) = x$;
(e) $td_p(G)$ and $td(G)$ are subgroups of G.

Hint. (a) If $U \in \mathcal{V}_\tau(0)$, pick a sequence $\{k_n\}_{n\in\mathbb{N}}$ in $\mathbb{Z}$ in which every integer appears infinitely many times. Then $k_n p^n \to 0$ yields $k_n p^n \in U$ for all sufficiently large n, so $p^n\mathbb{Z} \subseteq U$ for some $n \in \mathbb{N}$. This proves $\tau \leq \nu_{\mathbb{Z}}^p$. Now apply Exercise 3.5.25. For (b), apply the argument from the hint of (a) to deduce that for every $U \in \mathcal{V}_\tau(0)$ there exists $n \in \mathbb{N}_+$ such that $n!\mathbb{Z} \subseteq U$. Now apply Exercise 3.5.24 to conclude that τ is linear.

For (c_1) and (c_2), use items (a) and (b). Deduce (d) from (c_1) and (c_2) (or from (a) and (b)). Finally, (e) can be deduced from (c).

Exercise 5.4.14. For a Hausdorff abelian group G and a prime p, show that:

(a) if H is a subgroup of G, then $td(H) = td(G) \cap H$, $H_p = G_p \cap H$, and $td_p(H) = td_p(G) \cap H$;
(b) if $f: G \to H$ is a continuous homomorphism into a Hausdorff abelian group H, then $f(td(G)) \subseteq td(H)$, $f(G_p) \subseteq H_p$, and $f(td_p(G)) \subseteq td_p(H)$;
(c) $td(\prod_{i \in I} G_i) = \prod_{i \in I} td(G_i)$, $(\prod_{i \in I} G_i)_p = \prod_{i \in I} (G_i)_p$, and $td_p(\prod_{i \in I} G_i) = \prod_{i \in I} td_p(G_i)$ for every family $\{G_i : i \in I\}$ of Hausdorff abelian groups;
(d) the sum $wtd(G) := \sum_{p \in \mathbb{P}} td_p(G)$ is direct.

Hint. (a)–(c) Use (c_1) or (d) of the previous exercise.

(d) To see that the sum $\sum_{p \in \mathbb{P}} td_p(G)$ is direct, assume that $0 \neq g \in td_p(G) \cap \sum_{i=1}^r td_{q_i}(G)$, where all primes $p, q_1, \dots, q_r$ are distinct. For every $m \in \mathbb{N}_+$, the numbers $a_m = (q_1 \cdots q_r)^m$ and p^m are coprime, so $1 = u_m p^m + v_m a_m$ for some $u_m, v_m \in \mathbb{Z}$. Since $g \in \sum_{i=1}^r td_{q_i}(G)$ implies that $v_m a_m g \to 0$, and $g \in td_p(G)$ implies that $u_m p^m g \to 0$, we deduce that $g = u_m p^m g + v_m a_m g \to 0$, a contradiction.

Exercise 5.4.15.* Show that $\mathbb{T}_p = td_p(\mathbb{T}) = \mathbb{Z}(p^\infty)$ and $wtd(\mathbb{T}) = td(\mathbb{T}) = \mathbb{Q}/\mathbb{Z}$.

Hint. The proof of the equality $\mathbb{T}_p = \mathbb{Z}(p^\infty)$ can be found in [8] and it implies the remaining ones, except for $td(\mathbb{T}) = \mathbb{Q}/\mathbb{Z}$. To check that, it suffices to verify that for every $x \in \mathbb{T} \setminus \mathbb{Q}/\mathbb{Z}$ the subgroup $\langle x \rangle$ is strongly monothetic, so $x \notin td(\mathbb{T})$.

Exercise 5.4.16. Show that, for some topological group G and a prime p, the inclusion $td_p(G) \subseteq G_p$ may be proper.

Hint. By Exercise 7.3.2, $G = (\mathbb{Z}, \tau_{\{p^n\}})$ is not metrizable, so $td_p(G) = \{0\} \subsetneq \mathbb{Z} = G_p$.

Exercise 5.4.17. Let $G = H \times N$, where H and N are abelian groups. Prove that if $a_n = (a, b_n) \in G$ with $mb_n = 0$ for all $n \in \mathbb{N}$, while $ma \neq 0$, then $\{a_n\}_{n \in \mathbb{N}}$ is not a T-sequence of G.

Hint. Use the fact that if $\{a_n\}_{n \in \mathbb{N}}$ is a T-sequence of G, then so is the sequence $\{ma_n\}_{n \in \mathbb{N}}$ for every $m \in \mathbb{N}$.

Exercise 5.4.18. Let $\alpha \in \mathbb{R}$ be an irrational number and let $\frac{p_n}{q_n}$ be the convergents of the continued fraction representing α. Prove that $\{q_n\}_{n \in \mathbb{N}}$ is a T-sequence.

Hint. From the known inequality $|\alpha - \frac{p_n}{q_n}| \leq \frac{1}{q_n q_{n+1}}$ deduce that $\|q_n \alpha\| < \frac{1}{q_{n+1}}$ for the norm in $\mathbb{T}$ and conclude that $\{q_n \alpha + \mathbb{Z}\}_{n \in \mathbb{N}}$ is a null sequence in $\mathbb{T}$.

Exercise 5.4.19. (a)* Prove that there exists a T-sequence $\{a_n\}_{n \in \mathbb{N}}$ in $\mathbb{Z}$ with $\lim_{n \to \infty} a_{n+1}/a_n = 1$ (see [240] and also Example 14.4.17).
(b)* Every sequence $\{a_n\}_{n \in \mathbb{N}}$ in $\mathbb{Z}$ with $\lim_{n \to \infty} \frac{a_{n+1}}{a_n} = +\infty$ is a T-sequence (see [21, 240] and Theorem 14.4.6).
(c)* Every sequence $\{a_n\}_{n \in \mathbb{N}}$ in $\mathbb{Z}$ such that $\lim_{n \to \infty} \frac{a_{n+1}}{a_n} \in \mathbb{R}$ is transcendental is a T-sequence (see [240]).

Exercise 5.4.20. Show that $td(\mathbb{T}) = \mathbb{Q}/\mathbb{Z}$ is a proper subgroup of $\mathbb{T}!$ and that $\mathbb{T}!$ is a proper subgroup of $\mathbb{T}$.

Hint. If $z \in \mathbb{T} \setminus \mathbb{Q}/\mathbb{Z}$, then no nonproper subgroup of $\langle z \rangle$ can be open, so $z \notin td(\mathbb{T})$.

The elements $\bar{x}, \bar{y}$ of $\mathbb{T}$ determined by $x = e - 2 = \sum_{n=2}^{\infty} \frac{1}{n!} \in \mathbb{R}$ and $y = \frac{x+1}{2} \in \mathbb{R}$ satisfy $\bar{x} \in \mathbb{T}! \setminus \mathbb{Q}/\mathbb{Z}$ and $\bar{y} \in \mathbb{T} \setminus \mathbb{T}!$.

6 Connectedness in topological groups

This chapter is dedicated to the notions of connectedness and hereditary disconnectedness for topological groups, and moreover, we consider the connected component, arc component, quasicomponent, and the fourth component given by the intersection of all open subgroups.

6.1 Connected and hereditarily disconnected groups

We start with an elementary property of connected sets in a topological group.

Lemma 6.1.1. *Let G be a topological group.*
(a) *If $C_1, \dots, C_n$ are connected subsets of G, then also $C_1 \cdots C_n$ is connected.*
(b) *If C is a connected subset of G, then C^{-1}, as well as the subgroup $\langle C\rangle$ generated by C, is connected.*

Proof. (a) Consider the case $n = 2$, the general case easily follows by induction. The subset $C_1 \times C_2$ of $G \times G$ is connected. Now the multiplication $\mu: G \times G \to G$, $(x, y) \mapsto xy$, is continuous and $\mu(C_1 \times C_2) = C_1C_2$. So, by Lemma B.6.4(a), also C_1C_2 is connected.

(b) Since C is connected and C^{-1} is the continuous image of C under the inversion map $\iota: G \to G$, $x \mapsto x^{-1}$, we conclude that C^{-1} is connected as well, by Lemma B.6.4(a). To prove the second assertion, consider the set $C_1 = CC^{-1}$, which is connected by item (a). Obviously, $e_G \in C_1$, so $C \cup C^{-1} \subseteq C_1C_1 = C_1^2$. Since $\langle C\rangle = \langle C_1\rangle = \bigcup_{n\in\mathbb{N}} C_1^n$, and each set C_1^n is connected by item (a), we conclude that $\langle C\rangle$ is connected, by Lemma B.6.5. □

We see that connectedness and hereditary disconnectedness are properties stable under extension:

Proposition 6.1.2. *Let G be a topological group and N a closed normal subgroup of G. If N and G/N are connected (respectively, hereditarily disconnected), then also G is connected (respectively, hereditarily disconnected).*

Proof. Let $q: G \to G/N$ be the canonical projection.

Assume that N and G/N are connected and let A be a nonempty clopen set of G. As every coset aN with $a \in G$ is connected in view of Lemma B.6.4(a) and Lemma 2.1.6, one has either $aN \subseteq A$ or $aN \cap A = \emptyset$. Hence, $aN \subseteq A$ for every $a \in A$, and so $A = q^{-1}(q(A))$. This implies that $q(A)$ is a nonempty clopen set of the connected group G/N. Thus, $q(A) = G/N$, and consequently, $A = G$.

Now suppose that N and G/N are hereditarily disconnected. Assume that C is a connected subset of G. Then $q(C)$ is a connected subset of G/N by Lemma B.6.4(a), so $q(C)$ is a singleton by our hypothesis. This means that C is contained in some coset xN. Since xN is hereditarily disconnected as well, we conclude that C is a singleton. This proves that G is hereditarily disconnected. □

https://doi.org/10.1515/9783110654936-006

Corollary 6.1.3. *A topological group G is connected if and only if $\mathfrak{h}G$ is connected.*

Proof. Since $\mathfrak{h}G = G/\mathrm{core}(G)$ and $\mathrm{core}(G)$ is connected (being indiscrete), Proposition 6.1.2 and Lemma B.6.4(a) apply. □

Corollary 6.1.4. *If a connected group G is not indiscrete, then $|\mathfrak{h}G| \geq \mathfrak{c}$.*

Proof. In view of Corollary 6.1.3, we can assume without loss of generality that $G = \mathfrak{h}G$ is Hausdorff. Then G is also a Tichonov space by Theorem 5.2.14, so we can arrange for a nonconstant continuous function $f: G \to [0,1]$. Since G is connected, the image $f(G)$ is connected by Lemma B.6.4(a), so a subinterval of $[0,1]$. Since $f(G)$ is not a singleton, we deduce that $|f(G)| = \mathfrak{c}$. Hence, $|G| \geq \mathfrak{c}$. □

This corollary shows that certain groups cannot carry a connected Hausdorff group topology. Obviously, these are all nontrivial groups of size $< \mathfrak{c}$. But one can get also less trivial examples like those pointed out below.

Example 6.1.5. (a) Let $G = \mathbb{Z}(2) \times \mathbb{Z}(3)^{\omega}$; clearly, $|G| = \mathfrak{c}$. Assume that τ is a connected Hausdorff group topology on G. Then the multiplication by 3 defines a continuous endomorphism $\mu_3: G \to G$, and so $\mu_3(G) = 3G$ must be connected by Lemma B.6.4(a). But $|3G| = 2$, so $3G$ cannot carry any connected Hausdorff group topology in view of Corollary 6.1.4, a contradiction.

(b) Obviously, the argument from item (a) produces similar examples of abelian groups that admit no connected Hausdorff group topology, that is, when the starting abelian group G has the property that, for some $m \in \mathbb{N}_+$, mG is nontrivial and $|mG| < \mathfrak{c}$ (e. g., $G = H \times \mathbb{Z}(m)^{\omega}$ with $1 < |mH| < \mathfrak{c}$), G cannot carry any connected Hausdorff group topology.

(c) Surprisingly, the necessary condition for having a connected Hausdorff group topology from item (b) turns out to be also sufficient: every abelian group G such that for every $m \in \mathbb{N}_+$ the subgroup mG is either trivial or has size $\geq \mathfrak{c}$ admits a connected Hausdorff group topology (see also Chapter 4). The proof given in [107] makes substantial use of a construction by Hartman and Mycielcki [168] of a group $HM(G)$ defined for every abelian group G that carries a connected group topology and is algebraically isomorphic to $G^{(\mathfrak{c})}$ (see also [70]). In particular, the group $\mathbb{Z}(m)^{(\mathfrak{c})}$ carries a connected group topology for every $m \in \mathbb{N}$ with $m > 1$.

In the next remark we discuss zero-dimensionality. We recall that in general, under the axiom T_0, zero-dimensionality implies total disconnectedness, which yields, in turn, hereditary disconnectedness (see §B.6).

Remark 6.1.6. (a) A topological group G is zero-dimensional if and only if $\mathfrak{h}G$ is zero-dimensional. On the other hand, a zero-dimensional group G is totally disconnected if and only if G is Hausdorff.

(b) Proposition 3.1.7(b) yields that every linearly topologized group G is zero-dimensional; in particular, when G is Hausdorff, it is totally disconnected and so also hereditarily disconnected.

The next statement is true for regular spaces as well, but not for countable Hausdorff ones.

Proposition 6.1.7. *Every countable Hausdorff group G is zero-dimensional.*

Proof. Using the Tichonov separation axiom, for every $U \in \mathcal{V}(e_G)$, there exists a continuous function $f: G \to [0,1]$ such that $f(e_G) = 1$ and $f(G \setminus U) = \{0\}$. The subset $X = f(G)$ of $[0,1]$ is countable, hence there exists $a \in [0,1] \setminus f(G)$. Then $W = (a,1] \cap f(G)$ is a clopen set of $f(G)$. Therefore, $f^{-1}(W) \subseteq U$ is a clopen set of G containing e_G. This shows that G has a base of clopen sets. □

Example 6.1.8. (a) The countable Hausdorff group $G = \mathbb{Q}/\mathbb{Z}$ is zero-dimensional yet has no proper open subgroups (see Exercise 6.3.1), so its topology is not linear.

(b) Every proper subgroup H of $\mathbb{T}$ is zero-dimensional. Indeed, H is either finite or dense, in view of Example 3.1.14(a). If H is finite, then it is clearly zero-dimensional. If H is dense, then for any fixed $a \in \mathbb{T} \setminus H$ also $a + H$ is dense and disjoint with H. Hence, $\{\Gamma_{b,c} \cap H: b, c \in a + H\}$, where $\Gamma_{b,c}$ is an open arc with ends b, c is a base of the induced topology on H consisting of clopen sets of H.

Van Dantzig theorem 8.5.1 will show that every zero-dimensional locally compact group G has a local base at e_G consisting of open subgroups. Example 6.1.8(a) shows that local compactness is essential.

6.2 The four components

The first of the four components that we recall in this section is the connected component, that we will use also in the sequel.

Definition 6.2.1. For a topological group G, denote by $c(G)$ the connected component C_{e_G} of e_G and call it briefly the *connected component of G*.

Proposition 6.2.2. *The connected component $c(G)$ of a topological group G is a closed normal subgroup of G. The connected component of an element $x \in G$ is simply the coset $xc(G) = c(G)x$.*

Proof. To prove that $c(G)$ is stable under multiplication, it suffices to note that $c(G)c(G)$ is still connected by Lemma 6.1.1(a) and contains e_G, so must be contained in the connected component $c(G)$. Similarly, an application of Lemma 6.1.1(b) implies that $c(G)$ is stable with respect to the inversion map $\iota: G \to G$, $x \mapsto x^{-1}$, so $c(G)$ is a subgroup of G. Moreover, for every $a \in G$, the image $ac(G)a^{-1}$ under the conjugation ϕ_a by a is connected by Lemmas 2.1.6 and B.6.4(a), and contains e_G, so must be contained in the connected component $c(G)$. Therefore, $c(G)$ is a normal subgroup of G.

The fact that $c(G)$ is closed follows from the fact that the closure of $c(G)$ is a connected subgroup of G, by Lemmas 3.1.1(b) and B.6.4(b).

To prove the last assertion, it suffices to recall that, for all $x \in G$, the translations ${}_xt\colon y \mapsto xy$ and $t_x\colon y \mapsto yx$ are homeomorphisms, by Lemma 2.1.6. □

Our next aim is to see that the quotient $G/c(G)$ is hereditarily disconnected.

Lemma 6.2.3. *For every topological group G, the quotient $G/c(G)$ is hereditarily disconnected.*

Proof. Let $q\colon G \to G/c(G)$ be the canonical projection and $H = q^{-1}(c(G/c(G)))$. Apply Proposition 6.1.2 to H and its connected quotient $H/c(G) \cong c(G/c(G))$ to conclude that H is connected. Since H contains $c(G)$, we have $H = c(G)$. Hence, $G/c(G)$ is hereditarily disconnected. □

Next comes the second component that we introduce for every topological group.

Definition 6.2.4. For a topological group G, denote by $a(G)$ the set of points $x \in G$ connected to e_G by an arc, i. e., there exists a continuous map $f\colon [0,1] \to G$ such that $f(0) = e_G$ and $f(1) = x$. We call *arc* the image $f([0,1])$ in G and *arc component* the subset $a(G)$. The group G is called *arcwise connected* if $a(G) = G$.

Obviously, all points of $f([0,1])$ belong to $a(G)$.

The following result can be proved in analogy to Proposition 6.2.2.

Proposition 6.2.5. *For a topological group G, the arc component $a(G)$ of G is a normal subgroup of G.*

Proof. By Exercise 6.3.3(a) and the continuity of the multiplication $\mu\colon G \times G \to G$, $(x,y) \mapsto xy$, we get $a(G)a(G) \subseteq a(G)$. Analogously, using the continuity of the inversion map $\iota\colon G \to G$, $x \mapsto x^{-1}$, we get that $a(G)^{-1} \subseteq a(G)$. Then $a(G)$ is a subgroup of G. To show that $a(G)$ is stable under conjugation, use Exercise 6.3.3(b) and the continuity of the conjugation from Lemma 2.1.6. □

Clearly, $a(G) = a(c(G)) \subseteq c(G)$ for any topological group G, so it makes sense to study the subgroup $a(G)$ for connected groups G. In general, $a(G)$ need not be closed in G. Actually, the subgroup $a(K)$ is dense in $c(K)$ for every compact abelian group K (see Exercise 13.7.13).

The third component in our list is the quasicomponent, which always contains the connected component.

Definition 6.2.6. For a topological group G, denote by $Q(G)$ the quasicomponent of the neutral element e_G of G (i. e., $Q(G)$ is the intersection of all clopen sets of G containing e_G) and call it the *quasicomponent* of G.

Proposition 6.2.7. *For a topological group G, the quasicomponent $Q(G)$ is a closed normal subgroup of G. The quasicomponent Q_x of $x \in G$ coincides with the coset $xQ(G) = Q(G)x$.*

Proof. Let $x, y \in Q(G)$. To prove that $xy \in Q(G)$, we need to verify that $xy \in O$ for every clopen set O of G containing e_G. Let O be such a set, then $x, y \in O$. Obviously, Oy^{-1} is a clopen set containing e_G, hence $x \in Oy^{-1}$. This implies $xy \in O$, and so that $Q(G)$ is stable under multiplication. For every clopen set O of G containing e_G, the set O^{-1} has the same propriety, hence $Q(G)$ is stable also with respect to the inversion $\iota: G \to G$, $x \mapsto x^{-1}$. This implies that $Q(G)$ is a subgroup of G. Moreover, for every $a \in G$ and for every clopen set O of G containing e_G its image aOa^{-1} under the conjugation by $a \in G$ is a clopen set containing e_G, by Lemma 2.1.6. So, $Q(G)$ is stable also under conjugation, hence $Q(G)$ is a normal subgroup of G. As an intersection of clopen sets, $Q(G)$ is closed. □

Clearly, for a topological group G,

$$a(G) \subseteq c(G) \subseteq Q(G).$$

According to the general Lemma B.6.11, $c(K) = Q(K)$ for every compact group K. Actually, this remains true also in the case of locally compact groups G (see Corollary 8.5.4), as well as for countably compact groups (see [75]).

The fourth component that we consider is the following.

Definition 6.2.8. For a topological group G, denote by $o(G)$ the intersection of all open subgroups of G.

Clearly, $\mathrm{core}(G) \subseteq o(G)$ for every topological group G.

Since the conjugate of an open subgroup is still open, $o(G)$ is a closed normal subgroup of G and

$$Q(G) \subseteq o(G).$$

This inclusion may be proper as $G = \mathbb{Q}/\mathbb{Z}$ shows: here $Q(G) = \{0\}$, while $o(G) = G$ (see also Example 6.1.8(a), where the weaker property that G is not linearly topologized was pointed out).

For a Hausdorff group G, always $o(G) = \{e_G\}$ whenever G has a linear topology, or more generally, when $\mathcal{V}(e_G)$ has a base of open subgroups.

6.3 Exercises

Exercise 6.3.1. Show that the group $\mathbb{Q}/\mathbb{Z}$ is zero-dimensional but has no proper open subgroups.

Exercise 6.3.2. For G a connected group and H a Hausdorff group, prove that:

(a) if $h: G \to H$ is a continuous homomorphism, then h is trivial whenever $\ker h$ has nonempty interior;

(b) if $f_1, f_2: G \to H$ are continuous homomorphisms that coincide on some neighborhood of e_G in G, then $f_1 = f_2$.

Hint. (a) Use that $\ker f$ is an open subgroup of G, so must coincide with G.

(b) In case G, H are abelian, apply (a) to the homomorphism $h = f_1 - f_2: G \to H$. Otherwise, use that the subgroup $A = \{a \in G: f_1(a) = f_2(a)\}$ is open, so $H = G$.

Exercise 6.3.3. (a) If G, H are topological groups, prove that $c(G \times H) = c(G) \times c(H)$, $a(G \times H) = a(G) \times a(H)$, $Q(G \times H) = Q(G) \times Q(H)$, and $o(G \times H) = o(G) \times o(H)$.

(b) If $f: G \to H$ is a continuous map of topological groups with $f(e_G) = e_H$, then $f(c(G)) \subseteq c(H)$, $f(a(G)) \subseteq a(H)$, and $f(Q(G)) \subseteq Q(H)$. Moreover, if f is also a homomorphism, prove that f induces continuous homomorphisms $\bar{f}: G/c(G) \to H/c(H)$ and $\tilde{f}: G/Q(G) \to H/Q(H)$ commuting with the respective projections $G \to G/c(G)$, $H \to H/c(H)$, and $G \to G/Q(G)$, $H \to H/Q(H)$.

(c)* Let G be a topological abelian group and $l(G)$ the set of elements $x \in G$ such that there exists a continuous homomorphism $f: \mathbb{R} \to G$ with $f(1) = x$. Check that $l(G)$ is a subgroup of G contained in $a(G)$. If G is also locally compact, $a(G) = l(G)$.

(d) Can (a) be extended to arbitrary products?

Hint. (c) The last assertion is not trivial, a proof can be found in [123].

Exercise 6.3.4. Prove that the group $SO_3(\mathbb{R})$ of all rotations of $\mathbb{R}^3$ is connected. Is $GL_3(\mathbb{R})$ connected? What about $GL_2(\mathbb{R})$?

Exercise 6.3.5.* Prove that for an abelian group G the following are equivalent:

(a) $(G, \mathfrak{Z}_G)$ is connected;

(b) for every $m \in \mathbb{N}$, the subgroup mG is either trivial or infinite;

(c) for every $m \in \mathbb{N}_+$, $G[m]$ either coincides with G or has infinite index.

Hint. (a)⇒(b) If $(G, \mathfrak{Z}_G)$ is connected, then mG is connected, by Exercise 4.5.11. Hence, mG must be a singleton (necessarily $\{0\}$), if it is finite.

(b)⇔(c) For every $m \in \mathbb{N}_+$, $G[m] = \ker \mu_m$, so $G/G[m] \cong mG$.

For a proof of the implication (c)⇒(a), see [104].

Exercise 6.3.6.* Deduce from Exercise 6.3.3(b) that:

(a) the assignment $G \mapsto c(G)$ (respectively, $G \mapsto a(G)$), along with the inclusion $c(G) \to G$ (respectively, $a(G) \to G$), defines a monocoreflection from the category **TopGrp** to its full subcategory of all connected (respectively, arcwise connected) topological groups;

(b) the assignment $G \mapsto G/c(G)$ (respectively, $G \mapsto G/Q(G)$), along with the canonical projection $G \to G/c(G)$ (respectively, $G \to G/Q(G)$), defines an epireflection from the category **TopGrp** to its full subcategory of all hereditarily (respectively, totally) disconnected groups.

Does the assignment $G \mapsto Q(G)$ define a coreflection?

Hint. To negatively answer the question, note that $Q(G) = G$ precisely when G is connected. Next use the fact that there exist topological groups G such that $Q(G)$ is not connected (see [74]).

7 Completeness and completion

In this chapter we explicitly construct the Raĭkov completion using Cauchy nets, we discuss it also by means of filters, and compare it with the Weil completion.

7.1 Completeness and completion via Cauchy nets

7.1.1 Cauchy nets and completeness

Definition 7.1.1. A net $\{g_\alpha\}_{\alpha\in A}$ in a topological group G is a *Cauchy net* if for any $U \in \mathcal{V}_G(e_G)$ there exists $\alpha_0 \in A$ such that $g_\alpha^{-1}g_\beta \in U \ni g_\beta g_\alpha^{-1}$ for every $\alpha,\beta > \alpha_0$.

Clearly, every convergent net is a Cauchy net, but the converse does not hold true:

Example 7.1.2. Consider $\mathbb{Q}$ with the subgroup topology induced by the Euclidean topology of $\mathbb{R}$. Then $\{(1+1/n)^n\}_{n\in\mathbb{N}_+}$ is a Cauchy net in $\mathbb{Q}$ not converging in $\mathbb{Q}$.

Remark 7.1.3. It is easy to see that if H is a subgroup of a topological group G, then a net $\{h_\alpha\}_{\alpha\in A}$ in H is Cauchy if and only if it is a Cauchy net of G. In other words, this is an intrinsic property of the net and it does not depend on the topological group where the net is considered.

Consequently, a net $\{h_\alpha\}_{\alpha\in A}$ is Cauchy in H if and only if it is a Cauchy net of the subgroup $\langle h_\alpha : \alpha \in A\rangle$ of H.

By Exercise 7.3.1(a), a net $\{g_\alpha\}_{\alpha\in A}$ in a topological group G is a Cauchy net whenever it converges in some larger topological group H containing G as a topological subgroup. Our aim in this subsection is to see that all Cauchy nets of G arise in this way (see Theorem 7.1.10); moreover, there is a topological group H, in which G is dense, witnessing this simultaneously for all Cauchy nets of G.

Lemma 7.1.4. *Let G be a topological group.*
(a) *A Cauchy net $\{g_\alpha\}_{\alpha\in A}$ of G is convergent if and only if it has a convergent subnet.*
(b) *If $\{g_\alpha\}_{\alpha\in A}$ is a Cauchy net of G, then also $\{g_\alpha^{-1}\}_{\alpha\in A}$ is a Cauchy net of G.*
(c) *If $x' = \{x_\alpha\}_{\alpha\in A}$ and $y' = \{y_\alpha\}_{\alpha\in A}$ are Cauchy nets of G, then also $\{x_\alpha y_\alpha\}_{\alpha\in A}$ is a Cauchy net of G.*

Proof. (a) Let $\{g_{\alpha_\gamma}\}_{\gamma\in\Gamma}$ be a subnet of $\{g_\alpha\}_{\alpha\in A}$ with $g_{\alpha_\gamma} \to x \in G$. We prove that $g_\alpha \to x$. Let $U \in \mathcal{V}_G(e_G)$ and let $V \in \mathcal{V}(e_G)$ be symmetric and such that $VV \subseteq U$. Since $g_{\alpha_\gamma} \to x$, there exists $\gamma_0 \in \Gamma$ such that $g_{\alpha_\gamma} \in Vx$ for every $\gamma \geq \gamma_0$. Moreover, there exists $\alpha_0 \in A$ such that $\alpha_0 \geq \alpha_{\gamma_0}$ and $g_\alpha g_\beta^{-1} \in V$ for every $\alpha,\beta \geq \alpha_0$ in A. Let $\beta = \alpha_{\gamma_0}$. Then $g_\alpha \in VVx \subseteq Ux$ for every $\alpha \geq \alpha_0$, that is, $g_\alpha \to x$.

(b) is clear from the definition.

https://doi.org/10.1515/9783110654936-007

(c) Let $U_0 \in \mathcal{V}_G(e_G)$ and let $U \in \mathcal{V}_G(e_G)$ with $UU \subseteq U_0$. Moreover, let $V \in \mathcal{V}_G(e_G)$ be symmetric and such that $VVV \subseteq U$. There exists $\alpha_0 \in A$ such that $x_\beta x_\alpha^{-1} \in V$ for every $\alpha, \beta \geq \alpha_0$. Let $W \in \mathcal{V}_G(e_G)$ with $x_{\alpha_0} W x_{\alpha_0}^{-1} \subseteq V$ and $W \subseteq U$. Then, for every $\alpha \geq \alpha_0$,

$$x_\alpha W x_\alpha^{-1} = (x_\alpha x_{\alpha_0}^{-1})(x_{\alpha_0} W x_{\alpha_0}^{-1})(x_{\alpha_0} x_\alpha^{-1}) \subseteq VVV \subseteq U.$$

By hypothesis, there exists $\alpha_1 \in A$ such that $\alpha_1 \geq \alpha_0$ and $x_\alpha x_\beta^{-1} \in W$ and $y_\alpha y_\beta^{-1} \in W$ for every $\alpha, \beta \geq \alpha_1$. Therefore, for every $\alpha, \beta \geq \alpha_1$,

$$x_\alpha y_\alpha y_\beta^{-1} x_\beta^{-1} \in x_\alpha W x_\alpha^{-1} x_\alpha x_\beta^{-1} \subseteq U x_\alpha x_\beta^{-1} \subseteq UW \subseteq UU \subseteq U_0.$$

Analogously, there exists $\alpha' \in A$ such that, for all $\alpha, \beta \geq \alpha'$, $y_\beta^{-1} x_\beta^{-1} x_\alpha y_\alpha \in U_0$. □

Definition 7.1.5. A topological group G is *complete (in the sense of Raĭkov)* (or, *Raĭkov complete*) if every Cauchy net of G converges in G.

Example 7.1.6. (a) A discrete group G is complete, since every Cauchy net of G is eventually constant and so convergent.
(b) If G is an abelian group and $\{a_n\}_{n\in\mathbb{N}}$ is a T-sequence of G, then $(G, \tau_{\{a_n\}})$ is complete (see [240]) and nonmetrizable (see Exercise 7.3.2).

Of course, one can define Cauchy nets and completeness also in non-Hausdorff groups. In such a case the Hausdorff reflection provides a nice connection between Cauchy and convergent nets of G and its Hausdorff reflection $\mathfrak{h}G$:

Proposition 7.1.7. *Let G be a topological group and $q\colon G \to \mathfrak{h}G$ its Hausdorff reflection. Then:*
(a) *a net $\{x_\alpha\}_{\alpha\in A}$ in G is convergent (respectively, Cauchy) if and only if $\{q(x_\alpha)\}_{\alpha\in A}$ is convergent (respectively, Cauchy) in $\mathfrak{h}G$;*
(b) *G is complete if and only if $\mathfrak{h}G$ is complete.*

Proof. (a) follows from Exercise 7.3.1(b) and Lemma B.3.2; (b) follows from (a). □

The class of complete Hausdorff groups is closed under taking closed subgroups and products:

Proposition 7.1.8. *Let G be a complete Hausdorff group and H a subgroup of G. Then H is complete if and only if H is closed.*

Proof. Assume that H is a closed subgroup of the complete Hausdorff group G and let $\{h_\alpha\}_{\alpha\in A}$ be a Cauchy net of H. Since G is complete, $\{h_\alpha\}_{\alpha\in A}$ converges to some $g \in G$. Since H is closed, $g \in H$. This proves that H is complete. Conversely, suppose that H is complete and let $\{h_\alpha\}_{\alpha\in A}$ be a net in H that is convergent to some $x \in G$. Since $\{h_\alpha\}_{\alpha\in A}$ is a Cauchy net of H, necessarily there exists $h = \lim_{\alpha\in A} h_\alpha \in H$. As G is Hausdorff, $x = h \in H$. Hence, H is closed. □

Proposition 7.1.9. *Let $\{G_i : i \in I\}$ be a family of topological groups. Then $G = \prod_{i \in I} G_i$ is complete if and only if G_i is complete for every $i \in I$.*

Proof. Assume that G_i is complete for every $i \in I$ and let $\{x_\alpha\}_{\alpha \in A}$ be a Cauchy net of $G = \prod_{i \in I} G_i$. Since for every $i \in I$ the projection $p_i : G \to G_i$ is continuous, the net $\{p_i(x_\alpha)\}_{\alpha \in A}$ is a Cauchy net of G_i by Exercise 7.3.1(b), and hence $p_i(x_\alpha) \to y_i \in G_i$. Therefore, $\{x_\alpha\}_{\alpha \in A}$ converges to $(y_i)_{i \in I} \in G$.

Now suppose that G is complete. Let $\{x_\alpha\}_{\alpha \in A}$ be a Cauchy net in G_i and denote by $\iota_i : G_i \to G$ the canonical embedding. Then $\{\iota_i(x_\alpha)\}_{\alpha \in A}$ is a Cauchy net in G, which converges by assumption. Hence, $\{p_i(\iota_i(x_\alpha))\}_{\alpha \in A} = \{x_\alpha\}_{\alpha \in A}$ converges, so G_i is complete. □

Combining Propositions 7.1.8 and 7.1.9, one can prove that for every topological group G there exist a complete Hausdorff group cG and a continuous homomorphism $f : G \to cG$ such that $f(G)$ is dense in cG, and that the assignment $G \mapsto cG$ induces a functor (a reflection) from the category **TopGr** to its full subcategory of complete Hausdorff groups. But at this point we cannot say much about the map $f : G \to cG$. It is proved in the next section that it is an embedding when G is Hausdorff, and then cG will be given the name *completion* of G.

7.1.2 Completion via Cauchy nets

The proof of the next general theorem is deferred to the end of this section. In order to make that proof easily accessible to the reader, we precede it with the much easier proofs in the metrizable and abelian cases.

Theorem 7.1.10. *For every Hausdorff group G, there exist a complete Hausdorff group $\widetilde{G}$ and a topological embedding $\iota : G \hookrightarrow \widetilde{G}$ such that $\iota(G)$ is dense in $\widetilde{G}$.*

For a Hausdorff group G, the pair $(\widetilde{G}, \iota)$ given by Theorem 7.1.10 is called *(Raĭkov) completion* of G. We see below that it is unique up to topological isomorphisms.

We use Lemma 7.1.12, based on the next notion, which makes sense since a local base $\mathcal{B}$ at e_G in a topological group G is a directed set when endowed with the containment order $(\mathcal{B}, \supseteq)$ (see also Remark 3.1.4).

Definition 7.1.11. Let G be a topological group and $\mathcal{B}$ a local base at e_G. A *$\mathcal{B}$-net* in G is a net $\{x_U\}_{U \in \mathcal{B}}$ in G.

Lemma 7.1.12. *Let G be a topological group and $\mathcal{B}$ a local base at e_G. Then G is complete if and only if every Cauchy $\mathcal{B}$-net of G converges in G.*

Proof. Let $\{x_\alpha\}_{\alpha \in A}$ be a Cauchy net of G. For every $U \in \mathcal{B}$, there exists $\alpha_U \in A$ such that, for every $\alpha, \beta \geq \alpha_U$, $x_\alpha^{-1} x_\beta \in U$ and $x_\beta x_\alpha^{-1} \in U$. We prove that $\{x_{\alpha_U}\}_{U \in \mathcal{B}}$ is a Cauchy net of G. In fact, let $W \in \mathcal{B}$ and let $W_0 \in \mathcal{B}$ be such that $W_0 W_0 \subseteq W$. Let $U, V \in \mathcal{B}$ be contained in W_0; for every $\alpha \in A$ with $\alpha \geq \alpha_U$, $\alpha \geq \alpha_V$,

$$x_{\alpha_U} x_{\alpha_V}^{-1} = x_{\alpha_U} x_\alpha^{-1} x_\alpha x_{\alpha_V}^{-1} \in UV \subseteq W_0 W_0 \subseteq W,$$
$$x_{\alpha_V}^{-1} x_{\alpha_U} = x_{\alpha_V}^{-1} x_\alpha x_\alpha^{-1} x_{\alpha_U} \in VU \subseteq W_0 W_0 \subseteq W.$$

Since the $\mathcal{B}$-net $\{x_{\alpha_U}\}_{U \in \mathcal{B}}$ is a Cauchy net of G, it is convergent by hypothesis, so let $x \in G$ be a limit of $\{x_{\alpha_U}\}_{U \in \mathcal{B}}$. Now, according to Lemma 7.1.4(a), $x_\alpha \to x$, too. □

The use of $\mathcal{B}$-nets allows us to use only sequences in metrizable groups:

Lemma 7.1.13. *A metrizable group is complete if and only if every Cauchy sequence of G converges in G.*

Proof. By Birkhoff–Kakutani theorem 5.2.17, there exists a countable base $\mathcal{B} = \{U_n : n \in \mathbb{N}\}$ of $\mathcal{V}_G(e_G)$. For every $n \in \mathbb{N}$ let $V_n = U_0 \cap \cdots \cap U_n$. Then $\mathcal{B}' = \{V_n : n \in \mathbb{N}\}$ is a countable base of $\mathcal{V}_G(e_G)$ and $V_{n+1} \subseteq V_n$ for every $n \in \mathbb{N}$. By Lemma 7.1.12, it suffices to consider $\mathcal{B}'$-nets of G, and these are sequences. □

In the next theorem we offer a proof of Theorem 7.1.10 for metrizable groups; as a consequence, we obtain a proof also for abelian groups.

Theorem 7.1.14. *The completion $\widetilde{G}$ exists for every metrizable group G.*

Proof. Fix a countable base $\mathcal{B} = \{U_n : n \in \mathbb{N}\}$ of $\mathcal{V}_G(e_G)$ (see Birkhoff–Kakutani theorem 5.2.17). Let $G^{\mathbb{N}}$ be the group of all sequences of elements of G. Consider the subgroup G_0 of $G^{\mathbb{N}}$ consisting of all null sequences in G (i. e., all sequences in G converging to e_G) and let G_C be the larger subgroup of $G^{\mathbb{N}}$ consisting of all Cauchy sequences of G. There exists an injective homomorphism $\zeta : G \to G_C$, $g \mapsto \{g, g, \ldots, g, \ldots\}$. For every $k \in \mathbb{N}$, let

$$U_k^\sim = \{\{x_n\}_{n\in\mathbb{N}} \in G_C : \exists n_0 \in \mathbb{N}, \forall n \geq n_0, x_n \in U_k\}.$$

Then the filter base $\mathcal{B}^\sim = \{U_n^\sim : n \in \mathbb{N}\}$ on G_C satisfies (gt1), (gt2), and (gt3), so it can be taken as a base of $\mathcal{V}_{G_C}(\zeta(e_G))$ in a group topology τ on G_C. Clearly:

(i) $\zeta(G)$ is dense in (G_C, τ) (for $x^\sim = \{x_n\}_{n\in\mathbb{N}} \in G_C$ and $U_k \in \mathcal{B}$, there exists $n_0 \in \mathbb{N}$ such that $x_{n_0} x_n^{-1} \in U_k$ for all $n \geq n_0$; then $\zeta(x_{n_0}) \in U_k^\sim x^\sim$);
(ii) $\operatorname{core}(G_C, \tau) = G_0$;
(iii) for $n \in \mathbb{N}$, $\zeta(U_n) = U_n^\sim \cap \zeta(G)$, so $\zeta : G \to \zeta(G)$ is a topological isomorphism;
(iv) a sequence $\{x_n\}_{n\in\mathbb{N}}$ in G is a Cauchy sequence of G if and only if $\{\zeta(x_n)\}_{n\in\mathbb{N}}$ is a Cauchy sequence of G_C;
(v) $\zeta(G) \cap G_0 = \{\zeta(e_G)\}$.

We show that G_C/G_0 is the desired completion $\widetilde{G}$. To see that G_C is complete, let $\{x_n^\sim\}_{n\in\mathbb{N}}$ be a Cauchy sequence of G_C. By the density of $\zeta(G)$ in G_C, for every $n \in \mathbb{N}$ there exists $g_n \in G$ such that $\zeta(g_n) \in x_n^\sim U_n^\sim$. Then, for every $n \in \mathbb{N}$, $y_n^\sim = x_n^{\sim\,-1} \zeta(g_n) \in U_n^\sim$, and hence $\{y_n^{\sim}\}_{n\in\mathbb{N}}$ is a null sequence of G_C. Since $\zeta(g_n) = x_n^{\sim} y_n^{\sim}$ for every $n \in \mathbb{N}$, being a product of Cauchy sequences of G_C, $\{\zeta(g_n)\}_{n\in\mathbb{N}}$ is in its own turn a Cauchy sequence of G_C. This implies that $\{g_n\}_{n\in\mathbb{N}}$ is a Cauchy sequence of G in view of (iv), namely, $\{g_n\}_{n\in\mathbb{N}} \in G_C$. We verify that $x^{\sim} := \{g_n\}_{n\in\mathbb{N}} \in \lim_{n\in\mathbb{N}} \zeta(g_n)$: fix $U_k^{\sim} \in \mathcal{B}^{\sim}$; since $\{g_n\}_{n\in\mathbb{N}}$ is a Cauchy sequence of G, there exists $n_0 \in \mathbb{N}$ such that, for every $n, m \geq n_0$, $g_m \in g_n U_k$; thus, for every $m \geq n_0$, $\zeta(g_m) \in \{g_n\}_{n\in\mathbb{N}} U_k^{\sim} = x^{\sim} U_k^{\sim}$, namely, $\zeta(g_n) \to x^{\sim}$. Hence $\lim_{n\in\mathbb{N}} \zeta(g_n) y_n^{\sim -1} = \lim_{n\in\mathbb{N}} x_n^{\sim}$, so G_C is complete, by Lemma 7.1.13.

Let $q\colon G_C \to G_C/G_0 = \mathfrak{h}G_C$ be the Hausdorff reflection of G_C. Then $\widetilde{G} := G_C/G_0 = \mathfrak{h}G_C$ is Hausdorff, and $\widetilde{G}$ is complete by Proposition 7.1.7. Let $\iota = q \circ \zeta$:

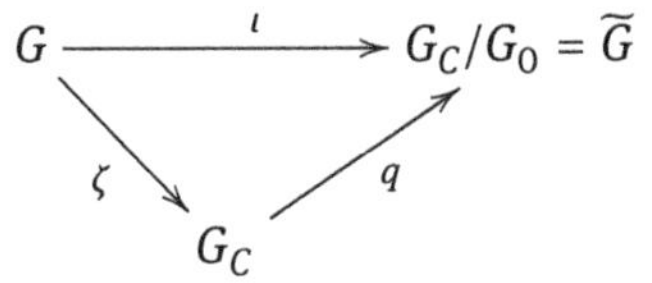

Since G_0 is indiscrete, $G_0 \cap \zeta(G) = \{\zeta(e_G)\}$ is dense in G_0, so $q \restriction_{\zeta(G)}\colon \zeta(G) \to q(\zeta(G))$ is open by Theorem 3.2.9, hence $\iota\colon G \to \widetilde{G}$ is a topological embedding. □

An easy example shows that the sets $U_k^{\sim}$ in the above proof need not be open.

Example 7.1.15. Let $\mathbb{Q}$ be endowed with the usual topology. For every $k \in \mathbb{N}_+$, let $U_k = (-\pi/k, \pi/k) \cap \mathbb{Q}$; clearly, $\{U_k : k \in \mathbb{N}_+\}$ is a base of $\mathcal{V}_{\mathbb{Q}}(0)$. Let $\{x_n\}_{n\in\mathbb{N}}$ be a sequence of $\mathbb{Q}$ such that $x_n \in U_1$ for every $n \in \mathbb{N}_+$ and $x_n \to \pi$ in $\mathbb{R}$; so $\{x_n\}_{n\in\mathbb{N}}$ is a Cauchy sequence of $\mathbb{Q}$ that does not converge in $\mathbb{Q}$.

Now $\{x_n\}_{n\in\mathbb{N}} \in U_1^{\sim}$, but there exists no $m \in \mathbb{N}_+$ such that $\{x_n\}_{n\in\mathbb{N}} + U_m^{\sim} \subseteq U_1^{\sim}$.

Proposition 7.1.16. *If a topological group G is metrizable, then its completion $\widetilde{G}$ is metrizable as well.*

Proof. Since $\widetilde{G}$ is Hausdorff, by Birkhoff–Kakutani theorem 5.2.17 it suffices to prove that in case there exists a countable base $\{U_n : n \in \mathbb{N}\}$ of $\mathcal{V}_G(e_G)$, there exists a countable base of $\mathcal{V}_{\widetilde{G}}(e_{\widetilde{G}})$. This countable base is $\{\overline{U_n}^{\widetilde{G}} : n \in \mathbb{N}\}$, by Lemma 5.1.3. □

This can be deduced also from the proof of Theorem 7.1.14, where we furnish a countable local base at e_{G_C}, which gives a countable local base at $e_{\widetilde{G}}$.

Corollary 7.1.17. *The completion $\widetilde{G}$ exists for every Hausdorff abelian group G.*

Proof. According to Corollary 5.2.19, G is isomorphic to a subgroup of a product $\prod_{i\in I} M_i$, where each abelian group M_i is metrizable. By Theorem 7.1.14, every M_i has a completion $\widetilde{M_i}$. Then $P = \prod_{i\in I} \widetilde{M_i}$ is complete by Proposition 7.1.9. The closure $\overline{G}^P$ is complete by Proposition 7.1.8, so $\overline{G}^P$ is the completion of G. □

As promised at the beginning of this section, now we are in position to prove the general theorem showing the existence of the completion.

Proof of Theorem 7.1.10. Fix a base $\mathcal{B}$ of $\mathcal{V}_G(e_G)$ and let $G^\sim$ be the family of all Cauchy $\mathcal{B}$-nets of G. For $x^\sim = \{x_U\}_{U\in\mathcal{B}}, y^\sim = \{y_U\}_{U\in\mathcal{B}} \in G^\sim$, let $x^\sim y^\sim = \{x_U y_U\}_{U\in\mathcal{B}}$. According to Lemma 7.1.4, $x^\sim y^\sim \in G^\sim$. For $x^\sim = \{x_U\}_{U\in\mathcal{B}} \in G^\sim$, also $\{x_{\tilde U}^{-1}\}_{U\in\mathcal{B}} \in G^\sim$ by Lemma 7.1.4, so it is the inverse of $x^\sim$. Then $G^\sim$ is a group with identity the constant net $\{e_G\}_{U\in\mathcal{B}}$. Let $\zeta: G \to G^\sim$ be the injective homomorphism sending each $g \in G$ to the constant Cauchy net $\zeta(g)$.

For every $U \in \mathcal{B}$, let

$$U^\sim = \{\{x_U\}_{U\in\mathcal{B}} \in \widetilde{G}: \exists V_0 \in \mathcal{B}, \forall V \subseteq V_0, x_V \in U\} \subseteq G^\sim.$$

One can check that the filter base $\mathcal{B}^\sim = \{U^\sim : U \in \mathcal{B}\}$ satisfies (gt1), (gt2), and (gt3), so it can be taken as a base of the filter of the neighborhoods of the neutral element $\zeta(e_G)$ of $G^\sim$ in a group topology on $G^\sim$. Since $\zeta(U) = U^\sim \cap \zeta(G)$ for every $U \in \mathcal{B}$, the homomorphism $\zeta: G \to \zeta(G)$ is a topological isomorphism. Moreover, $\zeta(G)$ is dense in $G^\sim$. To check it, pick $x^\sim = \{x_W\}_{W\in\mathcal{B}} \in G^\sim$ and a neighborhood $U^\sim \in \mathcal{B}^\sim$ of $e_{G^\sim}$. There exists $V_0 \in \mathcal{B}$ such that $x_V x_O^{-1} \in U$ for all $O, V \in \mathcal{B}$, with $O, V \subseteq V_0$. Let $g = x_{V_0}$. Then $g x_O^{-1} \in U$ for all $O \in \mathcal{B}$ with $O \subseteq V_0$. This proves that $\zeta(g) \in U^\sim x^\sim$.

Since $(\mathcal{B}, \supseteq) \to (\mathcal{B}^\sim, \supseteq)$, $U \mapsto U^\sim$, is a monotone bijection, we can consider $\mathcal{B}$-nets of $G^\sim$ instead of $\mathcal{B}^\sim$-nets. We use below that a $\mathcal{B}$-net $\{x_U\}_{U\in\mathcal{B}}$ of G is a Cauchy $\mathcal{B}$-net of G if and only if $\{\zeta(x_U)\}_{U\in\mathcal{B}}$ is a Cauchy $\mathcal{B}$-net of $G^\sim$.

To see that $G^\sim$ is complete, it is enough to see that every Cauchy $\mathcal{B}$-net $\{x_{\tilde U}\}_{U\in\mathcal{B}}$ of $G^\sim$ converges (see Lemma 7.1.12). By the density of $\zeta(G)$ in $G^\sim$, for every $U \in \mathcal{B}$ there exists $g_U \in G$ with $\zeta(g_U) \in U^\sim x_{\tilde U}$. Then $y_{\tilde U} = \zeta(g_U) x_{\tilde U}^{-1} \in U^\sim$, hence $\{y_{\tilde U}\}_{U\in\mathcal{B}}$ is a null $\mathcal{B}$-net of $G^\sim$, and therefore a Cauchy $\mathcal{B}$-net of $G^\sim$. We conclude that since $\{\zeta(g_U)\}_{U\in\mathcal{B}} = \{y_{\tilde U}\, x_{\tilde U}\}_{U\in\mathcal{B}}$ is a product of Cauchy $\mathcal{B}$-nets of $G^\sim$, $\{\zeta(g_U)\}_{U\in\mathcal{B}}$ is in its own turn a Cauchy $\mathcal{B}$-net of $G^\sim$.

This implies that $\{g_U\}_{U\in\mathcal{B}}$ is a Cauchy $\mathcal{B}$-net of G. We verify that $x^\sim := \{g_U\}_{U\in\mathcal{B}} \in \lim_{U\in\mathcal{B}} \zeta(g_U)$: fix $W^\sim \in \mathcal{B}^\sim$; since $\{g_U\}_{U\in\mathcal{B}}$ is a Cauchy $\mathcal{B}$-net of G, there exists $U_0 \in \mathcal{B}$ such that $g_{U_1} \in g_{U_2} W$ for all $U_1, U_2 \in \mathcal{B}$ contained in U_0; thus, for every $U_1 \in \mathcal{B}$ contained in U_0, $\zeta(g_{U_1}) \in \{g_U\}_{U\in\mathcal{B}} W^\sim = x^\sim W^\sim$. This yields $\zeta(g_U) \to x^\sim$. Hence $x^\sim \in \lim_{U\in\mathcal{B}} y_{\tilde U}^{-1} \zeta(g_U) = \lim_{U\in\mathcal{B}} x_{\tilde U}$, so $G^\sim$ is complete.

Let $q: G^\sim \to \mathfrak{h}G^\sim = G^\sim/\mathrm{core}(G^\sim)$ be the Hausdorff reflection of $G^\sim$. Then $\widetilde{G} := \mathfrak{h}G^\sim$ is Hausdorff and $\widetilde{G}$ is complete by Proposition 7.1.7. Since $\mathrm{core}(G^\sim) = \bigcap_{U\in\mathcal{B}} U^\sim$ coincides with the subgroup of $G^\sim$ of all null $\mathcal{B}$-nets of G, $\mathrm{core}(G^\sim) \cap \zeta(G) = \{\zeta(e_G)\}$. Let $\iota = q \circ \zeta$:

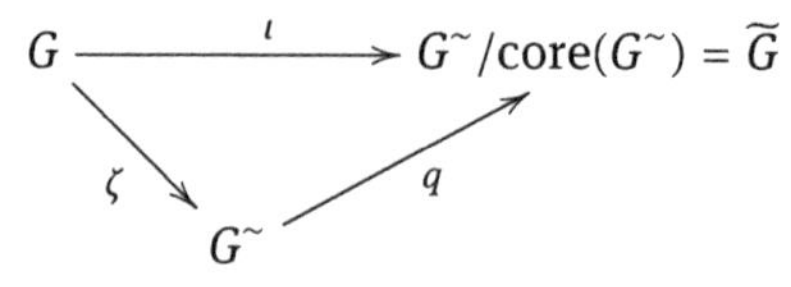

Since $\mathrm{core}(G^\sim)$ is indiscrete, $\mathrm{core}(G^\sim) \cap \zeta(G) = \{\zeta(e_G)\}$ is dense in $\mathrm{core}(G^\sim)$, so $q\restriction_{\zeta(G)}: \zeta(G) \to q(\zeta(G))$ is open by Theorem 3.2.9, hence $\iota: G \to \widetilde{G}$ is a topological embedding. □

The complete group $\widetilde{G}$ has an important universal property.

Theorem 7.1.18. *Let G be a Hausdorff group and let $\widetilde{G}$ be a complete Hausdorff group together with a topological embedding $\iota: G \hookrightarrow \widetilde{G}$ such that $\iota(G)$ is dense in $\widetilde{G}$. If $f: G \to H$ is a continuous homomorphism, where H is a complete Hausdorff group, then there is a unique continuous homomorphism $\widetilde{f}: \widetilde{G} \to H$ with $f = \widetilde{f} \circ \iota$:*

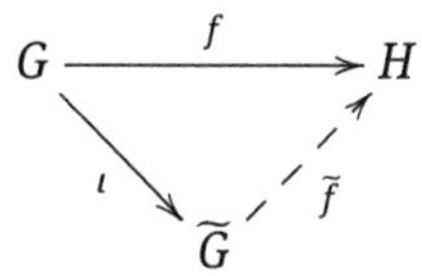

Proof. Let $g \in \widetilde{G}$. There exists a net $\{g_\alpha\}_{\alpha\in A}$ in G such that $g = \lim_{\alpha\in A} g_\alpha$. Then $\{g_\alpha\}_{\alpha\in A}$ is a Cauchy net, hence $\{f(g_\alpha)\}_{\alpha\in A}$ is a Cauchy net of H by Exercise 7.3.1(b). By the completeness of H, it must be convergent.

Put $\widetilde{f}(g) = \lim_{\alpha\in A} f(g_\alpha)$. To see that this limit does not depend on the choice of the net $\{g_\alpha\}_{\alpha\in A}$ with $g = \lim_{\alpha\in A} g_\alpha$, pick another net $\{x_\alpha\}_{\alpha\in A}$ in G such that $g = \lim_{\alpha\in A} x_\alpha$. Then again $\{f(x_\alpha)\}_{\alpha\in A}$ is a Cauchy net of H, so there exists $h \in H$ such that $f(x_\alpha) \to h$. On the other hand, $\{x_\alpha^{-1} g_\alpha\}_{\alpha\in A}$ converges to $e_{\widetilde{G}}$ in $\widetilde{G}$, hence in G as well. Therefore, $f(x_\alpha^{-1} g_\alpha) \to e_H$, by the continuity of f. Since $f(x_\alpha^{-1}) = f(x_\alpha)^{-1} \to h^{-1}$ in H, we deduce that $f(x_\alpha^{-1} g_\alpha) \to h^{-1} \lim_{\alpha\in A} f(g_\alpha)$. By the uniqueness of the limit, $h^{-1} \lim_{\alpha\in A} f(g_\alpha) = e_H$, so $h = \lim_{\alpha\in A} f(g_\alpha)$.

To see that $\widetilde{f}$ is continuous, pick $W, V \in \mathcal{V}_H(e_H)$ with $VV \subseteq W$. By the continuity of f, there exists $U \in \mathcal{V}_G(e_G)$ with $f(U) \subseteq V$. It is enough to check that $\widetilde{f}(\overline{U}) \subseteq W$, where the closure is taken in $\widetilde{G}$. If $u \in \overline{U}$, then $u = \lim_{\beta\in B} u_\beta$ with $\{u_\beta\}_{\beta\in B}$ a net in U. Then

$$\widetilde{f}(u) = \lim_{\beta\in B} f(u_\beta) \in \overline{f(U)} \subseteq f(U)V \subseteq VV \subseteq W.$$

This proves that $\widetilde{f}(\overline{U}) \subseteq W$.

To verify the uniqueness of the extension $\widetilde{f}$, suppose that $f': \widetilde{G} \to H$ is another continuous homomorphism with $f = f' \circ \iota$. This means that f' and $\widetilde{f}$ coincide on the dense subgroup $\iota(G)$, so $\widetilde{f} = f'$. □

From the two theorems above, one can deduce that every Hausdorff group has a unique, up to topological isomorphisms, (Raĭkov) completion $(\widetilde{G}, \iota)$, and we can assume that G is simply a dense subgroup of $\widetilde{G}$.

Corollary 7.1.19. *Let $\{G_i : i \in I\}$ be a family of Hausdorff groups and $G = \prod_{i\in I} G_i$. Then $\widetilde{G} \cong \prod_{i\in I} \widetilde{G}_i$.*

Proof. By Proposition 7.1.9, the product $\prod_{i\in I} \widetilde{G}_i$ is complete. Since G is a dense subgroup of $\prod_{i\in I} \widetilde{G}_i$, Theorem 7.1.18 applies. □

The following is a direct consequence of Theorem 7.1.18.

Corollary 7.1.20. *For Hausdorff groups G, H and a continuous homomorphism $f: G \to H$, there exists a continuous homomorphism $\widetilde{f}: \widetilde{G} \to \widetilde{H}$ extending f.*

This corollary shows that the assignments $G \mapsto \widetilde{G}$ and $f \mapsto \widetilde{f}$ define a reflector from the category **TopGr₂** to its full subcategory of complete Hausdorff groups.

Remark 7.1.21. Obviously, composing the completion functor with the functor $G \mapsto \mathfrak{h}G$ of the Hausdorff reflection, one obtains a reflector from the category **TopGr** to the full subcategory of complete Hausdorff groups, but now the maps $G \to \widetilde{\mathfrak{h}G}$ are not injective any more.

The next proposition gives a useful characterization of completeness in terms of "absolute closedness". (Recall that *H-closed spaces* were introduced by Alexandrov and Urysohn as those Hausdorff spaces which are always closed whenever embedded in some Hausdorff space.)

Proposition 7.1.22. *A Hausdorff group G is complete if and only if for every embedding $j: G \hookrightarrow H$ into a Hausdorff group H the subgroup $j(G)$ of H is closed.*

Proof. Assume that there exists an embedding $j: G \hookrightarrow H$ into a Hausdorff group H such that $j(G)$ is not a closed subgroup of H. Then there exists a net $\{y_\alpha\}_{\alpha\in A}$ in $j(G)$ converging to some element $h \in H$ that does not belong to $j(G)$. By Remark 7.1.3, $\{y_\alpha\}_{\alpha\in A}$ is a Cauchy net of $j(G)$. Since it converges to $h \in H \setminus j(G)$, and H is a Hausdorff group, we conclude that this net does not converge in $j(G)$. Since $j: G \to j(G)$ is a topological isomorphism, this provides a nonconvergent Cauchy net of G. Hence, G is not complete.

Now assume that G is not complete and consider the dense inclusion $\iota: G \hookrightarrow \widetilde{G}$. Since $G = \iota(G)$ is a proper dense subgroup of $\widetilde{G}$, we conclude that $\iota(G)$ is not closed in $\widetilde{G}$. □

Some compactness-like properties (like local compactness) imply completeness (see Remark 7.1.28 and Proposition 8.2.6). Here is another obviously stronger property:

Definition 7.1.23 ([115]). A Hausdorff group G is *h-complete* if every continuous homomorphic image of G is complete.

For example, discrete groups are complete but not h-complete, so this completeness property is strictly stronger. One can prove that abelian (actually, nilpotent) h-complete groups are compact (see [115]).

On the other hand, other relatively strong compactness-like properties (as countable compactness) do not imply completeness. To face this phenomenon, a weaker form of completeness was proposed in [110, 111]:

Definition 7.1.24. A topological group G is *sequentially complete* if every Cauchy sequence $\{g_n\}_{n\in\mathbb{N}}$ of G is convergent.

A topological group G is sequentially complete if and only if G is sequentially closed in its completion. We shall see in the sequel that countably compact groups are sequentially complete, although they need not be complete in general (see Remark 15.2.6).

Groups without nontrivial convergent sequences are sequentially complete (see Exercise 7.3.7), so for every abelian group G the group $G^{\#}$ is sequentially complete (since it has no nontrivial convergent sequences, in view of Glicksberg theorem 11.6.11 – see Theorem 13.4.9 for a proof in the present case), yet it is quite far from being complete, as we shall see in the sequel.

Plenty of results on the remarkable class of sequentially complete groups can be found in [80, 93, 102, 110, 111].

7.1.3 Weil completion

Historically the following concepts (see also Definition 7.1.27) appeared before that in §7.1.1.

Definition 7.1.25. A net $\{g_\alpha\}_{\alpha\in A}$ in a topological group G is a *left* (respectively, *right*) *Cauchy net* if for every neighborhood U of e_G in G there exists $\alpha_0 \in A$ such that $g_\alpha^{-1}g_\beta \in U$ (respectively, $g_\beta g_\alpha^{-1} \in U$) for every $\alpha, \beta > \alpha_0$.

Clearly, a net is Cauchy if and only if it is both a left and a right Cauchy net.

Lemma 7.1.26. *Let G be a Hausdorff group. Every left (respectively, right) Cauchy net of G with a convergent subnet is convergent.*

Proof. Let $\{g_\alpha\}_{\alpha\in A}$ be a left Cauchy net of G and let $\{g_\beta\}_{\beta\in B}$ be a subnet convergent to $x \in G$, where B is a cofinal subset of A. Let U be a neighborhood of e_G in G and V a symmetric neighborhood of e_G in G such that $VV \subseteq U$. Since $g_\beta \to x$, there exists $\beta_0 \in B$ such that $g_\beta \in xV$ for every $\beta > \beta_0$. On the other hand, there exists $\alpha_0 \in A$ such that $\alpha_0 \geq \beta_0$ and, for every $\alpha, \gamma > \alpha_0$, $g_\gamma^{-1}g_\alpha \in V$, that is, $g_\alpha \in g_\gamma V \subseteq xVV \subseteq xU$. Therefore, for every $\alpha > \alpha_0$, $g_\alpha \in xU$, namely, $g_\alpha \to x$. □

Definition 7.1.27. A topological group G is *complete in the sense of Weil* (or *Weil complete*) if every left Cauchy net of G converges in G.

Remark 7.1.28. (a) For every left Cauchy net $\{g_\alpha\}_{\alpha\in A}$ of a topological group G, the net $\{g_\alpha^{-1}\}_{\alpha\in A}$ is right Cauchy. Therefore, if every left Cauchy net of G converges in G, then the same applies to all right Cauchy nets of G. Hence, a topological group G is Weil complete if and only if every right Cauchy net of G converges in G.

(b) We shall see in the sequel that locally compact groups are Weil complete (see Proposition 8.2.6).

Obviously, every Weil complete group is also Raĭkov complete, but the converse does not hold in general (see Proposition 7.1.29). Clearly, these two concepts coincide for abelian groups.

It is possible to define the *Weil completion* of a Hausdorff group in analogy with the Raĭkov completion. If a Hausdorff group G admits a Weil completion H, then $H = \widetilde{G}$ is the Raĭkov completion of G, in view of Exercise 7.3.3. The converse does not hold true:

Proposition 7.1.29. *For an infinite set X, let $S(X)$ be equipped with $\mathcal{T}_X$:*
(a) *$S(X)$ is Raĭkov complete;*
(b) *$S(X)$ admits no Weil completion (and $S(X)$ is not Weil complete).*

Proof. (a) Let $\{f_\alpha\}_{\alpha\in A}$ be a Cauchy net of $S(X)$. For every finite subset E of X, there exists $\alpha_0 \in A$ such that for every $\alpha, \beta \geq \alpha_0$, $f_\beta^{-1} f_\alpha \in S_E$ and $f_\alpha f_\beta^{-1} \in S_E$; in particular, $f_\alpha \restriction_E = f_\beta \restriction_E$ and $f_\alpha^{-1} \restriction_E = f_\beta^{-1} \restriction_E$. Taking $E = \{x\}$ for $x \in X$, this means that $\{f_\alpha(x)\}_{\alpha\in A}$ and $\{f_\alpha^{-1}(x)\}_{\alpha\in A}$ are convergent in the discrete space X; so, let

$$f(x) = \lim_{\alpha\in A} f_\alpha(x) \quad \text{and} \quad g(x) = \lim_{\alpha\in A} f_\alpha^{-1}(x).$$

Hence, we have defined two maps $f, g: X \to X$ such that $f \circ g = g \circ f = id_X$, so $f \in S(X)$, and $f_\alpha \to f$ in $\mathcal{T}_X$. This proves that $S(X)$ is complete.

(b) To see that $S(X)$ is not Weil complete, we produce a left Cauchy net that does not converge. Let $Y = \{x_n : n \in \mathbb{N}_+\}$ be a subset of X (with pairwise distinct elements). For $n \in \mathbb{N}_+$, define $f_n: X \to X$ by

$$\begin{cases} f_n(x_i) = x_{i+1} & \text{for } i \in \{1, \dots, n-1\}, \\ f_n(x_n) = x_1, \\ f_n(x) = x & \text{for } x \in X \setminus \{x_1, \dots, x_n\}. \end{cases}$$

Then $\{f_n\}_{n\in\mathbb{N}_+}$ is a left Cauchy net of $S(X)$: for a finite subset E of X, let $k \in \mathbb{N}_+$ with $E \cap Y \subseteq \{x_1, \dots, x_k\}$; so, for $m, n > k$, $f_n \restriction_E = f_m \restriction_E$, that is, $f_m^{-1} f_n \in S_E$. We see that $\{f_n\}_{n\in\mathbb{N}_+}$ does not converge in $S(X)$. Let $f: X \to X$ be defined by

$$\begin{cases} f(x_i) = x_{i+1} & \text{for every } i \in \mathbb{N}_+, \\ f(x) = x & \text{for every } x \in X \setminus Y. \end{cases}$$

Then $f_n \to f$ in X^X endowed with the pointwise convergence topology, but $f(X) = X \setminus \{x_1\}$, so $f \notin S(X)$. It follows that $\{f_n\}_{n\in\mathbb{N}_+}$ does not converge in $S(X)$. Thus, $S(X)$ is not Weil complete.

By Exercise 7.3.3, $S(X)$ cannot admit a Weil completion, otherwise the above left Cauchy sequence $\{f_n\}_{n\in\mathbb{N}}$ would be also a right Cauchy sequence, and so it would converge by item (a). □

7.2 Completeness via filters

7.2.1 Cauchy filters

The proof of Theorem 7.1.10 becomes particularly involved when G is not metrizable. This is why many authors prefer to avoid the Cauchy nets for the construction of the completion. This can be done by means of the following notion.

Definition 7.2.1. A filter $\mathcal{F}$ on a Hausdorff group G is *Cauchy* if for every $U \in \mathcal{V}_G(e_G)$ there exists $F \in \mathcal{F}$ such that $FF^{-1} \cup F^{-1}F \subseteq U$.

One can define a Cauchy filter base in the same way, or, equivalently, by asking that the filter generated by this filter base is Cauchy. If $\mathcal{F}, \mathcal{H}$ are filters on G, we denote by $\mathcal{F} \cdot \mathcal{H}$ the filter on G generated by the filter base $\{FH: F \in \mathcal{F}, H \in \mathcal{H}\}$.

Lemma 7.2.2. *Let G be a Hausdorff group.*

(a) *For a filter $\mathcal{F}$ on G, the following conditions are equivalent:*
   (a_1) *the filter $\mathcal{F}$ is Cauchy;*
   (a_2) *the filter $\mathcal{F}^{-1} := \{F^{-1}: F \in \mathcal{F}\}$ is Cauchy;*
   (a_3) *the filters $\mathcal{F}^{-1} \cdot \mathcal{F}$ and $\mathcal{F} \cdot \mathcal{F}^{-1}$ converge to e_G;*
   (a_4) *for every $U \in \mathcal{V}_G(e_G)$, there exists $g \in G$ such that $Ug \in \mathcal{F}$ and $gU \in \mathcal{F}$.*
   (a_5) *for every $U \in \mathcal{V}_G(e_G)$, there exist $g, h \in G$ such that $Ug \in \mathcal{F}$ and $Uh \in \mathcal{F}$.*

(b) *If $\mathcal{F}$ is a Cauchy filter on G and $x_F \in F$ for every $F \in \mathcal{F}$, then the net $\{x_F\}_{F\in\mathcal{F}}$ is a Cauchy net (here $\mathcal{F}$ is considered as a directed partially ordered set with respect to inclusion).*

(c) *If $\{x_\alpha\}_{\alpha\in A}$ is a Cauchy net of G and $F_\alpha = \{x_\beta: \beta \in A, \beta \geq \alpha\}$, then the family $\{F_\alpha: \alpha \in A\}$ is a Cauchy filter base on G.*

(d) *If $\mathcal{F}, \mathcal{H}$ are Cauchy filters on G, then the filter $\mathcal{F} \cdot \mathcal{H}$ is Cauchy.*

Proof. The verification of (a)–(c) is a straightforward application of the definitions.

(d) Pick $U \in \mathcal{V}_G(e_G)$ and $V \in \mathcal{V}_G(e_G)$ with $VV \subseteq U$. According to (a_4), there exist $H \in \mathcal{H}$ and $h \in G$ such that $H \subseteq hV$. Since $hVh^{-1} \in \mathcal{V}_G(e_G)$, according to ($a_4$) there exist $F \in \mathcal{F}$ and $g \in G$ such that $F \subseteq ghVh^{-1}$. Now

$$FH \subseteq ghVh^{-1}hV \subseteq ghVV \subseteq ghU.$$

Similarly, one can prove that there exist $g', h' \in G$, $F' \in \mathcal{F}$ and $H' \in \mathcal{H}$, such that $F'H' \subseteq Uh'g'$. We conclude that $\mathcal{F} \cdot \mathcal{H}$ is a Cauchy filter. □

We immediately deduce from Lemma 7.2.2 the following characterization of complete Hausdorff groups by means of filters.

Proposition 7.2.3. *A Hausdorff group G is complete if and only if every Cauchy filter on G converges.*

The proof of Theorem 7.1.10 uses essentially Cauchy nets, whereas the construction of the completion in Theorem 7.2.10 is based on Cauchy filters. Nevertheless, it is important at this stage to realize that the use of filters does not lead to a new notion of completeness, since the completion is unique up to topological isomorphisms, by Theorem 7.1.18 (see also Lemma 7.2.9). Our motivation to rigorously carry out the Raĭkov completion procedure also in terms of Cauchy filters is due to its simplicity with respect to its counterpart based on the use of Cauchy nets. For example, the completion procedure with Cauchy filters in Theorem 7.2.10 does not require the "second step", i. e., taking a quotient of the group of minimal Cauchy filters, in order to obtain a Hausdorff group.

One can define left and right Cauchy filters on a Hausdorff group in analogy with left and right Cauchy nets and Definition 7.2.1, so that the Cauchy filters are those filters that are simultaneously left and right Cauchy. As observed above for the Raĭkov completeness and the Raĭkov completion, developing the Weil completeness and the Weil completion in terms of left (or right) Cauchy filters in full details gives nothing new with respect to what we already did in terms of left (or right) Cauchy nets. This is why we are not going to do that.

7.2.2 Minimal Cauchy filters

Let us see that one can relax the criterion for completeness by imposing convergence only on a much smaller family of Cauchy filters introduced as follows.

Definition 7.2.4. A Cauchy filter $\mathcal{F}$ on a Hausdorff group G is *minimal* if for every $F \in \mathcal{F}$ there exist $F' \in \mathcal{F}$ and $U \in \mathcal{V}_G(e_G)$ such that $UF'U \subseteq F$.

A leading example of a minimal Cauchy filter are the neighborhood filters $\mathcal{V}_G(x)$ (see Proposition 7.2.6). Clearly, the minimal Cauchy filters are *open* (i. e., have a base of open sets). Nevertheless, an open Cauchy filter need not be minimal even if it is convergent.

Example 7.2.5. Let $\mathbb{R}$ be equipped with the usual Euclidean topology, and let $\mathcal{B} = \{(0, 1/n): n \in \mathbb{N}_+\}$, which is a filter base consisting of open sets of $\mathbb{R}$. Since $\mathcal{B}$ converges to 0, it generates an open Cauchy filter $\mathcal{F}$. But $\mathcal{F}$ is not minimal, as $F = (0,1) \in \mathcal{F}$ does not contain any set of the form $F' + U$, where $F' \in \mathcal{F}$ and $U = (-1/m, 1/m)$ for some $m \in \mathbb{N}_+$. This shows also that a convergent open filter need not be minimal.

As the next proposition shows, the minimal Cauchy filters on a topological group G are precisely the minimal elements in the poset of all Cauchy filters on G ordered by inclusion; in particular, if $\mathcal{F} \neq \mathcal{G}$ are minimal Cauchy filters on G they are not comparable, that is, neither $\mathcal{F} \subseteq \mathcal{G}$ nor $\mathcal{F} \supseteq \mathcal{G}$. This explains the use of the term *minimal*, introduced by Bourbaki [38] in the framework of uniform spaces.

Proposition 7.2.6. *Let G be a Hausdorff group and $\mathcal{F}$ a Cauchy filter on G. The following conditions are equivalent:*

(a) $\mathcal{F}$ *is minimal;*
(b) *if* $\mathcal{F}_1$ *is a Cauchy filter on* G *such that* $\mathcal{F}_1 \subseteq \mathcal{F}$, *then* $\mathcal{F}_1 = \mathcal{F}$;
(c) $\mathcal{F} = \mathcal{F} \cdot \mathcal{V}_G(e_G) = \mathcal{V}_G(e_G) \cdot \mathcal{F}$.

In particular, $\mathcal{V}_G(x)$ *is a minimal Cauchy filter for every* $x \in G$.

Proof. (a)⇒(b) Let $F \in \mathcal{F}$. By the minimality of $\mathcal{F}$, there exist $F' \in \mathcal{F}$ and $U \in \mathcal{V}(e_G)$ such that $UF'U \subseteq F$. Moreover, there exists $F_1 \in \mathcal{F}_1$ such that $F_1F_1^{-1} \cup F_1^{-1}F_1 \subseteq U$. Since $F_1 \in \mathcal{F}_1 \subseteq \mathcal{F} \ni F'$, clearly $F_1 \cap F' \neq \emptyset$; pick $x \in F_1 \cap F'$. Then $x^{-1}F_1 \subseteq F_1^{-1}F_1 \subseteq U$, and so $F_1 \subseteq xU \subseteq F'U \subseteq UF'U \subseteq F$. We can conclude that $F \in \mathcal{F}_1$.

(b)⇒(c) By Lemma 7.2.2(d), $\mathcal{F} \cdot \mathcal{V}_G(e_G)$ and $\mathcal{V}_G(e_G) \cdot \mathcal{F}$ are Cauchy filters on G contained in $\mathcal{F}$.

(c)⇒(a) Our hypothesis implies $\mathcal{F} = \mathcal{V}_G(e_G) \cdot \mathcal{F} \cdot \mathcal{V}_G(e_G)$. To prove that $\mathcal{F}$ is minimal, pick an $F \in \mathcal{F}$ and using this equality find $U \in \mathcal{V}_G(e_G)$ and $E \in \mathcal{F}$ such that $UEU \subseteq F$. □

Lemma 7.2.7. *Let* G *be a topological subgroup of a Hausdorff group* H. *Let* $\mathcal{F}$ *be a Cauchy filter on* H *such that the restriction* $\mathcal{F}\restriction_G := \{G \cap U : U \in \mathcal{F}\}$ *is a filter base on* G. *Then* $\mathcal{F}\restriction_G$ *is a Cauchy filter on* G.
(a) *If* $\mathcal{F}$ *is minimal, then* $\mathcal{F}\restriction_G$ *is minimal.*
(b) *If* G *is dense in* H *and* $\mathcal{F}$ *is open, then* $\mathcal{F}\restriction_G$ *is open.*

Proof. To verify that $\mathcal{F}\restriction_G$ is Cauchy, let $V \in \mathcal{V}_G(e_G)$; so $V = U \cap G$ for some $U \in \mathcal{V}_H(e_H)$. Let $B \in \mathcal{F}$ such that $BB^{-1} \cup B^{-1}B \subseteq U$. Then $C = B \cap G \in \mathcal{F}\restriction_G$ and $CC^{-1} \cup C^{-1}C \subseteq V$.

(a) This is straightforward as above.

(b) Clearly, if $\mathcal{B}$ is a base of $\mathcal{F}$ consisting of open sets of H, then $\{B \cap G : B \in \mathcal{B}\}$ is a base of $\mathcal{F}\restriction_G$ consisting of open sets of G. □

Lemma 7.2.8. *Let* G *be a topological group. If* $\mathcal{F}, \mathcal{H}$ *are minimal Cauchy filters on* G, *then also* $\mathcal{F} \cdot \mathcal{H}$ *and* $\mathcal{F}^{-1}$ *are minimal Cauchy filters on* G.

Proof. We already noticed that $\mathcal{F} \cdot \mathcal{H}$ and $\mathcal{F}^{-1}$ are Cauchy filters in Lemma 7.2.2. The proof that they are minimal is straightforward. □

Lemma 7.2.9. *If* G *is a topological subgroup of a Hausdorff group* H *and* $h \in \overline{G}^H$, *then* $\mathcal{F} = \mathcal{V}_H(h)\restriction_G$ *is a minimal Cauchy filter on* G. *Consequently, a Hausdorff group* G *is complete if and only if every minimal Cauchy filter on* G *converges.*

Proof. The first assertion follows from Lemma 7.2.7 applied to $\mathcal{F} = \mathcal{V}_H(h)$.

To prove the second assertion, in view of Proposition 7.2.3 we only need to check that if every minimal Cauchy filter on G converges then G is complete. To this end, argue by contradiction and let G be a proper (dense) subgroup of its completion $\widetilde{G}$. By the first assertion, any element $h \in \widetilde{G} \setminus G$ gives rise to a minimal Cauchy filter $\mathcal{V}_{\widetilde{G}}(h)\restriction_G$ on G, which is not convergent in G. □

It follows from Lemma 7.2.9 that the Raĭkov completion $\widetilde{G}$ of a Hausdorff group G can be built also by using the minimal Cauchy filters on G. More precisely, one imposes on the extension $\widetilde{G}$ of G these two conditions: G is dense in $\widetilde{G}$ and every minimal Cauchy filter on $\widetilde{G}$ converges in $\widetilde{G}$ (i. e., $\widetilde{G}$ is complete, according to Lemma 7.2.9).

Theorem 7.2.10. *Let (G, τ) be a Hausdorff group, let*

$$\overline{G} = \{\mathcal{F} : \mathcal{F} \text{ minimal Cauchy filter on } G\} \quad \text{and} \quad \iota: G \to \overline{G},\ x \mapsto \mathcal{V}_G(x).$$

Then $\overline{G}$ with the binary operation $(\mathcal{F}, \mathcal{H}) \mapsto \mathcal{F} \cdot \mathcal{H}$ is a group that admits a complete Hausdorff group topology $\overline{\tau}$ such that $\iota(G)$ is a dense subgroup of $\overline{G}$ topologically isomorphic to G. Therefore, $(\overline{G}, \iota)$ is (topologically isomorphic to) the Raĭkov completion of G.

Proof. Due to Lemma 7.2.8, $\overline{G}$ is a group with binary operation $(\mathcal{F}, \mathcal{H}) \mapsto \mathcal{F} \cdot \mathcal{H}$. Moreover, $\iota(G)$ is a subgroup of $\overline{G}$ and ι is injective.

For every open set V of G, let $\widetilde{V} = \{\mathcal{F} \in \overline{G} : V \in \mathcal{F}\}$. Then

$$\mathcal{B} = \{\widetilde{V} : V \in \mathcal{V}_G(e_G),\ V \text{ open}\}$$

satisfies (gt1), (gt2), and (gt3), so it is a filter of neighborhoods of $e_{\overline{G}}$ in a group topology $\overline{\tau}$ on $\overline{G}$. Let us see that $\mathcal{B}$ satisfies also (gt4). To this end, let $\widetilde{U} \in \mathcal{B}$ and $\mathcal{F} \in \widetilde{U}$. Since $U \in \mathcal{F}$ and $\mathcal{F}$ is minimal, there exist an open $V \in \mathcal{V}_G(e_G)$ and $F \in \mathcal{F}$ such that $VFV \subseteq U$. If $\mathcal{G} \in \widetilde{V}$, then $\mathcal{G} \cdot \mathcal{F} \in \overline{G}$, in view of Lemma 7.2.8. Moreover, $\mathcal{G} \cdot \mathcal{F} \in \widetilde{U}$, since $VF \in \mathcal{G} \cdot \mathcal{F}$ and $VF \subseteq VFV \subseteq U$. This means that $\widetilde{V} \cdot \mathcal{F} \subseteq \widetilde{U}$. Then $\mathcal{B}$ satisfies (gt4), so $\mathcal{B} \subseteq \overline{\tau}$ by Remark 2.1.14. In other words, $\{\widetilde{V} : V \in \tau\}$ is a base of $\overline{\tau}$. (Observe that for $x \in G$ and $V \in \mathcal{V}_G(e_G)$ open, $\widetilde{xV} = \mathcal{V}_G(x) \cdot \widetilde{V}$ holds.)

To see that $\overline{\tau}$ induces on $\iota(G)$ the topology $\iota(\tau)$, let V be an open set of G; we verify that

$$\widetilde{V} \cap \iota(G) = \iota(V). \tag{7.1}$$

If $x \in V$, then $V \in \mathcal{V}_G(x)$, so $\mathcal{V}_G(x) \in \widetilde{V}$, namely, $\iota(x) \in \widetilde{V} \cap \iota(G)$. Vice versa, let $\iota(x) = \mathcal{V}_G(x) \in \widetilde{V}$; then $x \in V$.

By the definition of $\overline{\tau}$ and (7.1), every nonempty $\overline{\tau}$-open set of $\overline{G}$ hits $\iota(G)$, therefore $\iota(G)$ is dense in $\overline{G}$.

To see that $(\overline{G}, \overline{\tau})$ is Hausdorff, let $\mathcal{F} \in \overline{G}$ with $\mathcal{F} \neq e_{\overline{G}} = \mathcal{V}_G(e_G)$. Since $\mathcal{F}$ and $\mathcal{V}_G(e_G)$ are minimal Cauchy filters, this implies that $\mathcal{V}_G(e_G) \nsubseteq \mathcal{F}$ by Proposition 7.2.6(b). Then there exists an open $W \in \mathcal{V}_G(e_G)$ such that $W \notin \mathcal{F}$. Therefore, $\mathcal{F} \notin \widetilde{W} \in \mathcal{V}_{\overline{G}}(e_{\overline{G}})$. This proves that $\overline{G}$ is Hausdorff.

Let us see that for a minimal Cauchy filter $\mathcal{F}$ on G, the filter $\mathfrak{G}$ generated by the filter base $\iota(\mathcal{F})$ on $\overline{G}$ converges to $\mathcal{F} \in \overline{G}$. We have to verify that

$$\mathcal{V}_{\overline{G}}(\mathcal{F}) \subseteq \mathfrak{G}. \tag{7.2}$$

A basic member of $\mathcal{V}_{\overline{G}}(\mathcal{F})$ is a $\overline{\tau}$-open set of the form $\widetilde{V} \ni \mathcal{F}$, where V is an open set of G. This means that $V \in \mathcal{F}$. Since $\widetilde{V} \supseteq \iota(V)$ by (7.1), and $\iota(V) \in \iota(\mathcal{F})$ as $V \in \mathcal{F}$, we conclude that $\widetilde{V} \in \mathfrak{G}$. This proves the inclusion (7.2), which means that $\mathfrak{G}$ converges to $\mathcal{F}$.

To show that $(\overline{G}, \overline{\tau})$ is complete, take a minimal Cauchy filter $\mathfrak{F}$ on $\overline{G}$. By Lemma 7.2.7, $\mathfrak{F}\restriction_{\iota(G)}$ is a minimal Cauchy filter on $\iota(G)$, hence of the form $\iota(\mathcal{F})$ for a suitable minimal Cauchy filter $\mathcal{F}$ on G. Since $\mathfrak{F}$ is a minimal Cauchy filter, it has a base $\mathcal{C}$ consisting of open sets. So, the filter $\mathfrak{G}$ generated by $\iota(\mathcal{F})$ is the filter generated by the sets $\{\mathbf{U} \cap \iota(G) : \mathbf{U} \in \mathcal{C}\}$. Denote by $\overline{\mathfrak{G}}$ the filter generated by the closures of the elements of $\mathfrak{G}$. Then we obtain that $\overline{\mathfrak{G}}$ is the filter generated by the sets $\{\overline{\mathbf{U} \cap \iota(G)} : \mathbf{U} \in \mathcal{C}\} = \{\overline{\mathbf{U}} : \mathbf{U} \in \mathcal{C}\} \subseteq \mathfrak{F}$ (where we apply Lemma B.1.19, since $\iota(G)$ is dense in $\overline{G}$). This implies $\overline{\mathfrak{G}} \subseteq \mathfrak{F}$.

By (7.2), $\mathfrak{G}$ contains $\mathcal{V}_{\overline{G}}(\mathcal{F})$. Since $\overline{\tau}$ is regular, the filter $\mathcal{V}_{\overline{G}}(\mathcal{F})$ has a base of $\overline{\tau}$-closed sets of $\overline{G}$; hence, also the filter $\overline{\mathcal{V}_{\overline{G}}}(\mathcal{F})$, generated by the closures of the members of $\mathcal{V}_{\overline{G}}(\mathcal{F})$, equals $\mathcal{V}_{\overline{G}}(\mathcal{F})$. Combining the results, we obtain $\mathcal{V}_{\overline{G}}(\mathcal{F}) = \overline{\mathcal{V}_{\overline{G}}}(\mathcal{F}) \subseteq \overline{\mathfrak{G}} \subseteq \mathfrak{F}$. This shows that $\mathfrak{F}$ converges to $\mathcal{F}$.

In view of Lemma 7.2.9, we conclude that $(\overline{G}, \overline{\tau})$ is complete. Hence, $\overline{G}$ (with ι) is the completion of G, by Theorem 7.1.18. □

7.2.3 Completeness of the linearly topologized groups

Proposition 7.2.11. *Let G be a Hausdorff linearly topologized group and let $\{N_i : i \in I\}$ be a base of $\mathcal{V}_G(e_G)$ consisting of open normal subgroups of G. Then the completion $\widetilde{G}$ of G is isomorphic to the inverse limit $\varprojlim_{i\in I} G/N_i$ of the discrete quotients G/N_i. Moreover, $\widetilde{G} \cong \varprojlim_{i\in I} G/N_i$ is compact if and only if all N_i have finite index in G.*

Proof. Since $\bigcap_{i\in I} N_i$ is trivial, there is a natural embedding of G in the product $P = \prod_{i\in I} G/N_i$ of the discrete quotients G/N_i. Clearly, P is complete, by Proposition 7.1.9. Hence, the closure $\overline{G}^P$ is complete. It is easy to realize that $\overline{G}^P$ coincides with $\varprojlim_{i\in I} G/N_i$, in view of Remark 3.3.7. The last assertions is obvious. □

Now we discuss a property stronger than completeness in the class of Hausdorff linearly topologized abelian groups.

Definition 7.2.12. A Hausdorff linearly topologized abelian group G is *linearly compact* if every collection of closed cosets of subgroups of G with the finite intersection property has nonempty intersection.

Lemma 7.2.13. *Closed subgroups and continuous homomorphic images (provided they are Hausdorff and linearly topologized) of a linearly compact abelian group G are linearly compact.*

Proof. Obviously, closed subgroups of linearly compact groups are linearly compact.

Assume that H is a Hausdorff linearly topologized abelian group and that $f : G \to H$ is a continuous surjective homomorphism. If $\mathcal{F}$ is a filter base of closed cosets of subgroups of H, then $\mathcal{F}^* = \{f^{-1}(F) : F \in \mathcal{F}\}$ is a filter base of closed cosets of subgroups of G. Since G is linearly compact, $\mathcal{F}^*$ has nonempty intersection. Therefore, $\mathcal{F}$ has a nonempty intersection as well. □

Theorem 7.2.14. *A linearly compact abelian group G is complete.*

Proof. To see that G is complete, it suffices to check that every minimal Cauchy filter $\mathcal{F}$ on G converges. From the definition of minimal Cauchy filter and the fact that G is linearly topologized, we deduce that $\mathcal{F}$ has a base consisting of cosets of open subgroups. Now the linear compactness of G implies that $\mathcal{F}$ is fixed; let $x \in \bigcap \mathcal{F}$. Since $\mathcal{F}$ has a base of open sets, this means that $\mathcal{F} \subseteq \mathcal{V}_G(x)$. Since $\mathcal{V}_G(x)$ is a minimal Cauchy filter, this yields $\mathcal{V}_G(x) = \mathcal{F}$, and so $\mathcal{F} \to x$. □

In Exercises 7.3.12 and 7.3.13 we describe the linearly compact abelian groups.

7.3 Exercises

Exercise 7.3.1. Let G be a topological group. Prove that:

(a) if G is a subgroup of a topological group L, and $\{g_\alpha\}_{\alpha\in A}$ is a net in G that converges to some element $l \in L$, then $\{g_\alpha\}_{\alpha\in A}$ is a Cauchy net;

(b) if H is another topological group, $f: G \to H$ is a continuous homomorphism, and $\{g_\alpha\}_{\alpha\in A}$ is a (respectively, left, right) Cauchy net of G, then $\{f(g_\alpha)\}_{\alpha\in A}$ is a (respectively, left, right) Cauchy net of H.

Exercise 7.3.2. Let G be an abelian group and $\{a_n\}_{n\in\mathbb{N}}$ a T-sequence of G. Prove that $(G, \tau_{\{a_n\}})$ is not metrizable.

Hint. According to Lemma 5.3.11, the subgroup H of G generated by the countable set $\{a_n : n \in \mathbb{N}\}$ is $\tau_{\{a_n\}}$-open. Therefore, it suffices to prove that $(H, \tau_{\{a_n\}})$ is not metrizable. Since $(G, \tau_{\{a_n\}})$ is complete by Example 7.1.6(b), $(H, \tau_{\{a_n\}})$ is complete as well. Since complete metrizable groups are Baire spaces by the Baire category theorem B.5.17, and since a countable topological group that is a Baire space must be discrete (see Corollary 2.1.7), we deduce that $(H, \tau_{\{a_n\}})$ is not metrizable (as it is not discrete).

Exercise 7.3.3. Prove that if a Hausdorff group G admits a Weil completion, then in G the left and right Cauchy nets coincide, so they are simply the Cauchy nets.

Exercise 7.3.4. Let X be an infinite set and let $G = (S(X), \mathcal{T}_X)$. Prove that:

(a) a net $\{f_\alpha\}_{\alpha\in A}$ in G is left Cauchy if and only if there exists a (not necessarily bijective) map $f: X \to X$ such that $f_\alpha \to f$ in X^X, and such an f must necessarily be injective;

(b) a net $\{f_\alpha\}_{\alpha\in A}$ in G is right Cauchy if and only if there exists a (not necessarily bijective) map $g: X \to X$ such that $f_\alpha^{-1} \to g$ in X^X, and such a g must necessarily be injective.

Exercise 7.3.5. Prove that:

(a) if G is a Weil complete group and H is a subgroup of G, then H is Weil complete if and only if H is closed;

(b) for a family $\{G_i : i \in I\}$, the product $\prod_{i\in I} G_i$ is Weil complete if and only if G_i is Weil complete for every $i \in I$.

Exercise 7.3.6. Prove that a filter $\mathcal{F}$ on a Hausdorff group G is Cauchy if and only if for any $U \in \mathcal{V}_G(e_G)$ there exists $g \in G$ such that $gU \in \mathcal{F} \ni Ug$.

Exercise 7.3.7. Prove that if G is a Hausdorff group without nontrivial convergent sequences, then G is sequentially complete.

Exercise 7.3.8. Let G be a discrete group. Prove that $C^*(G)$ equipped with the topology of uniform convergence is complete.

Exercise 7.3.9. Let U, V be vector spaces over a field K. Prove that the group $\mathrm{Hom}(V, U)$, equipped with the finite topology τ_{fin}, is complete.

Exercise 7.3.10. (a) Let p be a prime number. Prove that the completion of $(\mathbb{Z}, \nu_{\mathbb{Z}}^p)$ is the compact group $\mathbb{J}_p$ of p-adic integers.
(b) Prove that the completion of $(\mathbb{Z}, \nu_{\mathbb{Z}})$ is isomorphic to $\prod_{p\in\mathbb{P}} \mathbb{J}_p$.

Exercise 7.3.11. Let p be a prime number. Prove that:
(a) $\mathbb{Z}$ admits a finest group topology τ such that $p^n \to 0$ in τ (this is $\tau_{\{p^n\}}$ in the notation of §5.3);
(b)* ([240, 241]) $(\mathbb{Z}, \tau)$ is complete;
(c) conclude that τ is not metrizable.

Exercise 7.3.12. Let G be a linearly compact abelian group. Prove that:
(a) if G is discrete, then G is torsion and contains no infinite direct sums;
(b) conclude, from (a), that when G is discrete, G is isomorphic to a subgroup of $\prod_{i=1}^n \mathbb{Z}(p_i^\infty)$, where the primes $p_1, \dots, p_n$ are not necessarily distinct;
(c)* for every linearly topologized abelian group H, the projection $p: G \times H \to H$ sends closed subgroups of $G \times H$ to closed subgroups of H.

Exercise 7.3.13. Prove that:
(a) the groups G of the form described in Exercise 7.3.12(b) are precisely those in which every descending chain of subgroups stabilizes (briefly called *DCC groups*);
(b) every DCC group G is linearly compact in the discrete topology;
(c) products and inverse limits of linearly compact groups are linearly compact;
(d) a linearly topologized abelian group G is linearly compact if and only if G is complete and G/U is a DCC group for every open subgroup U of G;
(e) for a linearly topologized abelian group G the completion $\widetilde{G}$ is linearly compact if and only if G/U is a DCC group for every open subgroup U of G.

Exercise 7.3.14. Prove that a dense subgroup H of a topological abelian group G is dually embedded.

Hint. Since $\mathbb{T}$ is complete, the continuous characters of H can be extended to continuous characters of G, by Theorem 7.1.18.

8 Compactness and local compactness – a first encounter

A topological group G is locally compact if there exists a compact neighborhood of e_G in G (compare with Definition B.5.5(v)). Since a group G is (locally) compact if and only if its Hausdorff reflection $\mathfrak{h}G$ is (locally) compact, we assume without explicitly mentioning it that all locally compact groups are Hausdorff.

8.1 Examples

Obviously, $\mathbb{T} \cong \mathbb{S}$ is compact, so as an immediate consequence of the Tichonov theorem, we obtain the following generic example of a compact abelian group:

Example 8.1.1. Every power $\mathbb{T}^I$ of $\mathbb{T}$, as well as every closed subgroup of $\mathbb{T}^I$, is compact. It becomes clear in the sequel that this is the most general instance of a compact abelian group: *every compact abelian group is topologically isomorphic to a closed subgroup of a power of* $\mathbb{T}$ (see Corollary 11.5.2).

The above example helps us to produce another important one.

Example 8.1.2. For every abelian group G, the group $G^* = \operatorname{Hom}(G, \mathbb{T})$ of all characters of G is closed in the product $\mathbb{T}^G$. In fact, considering the projections $\pi_x : \mathbb{T}^G \to \mathbb{T}$ for every $x \in G$,

$$\begin{aligned} G^* &= \bigcap_{h,g\in G} \{f \in \mathbb{T}^G : f(h+g) = f(h) + f(g)\} \\ &= \bigcap_{h,g\in G} \{f \in \mathbb{T}^G : \pi_{h+g}(f) = \pi_h(f) + \pi_g(f)\} \\ &= \bigcap_{h,g\in G} \{f \in \mathbb{T}^G : (\pi_h + \pi_g - \pi_{h+g})(f) = 0\} = \bigcap_{h,g\in G} \ker(\pi_h + \pi_g - \pi_{h+g}). \end{aligned}$$

Since π_x is continuous for every $x \in G$ and $\{0\}$ is closed in $\mathbb{T}$, all kernels $\ker(\pi_h + \pi_g - \pi_{h+g})$ are closed; so, G^* is closed, too. As $\mathbb{T}^G$ is compact by Example 8.1.1, G^* (endowed with the topology inherited from $\mathbb{T}^G$) is compact, too.

It becomes clear with Theorem 13.4.7 that this example is the most general one: *every compact abelian group is topologically isomorphic to some compact abelian group of the form* G^*.

The next lemma, based on the previous two examples, contains a useful fact: the existence of a "diagonal" convergent subnet of any given net of characters.

Lemma 8.1.3. *Let G be an abelian group and $N = \{\chi_\alpha\}_{\alpha\in A}$ a net in G^*. Then there exist $\chi \in G^*$ and a subnet $S = \{\chi_{\alpha_\beta}\}_{\beta\in B}$ of N such that $\chi_{\alpha_\beta}(x) \to \chi(x)$ for every $x \in G$. If G is countable, then every sequence in G has a convergent subsequence.*

https://doi.org/10.1515/9783110654936-008

Proof. By Example 8.1.1, the group $\mathbb{T}^G$ endowed with the product topology is compact. Since G^* is a topological subgroup of $\mathbb{T}^G$, there exist $\chi \in \mathbb{T}^G$ and a subnet $S = \{\chi_{\alpha_\beta}\}_{\beta\in B}$ of N that converges to χ (see Lemma B.5.7(c)). Therefore, $\chi_{\alpha_\beta}(x) \to \chi(x)$ for every $x \in G$ and $\chi \in G^*$, because G^* is closed in $\mathbb{T}^G$ by Example 8.1.2. To prove the second assertion, it suffices to note that $\mathbb{T}^G$ is also metrizable when G is countable. □

An example of a nonabelian compact group can be obtained as a topological subgroup of the full linear group $\mathrm{GL}_n(\mathbb{C})$ considered in Example 2.1.4:

Example 8.1.4. For $n \in \mathbb{N}_+$ the set $U(n)$ of all $n \times n$ unitary matrices over $\mathbb{C}$ (a matrix is *unitary* if its inverse coincides with its conjugate transposed) is a subgroup of $\mathrm{GL}_n(\mathbb{C})$. As a subset of $\mathbb{C}^{n^2}$, $U(n)$ is closed and bounded. So, $U(n)$ is compact, by Example B.5.6. It is easy to see that $U(1) \cong \mathbb{S}$.

Clearly, $\mathbb{U} := \prod_{n\in\mathbb{N}_+} U(n)$ is compact, as well as all powers $\mathbb{U}^I$ and all closed subgroups of $\mathbb{U}^I$. It is a remarkable fact that *every compact group is isomorphic to a closed subgroup of a power of* $\mathbb{U}$ (see Corollary 10.3.4).

Here we collect examples of locally compact groups.

Example 8.1.5. (a) Obviously, every discrete group is locally compact. Vice versa, every countable locally compact group G is discrete; indeed, G is of second category by Theorem B.5.20, and since G is countable, G is discrete by Corollary 2.1.7.

(b) For every $n \in \mathbb{N}$, the group $\mathbb{R}^n$ is locally compact (and not compact).

(c) A topological group with a compact open subgroup is locally compact.

(d) Since finite products preserve local compactness (see Theorem B.5.15(b)), it follows from (a) and (b) that every group of the form $\mathbb{R}^n \times G$, where G has a compact open subgroup, is necessarily locally compact. According to Theorem 14.2.18, every locally compact abelian group has this form.

(e) Let $\{G_i : i \in I\}$ be a family of topological groups and let K_i be a compact open subgroup of G_i for every $i \in I$ (so each G_i is locally compact). The *local direct product* of $\{G_i : i \in I\}$ modulo $\{K_i : i \in I\}$ is the subgroup

$$\prod_{i\in I}^{\mathrm{loc}}(G_i, K_i) := \left\{(x_i)_{i\in I} \in \prod_{i\in I} G_i : x_i \in K_i \text{ for all but finitely many } i \in I\right\}$$

of $\prod_{i\in I} G_i$ endowed with the topology with respect to which $K = \prod_{i\in I} K_i$ is open and equipped with the (compact) product topology. This makes $\prod^{\mathrm{loc}}_{i\in I}(G_i, K_i)$ a locally compact group. If all groups G_i are abelian, $\prod^{\mathrm{loc}}_{i\in I}(G_i, K_i) = K + \bigoplus_{i\in I} G_i$.

Example 8.1.6. The Hilbert space $(\ell_2, \|-\|)$ of square summable real sequences is not locally compact. Indeed, the closed unit disk is not compact: since ℓ_2 is metrizable, it is enough to observe that the sequence $\{e_n\}_{n\in\mathbb{N}}$ of the vectors of the canonical base has no Cauchy subsequences (so no convergent subsequences), as $\|e_n - e_m\| = \sqrt{2}$ for all $n \neq m$ in $\mathbb{N}$. A similar argument shows that no neighborhood of 0 is compact. More generally, every locally compact real or complex normed space is finite-dimensional.

8.2 Specific properties of compactness and local compactness

Lemma 8.2.1. *Let G be a topological group and C, K closed sets of G.*
(a) *If K is compact, then both CK and KC are closed.*
(b) *If both C and K are compact, then CK and KC are compact.*
(c) *If K is compact and $K \subseteq U$ for an open set U of G, then there exists an open neighborhood V of e_G such that $KV \subseteq U$.*

Proof. (a) Let $\{x_\alpha\}_{\alpha\in A}$ be a net in CK such that $x_\alpha \to x_0 \in G$. One has to show that $x_0 \in CK$. For every $\alpha \in A$, there exist $y_\alpha \in C$ and $z_\alpha \in K$ such that $x_\alpha = y_\alpha z_\alpha$. Since K is compact, there exist $z_0 \in K$ and a subnet $\{z_{\alpha_\beta}\}_{\beta\in B}$ such that $z_{\alpha_\beta} \to z_0$ (see Lemma B.5.7(c)). Thus, $\{(x_{\alpha_\beta}, z_{\alpha_\beta})\}_{\beta\in B}$ is a net in $G \times G$ which converges to (x_0, z_0). For every $\beta \in B$, let $y_{\alpha_\beta} = x_{\alpha_\beta} z_{\alpha_\beta}^{-1}$; then $y_{\alpha_\beta} \to x_0 z_0^{-1}$, because the map $G \times G \to G$, $(x, y) \mapsto xy^{-1}$, is continuous. Since $y_{\alpha_\beta} \in C$ for every $\beta \in B$ and C is closed, $x_0 z_0^{-1} \in C$, and so $x_0 = (x_0 z_0^{-1}) z_0 \in CK$. Analogously, KC is closed.

(b) The product $C \times K$ is compact by the Tichonov theorem and the continuous map $\mu: G \times G \to G$, $(x, y) \mapsto xy$, sends $C \times K$ onto CK. Thus, CK is compact.

(c) Let $C = G \setminus U$. Then C is a closed set of G with $C \cap K = \emptyset$. Since the map $\iota: G \to G$, $x \mapsto x^{-1}$, is continuous, K^{-1} is compact. Moreover, $e_G \notin K^{-1}C$ as $C \cap K = \emptyset$. By (a), $K^{-1}C$ is closed, so there exists a symmetric neighborhood V of e_G that misses $K^{-1}C$. Then $KV \cap C = \emptyset$, and consequently $KV \subseteq U$. □

Compactness of K cannot be omitted in Lemma 8.2.1(a). Indeed, $K = \mathbb{Z}$ and $C = \langle \sqrt{2} \rangle$ are closed subgroups of $\mathbb{R}$, but the subgroup $K + C$ of $\mathbb{R}$ is dense (see Proposition 3.1.11 or Proposition 9.4.6).

Now we see that the open canonical projection $q: G \to G/K$ from a topological group G onto its quotient G/K over a closed normal subgroup K of G is also closed in case K is compact.

Lemma 8.2.2. *Let G be a topological group and K a compact normal subgroup of G. Then the canonical projection $q: G \to G/K$ is closed.*

Proof. Let C be a closed set of G. As $q^{-1}(q(C)) = CK$ is closed by Lemma 8.2.1(a), we may conclude that $q(C)$ is closed. □

This lemma says that the canonical projection $q: G \to G/K$ is a perfect map when K is a compact normal subgroup of G. (Recall that a map $f: X \to Y$ between topological spaces is *perfect* if f is closed and $f^{-1}(y)$ is compact for all $y \in Y$.)

Lemma 8.2.3. *Let H be a closed normal subgroup of a topological group G.*
(a) *If G is compact, then G/H is compact.*
(b) *If H and G/H are compact, then G is compact.*

Proof. (a) is obvious, since the canonical projection $q: G \to G/H$ is continuous.

(b) Let $\mathcal{F} = \{F_\alpha : \alpha \in A\}$ be a family of closed sets of G with the finite intersection property. Then $q(\mathcal{F})$ is a family of closed sets of G/H with the finite intersection property, by Lemma 8.2.2. By the compactness of G/H, $\bigcap_{\alpha\in A} q(F_\alpha) \neq \emptyset$ (see Lemma B.5.7(a)), so there exists $x \in G$ such that $q(x) \in q(F_\alpha)$ for every $\alpha \in A$. Then $F_\alpha^* = F_\alpha \cap xH \neq \emptyset$ for every $\alpha \in A$. It follows that the family $\{F_\alpha^* : \alpha \in A\}$ of closed sets of the compact set xH has the finite intersection property (again by Lemma B.5.7(a)). Thus, $\bigcap_{\alpha\in A} F_\alpha^* \neq \emptyset$, and so also $\bigcap_{\alpha\in A} F_\alpha \neq \emptyset$. □

Another proof of Lemma 8.2.3(b) is provided by Lemma 8.2.2, as inverse images of compact sets under a perfect map are compact, hence $G = q^{-1}(G/H)$ is compact whenever H and G/H are compact.

Remark 8.2.4. Every closed subgroup of a locally compact group is locally compact. The counterpart regarding products is more delicate.
(a) Finite products of locally compact groups are locally compact.
(b) An infinite product of locally compact groups is locally compact if and only if all but finitely many of them are compact. In particular, $G^{\mathbb{N}}$ is locally compact if and only if G is compact (so $\mathbb{R}^{\mathbb{N}}$ is not locally compact).

Lemma 8.2.5. *Let G be a locally compact group, H a closed normal subgroup of G, and $q: G \to G/H$ the canonical projection. Then:*
(a) *G/H is locally compact, too;*
(b) *if $C \subseteq G/H$ is compact, there exists $K \subseteq G$ compact with $q(K) = C$.*

Proof. Let U be an open neighborhood of e_G in G with compact closure.

(a) Consider the open neighborhood $q(U)$ of $e_{G/H}$ in G/H. By the continuity of q, $q(\overline{U}) \subseteq \overline{q(U)}$ and $q(\overline{U})$ is compact in G/H. Since G/H is Hausdorff by Lemma 3.2.10(b), $q(\overline{U})$ is closed and contains $q(U)$. So, $\overline{q(U)} = q(\overline{U})$ is compact.

(b) Since q is open, $\{q(sU) : s \in G\}$ is an open cover of G/H. Since C is compact, there exists a finite subcover $\{q(s_iU) : i \in \{1, \dots, m\}\}$ of C. The compact set $K = (s_1\overline{U} \cup \dots \cup s_m\overline{U}) \cap q^{-1}(C) \subseteq G$ obviously satisfies $q(K) = C$. □

Proposition 8.2.6. *A locally compact group G is Weil complete.*

Proof. Let U be a neighborhood of e_G in G with compact closure and let $\{g_\alpha\}_{\alpha\in A}$ be a left Cauchy net of G. Then there exists $\alpha_0 \in A$ such that $g_\alpha^{-1} g_\beta \in U$ for every $\alpha, \beta \geq \alpha_0$. In particular, $g_\beta \in g_{\alpha_0} U$ for every $\beta \geq \alpha_0$. By the compactness of $g_{\alpha_0}\overline{U}$, there exists a convergent subnet $\{g_{\alpha_\beta}\}_{\beta\in B}$ (for some cofinal $B \subseteq A$) such that $g_{\alpha_\beta} \to g \in G$. Then also $g_\alpha \to g$, by Lemma 7.1.26. □

Consequently, every locally compact abelian group is (Raĭkov) complete.

Proposition 8.2.7. *The character and the pseudocharacter of a locally compact group G coincide.*

Proof. Clearly, $\psi(G) \leq \chi(G)$.

Let U be an open neighborhood of e_G such that $\overline{U}$ is compact. To prove that $\chi(G) \leq \psi(G)$, pick a family $\mathcal{B} = \{V_i : i \in I\}$ of neighborhoods of e_G with $\bigcap_{i\in I} V_i = \{e_G\}$ and $|I| = \psi(G)$ (in case G is discrete $|I| = 1$, otherwise I is infinite). Since G is a regular space, we can assume that $\overline{V_i} \subseteq U$ for every $i \in I$ and actually $\bigcap_{i\in I} \overline{V_i} = \{e_G\}$. Now pick an arbitrary open neighborhood W of e_G contained in U. Then $\bigcap_{i\in I} \overline{V_i} = \{e_G\} \subseteq W$ and $Y := \overline{U} \setminus W$ is compact. According to Lemma B.5.9, there exists a finite subset J of I such that $\bigcap_{i\in J} \overline{V_i} \cap Y = \emptyset$, that is, $\bigcap_{i\in J} \overline{V_i} \subseteq W$. This shows that the family $\mathcal{B}_1$ of all finite intersections of members of $\mathcal{B}$ forms a local base at e_G in G. Since $|\mathcal{B}_1| = |\mathcal{B}|$, we conclude that $\chi(G) \leq |\mathcal{B}_1| = |\mathcal{B}| = |I| = \psi(G)$. □

For compact groups we can say even more.

Corollary 8.2.8. *Any infinite compact group G satisfies $\psi(G) = \chi(G) = w(G)$.*

Proof. We are going to prove that $d(G) \leq \chi(G)$. Let $\mathcal{B}$ be a local base at e_G of cardinality $\chi(G)$ consisting of symmetric sets. For every $U \in \mathcal{B}$, choose a finite subset F_U of G such that $G = F_U U$. Then $\langle F_U : U \in \mathcal{B}\rangle$ is dense in G. Indeed, let O be a nonempty open set of G and let $x \in O$. Choose $U \in \mathcal{B}$ such that $U \subseteq x^{-1}O$ and $y_U \in F_U$ such that $x \in y_U U$. Then $y_U \in xU \subseteq O$. Since $|\langle F_U \colon U \in \mathcal{B}\rangle| = |\mathcal{B}| = \chi(G)$, we have shown that $d(G) \leq \chi(G)$.

Combining Proposition 8.2.7 with Lemma 5.1.7 and Remark 5.1.6(b) yields

$$\psi(G) = \chi(G) \leq w(G) = \chi(G) \cdot d(G) \leq \chi(G).$$

This implies $\psi(G) = \chi(G) = w(G)$. □

8.3 Compactly generated locally compact groups

Now we introduce a special class of σ-compact groups that plays an essential role in determining the structure of locally compact abelian groups.

Definition 8.3.1. A topological group G is *compactly generated* if there exists a compact subset K of G which generates G, that is, $G = \langle K\rangle = \bigcup_{n\in\mathbb{N}_+} (K \cup K^{-1})^n$.

Lemma 8.3.2. *If G is a compactly generated group, then G is σ-compact.*

Proof. There exists a compact subset K of G such that $G = \bigcup_{n\in\mathbb{N}_+} (K \cup K^{-1})^n$. Since K is compact, $(K \cup K^{-1})^n$ is compact for every $n \in \mathbb{N}_+$. □

While σ-compactness is a purely topological property, being compactly generated involves essentially the algebraic structure of the group. For example, the discrete σ-compact groups are simply the countable discrete groups, while the discrete compactly generated groups are the finitely generated discrete groups.

Corollary 8.3.3. *A locally compact group is a normal space.*

Proof. By hypothesis and Lemma 8.3.2, G contains a σ-compact open subgroup N. Then N is Lindelöff by Lemma B.5.18, so a normal space by Theorem B.5.10(b). Now $G = \bigsqcup_{g\in G} gN$ is a normal space as well. □

Lemma 8.3.4. *Let G be a locally compact group.*
(a) *If K is a compact subset of G and U is an open set of G such that $K \subseteq U$, then there exists an open neighborhood V of e_G in G such that $(KV)\cup(VK) \subseteq U$ and $\overline{(KV)\cup(VK)}$ is compact.*
(b) *If G is compactly generated, then there exists an open neighborhood V of e_G in G such that $\overline{V}$ is compact and V generates G.*

Proof. (a) By Lemma 8.2.1(c), there exists an open neighborhood V of e_G in G such that $(KV) \cup (VK) \subseteq U$. Since G is locally compact, one can choose V with compact closure. Thus, $K\overline{V}$ is compact by Lemma 8.2.1(b). Since $KV \subseteq K\overline{V}$, one obtains $\overline{KV} \subseteq K\overline{V}$, and so $\overline{KV}$ is compact. Analogously, $\overline{VK}$ is compact, so $\overline{(KV)\cup(VK)} = \overline{KV} \cup \overline{VK}$ is compact.

(b) Let K be a compact subset of G such that K generates G. So, $K \cup \{e_G\}$ is compact and, by (a) applied with $U = G$, there exists an open neighborhood V of e_G in G such that $K \cup \{e_G\} \subseteq V$, $\overline{V}$ is compact, and V generates G since $K \subseteq V$. □

For first countable topological groups Fujita and Shakmatov [143] described the precise relationship between σ-compactness and the stronger property of being compactly generated:

Theorem 8.3.5 ([143]). *A metrizable group G is compactly generated if and only if G is σ-compact and, for every open subgroup H of G, there exists a finite subset F of G such that $F \cup H$ algebraically generates G.*

Corollary 8.3.6. *A σ-compact metrizable group G is compactly generated in each of the following cases:*
(a) *G has no proper open subgroups;*
(b) *$\widetilde{G}$ is connected;*
(c) *G is totally bounded (see Definition 10.2.1).*

Theorem 8.3.7 ([143]). *A countable metrizable group is compactly generated if and only if it is algebraically generated by a null sequence (possibly eventually constant).*

Examples showing that the various conditions above cannot be omitted can be found in [143]. The question of when a topological group contains a compactly generated dense subgroup is considered in [142].

8.4 The open mapping theorem

Theorem 8.4.1 (Open mapping theorem). *Let G, H be locally compact groups and $f: G \to H$ a continuous homomorphism. If G is σ-compact and f is surjective, then f is open.*

Proof. Let U be a neighborhood of e_G in G. There exists a symmetric open neighborhood V of e_G in G such that $\overline{V}\,\overline{V} \subseteq U$ and $\overline{V}$ is compact. Since $G = \bigcup_{x \in G} xV$ and G is Lindelöff by Lemma B.5.18, there exists $\{x_n : n \in \mathbb{N}\} \subseteq G$ such that $G = \bigcup_{n \in \mathbb{N}} x_n V$. Therefore, $H = \bigcup_{n \in \mathbb{N}} f(x_n \overline{V})$, because f is surjective. Put $y_n = f(x_n)$ for every $n \in \mathbb{N}$; hence $H = \bigcup_{n \in \mathbb{N}} y_n f(\overline{V})$, where $f(\overline{V})$ is compact and so closed in H. Since H is locally compact, Theorem B.5.20 yields that there exists $n \in \mathbb{N}$ such that $\operatorname{Int}(y_n f(\overline{V}))$ is not empty, so there exists a nonempty open set W of H such that $W \subseteq f(\overline{V})$. If $w \in W$, then $w \in f(\overline{V})$, and so $w = f(v)$ for some $v \in \overline{V} = (\overline{V})^{-1}$. Hence,

$$e_G \in w^{-1}W \subseteq w^{-1}f(\overline{V}) = f(v^{-1})f(\overline{V}) \subseteq f(\overline{V}\,\overline{V}) \subseteq f(U),$$

and this implies that $f(U)$ is a neighborhood of e_G in H. By Lemma 2.1.20, this proves that f is open. □

The following immediate corollary is frequently used.

Corollary 8.4.2. *If G, H are Hausdorff groups, $f: G \to H$ is a continuous surjective homomorphism, and G is compact, then f is open.*

The topological groups for which the open mapping theorem holds are known also under the name *totally minimal*. Compact groups are totally minimal, by Corollary 8.4.2. More precisely, one has the following pair of concepts.

Definition 8.4.3. A Hausdorff group G is:
(i) *totally minimal* if every continuous surjective homomorphism of G onto a Hausdorff group H is open;
(ii) *minimal* if every continuous isomorphism of G to a Hausdorff group H is open.

Remark 8.4.4. Clearly, totally minimal groups are minimal, and a Hausdorff group is totally minimal if and only if all its Hausdorff quotients are minimal.

Here we provide only a couple of examples (compare with Exercise 8.7.9 and §8.8).

Example 8.4.5 ([93]). The question when an infinite abelian group G may carry a minimal group topology has been studied thoroughly. It is known that none of the groups $\mathbb{Q}^n$, $\mathbb{Z}(p^\infty)$, $\mathbb{Z}(p^\infty)^n$, $\mathbb{Z}(p_1^\infty) \oplus \cdots \oplus \mathbb{Z}(p_n^\infty)$, where $n \in \mathbb{N}_+$ and $p, p_1, \ldots, p_n$ are primes, carries a minimal group topology, while the group $\mathbb{Q}_{(p)} := \{a/b \in \mathbb{Q} : (b, p) = 1\}$ (i. e., the additive group of the localization of the ring $\mathbb{Z}$ at the prime ideal (p)) admits a unique minimal group topology.

Example 8.4.6. In view of Corollary 4.2.14 and Exercise 8.7.7, the groups $(S(X), \mathcal{T}_X)$, where X is an infinite set, are totally minimal.

8.5 Compactness vs connectedness

As already observed in Remark 6.1.6(b), a linear Hausdorff group is zero-dimensional, so hereditarily disconnected. Linearity and hereditary disconnectedness coincide for compact groups and for locally compact abelian groups:

Theorem 8.5.1 (van Dantzig theorem [276]). *Every hereditarily disconnected locally compact group G has a local base at e_G consisting of compact open subgroups. Moreover, a hereditarily disconnected locally compact group that is either abelian or compact has a linear topology.*

This can be derived from the following more precise result.

Theorem 8.5.2. *Let G be a locally compact group. Then:*
(a) *if G is hereditarily disconnected, every neighborhood of e_G contains a compact open subgroup of G;*
(b) *$c(G)$ coincides with the intersection of all open subgroups of G.*

If G is compact, then the open subgroups in items (a) and (b) can be chosen normal.

Proof. (a) By Vedenissov theorem B.6.10, there is a neighborhood base $\mathcal{O}$ at e_G consisting of compact symmetric clopen sets. Let $U \in \mathcal{O}$. Then, by Lemma 3.1.1(a),

$$U = \overline{U} = \bigcap\{UV : V \in \mathcal{O}, V \subseteq U\},$$

where every set UV is compact by Lemma 8.2.1(b), hence closed. Since U is open, $\{UV \setminus U : V \in \mathcal{O}, V \subseteq U\}$ is a family of closed compact subsets with empty intersection contained in the compact set UU. So Lemma B.5.9 implies that there exist $V_1, \dots, V_n \in \mathcal{O}$ such that $\bigcap_{k=1}^n UV_k \subseteq U$, so $U = \bigcap_{k=1}^n UV_k$. Then, for

$$V = U \cap \left(\bigcap_{k=1}^n V_k\right) \subseteq U,$$

one has $UV = U$. In particular, $V^n \subseteq U$ for every $n \in \mathbb{N}_+$. So, since V is symmetric, the subgroup $H = \langle V \rangle$ is contained in U as well. From $V \subseteq H$, one can deduce that H is open, so closed by Proposition 3.1.7(b), and hence compact as $H \subseteq U$ and U is compact.

In case G is compact, the normal core $H_G := \bigcap_{x \in G} x^{-1}Hx$ of H in G is a closed normal subgroup of G. As the number of distinct conjugates $x^{-1}Hx$ of H in G is finite (being equal to $[G : N_G(H)] \leq [G : H] < \infty$), H_G is an open normal subgroup of G contained in H, so also in U.

(b) Letting $C = c(G)$, the quotient G/C is hereditarily disconnected by Lemma 6.2.3, hence $\{e_{G/C}\}$ is the intersection of all open (respectively, open normal, in case G is compact) subgroups of G/C by item (a). The intersection of the inverse images of these subgroups with respect to the canonical projection $G \to G/C$ coincides with C. □

According to Example 6.3.1, none of the items (a) and (b) of Theorem 8.5.2 remains true without the hypothesis "locally compact". The next example shows that the last assertion of Theorem 8.5.2 (as well as that of van Dantzig theorem 8.5.1) fails too, if the group is not compact or abelian.

Example 8.5.3. Let p be a prime and $G = \mathbb{Q}_p \rtimes \mathbb{Z}$ such that $\mathbb{Z} \cong \{p^n : n \in \mathbb{Z}\}$ acts on $\mathbb{Q}_p$ by multiplication by p, and the subgroup $O = \mathbb{Q}_p \rtimes \{1\}$ of G is taken to be open carrying its natural p-adic topology. Then G is a hereditarily disconnected locally compact group. It has as a local base of neighborhoods at e_G the family of compact open subgroups $U_n = p^n \mathbb{J}_p \rtimes \{1\}$, with $n \in \mathbb{N}$.

Nevertheless, the only compact open normal subgroups of G are those containing $O = \mathbb{Q}_p \rtimes \{1\}$. Indeed, if V is a compact open subgroup of G, there exists $n \in \mathbb{N}$ such that $U_n \subseteq V$. Since the normal closure of U_n (i. e., the smallest normal subgroup of G containing U_n) is O, we deduce that $V = O$. This contradicts the fact that $\mathbb{Q}_p$ is not compact.

Corollary 8.5.4. *Let G be a locally compact group. Then $o(G) = Q(G) = c(G)$. So, G is hereditarily disconnected if and only if it is totally disconnected.*

Proof. It is always true that $c(G) \subseteq Q(G) \subseteq o(G)$. By Theorem 8.5.2(b), $c(G)$ is the intersection of open subgroups, so $c(G) \supseteq o(G)$. □

Remark 8.5.5. If G is a hereditarily disconnected compact group, then G has a local base $\{N_i : i \in I\}$ at e_G consisting of open normal subgroups, by Theorem 8.5.2(a). Then $G \cong \varprojlim G/N_i$, by Proposition 7.2.11 (with the inverse system $(G/N_i, \nu_{ji}, I)$ defined as there).

Definition 8.5.6. The topological groups that are inverse limits of finite groups (respectively, finite p-groups for a prime p) are named *profinite groups* (respectively, *pro-p-groups* or *pro-p-finite groups*).

The next equivalence follows from Remarks 8.5.5 and 3.3.7.

Corollary 8.5.7. *A compact group is hereditarily disconnected if and only if it is profinite.*

Remark 8.5.8. The (compact) topology of a profinite group is coarser than its profinite topology, and in general they need not coincide (see Exercise 8.7.10).

For every residually finite group G (i. e., the profinite topology ϖ_G on G is Hausdorff), the completion of (G, ϖ_G) is a profinite group, by Proposition 7.2.11.

In general, total disconnectedness is not preserved under taking quotients. The first example to this effect can be found in [7, Chapter 3, p. 21, Exercise 211], where $\mathbb{R}$ is presented as a quotient of a totally disconnected group. Later Kaplan [181] showed that $\mathbb{R}$ is also a quotient of a zero-dimensional metrizable group. Arhangel′skii [6] pushed this further by proving that every second countable topological group is a quotient of a second countable zero-dimensional group (see also [79]).

Corollary 8.5.9. *A quotient of a hereditarily disconnected locally compact group G is hereditarily disconnected.*

Proof. Let N be a closed normal subgroup of G. It follows from Theorem 8.5.2(a) that G has a local base at e_G formed by compact open subgroups. This yields that the quotient G/N has the same property. In particular, G/N is hereditarily disconnected, too. □

Corollary 8.5.10. *Let G, H be locally compact groups and $f: G \to H$ a continuous surjective homomorphism. If G is σ-compact, then:*
(a) *$f(c(G)) = c(H)$, provided $c(G)$ is compact;*
(b) *if G is hereditary disconnected, H is hereditary disconnected, too.*

Proof. (a) Since $c(G)$ is a compact normal subgroup of G, $f(c(G))$ is a compact (so, closed) normal subgroup of H, as f is surjective. The group $G/c(G)$ is hereditarily disconnected by Corollary 8.5.9 and f induces a continuous surjective homomorphism $\bar{f}: G/c(G) \to H/f(c(G))$. Since $G/c(G)$ is σ-compact, the open mapping theorem (Theorem 8.4.1) implies that $\bar{f}$ is open. Hence, by Corollary 8.5.9, $H/f(c(G))$ is hereditarily disconnected. Since $f(c(G)) \subseteq c(H)$ and $c(H)/f(c(G))$ is a connected subgroup of the hereditarily disconnected group $H/f(c(G))$, this implies $c(H) = f(c(G))$.

Item (b) follows from (a). □

Remark 8.5.11. (a) Item (a) of the above corollary fails when $c(G)$ is not compact, even when $H = c(H)$ is compact (see Example 14.3.12).
(b) If G, H are locally compact abelian groups and $f: G \to H$ a continuous surjective homomorphism, then $f(a(G)) = a(H)$ (see [177, Theorem 8.30(iv)]).

8.6 The Bohr compactification

Theorem 8.6.1. *For every topological group G, there exist a compact group bG and a continuous homomorphism $b_G: G \to bG$ with $\overline{b_G(G)} = bG$ such that for every continuous homomorphism $f: G \to K$, where K is a compact group, there exists a (necessarily unique) continuous homomorphism $f': bG \to K$ with $f' \circ b_G = f$:*

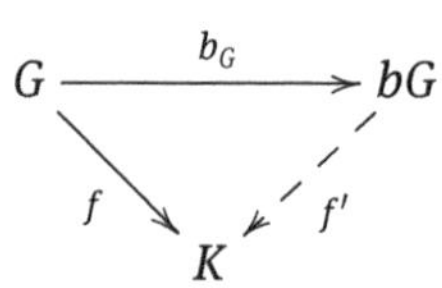

Moreover, $b_G: G \to bG$ is unique (up to isomorphism) with this property, i. e., if K is a compact group and $h: G \to K$ is a continuous homomorphism with $\overline{h(G)} = K$ and such

that for every continuous homomorphism $f: G \to C$, where C is a compact group, there exists a unique continuous homomorphism $f': K \to C$ with $f' \circ h = f$, then there exists a topological isomorphism $j: bG \to K$ such that $j \circ b_G = h$.

Proof. Let $\{N_j : j \in J\}$ be the family of all kernels of continuous homomorphisms $G \to C$ where C is a compact group. For every $j \in J$, let $q_j: G \to G/N_j$ be the canonical projection and let $\mathcal{F}_j = \{\tau_{(j,a)} : a \in A_j\}$ be the family of all group topologies on G/N_j coarser than the quotient topology of the topology on G such that the completion $K_{(j,a)}$ of $(G/N_j, \tau_{(j,a)})$ is compact. Let $I = \{(j,a) : j \in J,\ a \in A_j\}$ and, for every $i = (j,a) \in I$, let $h_i: G \to K_i$ be the composition of $q_j: G \to G/N_j$ with the inclusion $G/N_j \to K_{(j,a)} = K_i$.

Let $h: G \to \prod_{i \in I} K_i$ be the diagonal homomorphism of the family $\{h_i : i \in I\}$ and $bG := \overline{h(G)}$. For every $i \in I$, denote by $p_i: \prod_{s \in I} K_s \to K_i$ the canonical projection. Then the corestriction $b_G: G \to bG$ of h has the desired property.

In fact, let $f: G \to C$ be a continuous homomorphism where C is a compact group. Then $\ker f = N_j$ for some $j \in J$ and $f = \bar{f} \circ q_j$, where $\bar{f}: G/N_j \to C$ is a continuous injective homomorphism when G/N_j carries the quotient topology τ. Let τ' denote the initial topology induced by $\bar{f}$. Then $\tau' \le \tau$ holds. Since the completion of $(G/N_j, \tau')$ is compact (isomorphic to $\overline{\bar{f}(G/N_j)}$), one obtains $\tau' = \tau_{(j,a)}$ for some $a \in A_j$. So, $\bar{f}: (G/N_j, \tau_{(j,a)}) \to C$ is a topological embedding, and by Corollary 7.1.20 there exists a continuous extension $f_{i_0}: K_{i_0} \to C$ for $i_0 = (j,a)$ such that $f_{i_0} \circ h_{i_0} = f$. Let $f' = f_{i_0} \circ p_{i_0} \restriction_{bG}: bG \to C$. Now it is clear that, for every $x \in G$,

$$f'(b_G(x)) = f_{i_0}(p_{i_0}(b_G(x))) = f_{i_0}(p_{i_0}((h_i(x))_{i \in I})) = f_{i_0}(h_{i_0}(x)) = f(x).$$

Hence, $f' \circ b_G = f$, as required. The uniqueness of f' is a consequence of the density of $b_G(G)$ in bG.

To show the uniqueness of bG, assume that $h: G \to K$ is a continuous homomorphism with $\overline{h(G)} = K$ and such that for every continuous homomorphism $f: G \to C$, where C is a compact group, there exists a unique continuous homomorphism $f': K \to C$ with $f' \circ h = f$. In particular, there exists a continuous homomorphism $f^*: K \to bG$ such that $f^* \circ h = b_G$. Analogously, by the properties of b_G and bG, there exists a continuous homomorphism $h': bG \to K$ such that $h' \circ b_G = h$.

To prove that f^* and h' are topological isomorphisms, note that $id_K \circ h = h' \circ b_G = h' \circ f^* \circ h$. So, the homomorphisms id_K and $h' \circ f^*$ coincide on the dense subgroup $h(G)$ of K, and hence $id_K = h' \circ f^*$. Similarly, $id_{bG} = f^* \circ h'$. So, both f^* and h' are topological isomorphisms witnessing that $h: G \to K$ coincides, up to isomorphism, with $b_G: G \to bG$. □

Definition 8.6.2. For a topological group G, the compact group bG and the continuous homomorphism $b_G: G \to bG$ from Theorem 8.6.1 are called *Bohr compactification* of G. The *von Neumann kernel* of G is $n(G) := \ker b_G$.

A relevant property of the Bohr compactification (directly following from Theorem 8.6.1 and Definition 8.6.2) is that every $\chi \in \widehat{G}$ factorizes through $b_G: G \to bG$:

Corollary 8.6.3. *For a topological abelian group G, the group bG is abelian and for every character $\chi \in \widehat{G}$ there exists a character $\xi \in \widehat{bG}$ such that $\chi = \xi \circ b_G$.*

The term compactification used here substantially differs from the notion of compactification used in general topology to describe compact spaces K containing a given Tichonov space X as a dense subspace. Nevertheless, the Bohr compactification can be seen as an appropriate counterpart of the Čech–Stone compactification βX of a Tichonov space X, due to the property described in the above theorem, which says that the assignment $G \mapsto bG$ induces a functor (reflector) from the category **TopGrp** to its subcategory of all compact groups (exactly as the Čech–Stone compactification βX provides a functor from $\mathbf{Top}_{3,5}$ to **CompTop**; the analogy becomes complete if one defines the compact space βX also for non-Tichonov spaces X as the Čech–Stone compactification of the Tichonov reflection $\mathrm{T}_{3.5}X$ of X – see Remark C.2.11). From this point of view, the Bohr compactification bG of a topological group G is the compact group that *best approximates* G in the sense of Theorem 10.2.15.

According to J. von Neumann, we adopt the following terminology concerning the injectivity of the map b_G:

Definition 8.6.4. A topological group G is:
(i) *maximally almost periodic* (briefly, MAP) if b_G is injective (i. e., $n(G) = \{e_G\}$);
(ii) *minimally almost periodic* if bG is a singleton (i. e., $n(G) = G$).

Clearly, every compact group G is MAP as $bG = G$.

Example 8.6.5. According to Corollary A.2.6, every discrete abelian group G is MAP. Further examples will be given below, we prove in particular that bG coincides with the completion of $G^\#$ (see Theorem 10.2.15).

The terms maximally/minimally almost periodic are justified by the notion of almost periodic function (see Definition 12.1.1) and its connection with the Bohr compactification.

8.7 Exercises

Exercise 8.7.1. Prove that the group topology on $\mathbb{J}_p$ described in Example 2.1.5 is compact.

Exercise 8.7.2. (a) Prove item (c) of Lemma 8.2.1 directly, without making any recourse to item (a).
(b) Deduce item (a) of Lemma 8.2.1 from item (c).

Hint. (a) If U is an open set of G containing the compact set K, then for each $x \in K$ there exists an open $V_x \in \mathcal{V}(e_G)$ such that $xV_xV_x \subseteq U$, and moreover $K \subseteq \bigcup_{x\in K} xV_x$. Hence, there exist $x_1, \dots, x_n \in K$ such that $K \subseteq \bigcup_{k=1}^n x_k V_{x_k}$. For $V = \bigcap_{k=1}^n V_k$, show that $KV \subseteq U$.

(b) Argue as in the proof of Lemma 8.2.1(c): if $x \in G \setminus KC$, then for the compact subset K^{-1} of G one has $K^{-1}x \cap C = \emptyset$, so the compact set $K^{-1}x$ is contained in the open set $U = G \setminus C$ of G. By Lemma 8.2.1(c), there exists an open neighborhood V of e_G such that $K^{-1}xV \subseteq U$. Hence, $K^{-1}xV \cap C = \emptyset$, and consequently $xV \cap KC = \emptyset$. This proves that KC is closed.

Exercise 8.7.3. (a) Prove that if K is a compact abelian group and $m \in \mathbb{N}$, then the subgroup mK of K is closed.
(b) Show that (a) may fail if K is only locally compact.

Hint. (a) The image of the continuous endomorphism $\mu_m: K \to K$, $x \mapsto mx$, is mK.

(b) Consider the compact group $G = \mathbb{Z}(m^2)^{\mathbb{N}}$, its subgroups $H = \mathbb{Z}(m)^{\mathbb{N}}$ and $H_1 = \mathbb{Z}(m^2)^{(\mathbb{N})}$, and the local direct product $K = \prod_{n\in\mathbb{N}}^{\mathrm{loc}}(\mathbb{Z}(m^2), \mathbb{Z}(m))$, i.e., $K = H + H_1$. Then K is locally compact. Nevertheless, $mK = \mathbb{Z}(m)^{(\mathbb{N})}$ is a proper dense subgroup of H, so not closed in K.

Exercise 8.7.4. Let K be a compact torsion-free divisible abelian group. Prove that for every $r \in \mathbb{Q}\setminus\{0\}$, the multiplication $\mu_r: K \to K$, $x \mapsto rx$, is a topological automorphism.

Hint. Write $r = n/m$. The multiplication $\mu_m: K \to K$, $x \mapsto mx$, is a continuous automorphism. By the compactness of K and the open mapping theorem (Theorem 8.4.1), μ_m is a topological automorphism. In particular, its inverse $\mu_m^{-1}: K \to K$, $x \mapsto \frac{1}{m}x$, is a topological automorphism, too. Since $n \neq 0$, the multiplication $\mu_n: K \to K$, $x \mapsto nx$, is a topological isomorphism as well. Being the composition of the two topological automorphisms μ_m^{-1} and μ_n, also μ_r is a topological automorphism.

Exercise 8.7.5. (a) Give examples of σ-compact groups that are not compactly generated.
(b) Show that every connected locally compact group is compactly generated.

Exercise 8.7.6. A dense subgroup G of a Hausdorff group K is *essential* in K if every closed normal subgroup of K that trivially meets G is trivial. Prove that:
(a) if G is minimal, then G is essential;
(b) if K is compact and G is essential, then G is minimal.

Hint. Denote by τ the topology of G.

(a) Assume that G is minimal and N is a closed normal subgroup of K with $N \cap G = \{e_G\}$. Let $q: K \to K/N$ be the canonical projection. Since $N \cap G = \{e_G\}$, the restriction $q\restriction_G: G \to K/N$ is injective (and continuous). By the minimality of G, $q\restriction_G$ is open. By Theorem 3.2.9, $\overline{N \cap G} = N$, hence $N = \{e_G\}$.

(b) Assume that G is essential and $\tau' \leq \tau$ is a Hausdorff group topology on G. Then the identity map $id_G: (G,\tau) \to (G,\tau')$ is a continuous isomorphism. Consider its extension $\widetilde{f}: K \to \widetilde{(G,\tau')}$ and let $N = \ker\widetilde{f}$. Since $G \cap N = \{e_G\}$, we conclude by essentiality of G that $N = \{e_G\}$. Hence, $\widetilde{f}$ is a continuous injective homomorphism into $\widetilde{G}$. Since K is compact, we deduce that $\widetilde{f}: K \to \widetilde{f}(K)$ is a topological isomorphism and so is its restriction id_G. This proves that $\tau' = \tau$.

Exercise 8.7.7. Prove that $(S(X), \mathcal{T}_X)$ is topologically simple for an infinite set X.

Hint. We verify that a nontrivial closed normal subgroup N of $S(X)$ necessarily coincides with $S(X)$. By Exercise 8.7.6, $N_1 = S_\omega(X) \cap N$ is a nontrivial normal subgroup of $S_\omega(X)$, since $S_\omega(X)$ is a dense minimal subgroup of $S(X)$ in view of Theorem 4.2.11. Since $S_\omega(X) = \bigcup_{F\in[X]^{<\omega}} S(F)$, there exists a finite subset F_0 of X, with $|F_0| \geq 5$, such that $N_1 \cap S(F_0)$ is nontrivial. For all finite subsets F of X with $F \supseteq F_0$, $N_1 \cap S(F)$ is a nontrivial normal subgroup of $S(F)$, so N_1 contains the alternating group $A(F)$. This proves that $N_1 \supseteq A_\omega(X)$. By Exercise 4.5.16(b), $A_\omega(X)$ is dense and so $N_1 = S(X)$.

(For an alternative proof, see [99, Proposition 7.1.2(b)].)

Exercise 8.7.8. Making use of Theorem 3.2.9, prove that if G is a *totally dense* subgroup of a compact group K (i. e., $\overline{N \cap G} = N$ for every closed normal subgroup N of K), then G is totally minimal.

Exercise 8.7.9. Let p be a prime. Show that:
(a) $(\mathbb{Z}, \nu_{\mathbb{Z}}^p)$ is totally minimal, yet not compact;
(b) for $G = (\mathbb{Z}, \nu_{\mathbb{Z}}^p)$, the square $G \times G$ is not minimal, and conclude that minimality is not preserved by products;
(c) the subgroup $G = \mathbb{Q}/\mathbb{Z}$ of $\mathbb{T}$ is totally minimal and its subgroup $\operatorname{Soc}(G)$ is minimal, but not totally minimal; deduce that minimality is not preserved by taking quotients.

Hint. Apply Exercise 8.7.6. For (b), find a closed subgroup N of $\mathbb{J}_p \times \mathbb{J}_p$ with $N \cap (\mathbb{Z} \times \mathbb{Z}) = \{0\}$.

Exercise 8.7.10. Let $G = \mathbb{Z}(2)^{\mathbb{N}}$. Show that:
(a) G equipped with its product topology is profinite;
(b) ϖ_G of G is strictly finer than the product topology;
(c) no proper dense subgroup of G is minimal.

Hint. Apply Exercise 8.7.6.

Exercise 8.7.11. Prove that $td(K) = K! = K$ for a hereditarily disconnected compact abelian group K.

Hint. According to van Dantzig theorem 8.5.1, the topology of K is linear, hence for every $x \in K$ the group $\langle x \rangle$ is either finite or nondiscrete and carries a linear topology.

Exercise 8.7.12. If $G = td(G)$ for some topological abelian group G, then $td_p(G) = G_p$. In particular, $td_p(K) = K_p = K$ for a pro-p-finite group K.

Hint. Since the inclusion $td_p(G) \subseteq G_p$ is known, we have to prove $G_p \subseteq td_p(G)$. Pick an element $x \in G_p$ and put $C = \langle x \rangle$. If C is finite, then it is discrete, so $x \in G_p$ yields that x is p-torsion, therefore $x \in td_p(G)$. Assume now that C is infinite. Then our hypothesis $G = td(G)$ implies that $C \cong (\mathbb{Z}, \tau)$, where τ is a linear topology on $\mathbb{Z}$. If $\{m_n\mathbb{Z} : m_n \in \mathbb{N}_+, n \in \mathbb{N}\}$ is a local base at 0 of τ, then $x \in G_p$ implies that $p^n \to 0$ in $(\mathbb{Z}, \tau)$. Hence, for every fixed $k \in \mathbb{N}$ there exists n_0 such that $p^{n_0} \in m_k\mathbb{Z}$, i. e., $m_k \mid p^{n_0}$, so m_k is a power of p. This implies that $\tau \leq \nu_{\mathbb{Z}}^p$. By Exercise 3.5.25, $\tau = \nu_{\mathbb{Z}}^p$. Therefore, $x \in td_p(G)$.

8.8 Further readings, notes, and comments

Corollary 8.2.8 is also a consequence of the fact that a compact group G of infinite weight κ is a continuous image of $\{0,1\}^\kappa$ (so compact groups are *dyadic compacta*) in view of a theorem by Kuz'minov [192]. Moreover, $d(\{0,1\}^\kappa) = \log \kappa$ by Hewitt–Marczewski–Pondiczery theorem B.3.15, so by Lemma 5.1.9,

$$d(G) \leq \log \kappa \leq \kappa = \chi(G),$$

and thus $w(G) = d(G) \cdot \chi(G) = \chi(G) = \psi(G)$, by Lemma 5.1.7 and Proposition 8.2.7.

Minimal groups were introduced simultaneously and independently by Stephenson in [264] and by Doïtchinov in [126], where the first examples of noncompact minimal groups can be found (see also Exercise 8.7.9(a)). The first examples of minimal non-totally minimal groups can be found in [97] (see Exercise 8.7.9(c)), where the notion of a totally minimal group was explicitly given (it was introduced somewhat later also by Schwanengel [253]). Answering a question of Choquet, Doïtchinov [126] showed that minimality (unlike compactness) is not preserved even under finite direct products (see Exercise 8.7.9(b)). A complete description of the cases when minimality is preserved under (arbitrary) direct products can be found in [73]. The surveys [77] and [80] contain various information on minimal groups. The recent progress in this field is outlined in [93]. A crucial role in obtaining the results pointed out in Example 8.4.5 is played by the criterion for minimality of dense subgroups of compact groups due to Prodanov [231] and Stephenson [264] (see Exercise 8.7.6).

In the nonabelian case, complete (in particular, locally compact) minimal groups need not be compact (see [245] and §10.5 for a wealth of results in this direction). Stoyanov proved in [269] that the unitary group of an infinite-dimensional Hilbert space is totally minimal, complete, and noncompact.

Recall that a topological space is locally (arcwise) connected if every point admits a neighborhood base consisting entirely of (arcwise) connected open sets. Since $c(G)$ is open for a locally connected topological group G, it makes sense to study these properties of topological groups only for connected ones. Connected and locally arcwise connected spaces are obviously arcwise connected. In the opposite direction, Rickert [247] proved that locally compact arcwise connected groups are locally arcwise connected. This is why in the sequel we pay attention only to local connectedness and arcwise connectedness of the connected locally compact groups.

9 Properties of $\mathbb{R}^n$ and its subgroups

We denote by $e_1, \ldots, e_n$ the vectors of the canonical base of $\mathbb{R}^n$.

9.1 Lifting homomorphisms with domain $\mathbb{R}^n$

Every continuous homomorphism $f: \mathbb{R}^n \to H$, where H is a Hausdorff group, is uniquely determined by its restriction to any neighborhood of 0 in $\mathbb{R}^n$ (see Exercise 6.3.2(b)). Now we prove that every continuous map $f: B_\varepsilon(0) \to H$ defined only on some ball $B_\varepsilon(0)$ in $\mathbb{R}^n$ can be extended to a continuous homomorphism $f': \mathbb{R}^n \to H$ under a minor (necessary) additivity restraint.

Lemma 9.1.1. *Let $n \in \mathbb{N}_+$, H a topological abelian group, and $\varepsilon > 0$. Then every map $f: B_\varepsilon(0) \to H$ such that $f(x+y) = f(x) + f(y)$ whenever $x, y \in B_{\varepsilon/2}(0)$ can be uniquely extended to a homomorphism $f': \mathbb{R}^n \to H$. Moreover, f' is continuous if and only if f is continuous.*

Proof. Put $U = B_{\varepsilon/2}(0)$. For $x \in \mathbb{R}^n$, there exists $m \in \mathbb{N}_+$ such that $\frac{1}{m}x \in U$, and we put

$$f'(x) = mf\left(\frac{1}{m}x\right).$$

To see that this definition is correct, assume that $\frac{1}{k}x \in U$ as well and put $y = \frac{1}{km}x$. Then $iy \in U$ for all $i \le \max\{k, m\}$, and a simple inductive argument shows that $f(my) = mf(y)$ and $f(ky) = kf(y)$. So,

$$kf\left(\frac{1}{k}x\right) = kf(my) = kmf(y) = mf(ky) = mf\left(\frac{1}{m}x\right).$$

To verify that f' is a homomorphism, take $x, y \in \mathbb{R}^n$. There exists $m \in \mathbb{N}_+$ such that $\frac{1}{m}x, \frac{1}{m}y, \frac{1}{m}(x+y) \in U$. By our hypothesis,

$$f'(x+y) = mf\left(\frac{1}{m}(x+y)\right) = mf\left(\frac{1}{m}x\right) + mf\left(\frac{1}{m}y\right) = f'(x) + f'(y).$$

The uniqueness of f' follows from Exercise 6.3.2(b). For a direct proof, assume that $f'': \mathbb{R}^n \to H$ is another homomorphism extending f. Then for every $x \in \mathbb{R}^n$, there exists $m \in \mathbb{N}_+$ such that $y := \frac{1}{m}x \in U$. So, $f''(x) = f''(my) = mf''(y) = mf(y) = f'(x)$.

Since f' is a homomorphism, it suffices to check its continuity at 0, and this follows from the continuity of $f: U \to H$. □

The *global* structure of $\mathbb{R}^n$, and in particular the fact that it is torsion-free, played a prominent role in the above proof. The next example shows that the counterpart of the theorem with $\mathbb{T}$ in place of $\mathbb{R}^n$ fails.

https://doi.org/10.1515/9783110654936-009

Example 9.1.2. Let $U = (-\frac{1}{4}, \frac{1}{4}) + \mathbb{Z}$, that is, $U = B_{1/4}(0)$ in $\mathbb{T}$. Moreover, let $f: U \hookrightarrow (-\frac{1}{4}, \frac{1}{4}) \subseteq \mathbb{R}$ be the inclusion map, more precisely, f is the unique section of the canonical projection $q_0: \mathbb{R} \to \mathbb{T}$ restricted to U and with values in $(-\frac{1}{4}, \frac{1}{4})$. Then f satisfies the additivity restraint from Lemma 9.1.1, but f cannot be extended to $\mathbb{T}$.

The next lemma is used frequently in the sequel. Again the above example shows that $\mathbb{T}$ does not have the "lifting" property established for $\mathbb{R}^n$ (e. g., for the identity map $f = id_{\mathbb{T}}: \mathbb{T} \to \mathbb{T}$, the discrete subgroup $D = \mathbb{Z}$ of $\mathbb{R}$ and the canonical projection $q_0: \mathbb{R} \to \mathbb{T}$).

Lemma 9.1.3 (Lifting lemma). *Let H be a topological abelian group, D a discrete subgroup of H, and $p: H \to H/D$ the canonical projection. For every continuous homomorphism $f: \mathbb{R}^n \to H/D$, with $n \in \mathbb{N}_+$, there exists a unique continuous homomorphism $f': \mathbb{R}^n \to H$ such that $p \circ f' = f$:*

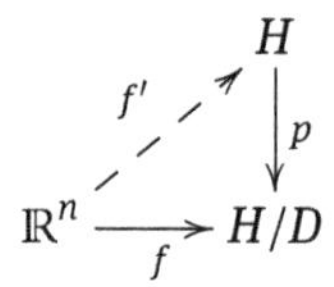

If f is open, then f' is open.

Proof. Let W be a symmetric open neighborhood of 0 in H such that $(W + W) \cap D = \{0\}$. (If f is open, then $f(\mathbb{R}^n)$ is open in H/D, so we pick W such that $p(W) \subseteq f(\mathbb{R}^n)$.) Then the restriction $p \restriction_W: W \to p(W)$ is a bijection; moreover, both $p \restriction_W$ and its inverse $\xi: p(W) \to W$ are homeomorphisms. Pick a symmetric open neighborhood W_1 of 0 in H such that $W_1 + W_1 \subseteq W$. Then

$$\xi(x + y) = \xi(x) + \xi(y) \quad \text{for every } x, y \in p(W_1). \tag{9.1}$$

For $U_0 = f^{-1}(p(W_1))$, pick an $\varepsilon > 0$ with $B_\varepsilon(0) \subseteq U_0$ and put

$$f^* = \xi \circ f \restriction_{B_\varepsilon(0)}: B_\varepsilon(0) \to H,$$

so that $f^*: B_\varepsilon(0) \to H$ is continuous, as a composition of the continuous mappings f and ξ. Moreover, (9.1) yields $f^*(x + y) = f^*(x) + f^*(y)$ whenever $x, y \in B_{\varepsilon/2}(0)$. By Lemma 9.1.1, the continuous map $f^*: B_\varepsilon(0) \to H$ can be uniquely extended to a continuous homomorphism $f': \mathbb{R}^n \to H$. Since $(p \circ f')(u) = (p \circ f^*)(u) = f(u)$ for every $u \in B_\varepsilon(0)$, the homomorphisms $p \circ f'$ and f coincide on $B_\varepsilon(0)$, hence $p \circ f' = f$, by Exercise 6.3.2(b).

Let $f'': \mathbb{R}^n \to H$ be a continuous homomorphism such that $p \circ f'' = p \circ f'$ holds. Then $g = f' - f''$ is a continuous homomorphism $\mathbb{R}^n \to H$ which satisfies $p \circ g = 0$, so $g(\mathbb{R}^n) \subseteq D$. Since D is discrete, $g^{-1}(\{0\})$ is an open subgroup of $\mathbb{R}^n$, hence $g^{-1}(\{0\}) = \mathbb{R}^n$. This shows that $g = 0$, or equivalently $f'' = f'$.

Assume that f is open. Then $f(U_0) = f(f^{-1}(p(W_1))) = p(W_1) \cap f(\mathbb{R}^n) = p(W_1)$, since $W_1 \subseteq W$ and $p(W) \subseteq f(\mathbb{R}^n)$. In order to prove that f' is open, it suffices to show that

$f'(V) = f^*(V)$ is open for every open neighborhood V of 0 contained in U_0. But this is clear, since $f^*(V) = \xi \circ f(V)$ is open, as $\xi\colon p(W) \to W$ is a homeomorphism and f was assumed to be open. □

9.2 The closed subgroups of $\mathbb{R}^n$

Our main goal here is to describe the closed subgroups of $\mathbb{R}^n$. In the next example we outline two important instances of such subgroups.

Example 9.2.1. Let $n, m \in \mathbb{N}_+$ and let $v_1, \dots, v_m$ be linearly independent in $\mathbb{R}^n$.
(a) The linear subspace $V = \mathbb{R}v_1 + \cdots + \mathbb{R}v_m \cong \mathbb{R}^m$ spanned by $v_1, \dots, v_m$ is a closed subgroup of $\mathbb{R}^n$.
(b) The subgroup $D = \langle v_1 \rangle + \cdots + \langle v_m \rangle = \langle v_1, \dots, v_m \rangle \cong \mathbb{Z}^m$ generated by $v_1, \dots, v_m$ is a discrete (hence, closed) subgroup of $\mathbb{R}^n$.

We prove that every closed subgroup of $\mathbb{R}^n$ is topologically isomorphic to a product $V \times D$ of a linear subspace $V \cong \mathbb{R}^s$ and a discrete subgroup $D \cong \mathbb{Z}^m$, with $s, m \in \mathbb{N}$ and $s + m \leq n$. More precisely:

Theorem 9.2.2. *Let $n \in \mathbb{N}_+$ and let H be a closed subgroup of $\mathbb{R}^n$. Then there exist $s, m \in \mathbb{N}$, with $s + m \leq n$, and linearly independent vectors $v_1, \dots, v_s, \dots, v_{s+m}$ such that $H = V \times D$, where $V \cong \mathbb{R}^s$ is the linear subspace of $\mathbb{R}^n$ spanned by $v_1, \dots, v_s$ and $D = \langle v_{s+1}, \dots, v_{s+m} \rangle \cong \mathbb{Z}^m$, so D is discrete and V is open in H.*

We give two proofs of this theorem. The first is relatively short and proceeds by induction. The second proof splits into several steps. Before starting the proofs, we note that the dichotomy imposed by Example 9.2.1 is reflected in the following topological dichotomy resulting from the theorem:
(i) the closed *connected* subgroups of $\mathbb{R}^n$ are always linear subspaces of $\mathbb{R}^n$, so isomorphic to $\mathbb{R}^s$ for some $s \leq n$;
(ii) the closed *hereditarily disconnected* subgroups D of $\mathbb{R}^n$ are free and have free-rank $r_0(D) \leq n$; in particular, they are discrete.

Remark 9.2.3. It follows from Theorem 9.2.2 that, for every closed subgroup H of $\mathbb{R}^n$, $H \cong V \times D$, where $c(H) \cong V \cong \mathbb{R}^s$ is open in H with $s \leq n$, and $D \cong \mathbb{Z}^m$ is discrete with $m \leq n - s$; in particular, H contains a discrete subgroup $D_1 \times D$ of free-rank $s + m$.

The next lemma prepares the inductive step in the first proof of Theorem 9.2.2.

Lemma 9.2.4. *If H is a closed subgroup of $\mathbb{R}^n$, $L \cong \mathbb{R}$ is a one-dimensional subspace of $\mathbb{R}^n$, and $H \cap L \neq \{0\}$, then denoting by $p\colon \mathbb{R}^n \to \mathbb{R}^n/L$ the canonical projection, $p(H)$ is a closed subgroup of $\mathbb{R}^n/L$.*

Proof. If $n = 1$, then $\mathbb{R}^n/L$ is trivial, so we are done.

Assume that $n > 1$ and consider the nonzero closed subgroup $H_1 = H \cap L$ of $L \cong \mathbb{R}$. If $H_1 = L$ (i. e., $L \subseteq H$), then the assertion follows from Theorem 3.2.8(b). Now assume that $H_1 \neq L \cong \mathbb{R}$. Then $H_1 = \langle a \rangle$ is cyclic, by Proposition 3.1.11. Making use of an appropriate linear automorphism α of $\mathbb{R}^n$ and replacing H by $\alpha(H)$, we may assume without loss of generality that

$$L = \mathbb{R} \times \{0\}^{n-1} \quad \text{and} \quad a = e_1 \text{ (i. e., } H_1 = \mathbb{Z} \times \{0\}^{n-1}).$$

Consider the canonical projection $\pi: \mathbb{R}^n \to \mathbb{R}^n/H_1$. Since H is a closed subgroup of $\mathbb{R}^n$ containing H_1, its image $\pi(H)$ is a closed subgroup of $\mathbb{R}^n/H_1 \cong \mathbb{T} \times \mathbb{R}^{n-1}$, by Theorem 3.2.8(b). Next observe that the projection $p: \mathbb{R}^n \to \mathbb{R}^{n-1}$ is the composition of π and the canonical projection $q: \mathbb{R}^n/H_1 \to \mathbb{R}^{n-1}$:

$$\begin{array}{ccc} \mathbb{R}^n & \xrightarrow{\quad p \quad} & \mathbb{R}^{n-1} \\ & {\scriptstyle \pi}\searrow \quad \nearrow{\scriptstyle q} & \\ & \mathbb{R}^n/H_1 \cong \mathbb{T} \times \mathbb{R}^{n-1} & \end{array}$$

Since $\ker q = L/H_1 \cong \mathbb{T}$ is compact and $\pi(H)$ is closed in $\mathbb{R}^n/H_1$, we conclude that $p(H) = q(\pi(H))$ is a closed subgroup of $\mathbb{R}^{n-1}$, by Lemma 8.2.2. □

We shall see in §9.3 that "closed" can be replaced by "discrete" in the conclusion of this lemma. Let us see that the hypothesis $H \cap L \neq \{0\}$ is relevant.

Example 9.2.5. Take the discrete (hence, closed) subgroup $H = \mathbb{Z}^2$ of $\mathbb{R}^2$ and the line $L = \mathbb{R}v$ in $\mathbb{R}^2$, where $v = (1, \sqrt{2})$. Then $L \cap H = \{0\}$, while $\mathbb{R}^2/L \cong \mathbb{R}$, so Proposition 3.1.11 yields that, denoting by $p: \mathbb{R}^2 \to \mathbb{R}^2/L \cong \mathbb{R}$ the canonical projection, the noncyclic image $p(H) \cong \mathbb{Z}^2$ of H in $\mathbb{R}$ is dense, so fails to be closed.

First proof of Theorem 9.2.2. We proceed by induction on $n \in \mathbb{N}_+$, and the case $n = 1$ is Proposition 3.1.11.

Assume that $n > 1$. If H is a linear subspace of $\mathbb{R}^n$, then $H = V$, and we are done. So, assume that H is not a linear subspace. Then there exists a nonzero $h \in H$ such that the line $L = \mathbb{R}h$ is not contained in H. Thus, the closed nonzero subgroup $H_1 = H \cap L$ of $L \cong \mathbb{R}$ is proper, hence cyclic, by Proposition 3.1.11; so, let $H_1 = \langle a \rangle$ for some $a \in H_1 \setminus \{0\}$. By Lemma 9.2.4, the canonical projection $p: \mathbb{R}^n \to \mathbb{R}^n/L \cong \mathbb{R}^{n-1}$ sends H to a closed subgroup $p(H)$ of $\mathbb{R}^{n-1}$.

By the inductive hypothesis, there exist $s, m \in \mathbb{N}$, with $s+m < n$, and $s+m$ linearly independent vectors $v'_1, \dots, v'_s, v'_{s+1}, \dots, v'_{s+m}$ in $\mathbb{R}^n/L \cong \mathbb{R}^{n-1}$ such that

$$p(H) = V' \times D', \quad \text{with } V' = \mathbb{R}v'_1 + \cdots + \mathbb{R}v'_s \cong \mathbb{R}^s \quad \text{and} \quad D' = \langle v'_{s+1}, \dots, v'_{s+m} \rangle \cong \mathbb{Z}^m.$$

Since both H and $p(H)$ are locally compact abelian groups by Remark 8.2.4 and H is also σ-compact (as a closed subgroup of $\mathbb{R}^n$), it follows from the open mapping theorem (Theorem 8.4.1) that the continuous surjective homomorphism $p\restriction_H: H \to p(H)$ is open, that is, $p(H) \cong H/H_1$ topologically. Since H_1 is discrete, we can apply Lemma 9.1.3 to obtain an open continuous homomorphism $f\colon V' \to H$ such that $p \circ f = j$ is the inclusion of $V' \cong \mathbb{R}^s$ in $p(H)$. Since $j = p \circ f$ is injective, so is f. Hence, letting $V = f(V')$, we get that $f\colon V' \to V$ is a topological isomorphism.

Let $v_i = f(v_i')$ for every $i \in \{1, \dots, s\}$. For every $j \in \{1, \dots, m\}$, find $v_{s+j} \in H$ such that $p(v_{s+j}) = v'_{s+j}$. Let $v_0 = a$. Since the canonical projection $p\colon \mathbb{R}^n \to \mathbb{R}^n/L$ is $\mathbb{R}$-linear, the vectors $v_0, v_1, \dots, v_s, v_{s+1}, \dots, v_{s+m}$ are linearly independent, so $D = \langle v_0, v_{s+1}, \dots, v_{s+m}\rangle \cong \mathbb{Z}^{m+1}$ is discrete (see Example 9.2.1(b)). From

$$p(H) = V' \times D' \quad \text{and} \quad \ker p\restriction_H = H \cap L = H_1 = \langle a\rangle,$$

we deduce that $H = V \times D \cong \mathbb{R}^s \times \mathbb{Z}^{m+1}$. □

Corollary 9.2.6. *For every $n \in \mathbb{N}_+$, the only compact subgroup of $\mathbb{R}^n$ is $\{0\}$.*

Proof. Let K be a compact subgroup of $\mathbb{R}^n$. By Theorem 9.2.2, $K = V \times D$, where, for some $s, m \in \mathbb{N}$, $V \cong \mathbb{R}^s$ is a linear subspace of $\mathbb{R}^n$ and $D \cong \mathbb{Z}^m$ is a discrete subgroup of $\mathbb{R}^n$. The compactness of K yields that both V and D are compact, and $\mathbb{R}^s$ is compact only for $s = 0$ while $\mathbb{Z}^m$ is compact only for $m = 0$. □

9.3 The second proof of Theorem 9.2.2

The second proof of Theorem 9.2.2 makes no recourse to induction, so from a certain point of view gives a better insight in the argument.

By Proposition 3.1.11, every discrete subgroup of $\mathbb{R}$ is cyclic. The first part of this proof consists in appropriately extending this property to discrete subgroups of $\mathbb{R}^n$ for every $n \in \mathbb{N}_+$ (see Proposition 9.3.2). The first step, namely, Lemma 9.3.1, is to prove directly that the free-rank $r_0(H)$ of a discrete subgroup H of $\mathbb{R}^n$ coincides with the dimension of the linear subspace of $\mathbb{R}^n$ generated by H. Note that the elements $x_1, \dots, x_k \in \mathbb{R}^n$ are independent if and only if they are $\mathbb{Q}$-linearly independent.

Lemma 9.3.1. *Let H be a discrete subgroup of $\mathbb{R}^n$. If the elements $v_1, \dots, v_m$ of H are independent, then they are also $\mathbb{R}$-linearly independent.*

Proof. Let $D = \langle v_1, \dots, v_m\rangle \cong \mathbb{Z}^m$, and let $V \cong \mathbb{R}^k$ be the linear subspace of $\mathbb{R}^n$ generated by H. We need to prove that $k \geq m$. We can assume without loss of generality that $V = \mathbb{R}^n$ (i. e., $k = n$).

Assume for a contradiction that $m > n$. Since $v_1, \dots, v_m$ generate the vector space $\mathbb{R}^n$, after possibly changing their enumeration, we can assume that the vectors $v_1, \dots, v_n$ are $\mathbb{R}$-linearly independent, so form a base of $\mathbb{R}^n$. Moreover, we can assume without loss of generality that $v_1 = e_1, \dots, v_n = e_n$. Indeed, as $v_1, \dots, v_n$ are $\mathbb{R}$-linearly independent, there exists an $\mathbb{R}$-linear isomorphism $\alpha\colon \mathbb{R}^n \to \mathbb{R}^n$ with $\alpha(v_i) = e_i$ for every $i \in \{1, \dots, n\}$. Clearly, $\alpha(H)$ is still a discrete subgroup of $\mathbb{R}^n$ and $\alpha(D) \cong D \cong \mathbb{Z}^m$.

Let $h = \alpha(v_{n+1})$; so,

$$\langle e_1, \dots, e_n, h\rangle \cong \mathbb{Z}^{n+1}. \tag{9.2}$$

As H is discrete, there exists $\varepsilon > 0$ with $\max\{|x_i|: i \in \{1,\dots,n\}\} \geq \varepsilon$ for every $x = (x_1,\dots,x_n) \in H \setminus \{0\}$. Pick an integer $M > 1/\varepsilon$ and let

$$C_i = \left[\frac{i}{M}, \frac{i+1}{M}\right] \quad \text{for all } i \in F := \{0, 1, \dots, M-1\}.$$

Further, let

$$C_{\bar{i}} = \prod_{k=1}^{n} C_{i_k} \quad \text{for } \bar{i} = (i_1, \dots, i_n) \in F^n.$$

These "small cubes" $C_{\bar{i}}$ represent the standard cube $C = [0,1]^n$ in $\mathbb{R}^n$ as a finite union $\bigcup_{\bar{i}\in F^n} C_{\bar{i}}$ such that each one of the small cubes has edges of length $< \varepsilon$.

As usual, for $r \in \mathbb{R}$ let $\{r\} = r - \lfloor r \rfloor$ and for $j \in \mathbb{Z}$ let

$$C \ni a_j = (\{jh_1\}, \dots, \{jh_n\}) = jh - (\lfloor jh_1 \rfloor, \dots, \lfloor jh_n \rfloor) \in jh + \mathbb{Z}^n \subseteq H.$$

Then $a_k \neq a_l$ for every pair $k \neq l$ in $\mathbb{Z}$, since otherwise $(k-l)h \in \mathbb{Z}^n = \langle e_1, \dots, e_n\rangle$ with $k - l \neq 0$ in contradiction with (9.2). Among the infinitely many points $a_j \in C$, there exist two elements $a_k \neq a_l$ of H belonging to the same small cube $C_{\bar{i}}$. Hence, $|\{kh_j\} - \{lh_j\}| < \varepsilon$ for every $j \in \{1,\dots,n\}$. So, $a_k - a_l \in H \cap (-\varepsilon, \varepsilon)^n = \{0\}$, and this contradicts $a_k \neq a_l$. □

Observe that it is essential to assume H to be closed: e. g., the elements $1, \sqrt{2}$ of the subgroup $H = \langle 1, \sqrt{2}\rangle$ of $\mathbb{R}$ are $\mathbb{Q}$-linearly independent, but not $\mathbb{R}$-linearly independent.

The aim of the next step is to see that the discrete subgroups of $\mathbb{R}^n$ are free.

Proposition 9.3.2. *For a discrete subgroup H of $\mathbb{R}^n$, H is free and $r_0(H) \leq n$.*

Proof. Since by Lemma 9.3.1 there are at most n independent vectors in H, we have $m := r_0(H) \leq n$. Then there exist m independent vectors $v_1, \dots, v_m$ of H. By Lemma 9.3.1, the vectors $v_1, \dots, v_m$ are also $\mathbb{R}$-linearly independent.

Let $V \cong \mathbb{R}^m$ be the linear subspace of $\mathbb{R}^n$ generated by $v_1, \dots, v_m$. Obviously, $H \subseteq V$, since H is contained in the $\mathbb{Q}$-linear subspace of $\mathbb{R}^n$ generated by the free subgroup $F = \langle v_1, \dots, v_m\rangle$ of H. As H is a discrete subgroup of V too, we can argue with V in place of $\mathbb{R}^n$, so we can assume without loss of generality that $r_0(H) = m = n$ and $V = \mathbb{R}^n$. Next we verify that H/F is finite; from this we obtain that H is finitely generated and torsion-free, so H must be free, by Theorem A.1.1.

Since the vectors $v_1, \dots, v_n$ are $\mathbb{R}$-linearly independent, we can assume without loss of generality that $F = \mathbb{Z}^n \subseteq H$. Indeed, let $\alpha: \mathbb{R}^n \to \mathbb{R}^n$ be the linear isomorphism with $\alpha(v_i) = e_i$ for every $i \in \{1,\dots,n\}$. Then $\alpha(H)$ is still a discrete subgroup of $\mathbb{R}^n$, $\mathbb{Z}^n = \alpha(F) \subseteq \alpha(H)$ and H/F is finite if and only if $\alpha(H)/\alpha(F) \cong H/F$ is finite.

To check that H/F is finite, consider the canonical projection $q: \mathbb{R}^n \to \mathbb{R}^n/\mathbb{Z}^n \cong \mathbb{T}^n$. According to Theorem 3.2.8(c), q sends the closed subgroup H onto a closed (hence, compact) subgroup $q(H)$ of $\mathbb{T}^n$; moreover, $H = q^{-1}(q(H))$, therefore the restriction of q to H is open and $q(H)$ is discrete. We conclude that $q(H) \cong H/F$ is both compact and discrete, so finite. □

The next lemma sharpens Lemma 9.2.4 to the case of discrete subgroups of $\mathbb{R}^n$.

Lemma 9.3.3. *Let H be a discrete subgroup of $\mathbb{R}^n$ and $L \cong \mathbb{R}$ a one-dimensional linear subspace of $\mathbb{R}^n$ with $H \cap L \neq \{0\}$. Then, denoting by $p: \mathbb{R}^n \to \mathbb{R}^n/L$ the canonical projection, $p(H)$ is a discrete subgroup of $\mathbb{R}^n/L$.*

Proof. If $n = 1$, then $L = \mathbb{R}$, so this case is trivial. Assume $n > 1$ in the sequel. Since $\{0\} \neq H_1 = H \cap L$ is a discrete subgroup of $L \cong \mathbb{R}$, we conclude that $H_1 = \langle a \rangle$ is cyclic, by Proposition 3.1.11. Making use of an appropriate linear automorphism $\alpha: \mathbb{R}^n \to \mathbb{R}^n$ and replacing H by $\alpha(H)$, we assume without loss of generality that

$$L = \mathbb{R} \times \{0\}^{n-1} \quad \text{and} \quad a = e_1,$$

where we consider L as a subgroup of $\mathbb{R}^n = \mathbb{R} \times \mathbb{R}^{n-1}$. Thus,

$$H_1 = \mathbb{Z} \times \{0\}^{n-1}. \tag{9.3}$$

For $\varepsilon > 0$, let $B_\varepsilon(0) = (-\varepsilon, \varepsilon)^n$ and $U_\varepsilon = B_\varepsilon(0) + L$. Let us prove that for some $\varepsilon > 0$ also

$$U_\varepsilon \cap H = \mathbb{Z} \times \{0\}^{n-1} \tag{9.4}$$

holds true. Assume for a contradiction that $U_{1/k} \cap H \not\subseteq L$ for every $k \in \mathbb{N}_+$, and pick $h_k := (x_k, y_k) \in U_{1/k} \cap H \subseteq \mathbb{R} \times \mathbb{R}^{n-1}$ with $y_k \neq 0$. Since $\mathbb{Z} \times \{0\}^{n-1} \subseteq H$ by (9.3), we can assume without loss of generality that $0 \leq x_k < 1$ for every $k \in \mathbb{N}_+$. Then there exists a converging subsequence $x_{k_l} \to z$, and so $h_{k_l} \to (z, 0) \in H$. Since H is discrete, this sequence is eventually constant, so $y_{k_l} = 0$ for all sufficiently large $k \in \mathbb{N}_+$, a contradiction. This proves that $U_\varepsilon \cap H \subseteq L \cap H = H_1$ and hence (9.4) holds for some $\varepsilon > 0$. Let $p: \mathbb{R}^n \to \mathbb{R}^{n-1}$ be the canonical projection along L. Then $U_\varepsilon = p^{-1}(p(B_\varepsilon(0)))$, so (9.4) implies that $p(B_\varepsilon(0)) \cap p(H) = \{0\}$ in $\mathbb{R}^{n-1}$. Thus, $p(H)$ is discrete. □

Exercise 9.5.1 provides a shorter alternative description of the discrete subgroups of $\mathbb{R}^n$, so of the first part of Theorem 9.2.2 carried out in Proposition 9.3.2.

Now we pass to the case of nondiscrete closed subgroups of $\mathbb{R}^n$.

Lemma 9.3.4. *If H is a nondiscrete closed subgroup of $\mathbb{R}^n$, then H contains a line through the origin (i. e., there exists $u_0 \in H$ such that $\mathbb{R}u_0 \subseteq H$).*

Proof. It is enough to prove that if H is a closed subgroup containing no line then H is discrete. In order to do so, we consider $S = \{x \in \mathbb{R}^n : \|x\|_2 = 1\}$ and the function $\varphi: S \to \mathbb{R}_{>0} \cup \{\infty\}$ defined as follows: for all $u \in S$ with $\mathbb{R}u \cap H = \{0\}$, we put $\varphi(u) = \infty$.

If $\mathbb{R}u \cap H = \langle au \rangle$ for some $a > 0$, we define $\varphi(u) = a$. We are going to show that $F_b := \varphi^{-1}((0, b])$ is closed in S for every $b \in \mathbb{R}_{>0}$. So, fix $b \in \mathbb{R}_{>0}$ and let $\{u_k\}_{k\in\mathbb{N}}$ be a sequence in F_b which converges to $\tilde{u} \in S$. We have to show that $\varphi(\tilde{u}) \le b$. By assumption, $b_k := \varphi(u_k) \in (0, b]$. So, the bounded sequence $\{b_k\}_{k\in\mathbb{N}}$ has a convergent subsequence. We may assume without loss of generality that $\{b_k\}_{k\in\mathbb{N}}$ converges to $\tilde{b} \in [0, b]$. If $\tilde{b} > 0$, since $\tilde{b}\tilde{u} = \lim_{k\to\infty} b_k u_k \in H$ (with H being closed by assumption), we obtain that $\varphi(\tilde{u}) \le \tilde{b} \le b$.

Let us assume now that $\tilde{b} = 0$, and fix $\varepsilon > 0$. We can assume without loss of generality that, for all $k \in \mathbb{N}$, $b_k \in (0, \varepsilon)$, and so we can choose $m_k \in \mathbb{N}$ such that $m_k b_k \in [\varepsilon, 2\varepsilon]$. Since $\{m_k b_k\}_{k\in\mathbb{N}}$ is bounded, we may assume that it converges to an element $\tilde{\varepsilon} \in [\varepsilon, 2\varepsilon]$. So, $\{m_k b_k u_k\}_{k\in\mathbb{N}}$ converges to $\tilde{\varepsilon}\tilde{u} \in H$, hence $0 < \varphi(\tilde{u}) = \tilde{\varepsilon} \le 2\varepsilon$. Since $\varepsilon > 0$ was arbitrary, we deduce that $\varphi(\tilde{u}) = 0$, a contradiction. So, the second case cannot occur. This shows that, for all $b \in \mathbb{R}_{>0}$, F_b is closed, hence its complement $V_b := \varphi^{-1}((b, \infty])$ is open. Since $S = \bigcup_{b\in\mathbb{R}_{>0}} V_b$, the compactness of S implies that there exists $b_0 > 0$ with $S = V_{b_0}$. By the definition of φ, $\|x\|_2 \ge b_0$ for every $x \in H \setminus \{0\}$. Hence, H is discrete. □

Second proof of Theorem 9.2.2. If H is a closed subgroup of $\mathbb{R}^n$ and V_1, V_2 are linear subspaces of $\mathbb{R}^n$ contained in H, then also the linear subspace $V_1 + V_2$ of $\mathbb{R}^n$ is contained in H. Therefore, H contains a largest linear subspace $\lambda(H)$ of $\mathbb{R}^n$. Since $\lambda(H)$ is a closed subgroup of $\mathbb{R}^n$ contained in H, the canonical projection $p\colon \mathbb{R}^n \to \mathbb{R}^n/\lambda(H) \cong \mathbb{R}^k$ (where $k = n - \dim \lambda(H)$) sends H to a closed subgroup $p(H)$ topologically isomorphic to $H/\lambda(H)$ of $\mathbb{R}^k$, by Theorem 3.2.8(b). Moreover, $p(H)$ contains no lines L, since such a line L would produce a linear subspace $p^{-1}(L)$ of $\mathbb{R}^n$ contained in H and properly containing $\lambda(H)$. By Lemma 9.3.4, $p(H)$ is discrete, and this means that $\lambda(H)$ is an open subgroup of H. Since $\lambda(H)$ is divisible, by Corollary A.2.7 it splits, so $H = \lambda(H) \times H'$, where H' is a discrete subgroup of H (and of $\mathbb{R}^n$). By Proposition 9.3.2, $H' \cong \mathbb{Z}^m$. □

9.4 Elementary LCA groups and the Kronecker theorem

9.4.1 Quotients of $\mathbb{R}^n$ and closed subgroups of $\mathbb{T}^k$

The next corollary of Theorem 9.2.2 describes the quotients of $\mathbb{R}^n$.

Corollary 9.4.1. *A quotient of $\mathbb{R}^n$ is isomorphic to $\mathbb{R}^k \times \mathbb{T}^m$, where $k+m \le n$. In particular, a compact quotient of $\mathbb{R}^n$ is isomorphic to $\mathbb{T}^m$ for some $m \le n$.*

Proof. Let H be a closed subgroup of $\mathbb{R}^n$. By Theorem 9.2.2, $H = V \times D$, with $V \cong \mathbb{R}^s$, $D \cong \mathbb{Z}^m$ discrete and $s + m \le n$. Let V_1 be the linear subspace of $\mathbb{R}^n$ spanned by D. Pick a complementing linear subspace V_2 of $\mathbb{R}^n$ for $V + V_1$; then $\dim V_2 = k$, where $k = n - (s + m)$, and $\mathbb{R}^n \cong V \times V_1 \times V_2$. Therefore, $\mathbb{R}^n/H \cong (V_1/D) \times V_2$. Moreover, $V_1/D \cong \mathbb{T}^m$, and hence $\mathbb{R}^n/H \cong \mathbb{T}^m \times \mathbb{R}^k$. □

This corollary implies, of course, that for $k, m \in \mathbb{N}$, a quotient of the group $\mathbb{R}^k \times \mathbb{T}^m$ is isomorphic to $\mathbb{R}^l \times \mathbb{T}^s$ for some $l, s \in \mathbb{N}$ with $l \le k$ and $l + s \le k + m$.

We introduce two classes of topological abelian groups that completely describe the closed subgroups of $\mathbb{T}^k$ and $\mathbb{R}^n \times \mathbb{T}^k$, respectively.

Definition 9.4.2. A topological abelian group is:
(i) *elementary compact* if it is topologically isomorphic to $\mathbb{T}^n \times F$, where $n \in \mathbb{N}$ and F is a finite abelian group;
(ii) *elementary locally compact* if it is topologically isomorphic to $\mathbb{R}^n \times \mathbb{Z}^m \times \mathbb{T}^s \times F$, where $n, m, s \in \mathbb{N}$ and F is a finite abelian group.

The class of elementary locally compact abelian groups is closed under taking quotients, closed subgroups, and finite products (see Exercise 9.5.2). The following result shows that the elementary (locally) compact abelian groups are very natural.

Corollary 9.4.3. *Every closed subgroup C of $\mathbb{R}^n \times \mathbb{T}^k$ is elementary locally compact. Consequently, every closed subgroup of $\mathbb{T}^s$ is elementary compact.*

Proof. Let $q: \mathbb{R}^{n+k} \to \mathbb{R}^n \times \mathbb{T}^k$ be the canonical projection. Then $H = q^{-1}(C)$ is a closed subgroup of $\mathbb{R}^{n+k}$. By Theorem 9.2.2, H is a direct product $H = V \times D$ with $V \cong \mathbb{R}^s$ and $D \cong \mathbb{Z}^m$ for some $s, m \in \mathbb{N}$. Since the restriction $q \restriction_H : H \to q(H) = C$ is open by Theorem 3.2.8(c), $q(V)$ is an open subgroup of C. Moreover, $q(V)$ is also divisible (being a quotient of the divisible abelian group V), so $C \cong q(V) \times B$ by Corollary A.2.7, where the subgroup B of C is discrete by Lemma 3.2.10(a).

As a quotient of $V \cong \mathbb{R}^s$, $q(V) \cong \mathbb{R}^l \times \mathbb{T}^h$ for some $l, h \in \mathbb{N}$ by Corollary 9.4.1. On the other hand, the chain of standard isomorphisms

$$B \cong \frac{C}{q(V)} \cong \frac{H/\ker q}{(V + \ker q)/\ker q} \cong \frac{H}{V + \ker q} \cong \frac{H/V}{(V + \ker q)/V} \cong \frac{D}{D_1},$$

where $D_1 = (V + \ker q)/V$, shows that B is isomorphic to a quotient of $D \cong \mathbb{Z}^m$, so $B \cong \mathbb{Z}^g \times F$ for some $g \in \mathbb{N}$ and some finite abelian group F. Therefore, $C \cong \mathbb{R}^l \times \mathbb{T}^h \times \mathbb{Z}^g \times F$. □

Also the converse implication is true, namely, every elementary locally compact abelian group can be embedded as a closed subgroup of $\mathbb{R}^n \times \mathbb{T}^k$ for some $n, k \in \mathbb{N}$ (see Exercise 9.5.4).

9.4.2 Closure in $\mathbb{R}^n$

Our next aim is the description of the closure of an arbitrary subgroup of $\mathbb{R}^n$.

Every base $(v_1, \dots, v_n)$ of $\mathbb{R}^n$ admits a dual base $(v_1', \dots, v_n')$ defined by the relations: for every $i, j \in \{1, \dots, n\}$, $(v_i \mid v_j') = \delta_{ij}$. For a subset X of $\mathbb{R}^n$, its *orthogonal subspace* is

$$X^o := \{u \in \mathbb{R}^n : \forall x \in X,\ (x \mid u) = 0\}.$$

If $X = \{v\} \neq \{0\}$ is a singleton, then X^o is the hyperspace of $\mathbb{R}^n$ orthogonal to v. So, X^o is always a linear subspace of $\mathbb{R}^n$, being an intersection of hyperspaces. If V is a linear subspace of $\mathbb{R}^n$, then V^o is the orthogonal complement of V, so $\mathbb{R}^n \cong V \times V^o$.

For a subgroup H of $\mathbb{R}^n$, its *associated subgroup* is

$$H^\dagger := \{u \in \mathbb{R}^n : \forall x \in H,\ (x \mid u) \in \mathbb{Z}\}.$$

Clearly, $H^o \subseteq H^\dagger$, $(\mathbb{Z}^n)^\dagger = \mathbb{Z}^n$, and $\{0\}^\dagger = \mathbb{R}^n$, while $H^\dagger = \mathbb{R}^n$ implies $H = \{0\}$. We describe $H^\dagger$ for every subgroup H of $\mathbb{R}^n$:

Lemma 9.4.4. *Let H, H_1 be subgroups of $\mathbb{R}^n$. Then:*
(a) *$H^\dagger$ is closed subgroup of $\mathbb{R}^n$ and the correspondence $H \mapsto H^\dagger$ is monotone decreasing;*
(b) *$(\overline{H})^\dagger = H^\dagger$;*
(c) *if H is a linear subspace of $\mathbb{R}^n$, $H^o = H^\dagger$;*
(d) *$(H + H_1)^\dagger = H^\dagger \cap H_1^\dagger$.*

Proof. The map $\mathbb{R}^n \times \mathbb{R}^n \to \mathbb{R}$ defined by $(x, y) \mapsto (x \mid y)$ is continuous.

(a) Let $q_0 : \mathbb{R} \to \mathbb{T} = \mathbb{R}/\mathbb{Z}$ be the canonical projection. For every $a \in \mathbb{R}^n$,

$$\chi_a = q_0 \circ (a \mid -) : \mathbb{R}^n \to \mathbb{T},\ x \mapsto q_0((a \mid x)),$$

is a continuous homomorphism, so $\chi_a^{-1}(0) = \{u \in \mathbb{R}^n : (a \mid u) \in \mathbb{Z}\}$ is closed in $\mathbb{R}^n$. Therefore, $H^\dagger = \bigcap_{h \in H} \chi_h^{-1}(0)$ is closed in $\mathbb{R}^n$, too. The same equality proves that the correspondence $H \mapsto H^\dagger$ is monotone decreasing.

(b) From the second part of (a) one has $(\overline{H})^\dagger \subseteq H^\dagger$. Suppose that $u \in H^\dagger$. For every $x \in \overline{H}$, by the continuity of the map $\mathbb{R}^n \to \mathbb{T}$, $x \mapsto \chi_x(u) = \chi_u(x)$, one can deduce that $\chi_u(x) = 0$, since $\chi_u(h) = 0$ for every $h \in H$. Hence, $u \in (\overline{H})^\dagger$.

(c) Let $y \in H^\dagger$. To prove that $y \in H^o$, take any $x \in H$ and assume that $m := (x \mid y) \neq 0$. Then $z := \frac{1}{2m} x \in H$ and $(z \mid y) = \frac{1}{2} \notin \mathbb{Z}$, a contradiction.

(d) The inclusion $(H + H_1)^\dagger \subseteq H^\dagger \cap H_1^\dagger$ follows from item (a). On the other hand, if $x \in H^\dagger \cap H_1^\dagger$, then obviously $x \in (H + H_1)^\dagger$. □

We study in the sequel the subgroup $H^\dagger$ associated to a closed subgroup H of $\mathbb{R}^n$.

Proposition 9.4.5. *For every subgroup H of $\mathbb{R}^n$, $\overline{H} = (H^\dagger)^\dagger$. In particular, H is dense in $\mathbb{R}^n$ if and only if $H^\dagger = \{0\}$.*

Proof. By Lemma 9.4.4(b), $(\overline{H})^\dagger = H^\dagger$, so we can assume without loss of generality that $H = \overline{H}$ is closed. According to Theorem 9.2.2, there exist a base $(v_1, \dots, v_n)$ of $\mathbb{R}^n$ and $k \leq n$ such that $H = V + L$, where V is the linear subspace generated by $v_1, \dots, v_s$ for some $s \in \{0, \dots, k\}$ and $L = \langle v_{s+1}, \dots, v_k \rangle$. Let $(v_1', \dots, v_n')$ be the dual base of $(v_1, \dots, v_n)$.

Let $V' = \mathbb{R}v_1' + \dots + \mathbb{R}v_s'$, $L' = \langle v_{s+1}', \dots, v_k' \rangle$ and $W' = \mathbb{R}v_{k+1}' + \dots + \mathbb{R}v_n'$. Then $V^\dagger = V^o = (\mathbb{R}v_{s+1}' + \dots + \mathbb{R}v_k') \oplus W'$ and $L^\dagger = V' \oplus L' \oplus W'$ since $(v_1', \dots, v_n')$ is the dual base of $(v_1, \dots, v_n)$. By Lemma 9.4.4(d), $H^\dagger = V^\dagger \cap L^\dagger = L' \oplus W'$. Similarly, $(H^\dagger)^\dagger = V + L = H$ since $(v_1, \dots, v_n)$ is the dual base of $(v_1', \dots, v_n')$. □

The next proposition is a particular case of the well-known Kronecker theorem.

Proposition 9.4.6. *Let $v_1, \dots, v_n \in \mathbb{R}$. Then, for $v = (v_1, \dots, v_n) \in \mathbb{R}^n$, the subgroup $\langle v\rangle + \mathbb{Z}^n$ of $\mathbb{R}^n$ is dense if and only if $v_0 = 1, v_1, \dots, v_n \in \mathbb{R}$ are linearly independent as elements of the vector space $\mathbb{R}$ over $\mathbb{Q}$.*

Proof. Assume that $v_0 = 1, v_1, \dots, v_n \in \mathbb{R}$ are linearly independent and let $H = \langle v\rangle + \mathbb{Z}^n$. Then $H^\dagger \subseteq \mathbb{Z}^n = (\mathbb{Z}^n)^\dagger$. Therefore, some $z \in \mathbb{Z}^n$ belongs to $(\langle v\rangle)^\dagger$ if and only if $z = 0$. Thus, $H^\dagger = \{0\}$. Consequently, H is dense in $\mathbb{R}^n$ by Proposition 9.4.5. Conversely, if $\sum_{i=0}^n k_i v_i = 0$ is a nontrivial linear combination with $k_i \in \mathbb{Z}$, then $k = (k_1, \dots, k_n) \in \mathbb{Z}^n$ is nonzero and $k \in H^\dagger$. Thus, $H^\dagger \neq \{0\}$, so H is not dense. □

Now we prove that $\mathbb{T}^n$ is monothetic for every $n \in \mathbb{N}_+$.

Corollary 9.4.7. *Let $q_0\colon \mathbb{R} \to \mathbb{T}$ be the canonical projection. For $n \in \mathbb{N}_+$ and $v_1, \dots, v_n \in \mathbb{R}$ such that $1, v_1, \dots, v_n \in \mathbb{R}$ are $\mathbb{Q}$-linearly independent in $\mathbb{R}$, $\langle (q_0(v_1), \dots, q_0(v_n))\rangle$ is dense in $\mathbb{T}^n$.*

Proof. By Proposition 9.4.6, with $v = (v_1, \dots, v_n) \in \mathbb{R}^n$, the subgroup $H = \langle v\rangle + \mathbb{Z}^n$ of $\mathbb{R}^n$ is dense. Consider the canonical projection $\pi\colon \mathbb{R}^n \to \mathbb{T}^n \cong \mathbb{R}^n/\mathbb{Z}^n$. Then $\pi(H) = \langle (q_0(v_1), \dots, q_0(v_n))\rangle$ is a dense subgroup of $\mathbb{T}^n$. □

We can finally verify with Proposition 9.4.6 that $\mathbb{T}^{\mathfrak{c}}$ has a dense cyclic subgroup:

Theorem 9.4.8. *The group $\mathbb{T}^{\mathfrak{c}}$ is monothetic.*

Proof. Let B be a Hamel base of $\mathbb{R}$ on $\mathbb{Q}$ that contains 1 and let $B_0 = B\setminus\{1\}$; in particular, $|B_0| = |B| = \mathfrak{c}$. To see that the element $x = (x_b)_{b\in B_0} \in \mathbb{T}^{B_0}$, defined by $x_b = b+\mathbb{Z} \in \mathbb{R}/\mathbb{Z} = \mathbb{T}$ for every $b \in B_0$, is a topological generator of the group $\mathbb{T}^{B_0}$, it suffices to check that for every finite subset B_1 of B_0 the projection $p\colon \mathbb{T}^{B_0} \to \mathbb{T}^{B_1}$ sends $H = \langle x\rangle$ to a dense subgroup of $\mathbb{T}^{B_1}$. This follows from Corollary 9.4.7. □

Clearly, $\mathbb{T}^\kappa$ is not monothetic in case $\kappa > \mathfrak{c}$, since a separable group has weight $\leq \mathfrak{c}$.

9.5 Exercises

Exercise 9.5.1. Prove by induction on $n \in \mathbb{N}_+$ that for every discrete (so closed) subgroup H of $\mathbb{R}^n$ there exist $m \leq n$ and m linearly independent vectors $v_1, \dots, v_m \in H$ such that $H = \langle v_1, \dots, v_m\rangle \cong \mathbb{Z}^m$.

Hint. The case $n = 1$ is Proposition 3.1.11. Assume $n > 1$. Pick any $h \in H \setminus \{0\}$ and let L be the line $\mathbb{R}h$ in $\mathbb{R}^n$. Since $H_1 = H \cap L \neq \{0\}$ is a discrete subgroup of $L \cong \mathbb{R}$, we conclude that $H = \langle a\rangle$ is cyclic, by Proposition 3.1.11. We can apply Lemma 9.3.3 to claim that the image $p(H)$ of H along the canonical projection $p\colon \mathbb{R}^n \to \mathbb{R}^n/L \cong \mathbb{R}^{n-1}$ is a discrete subgroup of $\mathbb{R}^{n-1}$. Then our inductive hypothesis yields $p(H) = \langle v_2', \dots, v_m'\rangle$ for some linearly independent vectors $v_2', \dots, v_m'$ in $\mathbb{R}^n/L$. Pick $v_i \in H$ such that $p(v_i) = v_i'$ for $i \in \{2, \dots, n\}$. Then, with $v_1 = a$, we have the desired presentation

$$H = \langle v_1, \dots, v_m\rangle = \langle v_1\rangle \oplus \langle v_2, \dots, v_m\rangle \cong \mathbb{Z} \oplus \mathbb{Z}^{m-1} \cong \mathbb{Z}^m.$$

Exercise 9.5.2. Prove that the class $\mathcal{EC}$ of elementary compact abelian groups is stable under taking closed subgroups, quotients and finite direct products.

Hint. Use Corollary 9.4.3.

Exercise 9.5.3. Prove that every elementary locally compact abelian group is a quotient of an elementary locally compact abelian group of the form $\mathbb{R}^n \times \mathbb{Z}^m$.

Exercise 9.5.4. Prove that every elementary locally compact abelian group can be embedded as a closed subgroup of $\mathbb{R}^n \times \mathbb{T}^k$ for some $n, k \in \mathbb{N}$.

Exercise 9.5.5. Determine for which of the following six possible choices of the vector $v \in \mathbb{R}^4$ the subgroup $\langle v \rangle + \mathbb{Z}^4$ of $\mathbb{R}^4$ is dense:

$$(\sqrt{2}, \sqrt{3}, \sqrt{5}, \sqrt{6}),\ (\sqrt{2}, \sqrt{3}, \sqrt{5}, \sqrt{7}),\ (\log 2, \log 3, \log 5, \log 6),\ (\log 2, \log 3, \log 5, \log 7)$$
$$(\log 3, \log 5, \log 7, \log 9), \quad \text{and} \quad (\log 5, \log 7, \log 9, \log 11).$$

Exercise 9.5.6. Let V be a hyperplane in $\mathbb{R}^n$ determined by $\sum_{i=1}^n a_i x_i = 0$, for $a_1, \dots, a_n \in \mathbb{R}$, such that there exists at least one coefficient $a_i = 1$. Prove that the subgroup $H = V + \mathbb{Z}^n$ of $\mathbb{R}^n$ is not dense if and only if all the coefficients a_i are rational.

Hint. Clearly, $V = (\mathbb{R}\alpha)^o = (\mathbb{R}\alpha)^\dagger$, where $\alpha = (a_1, \dots, a_n)$. Hence, $H^\dagger = (\mathbb{R}\alpha)^{\dagger\dagger} \cap (\mathbb{Z}^n)^\dagger = \mathbb{R}\alpha \cap \mathbb{Z}^n$, by Lemma 9.4.4(d). Clearly, $\mathbb{R}\alpha \cap \mathbb{Z}^n \neq \{0\}$ if and only if all coefficients a_i are rational. On the other hand, $H^\dagger = \mathbb{R}\alpha \cap \mathbb{Z}^n \neq \{0\}$ if and only if H is not dense in $\mathbb{R}^n$, by Proposition 9.4.5.

Exercise 9.5.7. (a) Prove that a subgroup H of $\mathbb{T}$ is dense if and only if H is infinite.
(b) Determine the minimal (with respect to inclusion) dense subgroups of $\mathbb{T}$.
(c)* Determine the minimal (with respect to inclusion) dense subgroups of $\mathbb{T}^2$.

Exercise 9.5.8. Prove that if $\phi\colon \mathbb{R}^n \to \mathbb{R}^n$ is a continuous endomorphism, then ϕ is also an $\mathbb{R}$-linear transformation.

10 Subgroups of compact groups

10.1 Big subsets of groups

Definition 10.1.1. A subset B of a group G is *left* (respectively, *right*) *big* if there exists a finite subset F of G such that $FB = G$ (respectively, $BF = G$). Moreover, B is *big* if it is simultaneously left and right big.

The notions of left and right big subsets coincide for abelian groups, but this need not be true in general.

Example 10.1.2. Let G be the free product of a cyclic group $A = \langle a \rangle$ of order 2 and an infinite cyclic group $C = \langle b \rangle$. Every element of $G \setminus C$ can be uniquely written as a product

$$w = b^{n_0} \cdot a \cdot b^{n_1} \cdot a \cdot b^{n_2} \cdots a \cdot b^{n_{k-1}} \cdot a \cdot b^{n_k}, \tag{10.1}$$

where $k \in \mathbb{N}_+$, and $n_0, \dots, n_k$ are integers such that if $k > 1$ then $n_1, \dots, n_{k-1}$ are nonzero, while n_0 and n_k may also have value 0. Obviously, the elements $w \in C$ can be obtained in the form (10.1) with $k = 0$ (and consequently, $w = b^{n_0}$). One refers to (10.1) as the *reduced form* of the element $w \in G$. The product $w_1 \cdot w_2$ of two words w_1 and w_2 can be brought to a reduced form after a finite number of cancelations, i. e., either $w_1 \cdot w_2$ is already in reduced form, or there exist factorizations $w_1 = \tilde{w}_1 \cdot u$ and $w_2 = u^{-1} \cdot \tilde{w}_2$ such that $w_1 \cdot w_2 = \tilde{w}_1 \cdot \tilde{w}_2$ is in reduced form.

Set

$$Y = \{w \in G : k \in \mathbb{N}_+ \text{ and } n_k = 0 \text{ in (10.1)}\} \quad \text{and} \quad X = G \setminus Y.$$

Note that $Y = Xa$, so that $X = Ya$, too. Thus, both X and Y are right big, since $G = X \cup Xa = Y \cup Ya$. Let us see now that neither X nor Y is left big. Since $Y = Xa$, it suffices to see that X is not left big.

Let us first note that the inverse of an element w as in (10.1) is given by

$$w^{-1} = b^{-n_k} \cdot a \cdot b^{-n_{k-1}} \cdot a \cdot b^{-n_{k-2}} \cdot a \cdots a \cdot b^{-n_1} \cdot a \cdot b^{-n_0}.$$

Assume that $G = \bigcup_{i=1}^{l} g_i X$ for some $g_1, \dots, g_l \in G$. There exist $n \neq m$ in $\mathbb{N}_+$ such that $b^n a \in g_i X$ and $b^m a \in g_i X$ for some $i \in \{1, \dots, l\}$. Then, for some $x \in X$, one has $b^n a = g_i x$, and so $g_i = b^n a x^{-1}$. Now $x = b^{n_0} a b^{n_1} a b^{n_2} a \cdots a b^{n_{k-1}} a b^{n_k}$ with either $x = 1$ or $n_k \neq 0$. In both cases the leading term of the reduced form of the word g_i is $b^n a$. An analogous argument shows that the leading term of the reduced form of the word g_i is $b^m a$, a contradiction.

On the other hand, a subset B of a group G is left big if and only if B^{-1} is right big; so, in case B is symmetric, B is left big precisely when B is right big.

Moreover, every nonempty subset of a finite group is big, while every big subset of an infinite group is necessarily infinite.

https://doi.org/10.1515/9783110654936-010

Example 10.1.3. An infinite subset B of $\mathbb{Z}$ is big if and only if:
(a) B is unbounded from above and from below; and
(b) if $B = \{b_n : n \in \mathbb{Z}\}$ is a one-to-one monotone enumeration of B, then the set $\{b_{n+1} - b_n : n \in \mathbb{Z}\}$ is bounded.

We list some basic properties of big subsets.

Lemma 10.1.4. (a) *Assume that B_j is a left big subset of the group G_j, for $j \in \{1, \dots, n\}$. Then $B_1 \times \cdots \times B_n$ is a left big subset of $G_1 \times \cdots \times G_n$.*
(b) *Let G, H be groups and $f: G \to H$ a surjective homomorphism. Then:*
(b_1) *if B is a left big subset of H, then $f^{-1}(B)$ is a left big subset of G;*
(b_2) *if B' is a left big subset of G, then $f(B')$ is a left big subset of H.*

Proof. (a) and (b_2) follow directly from the definition.

(b_1) Assume that there exists a finite subset F of H such that $FB = H$. Let E be a finite subset of G such that $f(E) = F$. Then we see that $G = Ef^{-1}(B)$, and so $f^{-1}(B)$ is left big in G. In fact, for $x \in G$, there exists $a \in F$ such that $f(x) \in aB$; for $c \in E$ such that $f(c) = a$, we get $f(x) \in aB = f(c)B$, so that $f(c^{-1}x) \in B$, and hence $c^{-1}x \in f^{-1}(B)$, that is, $x \in Ef^{-1}(B)$. □

If the group homomorphism $f: G \to H$ in Lemma 10.1.4(b) is not surjective, then the property may fail (e. g., taking the inclusion $f: G = 2\mathbb{Z} \hookrightarrow \mathbb{Z} = H$, then $B = 1 + 2\mathbb{Z}$ is big in H, yet $f^{-1}(B)$ is empty). The next proposition gives an easy remedy.

Proposition 10.1.5. *Let G be a group and B a left big subset of G. Then:*
(a) *for every subgroup H of G, $B^{-1}B \cap H$ is a big subset of H;*
(b) *for every $a \in G$, there exists $n \in \mathbb{N}_+$ such that $a^n \in B^{-1}B$.*

Proof. (a) Let F be a finite subset of G such that $FB = G$. For $f \in F$, if $fB \cap H \neq \emptyset$, choose $a_f \in fB \cap H$ and let $E = \{a_f : f \in F, fB \cap H \neq \emptyset\}$. For every $h \in H$, there exists $f \in F$ such that $h \in fB$; since $a_f \in fB$, $a_f^{-1}h \in B^{-1}B$. So, $H \subseteq E(B^{-1}B \cap H)$, that is, $B^{-1}B \cap H$ is left big in H. Since $B^{-1}B \cap H$ is symmetric, it is also right big.

(b) Take $H = \langle a \rangle$. If H is finite, there is nothing to prove as $a^n = e_G \in B^{-1}B$, where $n \in \mathbb{N}_+$ is the order of a. Otherwise, $H \cong \mathbb{Z}$ and, since $B^{-1}B \cap H$ is big in H by (a), there exists $n \in \mathbb{N}_+$ such that $a^n \in B^{-1}B$, by Example 10.1.3(a). □

Combining Proposition 10.1.5(a) with Lemma 10.1.4(b_1), we get

Corollary 10.1.6. *Let G, H be groups and $f: G \to H$ a group homomorphism. If B is a left big subset of H, then $f^{-1}(B^{-1}B)$ is a left big subset of G.*

Now we introduce a concept in the opposite direction of being big.

Definition 10.1.7. Call a nonempty subset S of an infinite group G *left* (respectively, *right*) *small* if there exist elements $\{g_n : n \in \mathbb{N}\}$ of G such that $g_n S \cap g_m S = \emptyset$ (respectively, $Sg_n \cap Sg_m = \emptyset$) whenever $m \neq n$ in $\mathbb{N}$.

Clearly, a nonempty subset S of an infinite group G is left small if and only if S^{-1} is right small.

Lemma 10.1.8. *Let G be an infinite group and S a subset of G.*
(a) *If S is finite, then S is left and right small.*
(b) *If S is left big, then S is not right small.*
(c) *If S is left small, then S is not right big.*
(d) *If S is symmetric and small, then S is not big.*
(e) *If SS^{-1} is not big, then S is left small.*

Proof. (a) obviously follows from (e).

(b) Let F be a finite set in G such that $G = FS$. To show that S^{-1} is not left small, pick an arbitrary faithfully indexed sequence $\{g_n\}_{n\in\mathbb{N}}$ in G. Then there exist $f \in F$ and two distinct $n, m \in \mathbb{N}$ such that $g_n \in fS \ni g_m$, so that $g_n = fs_1$ and $g_m = fs_2$ for some $s_1, s_1 \in S$. Hence, $g_m^{-1} g_n = s_2^{-1} s_1$ and consequently, $g_n s_1^{-1} = g_m s_2^{-1}$. Therefore, $g_n S^{-1} \cap g_m S^{-1} \neq \emptyset$. Hence, S^{-1} is not left small, i. e., S is not right small.

(c) follows from (b), while (d) follows from (c).

(e) Build the required infinite subset $\{g_n : n \in \mathbb{N}\}$ of G by induction, using the fact that $FSS^{-1} \neq G$ for every finite subset F of G, by hypothesis. Put $g_0 = e_G$ and assume that $F = \{g_0, \dots, g_n\}$ is already found. Then there exists $g_{n+1} \in G \setminus FSS^{-1}$, and it is straightforward to prove that $g_{n+1} S \cap g_i S = \emptyset$ for every $i \in \{0, \dots, n\}$. □

10.2 Precompact groups

10.2.1 Totally bounded and precompact groups

Using big subsets, we give the following fundamental definition.

Definition 10.2.1. A topological group G is *totally bounded* if every nonempty open set U of G is left big, and G is *precompact* if it is Hausdorff and totally bounded.

Clearly, compact groups are precompact.

Remark 10.2.2. The notions of total boundedness and precompactness defined by using *left* big subsets are only apparently asymmetric. Indeed, a topological group G is totally bounded if and only if every nonempty open set U of G is right big (see Exercise 10.4.1).

Lemma 10.2.3. *If $f: G \to H$ is a continuous surjective homomorphism of topological groups, then H is totally bounded whenever G is totally bounded. If G carries the initial* *topology of f and H is totally bounded, then also G is totally bounded. In particular, G is totally bounded if and only if $\mathfrak{h}G$ is precompact.*

Proof. To prove the first assertion, it suffices to recall that the homomorphic image of a left big subset under a surjective homomorphism is left big, by Lemma 10.1.4(b_2). The second assertion follows from the fact that the open sets of G are preimages of the open sets of H. So, Lemma 10.1.4(b_1) applies.

The last assertion follows from the first and second since $\mathfrak{h}G = G/\mathrm{core}(G)$ carries the initial topology with respect to the canonical projection $G \to G/\mathrm{core}(G)$. □

The simple connection between total boundedness and precompactness from Lemma 10.2.3 is frequently used in the sequel. Most often we deal with precompact groups, leaving the obvious counterpart for totally bounded groups to the reader, or vice versa.

Proposition 10.2.4. *If $\{G_i : i \in I\}$ is a family of topological groups, then $G = \prod_{i\in I} G_i$ is totally bounded if and only if each G_i is totally bounded.*

Proof. If G is totally bounded, then each G_i is totally bounded by Lemma 10.2.3.

Assume that each G_i is totally bounded and let U be a nonempty open set of G. Then there exist a finite subset J of I and a nonempty open set V of $G_J = \prod_{i\in J} G_i$ such that $p_J^{-1}(V) \subseteq U$, where $p_J : G \to G_J$ is the canonical projection. Since p_J is surjective and V is left big in G_J by Lemma 10.1.4(a), it follows from Lemma 10.1.4(b_1) that U is left big as well. □

Now we see that total boundedness is preserved under taking subgroups.

Proposition 10.2.5. *All subgroups of totally bounded groups are totally bounded. In particular, all subgroups of compact groups are precompact.*

Proof. Let H be a subgroup of G. If $U \in \mathcal{V}_H(e_G)$, there exists $W \in \mathcal{V}_G(e_G)$ such that $U = W \cap H$. Pick $V \in \mathcal{V}_G(e_G)$ such that $V^{-1}V \subseteq W$. Since V is left big in G, $(V^{-1}V) \cap H$ is left big in H, by Proposition 10.1.5(a). Since $(V^{-1}V) \cap H \subseteq W \cap H = U$, we conclude that U is left big in H. □

One can show that the precompact groups are precisely the subgroups of the compact groups. This requires two steps as the next theorem shows.

Theorem 10.2.6. (a) *A Hausdorff group G having a dense precompact subgroup H is necessarily precompact.*
(b) *The compact groups are precisely the complete precompact groups.*

Proof. (a) For every $U \in \mathcal{V}_G(e_G)$, choose an open $V \in \mathcal{V}_G(e_G)$ with $VV \subseteq U$. By the precompactness of H, there exists a finite subset F of H such that $H = F(V \cap H)$. Then $G = HV = F(V \cap H)V \subseteq FVV \subseteq FU$.

(b) Compact groups are complete by Proposition 8.2.6 and precompact.

To prove the other implication, take a complete precompact group G. To prove that G is compact, it suffices to verify that every ultrafilter $\mathcal{U}$ on G converges (see Lemma B.5.7(b)). First, we show that $\mathcal{U}$ is a Cauchy filter. Indeed, if $U \in \mathcal{V}_G(e_G)$, then U is a big subset of G and so there exist $g_1, \dots, g_n \in G$ such that $G = \bigcup_{i=1}^n g_i U$. Since $\mathcal{U}$ is an ultrafilter, $g_i U \in \mathcal{U}$ for some $i \in \{1, \dots, n\}$. Analogously, one can prove that there exists $g \in G$ such that $Ug \in \mathcal{U}$. By Lemma 7.2.2(a), $\mathcal{U}$ is a Cauchy filter. According to Proposition 7.2.3, we conclude that $\mathcal{U}$ converges. □

We have described the precompact groups internally (as the Hausdorff groups having big nonempty open sets), or externally (as the subgroups of the compact groups). Now we describe total boundedness in terms of small subsets.

Lemma 10.2.7. *For a topological group G, the following are equivalent:*
(a) *G is not totally bounded;*
(b) *G has a left small nonempty open set;*
(c) *G has a right small nonempty open set.*

Proof. (b)⇔(c) since a subset S of G is left small if and only if S^{-1} is right small, while (b)⇒(a) is a consequence of Lemma 10.1.8(c).

(a)⇒(b) If $U \in \mathcal{V}_G(e_G)$ is not left big, choose $V \in \mathcal{V}_G(e_G)$ such that $VV^{-1} \subseteq U$. Then V is left small by Lemma 10.1.8(e). □

The following result shows that totally bounded groups satisfy a much stronger version of the axiom (gt3).

Lemma 10.2.8. *If G is a totally bounded group, then for every $U \in \mathcal{V}(e_G)$, there exists $V \in \mathcal{V}(e_G)$ such that $g^{-1}Vg \subseteq U$ for all $g \in G$.*

Proof. Let $W \in \mathcal{V}(e_G)$ be symmetric and such that $WWW \subseteq U$. By hypothesis, $G = FW$ for some finite subset F of G. For every $a \in F$, pick $V_a \in \mathcal{V}(e_G)$ such that $a^{-1}V_a a \subseteq W$, and let $V = \bigcap_{a \in F} V_a$. Then $g^{-1}Vg \subseteq U$ for every $g \in G$: if $g \in aW$ for some $a \in F$, then $g = aw$ for some $w \in W$, and so

$$g^{-1}Vg = w^{-1}a^{-1}Vaw \subseteq w^{-1}a^{-1}V_a aw \subseteq w^{-1}Ww \subseteq U. \qquad \square$$

The next theorem reveals a remarkable dichotomy concerning monothetic locally compact groups.

Theorem 10.2.9 (Weil lemma). *Let G be a locally compact monothetic group. Then G is either compact or discrete (in the latter case, G is cyclic).*

Proof. By Corollary 3.1.23, the group G is abelian, so we denote it additively.

Let $x \in G$ be such that $\overline{\langle x \rangle} = G$. If G is finite, then G is both compact and discrete. We can suppose without loss of generality that $\langle x \rangle \cong \mathbb{Z}$ is infinite, and so also that $\mathbb{Z}$ is a subgroup of G (thus, $G = \overline{\mathbb{Z}}$).

If the induced topology τ on $\mathbb{Z}$ is discrete, then $\mathbb{Z}$ is closed by Proposition 3.1.17, and so $G = \mathbb{Z}$ is discrete. Suppose now that τ is not discrete. Our aim is to show that τ is precompact. Then, as G is locally compact and so complete by Proposition 8.2.6, the density of $\mathbb{Z}$ in G yields that $G = \widetilde{\mathbb{Z}} = \overline{\mathbb{Z}}$ is compact, by Theorem 10.2.6.

To verify that τ is totally bounded, we need the following property of $(\mathbb{Z}, \tau)$:

(P) every open set of $\mathbb{Z}$ has no maximal element, and so in particular every nonempty open set of $\mathbb{Z}$ contains positive elements.

Indeed, let U be an open set of $\mathbb{Z}$. Since the assertion is true for $U = \emptyset$, assume that $U \neq \emptyset$ and $0 \in U$. If U has a maximal element, then $-U$ is an open set of $\mathbb{Z}$ that contains 0 and it has a minimal element, so $U \cap -U$ is a finite open neighborhood of 0 in $\mathbb{Z}$; thus, τ is discrete against the assumption.

Pick a compact neighborhood U of 0 in G and a symmetric open neighborhood V of 0 in G with $V + V \subseteq U$. Since U is compact and V is open, there exist $g_1, \dots, g_m \in G$ such that $U \subseteq \bigcup_{i=1}^m (g_i + V)$. By (P), there exist positive $n_1, \dots, n_m \in \mathbb{Z}$ such that $n_i \in g_i + V$ for every $i \in \{1, \dots, m\}$; equivalently, $g_i \in n_i - V = n_i + V$. Thus,

$$U \subseteq \bigcup_{i=1}^m (g_i + V) \subseteq \bigcup_{i=1}^m (n_i + V + V) \subseteq \bigcup_{i=1}^m (n_i + U),$$

and this implies

$$U \cap \mathbb{Z} \subseteq \bigcup_{i=1}^m (n_i + U \cap \mathbb{Z}). \tag{10.2}$$

We show that $U \cap \mathbb{Z}$ is a big subset of $\mathbb{Z}$; more precisely, $(U \cap \mathbb{Z}) + F = \mathbb{Z}$, where $F = \{1, \dots, N\}$ and $N = \max\{n_1, \dots, n_m\}$. Let $t \in \mathbb{Z}$; since $U \cap \mathbb{Z}$ has no maximal element by (P), there exists $s \in U \cap \mathbb{Z}$ such that $s \geq t$. Define

$$s_t = \min\{s \in U \cap \mathbb{Z} : s \geq t\}.$$

By (10.2), $s_t = n_i + u_t$ for some $i \in \{1, \dots, m\}$ and $u_t \in U \cap \mathbb{Z}$. Since $n_i > 0$ for every $i \in \{1, \dots, m\}$, we conclude that $u_t < s_t$, and so $u_t < t$ by the choice of s_t. Hence, $u_t < t \leq s_t$. Then $t - u_t \leq s_t - u_t = n_i \in F$. Thus, $t = u_t + (t - u_t) \in (U \cap \mathbb{Z}) + F$ and consequently, $(U \cap \mathbb{Z}) + F = \mathbb{Z}$. Therefore, the topology τ is totally bounded. □

Corollary 10.2.10. *Let G be a locally compact group and $x \in G$. Then $\overline{\langle x \rangle}$ is either compact or discrete.*

Corollary 10.2.11. *Let G be a locally compact group such that M_G is dense in G. Then G is compact, connected and $w(G) \leq \mathfrak{c}$.*

Proof. The compactness of G and $w(G) \leq \mathfrak{c}$ follow, respectively, from Theorem 10.2.9 and Exercise 5.4.4 (the weaker hypothesis $M_G \neq \emptyset$ is sufficient). Now assume that G

is not connected. Then G has a nontrivial proper open subgroup U, by Theorem 8.5.2 (note that G cannot be discrete by the density of the proper subset M_G of G). By the density of M_G, the open set $U \setminus \{0\}$ meets M_G. Pick $0 \neq x \in U \cap M_G$. Then $\langle x\rangle \subseteq U$ cannot be dense in G, a contradiction. □

Remark 10.2.12. Let α be an infinite cardinal. A topological group G is *α-totally bounded* (shortly, *α-bounded*) if for every nonempty open set U of G there exists a subset F of G of size $< \alpha$ such that $FU = G$. Clearly, the ω-totally bounded groups are precisely the totally bounded groups. As in the case of totally bounded groups, the counterpart of this notion using the equality $UF = G$ does not lead to a different notion.

The ω_1-bounded groups are also known under the name *ω-narrow* groups; they were introduced by Guran [162] under the name *$\aleph_0$-bounded* groups. It is known that every ω-narrow group is topologically isomorphic to a subgroup of a product of second countable groups (see [162]).

10.2.2 A second (internal) approach to the Bohr compactification

Proposition 10.2.13. *For every topological group (G,τ), there exists the finest totally bounded group topology τ^+ on G coarser than τ.*

Proof. Let $\{\tau_i : i \in I\}$ be the family of all totally bounded group topologies on G coarser than τ, and let $\tau^+ = \sup\{\tau_i : i \in I\}$. Then (G,τ^+) is topologically isomorphic to the diagonal subgroup $\Delta_G = \{x = (x_i)_{i\in I} \in G^I : x_i = x_j \text{ for every } i,j \in I\}$ of $\prod_{i\in I}(G,\tau_i)$ (see Exercise 3.5.10), which is totally bounded, by Propositions 10.2.4 and 10.2.5. Hence, τ^+ is still totally bounded, and it is the finest totally bounded group topology on G coarser than τ. □

For a discrete group (G,δ_G), we denote δ_G^+ by $\mathcal{P}_G$. In case G is abelian, $\mathcal{P}_G$ is a precompact functorial topology, by Proposition 10.2.14.

Proposition 10.2.13 produces, for every topological group G, a “universal” precompact continuous surjective homomorphic image $q: G \to G^+$:

Proposition 10.2.14. *For every topological group (G,τ), the quotient group*

$$G^+ := G/\overline{\{e_G\}}^{\tau+} = \mathfrak{h}(G,\tau^+)$$

equipped with the quotient topology of τ^+ is precompact, and every continuous homomorphism $f: (G,\tau) \to P$, where P is a precompact group, factors through the canonical projection $q: G \to G^+$.

Proof. The precompactness of the quotient G^+ with the quotient topology of τ^+ follows from Lemma 10.2.3. Let τ_1 be the initial topology of G with respect to $f: G \to P$. According to Proposition 10.2.5, we may assume that f is surjective. Then $\tau_1 \le \tau$ and

τ_1 is totally bounded by Lemma 10.2.3, so Proposition 10.2.13 implies that $\tau_1 \le \tau^+$. Therefore, $f: (G, \tau^+) \to P$ is continuous as well. Now we can factorize f through the canonical projection $q: G \to G^+$ according to Lemma 3.4.1. □

According to Proposition 10.2.14, the assignment $G \mapsto G^+$ induces a functor from the category of all topological groups to the subcategory of all precompact groups. Now we see that $\widetilde{G^+} \cong bG$. To this end, let $j_G: G \to \widetilde{G^+}$ be the composition of the canonical homomorphism $q: G \to G^+$ and the inclusion of $G^+ \hookrightarrow \widetilde{G^+}$.

Theorem 10.2.15. *For every topological group G, there exists a topological isomorphism $i: bG \to \widetilde{G^+}$ such that $i \circ b_G = j_G$.*

Proof. In view of Theorem 8.6.1, it suffices to prove that $j_G: G \to \widetilde{G^+}$ has the universal property of $b_G: G \to bG$.

Let $f: G \to K$ be a continuous homomorphism, where K is a compact group. By Proposition 10.2.14, f factorizes through $q: G \to G^+$, i.e., there exists a continuous homomorphism $h: G^+ \to K$ such that $f = h \circ q$:

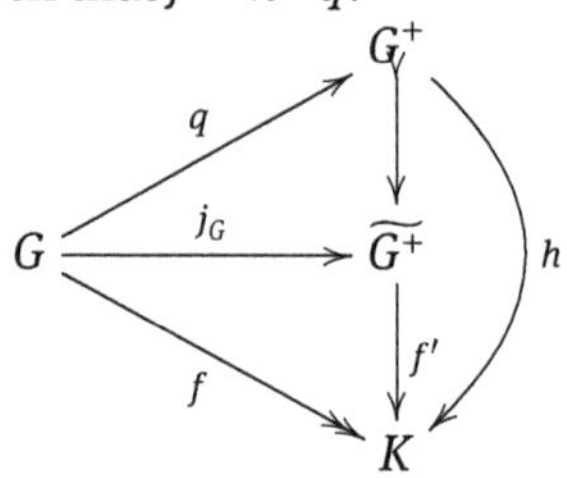

By Proposition 8.2.6, the group K is complete, being compact. In view of Corollary 7.1.20, we can extend h to a continuous homomorphism $f': \widetilde{G^+} \to K$ with $f' \circ j_G = f$. The uniqueness of f' follows from the fact that $j_G(G)$ is dense in $\widetilde{G^+}$. □

The above theorem implies that for (G, τ) one has $n(G) = \operatorname{core}(G, \tau^+) = \overline{\{e_G\}}^{\tau^+}$, so (G, τ) is MAP (respectively, minimally almost periodic) precisely when τ^+ is Hausdorff (respectively, τ^+ is indiscrete). In particular, precompact groups are MAP.

10.2.3 Precompactness of the topologies induced by characters

For a subset E of an abelian group G, we set $E_{(2)} = E - E$, $E_{(4)} = E - E + E - E$, $E_{(6)} = E - E + E - E + E - E$, and so on.

Now we adopt a different approach to describe the precompact abelian groups, based on the use of characters. Our first aim is to see that the group topologies induced by characters are always totally bounded.

Proposition 10.2.16. *If G is an abelian group, $\delta > 0$, and $\chi_1, \dots, \chi_s \in G^*$ with $s \in \mathbb{N}_+$, then $U(\chi_1, \dots, \chi_s; \delta)$ is a big subset of G. Moreover, for every $a \in G$, there exists $n \in \mathbb{N}_+$ such that $na \in U(\chi_1, \dots, \chi_s; \delta)$.*

Proof. Let $h: G \to \mathbb{S}^s$, $x \mapsto (\chi_1(x), \dots, \chi_s(x))$, and let

$$B = \left\{(z_1, \dots, z_s) \in \mathbb{S}^s : |\mathrm{Arg}(z_i)| < \frac{\delta}{2} \text{ for } i \in \{1, \dots, s\}\right\}$$
$$= \left\{z \in \mathbb{S} : |\mathrm{Arg}(z)| < \frac{\delta}{2}\right\}^s.$$

Then B is big in $\mathbb{S}^s$, as $\{z \in \mathbb{S} : |\mathrm{Arg}(z)| < \frac{\delta}{2}\}$ is big in $\mathbb{S}$ and

$$B^{-1}B \subseteq C := \{(z_1, \dots, z_s) \in \mathbb{S}^s : |\mathrm{Arg}(z_i)| < \delta \quad \text{for } i \in \{1, \dots, s\}\}.$$

Therefore, $U(\chi_1, \dots, \chi_s; \delta) = h^{-1}(C)$ is big in G, by Corollary 10.1.6.

The second statement follows from Proposition 10.1.5(b), since

$$U\left(\chi_1, \dots, \chi_s; \frac{\delta}{2}\right) - U\left(\chi_1, \dots, \chi_s; \frac{\delta}{2}\right) \subseteq U(\chi_1, \dots, \chi_s; \delta). \qquad \square$$

Corollary 10.2.17. *Let G be an abelian group and H a subgroup of G^*. Then $\mathcal{T}_H$ is totally bounded. Moreover, $\mathcal{T}_H$ is precompact if and only if H separates the points of G. In particular, $\mathfrak{B}_G$ is precompact.*

It requires a considerable effort to prove that, conversely, every totally bounded group topology on an abelian group G has the form $\mathcal{T}_H$ for some subgroup H of G (see Theorem 11.4.2).

Remark 10.2.18. (a) At this stage, since every abelian group G admits the finest totally bounded group topology $\mathcal{P}_G$ by Proposition 10.2.14, Corollary 10.2.17 gives so far only the inequality $\mathcal{P}_G \geq \mathfrak{B}_G$, which implies that also $\mathcal{P}_G$ is precompact.

(b) It follows easily from Corollary 10.2.17 and Proposition 10.2.16 that for an abelian group G and every neighborhood E of 0 in the Bohr topology $\mathfrak{B}_G$ (namely, a subset E of G containing a subset of the form $U(\chi_1, \dots, \chi_n; \varepsilon)$ with characters $\chi_i: G \to \mathbb{S}$ for $i \in \{1, \dots, n\}$ and $\varepsilon > 0$) there exists a big subset B of G such that $B_{(8)} \subseteq E$: just take $B = U(\chi_1, \dots, \chi_n; \varepsilon/8)$.
Surprisingly, the converse is also true. Namely, we shall obtain as a corollary of the Følner lemma that every subset E of G satisfying $B_{(8)} \subseteq E$ for some big subset B of G must be a neighborhood of 0 in $\mathfrak{B}_G$ (see Corollary 11.2.6). This means that $\mathcal{P}_G = \mathfrak{B}_G$ (see Corollary 11.2.7).

(c) According to a classical result of E. Følner, a topological abelian group G is MAP if and only if for every $a \in G \setminus \{0\}$ there exists a big subset B of G such that a does not belong to the closure of $B_{(4)}$. A weaker form of this theorem is proved in Følner theorem 11.3.5 (with the bigger set $B_{(10)}$ in place of $B_{(4)}$).

10.3 Unitary representations of locally compact groups

The nice structure theory of locally compact groups (see §11.6 for the abelian case) is due to the Haar integral and the Haar measure of locally compact groups. Every locally compact group G admits a *right Haar integral* (see §12.2). The Haar integral gives the possibility to obtain unitary representations of locally compact groups (one can see in §12.1.2 how these unitary representations arise in the case of compact abelian groups).

For a complex Hilbert space $\mathcal{H}$, denote by $U(\mathcal{H})$ the group of its unitary operators (briefly, unitary group of $\mathcal{H}$) equipped with the strong operator topology (this is, the coarsest topology on $U(\mathcal{H})$ such that, for each fixed $x \in \mathcal{H}$, the evaluation map $U(\mathcal{H}) \to \mathcal{H}$, $T \mapsto T(x)$, is continuous with respect to the norm topology of $\mathcal{H}$).

Definition 10.3.1. A *unitary representation* of a locally compact group G is a continuous homomorphism $V: G \to U(\mathcal{H})$, where $\mathcal{H}$ is a complex Hilbert space. A unitary representation $V: G \to U(\mathcal{H})$ of G is said to be *irreducible* if the only V-invariant closed subspaces of $\mathcal{H}$ are $\{0\}$ and $\mathcal{H}$.

Associated with any unitary representation V of G one has a cardinal number $d(V)$, called the *degree* of V, which is by definition the cardinality of an orthonormal base (briefly, the *dimension*) of the complex Hilbert space $\mathcal{H}$. If $d(V)$ is finite, then the unitary group $U(\mathcal{H}) = U(d(V))$ is compact (see Example 8.1.4).

Theorem 10.3.2 (Gel′fand–Raĭkov theorem). *For every locally compact group G and $a \in G \setminus \{e_G\}$, there exists an irreducible unitary representation $V: G \to U(\mathcal{H})$ of G by unitary operators of some Hilbert space $\mathcal{H}$ such that $V(a) \neq e_{U(\mathcal{H})}$.*

The proof of this theorem can be found in [174, 22.12].

If G is compact, all irreducible unitary representations of G have finite degree; this case is further discussed in Theorem 10.3.3. In particular, a topological group G is MAP if and only if the finite-degree irreducible unitary representations of G separate the points (see Corollary 10.3.6).

The smaller class of locally compact groups G with the property $d(V) < \infty$ for every irreducible unitary representation V of G was introduced and characterized by Moore [213] (later these locally compact groups were named *Moore groups* in [246, 248]). In the same paper, Moore characterized also the locally compact groups G such that all degrees $d(V)$, when V varies among all irreducible unitary representations of G, are bounded by some natural number.

The compact case of the Gel′fand–Raĭkov theorem is known as *Peter–Weyl–van Kampen theorem*, which shows that the compact groups are Moore groups:

Theorem 10.3.3 (Peter–Weyl–van Kampen theorem). *Let G be a compact group. For every $a \in G \setminus \{e_G\}$, there exist $n \in \mathbb{N}_+$ and a continuous homomorphism $f: G \to U(n)$ such that $f(a) \neq e_G$.*

The proof of the Peter–Weyl–van Kampen theorem can be found in [174, 22.13]. We give now several relevant consequences.

Corollary 10.3.4. *If G is a compact group, then G is isomorphic to a (closed) subgroup of some power $\mathbb{U}^I$ of the group $\mathbb{U}$.*

Proof. In view of Theorem 10.3.3, there exist a nonempty set I and a family $\{f_i: G \to \mathbb{U}: i \in I\}$ of continuous homomorphisms that separate the points of G. Then the diagonal map determined by $\{f_i: G \to \mathbb{U}: i \in I\}$ is a continuous injective homomorphism $G \hookrightarrow \mathbb{U}^I$, $x \mapsto (f_i(x))_{i\in I}$. By the compactness of G and the open mapping theorem (Theorem 8.4.1), this is the required embedding. □

Now we provide an alternative way to describe the Bohr compactification $b_G: G \to bG$ of a topological group G. By Theorem 10.2.15, $b_G: G \to bG$ has the property:

(Bc) every continuous homomorphism $f: G \to U(n)$ with $n \in \mathbb{N}$ factorizes through b_G by means of a unique continuous homomorphism $f': bG \to U(n)$ with $f' \circ b_G = f$:

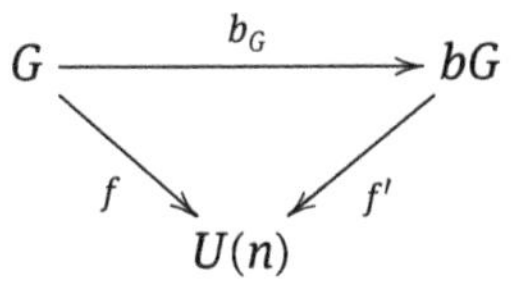

Our aim now is to show (without any recourse to Theorem 10.2.15) that a continuous homomorphism $h: G \to K$, with $K = \overline{h(G)}$ compact and satisfying (Bc), exists and coincides, up to isomorphism, with $b_G: G \to bG$.

Proposition 10.3.5. *The Bohr compactification of a topological group G can be equivalently defined by the property* (Bc).

Proof. First, we verify that if a continuous homomorphism $h: G \to K$, where K is a compact group, has the property (Bc) (i. e., every continuous homomorphism $f: G \to U(n)$ for some $n \in \mathbb{N}$ factorizes through h by means of a unique continuous homomorphism $f': K \to U(n)$ with $f' \circ h = f$), then $h: G \to K$ has also the (stronger) universal property of the Bohr compactification bG, so must coincide, up to isomorphism, with $b_G: G \to bG$, by Theorem 8.6.1.

Pick a continuous homomorphism $f: G \to C$ with C compact. By Corollary 10.3.4, we can assume that C is a subgroup of a product $P = \prod_{i\in I} U(n_i)$. Let $i \in I$ and let $p_i: P \to U(n_i)$ be the canonical projection. Then our hypothesis applied to $p_i \circ f: G \to U(n_i)$ provides a continuous homomorphism $f_i': K \to U(n_i)$ with $f_i' \circ h = p_i \circ f$. The diagonal homomorphism $f': K \to P$ of the family $\{f_i': i \in I\}$ satisfies $f' \circ h = f$ (compose with p_i and use the fact that $f_i' \circ h = p_i \circ f$).

The proof of the existence of a continuous homomorphism $h: G \to K$, with a compact group K, satisfying (Bc) is similar to the proof of Theorem C.2.10, so we only give a brief sketch. There exists a set of continuous homomorphisms $h_i: G \to U(n_i)$ with $i \in I$ so that every continuous homomorphism $f: G \to U(n)$ is isomorphic to some h_i. Take the diagonal homomorphism $d: G \to \prod_{i\in I} U(n_i)$, let $K = \overline{d(G)}$, and let $h: G \to K$ be the corestriction of d to K. This h does the job. □

Corollary 10.3.6. *A topological group G is MAP if and only if the continuous homomorphisms $G \to U(n)$ (with n varying in $\mathbb{N}$) separate the points of G, while G is minimally almost periodic if and only if for every $n \in \mathbb{N}$ the only continuous homomorphism $G \to U(n)$ is the trivial one.*

In the case of an abelian group G, the irreducible unitary representations have degree 1, so they simply are the continuous characters $G \to \mathbb{T}$. Hence:

Corollary 10.3.7 ([174, 22.17]). *A topological abelian group G is MAP if and only if the continuous characters $G \to \mathbb{T}$ separate the points of G, while G is minimally almost periodic if and only if $\widehat{G}$ is trivial.*

Using Peter–Weyl–van Kampen theorem 10.3.3, in the abelian case one can prove that every locally compact abelian group is MAP. The proof of this fact (see Theorem 11.6.3) requires several ingredients that we develop in Chapter 11.

Examples of minimally almost periodic groups are less easy to come by. The first known ones came from functional analysis, namely, the topological real vector spaces V having no nontrivial continuous linear functionals $V \to \mathbb{R}$, e. g., L_p with $0 < p < 1$. This is the space of equivalence classes of measurable functions $[0,1] \to \mathbb{R}$ with $\int_0^1 |f(x)|^p dx < \infty$, endowed with the topology induced by the metric $d_p: L_p \times L_p \to [0,\infty)$, $([f],[g]) \mapsto \int_0^1 |f(x) - g(x)|^p dx$. Day [64] showed that the null-function is the only continuous linear form of L_p and hence the trivial character is the only continuous character of L_p.

Minimally almost periodic group topologies on $\mathbb{Z}$ were built by Nienhuys [221]. Prodanov [230], unaware of the already existing examples, produced an elementary example of a minimally almost periodic abelian group (his idea was further developed in [55, 108, 257, 258]). Further examples were produced in [3, 240]. Remus [244] noticed that the connected group topologies on the bounded abelian groups produced by Markov [210] are necessarily minimally almost periodic (see Exercise 10.4.14). Comfort [54] raised the problem of the description of all abelian groups admitting a minimally almost periodic group topology. Gabriyelyan gave a solution in the countable case in [148, 149], the solution in the general case can be found in [106].

It is easier to encounter minimally almost periodic groups in the nonabelian case, e. g., $S(\mathbb{N})$ or $SL_2(\mathbb{R})$.

10.4 Exercises

Exercise 10.4.1. Prove that a topological group G is totally bounded if and only if every nonempty open set U of G is right big.

Hint. We can assume without loss of generality that $e_G \in U$. Take any open $V \in \mathcal{V}(e_G)$ such that $V^{-1} \subseteq U$. Since V must be left big in case G is totally bounded, V^{-1} is right big, so U is right big as well.

Exercise 10.4.2. Prove that for no infinite set X the group $S(X)$ admits a precompact group topology.

Hint. According to Theorem 4.2.2, if $S(X)$ admits a precompact group topology, then also $\mathcal{T}_X$ is precompact. Since the subgroups S_x of $S(X)$ are open in $\mathcal{T}_X$, this would imply that S_x has finite index, a contradiction.

Exercise 10.4.3. Prove that, for an infinite group G and a subgroup H of G, the following conditions are equivalent:
(a) H has infinite index;
(b) H is not left big;
(c) H is not right big;
(d) H is left small;
(e) H is right small.

Exercise 10.4.4.* Show that no infinite abelian group G is a finite union of small subsets.

Hint. Since G is abelian, it admits a finitely additive invariant (Banach) measure μ defined on the powerset of G and $\mu(G) = 1$. Clearly, every small subset of G has measure 0, while $\mu(G) = 1$. So, G cannot be a union of finitely many sets of measure 0.

For an elementary proof, due to U. Zannier, see [283] or [99, Exercise 1.6.20].

Exercise 10.4.5 ([100]).* Prove that every infinite abelian group has a small set of generators.

Exercise 10.4.6.* Describe the abelian groups G such that ν_G is precompact; the same with ν_G^p.

Exercise 10.4.7.* Let α be an infinite cardinal.
(a) Show that the class $\mathbf{B}_\alpha$ of α-totally bounded groups is stable under taking subgroups, continuous homomorphic images and products.
(b) Deduce from (a) that every group G admits the finest α-totally bounded group topology which is finer than the pro-α topology on G (having as basic open neighborhoods of e_G all subgroups of G of index $< \alpha$).

Exercise 10.4.8. Describe the abelian groups G such that ν_G is ω_1-totally bounded; the same with ν_G^p.

Exercise 10.4.9. Show that every separable group is ω_1-totally bounded.

Exercise 10.4.10. Show that for a metrizable group G the following conditions are equivalent:
(a) G is second countable;
(b) G is Lindelöff;
(c) G is separable;
(d) G is ω_1-totally bounded.

Hint. The implications (a)⇒(b)⇒(c) are well known and are actually equivalences. The implication (c)⇒(d) is Exercise 10.4.9. To prove the missing implication (d)⇒(c), fix a countable local base $\{U_n : n \in \mathbb{N}\}$ at e_G. For every $n \in \mathbb{N}$, there exists a countable set F_n such that $F_nU_n = G$. Then the countable set $F = \bigcup_{n\in\mathbb{N}} F_n$ satisfies $FU_n = G$ for every $n \in \mathbb{N}$. This means that F is dense in G, so G is separable.

Exercise 10.4.11. If G is a countably infinite Hausdorff abelian group, show that for every compact subset K of G the set $K_{(2n)}$ is big for no $n \in \mathbb{N}_+$.

Hint. By Lemma 8.2.1, every set $K_{(2n)}$ is compact. So, if $K_{(2n)}$ were big for some $n \in \mathbb{N}_+$, then G itself would be compact. Now Lemma 8.1.5(a) applies.

Exercise 10.4.12. Prove that:
(a) if $S = \{a_n\}_{n\in\mathbb{N}}$ is a one-to-one T-sequence of an abelian group G, then for every $n \in \mathbb{N}$ the set $S_{(2n)}$ is small in G;
(b)* the sequence $\{p_n\}_{n\in\mathbb{N}}$ of all prime numbers in $\mathbb{Z}$ is not a T-sequence.

Hint. (a) Consider the (countable) subgroup generated by S. If $a_n \to 0$ in some Hausdorff group topology τ on G, then $S \cup \{0\}$ is compact in τ, so Exercise 10.4.11 applies.

(b) Use (a) and the fact that there exists a constant $m \in \mathbb{N}_+$ such that every integer number is the sum of at most m primes. More precisely, according to the positive solution of the ternary Goldbach conjecture, there exists a constant $C \in \mathbb{N}_+$ such that every odd integer $\geq C$ is the sum of three primes (see [286] for further details on the Goldbach conjecture).

Exercise 10.4.13. Prove that, for a direct product $(G,\tau) = (G_1,\tau_1) \times (G_2,\tau_2)$ of topological groups, one has $n(G) = n(G_1) \times n(G_2)$ and $G^+ \cong G_1^+ \times G_2^+$. Hence, $bG \cong bG_1 \times bG_2$.

Hint. The group $(G_1 \times G_2, \tau_1^+ \times \tau_2^+)$ is totally bounded and for every totally bounded group topology $\sigma \leq \tau$ on G, one has $\sigma\restriction_{G_i} \leq \tau\restriction_{G_i} = \tau_i$, so $\sigma\restriction_{G_i} \leq \tau_i^+$ for $i = 1,2$. By the properties of the product topology on finite products, $\sigma \leq \tau_1^+ \times \tau_2^+$. This proves that $\tau_1^+ \times \tau_2^+ = \tau^+$, and consequently, $n(G) = n(G_1) \times n(G_2)$. Therefore, $G^+ = G/n(G) \cong G_1/n(G_1) \times G_2/n(G_2)$ and this is a topological isomorphism whenever G^+ and $G_i^+ = G_i/n(G_i)$, for $i = 1,2$, carry the quotient topology. Therefore, $G^+ \cong G_1^+ \times G_2^+$ and $bG \cong bG_1 \times bG_2$.

Exercise 10.4.14. Show that every infinite bounded connected abelian group is minimally almost periodic. Deduce that for a connected abelian group G and every $m \in N_+$ the quotient $G/\overline{mG}$ is MinAP.

Hint. Prove that for such a group G every continuous character $G \to \mathbb{T}$ is trivial and apply Corollary 10.3.7. The second assertion immediately follows from the first one.

10.5 Further readings, notes, and comments

The fact that every infinite abelian group has a small set of generators (see Exercise 10.4.5) was extended to arbitrary infinite groups in [100]. One can find in the literature also different (weaker) forms of smallness in [10, 29].

Next, we discuss a remarkable connection, established by Prodanov and Stoyanov, between two compactness-like properties that partially inverts the implication "compact ⇒ minimal" due to the open mapping theorem (see Corollary 8.4.2):

Theorem 10.5.1 ([239]). *Minimal abelian groups are precompact.*

A simplified proof of this theorem can be found in [99, Theorem 2.7.7]. Combining it with Exercise 8.7.6, one can deduce that minimal abelian groups are precisely the dense essential subgroups of compact abelian groups. A significant generalization of Theorem 10.5.1 was recently obtained by Banakh [15].

Theorem 10.5.1 was conjectured by Prodanov in 1971 and partial results for some classes of minimal abelian groups were obtained in the consequent twelve years; we list here some of them. He showed in [232] that countable minimal abelian groups are precompact. Stoyanov [267] pushed further this result to abelian groups G satisfying $|G/(\mathrm{div}(G) + t(G))| < \mathfrak{c}$. The totally minimal abelian groups were proved to be precompact in [233]. Motivated by this result, Dierolf and Schwanengel [69] found the first example of a nonprecompact totally minimal group: the symmetric group $S(X)$ of any infinite set X (see §4.2), which is also complete, by Proposition 7.1.29. (Actually, $S(X)$ admits no precompact topologies whatsoever – see Exercise 10.4.2.) Answering [77, Question 3.5(a)], it was shown by Megrelishvili and the second named author in [92, Corollary 5.5] that precompactness of minimal topologies fails even in nilpotent groups of class 2.

An original method to approach Theorem 10.5.1 for some abelian groups was proposed by Prodanov [234]. He observed that for every abelian group G the submaximal topology $\mathcal{M}_G$ is finer than any minimal topology on G. Hence, if $\mathcal{M}_G = \mathfrak{B}_G$, then every minimal topology on this group G is precompact. (Divisible abelian groups and finite-rank torsion-free abelian groups have this property, as $\mathcal{M}_G = \mathfrak{B}_G$ occurs precisely when $\mathcal{S}_G \leq \mathfrak{B}_G$, i. e., $[G : (nG + \mathrm{Soc}(G))] < \infty$ for every $n \in \mathbb{N}_+$ – see Exercise 2.4.15.) This technique enabled him to prove that a relevant subgroup of every minimal abelian group must be precompact.

As far as complementation in the lattice of group topologies is concerned (see §2.5), a precompact group topology τ on a group G admits no transversal topologies, since obviously $\tau \leq \mathfrak{B}_G \leq \mathcal{M}_G$, so the criterion from §2.5 applies.

It was proved by Freudenthal and Weil that a connected MAP locally compact group has the form $\mathbb{R}^n \times G$, where G is compact (and necessarily connected).

The terms Bohr topology and Bohr compactification have been chosen as a reward to Harald Bohr for his work [18] on almost periodic functions closely related to the Bohr compactification (see Theorems 12.1.9 and 12.1.12). Otherwise, the Bohr compactification is due to Weil [288]. More general results were obtained later by Holm [178] and Prodanov [238] concerning the Bohr compactification of rings and other universal algebras.

The standard exposition of the Pontryagin-van Kampen duality exploits the Haar measure for the proof of the Peter–Weyl–van Kampen theorem. Our aim here is to obtain a proof of the Peter–Weyl–van Kampen theorem in the abelian case without any recourse to the Haar integration and tools of functional analysis. This elementary approach, based on the Følner theorem mentioned above and ideas of Prodanov, can be found in [99, Chapter 1]. It makes no recourse to the Haar measure at all; on the contrary, after giving a self-contained elementary proof of the Peter–Weyl–van Kampen theorem in the abelian case, one obtains as an easy consequence the existence of the Haar measure on locally compact abelian groups (see Theorem 12.2.5 for the compact case and Theorem 12.2.9 for the locally compact one).

Now we reveal the subtle (virtual) connection between the Zariski topology, the von Neumann radical and minimal almost periodicity. Why are we using "virtual" becomes clear immediately from the first connection. Namely, it is proved in [106] that an abelian group admits a minimally almost periodic group topology if and only if it is connected in its Zariski topology. (For example $G = \mathbb{Z}(2) \times \mathbb{Z}(3)^{\mathbb{N}}$ does not admit minimally almost periodic group topologies, since $G[3] = 2G$ is a proper open subgroup of $(G, \mathfrak{Z}_G)$, so $(G, \mathfrak{Z}_G)$ is not connected.) In particular, every unbounded abelian group admits a minimally almost periodic group topology (see [104, Theorem 4.6(ii)]). This answers positively a question set by Comfort [54].

The second connection involves the "realization as the von Neumann kernel" in the following sense. Following [106], we shall say that a subgroup H of an abelian group G can be realized as the von Neumann kernel of G when there exists a Hausdorff group topology τ on G such that $n(G, \tau) = H$. Clearly, the whole group G can be realized as the von Neumann kernel of G precisely when G admits a minimally almost periodic group topology. Hence, the problem of realization as the von Neumann kernel in the above sense is more general than the previous one. It is proved in [106] that a subgroup H of an abelian group G can be realized as the von Neumann kernel of G if and only if H is contained in the connected component of zero of G with respect to the Zariski topology of G.

In the above example $G = \mathbb{Z}(2)\times\mathbb{Z}(3)^{\mathbb{N}}$, the connected component C of 0 in $(G, \mathfrak{Z}_G)$ coincides with the subgroup $G[3] = 2G = \{0\} \times \mathbb{Z}(3)^{\mathbb{N}}$ of G. Indeed, $2G$ is an open subgroup of G, so it contains C. On the other hand, $(2G, \mathfrak{Z}_{2G})$ is connected since $\mathfrak{Z}_{2G} = \gamma_{2G}$, by Exercise 4.5.6(b). By Exercise 4.5.12, $\mathfrak{Z}_{2G}$ coincides with $\mathfrak{Z}_G \restriction_{2G}$, hence $2G$ is connected in $(G, \mathfrak{Z}_G)$ as well. This proves that $C = 2G$. Therefore, the above mentioned general result implies that the subgroup $H = \mathbb{Z}(2) \times \{0\}$ of G cannot be realized as the von Neumann kernel of G. More precisely, the subgroups of G that can be realized as the von Neumann kernel of G are precisely the subgroups of $2G$.

11 The Følner theorem

The first half of this chapter is entirely dedicated to the Følner theorem. In the second half we prove the Peter–Weyl theorem in the abelian case by applying the Følner theorem. Moreover, we use the Prodanov lemma to describe the precompact topologies on abelian groups.

Recall that, for an abelian group G, we denote by $G^* = \mathrm{Hom}(G, \mathbb{S})$ the group of all characters of G. In case G is endowed with a group topology, we denote by $\widehat{G}$ the subgroup of G^* consisting of all continuous characters of G.

11.1 Fourier theory for finite abelian groups

In the sequel G is a finite abelian group, so $G^* \cong G$ (see Exercise 11.7.1), hence in particular $|G^*| = |G|$.

Here we recall some well-known properties of the scalar product[1] $(- \mid -)$ in finite-dimensional complex vector spaces $V = \mathbb{C}^n$. We normalize the scalar product in such a way to let the vector $(1, 1, \dots, 1)$ (i. e., the constant function 1) have norm 1, i. e., for $u = (u_1, \dots, u_n), v = (v_1, \dots, v_n) \in V$, we let

$$(u \mid v) := \frac{1}{n} \sum_{i=1}^{n} u_i \overline{v_i}.$$

By taking a finite group G of size n, we can consider $V = \mathbb{C}^G \cong \mathbb{C}^n$, so that we have also an action of G on V. In these terms, for every $f, g \in \mathbb{C}^G$,

$$(f \mid g) = \frac{1}{|G|} \sum_{x \in G} f(x)\overline{g(x)}.$$

Let us see that the elements of the subset $G^* = \mathrm{Hom}(G, \mathbb{S})$ of $V = \mathbb{C}^G$ are pairwise orthogonal and have norm 1:

Proposition 11.1.1. *Let G be a finite abelian group and $\varphi, \chi \in G^*$, $x, y \in G$. Then:*

(a) $(\varphi \mid \chi) = \begin{cases} 1 & \text{if } \varphi = \chi, \\ 0 & \text{if } \varphi \neq \chi; \end{cases}$

(b) $\frac{1}{|G^*|} \sum_{\chi \in G^*} \chi(x)\overline{\chi(y)} = \begin{cases} 1 & \text{if } x = y, \\ 0 & \text{if } x \neq y. \end{cases}$

1 Frequently $(- \mid -)$ is called also Hermitian product. We keep the same notation that we use for the standard scalar product of $\mathbb{R}^n$.

https://doi.org/10.1515/9783110654936-011

Proof. (a) If $\varphi = \chi$, then $\chi(x)\overline{\chi(x)} = \chi(x)\chi(x)^{-1} = 1$. If $\varphi \neq \chi$, there exists $z \in G$ such that $\varphi(z) \neq \chi(z)$. Therefore, the equalities

$$\sum_{x \in G} \varphi(x)\overline{\chi(x)} = \frac{\varphi(z)}{\chi(z)} \sum_{x \in G} \varphi(x-z)\overline{\chi(x-z)} = \frac{\varphi(z)}{\chi(z)} \sum_{x \in G} \varphi(x)\overline{\chi(x)}$$

imply that $\sum_{x \in G} \varphi(x)\overline{\chi(x)} = 0$.

(b) By Exercise 11.7.1, G^* is a finite group. For every $x \in G$, the evaluation map $G^* \to \mathbb{S}$, $\chi \mapsto \chi(x)$, is a homomorphism, so belongs to G^{**}. Now the assertion follows from (a). □

Definition 11.1.2. Let G be a finite abelian group and $f \in \mathbb{C}^G$. For every $\chi \in G^*$, the *Fourier coefficient* of f corresponding to χ is

$$c_\chi := (f \mid \chi) = \frac{1}{|G|} \sum_{x \in G} f(x)\overline{\chi(x)}.$$

For $f, g \in \mathbb{C}^G$, define the function $f * g$ by letting, for every $x \in G$,

$$(f * g)(x) := \frac{1}{|G|} \sum_{y \in G} g(x+y)\overline{f(y)} = (g_x \mid f),$$

where $g_x(y) = g(x+y)$ for all $y \in G$.

Observe that if f and g are real-valued (nonnegative) functions, so is $f * g$. Proposition 11.1.1(a) shows that G^* is an orthonormal subset of $\mathbb{C}^G$.

Proposition 11.1.3. *Let G be a finite abelian group and $f \in \mathbb{C}^G$ with Fourier coefficients $\{c_\chi : \chi \in G^*\}$. Then:*
(a) *G^* is an orthonormal base of $\mathbb{C}^G$;*
(b) *$f(x) = \sum_{\chi \in G^*} c_\chi \chi(x)$ for every $x \in G$;*
(c) *if g is another complex-valued function on G with Fourier coefficients $\{d_\chi : \chi \in G^*\}$, then $f * g$ has Fourier coefficients $\{\overline{c_\chi} d_\chi : \chi \in G^*\}$;*
(d) *$(f * f)(x) = \sum_{\chi \in G^*} |c_\chi|^2 \chi(x)$ and $\frac{1}{|G|} \sum_{x \in G} |f(x)|^2 = \sum_{\chi \in G^*} |c_\chi|^2$.*

Proof. (a) According to Proposition 11.1.1(a), G^* is an orthonormal set of cardinality $|G|$ in $\mathbb{C}^G$, hence an orthonormal base.

(b) Since G^* is an orthonormal base of $\mathbb{C}^G$ by item (a), we obtain for $f \in \mathbb{C}^G$ that $f = \sum_{\chi \in G^*} (f \mid \chi)\chi = \sum_{\chi \in G^*} c_\chi \chi$.

(c) The Fourier coefficient of $f * g$ corresponding to $\chi \in G^*$ is

$$\begin{aligned}(f * g \mid \chi) &= \frac{1}{|G|} \sum_{x \in G} (f * g)(x)\overline{\chi(x)} = \frac{1}{|G|^2} \sum_{x \in G} \sum_{y \in G} g(x+y)\overline{f(y)}\,\overline{\chi(x)} \\ &= \frac{1}{|G|^2} \sum_{z \in G} \sum_{y \in G} g(z)\overline{f(y)}\,\overline{\chi(z-y)}\end{aligned}$$

$$= \overline{\left(\frac{1}{|G|}\sum_{y\in G} f(y)\overline{\chi(y)}\right)}\left(\frac{1}{|G|}\sum_{z\in G} g(z)\overline{\chi(z)}\right) = \overline{c_\chi}d_\chi.$$

(d) The first equality follows from item (c) with $g = f$.

For $x = 0$, it becomes $(f * f)(0) = \sum_{\chi\in G^*} |c_\chi|^2$, while the definition gives $(f * f)(0) = \frac{1}{|G|}\sum_{y\in G} |f(y)|^2$, so we obtain the second equality of item (d). □

Corollary 11.1.4. *Let G be a finite abelian group, $f: G \to \mathbb{R}_{\geq 0}$ a function, and $E = \{x \in G: f(x) > 0\}$. Then, for $g := f * f$ and $x \in G$:*
(a) *$g(x) > 0$ if and only if $x \in E_{(2)}$;*
(b) *$g(x) = \sum_{\chi\in G^*} |c_\chi|^2\chi(x)$, where the c_χ are the Fourier coefficients of f.*

Proof. (a) For $x \in G$, by definition $g(x) > 0$ if and only if there exists $y \in E$ with $x+y \in E$, that is, $x \in E - E = E_{(2)}$.

(b) follows from Proposition 11.1.3(d). □

11.2 The Bogoliouboff and Følner lemmas

Lemma 11.2.1 (Bogoliouboff lemma). *If F is a finite abelian group and E is a nonempty subset of F, then there exist $\chi_1,\dots,\chi_m \in F^*$, where $m = \lfloor (\frac{|F|}{|E|})^2 \rfloor$, such that $U(\chi_1,\dots,\chi_m;\frac{\pi}{2}) \subseteq E_{(4)}$.*

Proof. Let $f: F \to \{0,1\} \subseteq \mathbb{C}$ be the characteristic function of E. By Proposition 11.1.3(b), for every $x \in F$,

$$f(x) = \sum_{\chi\in F^*} c_\chi\chi(x), \quad \text{with } c_\chi = \frac{1}{|F|}\sum_{x\in F} f(x)\overline{\chi(x)}. \tag{11.1}$$

Denote $g = f * f$ and $h = g * g$. By definition, f and g have real values, and by Corollary 11.1.4(b), for every $x \in F$,

$$g(x) = \sum_{\chi\in F^*} |c_\chi|^2\chi(x) \quad \text{and} \quad h(x) = \sum_{\chi\in F^*} |c_\chi|^4\chi(x); \tag{11.2}$$

by Corollary 11.1.4(a), $g(x) > 0$ if and only if $x \in E_{(2)}$, and analogously

$$h(x) > 0 \quad \text{if and only if} \quad x \in E_{(4)}. \tag{11.3}$$

By Proposition 11.1.3(d),

$$\sum_{\chi\in F^*} |c_\chi|^2 = \frac{1}{|F|}\sum_{x\in F} |f(x)|^2 = \frac{|E|}{|F|}. \tag{11.4}$$

Set $a = \frac{|E|}{|F|}$ and order the Fourier coefficients $\{c_{\chi_0}, c_{\chi_1},\dots,c_{\chi_k}\}$ of f so that

$$|c_{\chi_0}| \geq |c_{\chi_1}| \geq \cdots \geq |c_{\chi_k}|.$$

Taking into account the fact that f is the characteristic function of E, it easily follows from (11.1) that the maximum value of $|c_{\chi_i}|$ is attained for the trivial character $\chi_0 \equiv 1$, namely, $c_{\chi_0} = a$. For every $j \in \{0, \dots, k\}$, by (11.4),

$$(j+1)|c_{\chi_j}|^2 \leq \sum_{i=0}^{j} |c_{\chi_i}|^2 \leq \sum_{\chi \in F^*} |c_\chi|^2 = a,$$

and so

$$|c_{\chi_j}|^4 \leq \frac{a^2}{(j+1)^2}. \tag{11.5}$$

Now let $m = \min\{k-1, \lfloor \frac{1}{a^2} \rfloor\}$. We are going to show that with $\chi_1, \dots, \chi_m \in F^*$, for $x \in G$ we get

$$h(x) > 0 \quad \text{for every } x \in U\left(\chi_1, \dots, \chi_m; \frac{\pi}{2}\right). \tag{11.6}$$

Let $x \in U(\chi_1 \dots, \chi_m; \frac{\pi}{2})$. Clearly, $\mathrm{Re}(\chi_j(x)) \geq 0$ for $j \in \{1, \dots, m\}$, and thus,

$$\left| a^4 + \sum_{j=1}^{m} |c_{\chi_j}|^4 \chi_j(x) \right| \geq \mathrm{Re}\left(a^4 + \sum_{j=1}^{m} |c_{\chi_j}|^4 \chi_j(x) \right) \geq a^4. \tag{11.7}$$

On the other hand, using also the equality

$$\sum_{j=m+1}^{k} \frac{1}{j(j+1)} = \sum_{j=m+1}^{k} \left(\frac{1}{j} - \frac{1}{j+1} \right) = \frac{1}{m+1} - \frac{1}{k+1},$$

(11.5) yields

$$\sum_{j=m+1}^{k} |c_{\chi_j}|^4 \leq \sum_{j=m+1}^{k} \frac{a^2}{(j+1)^2} < a^2 \sum_{j=m+1}^{k} \frac{1}{j(j+1)} \leq \frac{a^2}{m+1}. \tag{11.8}$$

Since h has nonnegative real values, (11.2), (11.7), and (11.8) give

$$\begin{aligned} h(x) = |h(x)| = \left| a^4 + \sum_{j=1}^{k} |c_{\chi_j}|^4 \chi_j(x) \right| &\geq \left| a^4 + \sum_{j=1}^{m} |c_{\chi_j}|^4 \chi_j(x) \right| - \sum_{j=m+1}^{k} |c_{\chi_j}|^4 \\ &\geq a^4 - \frac{a^2}{m+1} \geq a^2 \left(a^2 - \frac{1}{m+1} \right) > 0. \end{aligned}$$

This proves (11.6). Therefore, by (11.3), $U(\chi_1 \dots, \chi_m; \frac{\pi}{2}) \subseteq E_{(4)}$. □

Remark 11.2.2. Let us note that the estimate in Lemma 11.2.1 for the number m of characters is certainly not optimal when E is too small compared to $k = |F|$. For example, when E is the singleton $\{0\}$, the upper bound given by the lemma is just $|F|^2$, while one can certainly find at most $m = k - 1$ characters $\chi_1, \dots, \chi_m$ (namely, all nontrivial characters $\chi_1, \dots, \chi_{k-1} \in F^*$) such that $U(\chi_1, \dots, \chi_m; \frac{\pi}{2}) = \{0\}$. (For certain groups, e. g., $F = \mathbb{Z}(2)^k$, one can find even a much smaller number, say $m = \log_2 |F|$.) Therefore, one can always assume (as implicitly done in the proof) that $m \leq k - 1$, since the $k - 1$ nontrivial characters $\chi_1, \dots, \chi_{k-1}$ of F give $U(\chi_1, \dots, \chi_{k-1}; \frac{\pi}{2}) = \{0\} \subseteq E_{(2)}$, for arbitrary $\emptyset \neq E \subseteq F$.

Nevertheless, in the cases relevant for the proof of the Følner theorem, namely, when the subset E is relatively large with respect to F, this estimate seems more reasonable.

The next technical lemma is needed in the following proofs.

Lemma 11.2.3. *Let A be an abelian group and $\{A_n\}_{n \in \mathbb{N}_+}$ a sequence of finite subsets of A such that, for every $a \in A$,*

$$\lim_{n \to \infty} \frac{|(A_n - a) \cap A_n|}{|A_n|} = 1.$$

If $k \in \mathbb{N}_+$ and V is a subset of A such that k translates of V cover A, then for every $\varepsilon > 0$ there exists $N > 0$ such that, for every $n \geq N$,

$$|V \cap A_n| > \left(\frac{1}{k} - \varepsilon\right) |A_n|. \tag{11.9}$$

Proof. Let $a_1, \dots, a_k \in A$ be such that $A = \bigcup_{i=1}^k (a_i + V)$. Let $\varepsilon > 0$. By hypothesis, there exists $N > 0$ such that, for every $n \geq N$ and every $i \in \{1, \dots, k\}$,

$$|(A_n - a_i) \cap A_n| > (1 - \varepsilon)|A_n|,$$

and consequently,

$$|(A_n - a_i) \setminus A_n| < \varepsilon |A_n|. \tag{11.10}$$

Since, for every $n \in \mathbb{N}$, $A_n = \bigcup_{i=1}^k (a_i + V) \cap A_n$, there exists $i_n \in \{1, \dots, k\}$ such that

$$\frac{1}{k}|A_n| \leq |(a_{i_n} + V) \cap A_n| = |V \cap (A_n - a_{i_n})|. \tag{11.11}$$

Since $V \cap (A_n - a_{i_n}) \subseteq (V \cap A_n) \cup ((A_n - a_{i_n}) \setminus A_n)$, (11.10) and (11.11) yield that for every $n \geq N$,

$$\frac{1}{k}|A_n| \leq |V \cap (A_n - a_{i_n})| \leq |V \cap A_n| + |(A_n - a_{i_n}) \setminus A_n| < |V \cap A_n| + \varepsilon |A_n|. \qquad \square$$

Lemma 11.2.4 (Bogoliouboff–Følner lemma). *Let A be a finitely generated abelian group and let $r = r_0(A)$. If $k \in \mathbb{N}_+$ and V is a subset of A such that k translates of V cover A, then there exist $\rho_1, \dots, \rho_s \in A^*$, where $s = 3^{2r}k^2$, such that $U_A(\rho_1, \dots, \rho_s; \frac{\pi}{2}) \subseteq V_{(4)}$.*

Proof. By Theorem A.1.1, we can identify A with the group $\mathbb{Z}^r \times F$, where F is a finite abelian group. For every $n \in \mathbb{N}_+$, define

$$A_n = (-n, n]^r \times F.$$

Fix arbitrarily $a = (a_1, \dots, a_r; f) \in A$, $n \in \mathbb{N}_+$ and $i \in \{1, \dots, r\}$. The set

$$J_{n,i} = (-n, n] \cap (-n - a_i, n - a_i] \cap \mathbb{Z}$$

satisfies $|J_{n,i}| = 2n - |a_i|$ when $2n \geq |a_i|$. In particular, for every $i \in \{1, \dots, r\}$,

$$J_{n,i} \neq \emptyset \quad \text{for every } n > n_0 := \max\{|a_i| : i \in \{1, \dots, r\}\}.$$

For all $n > n_0$, as $(A_n - a) \cap A_n = \prod_{i=1}^r J_{n,i} \times F$,

$$|(A_n - a) \cap A_n| \geq |F| \prod_{i=1}^{r} (2n - |a_i|).$$

Since $|A_n| = |F|(2n)^r$ for every $n \in \mathbb{N}_+$, we can apply Lemma 11.2.3. Thus, for every $\varepsilon > 0$, there exists $N_\varepsilon > 0$ with $N_\varepsilon \geq n_0$ such that (11.9) holds for all $n \geq N_\varepsilon$.

For $n \geq N_\varepsilon$, define $G_n = A/(6n\mathbb{Z}^r)$ and $E = q(V \cap A_n)$ where $q\colon A \to G_n$ is the canonical projection. Observe that $q \restriction_{A_n}$ is injective, as $(A_n - A_n) \cap \ker q = \{0\}$. Then (11.9) gives

$$|E| = |V \cap A_n| > \left(\frac{1}{k} - \varepsilon\right)|A_n| = \left(\frac{1}{k} - \varepsilon\right)(2n)^r|F|,$$

and so

$$\frac{|G_n|}{|E|} \leq \frac{(6n)^r|F|}{(\frac{1}{k} - \varepsilon)(2n)^r|F|} = \frac{3^r k}{1 - k\varepsilon}.$$

Fix $\varepsilon > 0$ sufficiently small to have $\left\lfloor \frac{3^{2r}k^2}{(1-k\varepsilon)^2} \right\rfloor = 3^{2r}k^2$ and pick $n \geq N_\varepsilon$. Now apply Lemma 11.2.1 to find $s = 3^{2r}k^2$ characters $\xi_{1,n}, \dots, \xi_{s,n} \in G_n^*$ such that

$$U_{G_n}\left(\xi_{1,n}, \dots, \xi_{s,n}; \frac{\pi}{2}\right) \subseteq E_{(4)}.$$

For $j \in \{1, \dots, s\}$, define $\varrho_{j,n} = \xi_{j,n} \circ q \in A^*$:

$$\begin{array}{ccc} A & \xrightarrow{\varrho_{j,n}} & \mathbb{S} \\ & {\scriptstyle q}\searrow \quad \nearrow{\scriptstyle \xi_{j,n}} & \\ & G_n & \end{array}$$

If $a \in A_n \cap U_A(\varrho_{1,n}, \ldots, \varrho_{s,n}; \frac{\pi}{2})$, then $q(a) \in U_{G_n}(\xi_{1,n}, \ldots, \xi_{s,n}; \frac{\pi}{2}) \subseteq E_{(4)}$, and so there exist $b_1, b_2, b_3, b_4 \in V \cap A_n$ and $c = (c_i)_{i=1}^r \in 6n\mathbb{Z}^r$ such that $a = b_1 - b_2 + b_3 - b_4 + c$. Now

$$c = a - b_1 + b_2 - b_3 + b_4 \in (A_n)_{(4)} + A_n$$

implies $|c_i| \leq 5n$ for each $i \in \{1, \ldots, r\}$. So, $c = 0$ as $6n$ divides c_i for each $i \in \{1, \ldots, r\}$. Thus, $a \in V_{(4)}$ and we conclude that, for every $n \geq N_\varepsilon$,

$$A_n \cap U_A\left(\varrho_{1,n}, \ldots, \varrho_{s,n}; \frac{\pi}{2}\right) \subseteq V_{(4)}. \tag{11.12}$$

By Lemma 8.1.3, there exist $\varrho_1, \ldots, \varrho_s \in A^*$ and a subsequence $\{n_l\}_{l \in \mathbb{N}}$ of $\{n\}_{n \in \mathbb{N}_+}$ such that, for every $i \in \{1, \ldots, s\}$,

$$\varrho_i(a) = \lim_{l \to \infty} \varrho_{i,n_l}(a) \quad \text{for every } a \in A. \tag{11.13}$$

We prove that

$$U_A\left(\varrho_1, \ldots \varrho_s; \frac{\pi}{2}\right) \subseteq V_{(4)}. \tag{11.14}$$

Take $a \in U_A(\varrho_1, \ldots, \varrho_s; \frac{\pi}{2})$. Since $A = \bigcup_{l=t}^{\infty} A_{n_l}$ for any $t \in \mathbb{N}$, and by (11.13) we can pick $l \in \mathbb{N}$ to have $n_l \geq N_\varepsilon$ such that $a \in A_{n_l}$ and $|\mathrm{Arg}(\varrho_{i,n_l}(a))| < \pi/2$ for every $i \in \{1, \ldots, s\}$, i. e., $a \in U_A(\varrho_{1,n_l}, \ldots, \varrho_{s,n_l}; \frac{\pi}{2}) \cap A_{n_l}$. Now (11.12), applied to n_l, yields $a \in V_{(4)}$, and this proves (11.14). □

Our next aim is to eliminate the dependence of the number m of characters on the free-rank of the group A in the Bogoliouboff–Følner lemma. The price to pay for this is taking $V_{(8)}$ instead of $V_{(4)}$.

Lemma 11.2.5 (Følner lemma). *Let A be an infinite abelian group. If $k \in \mathbb{N}_+$ and V is a subset of A such that k translates of V cover A, then there exist $\chi_1, \ldots, \chi_m \in A^*$, where $m = k^2$, such that $U_A(\chi_1, \ldots, \chi_m; \frac{\pi}{2}) \subseteq V_{(8)}$.*

Proof. First, we consider the case when A is finitely generated, and let $r = r_0(A)$. By Lemma 11.2.4, there exist $\varrho_1, \ldots, \varrho_s \in A^*$, where $s = 3^{2r}k^2$, such that

$$U_A\left(\varrho_1, \ldots, \varrho_s; \frac{\pi}{2}\right) \subseteq V_{(4)}.$$

As above, by Theorem A.1.1, we can identify A with $\mathbb{Z}^r \times F$, where F is a finite abelian group. For $t \in \{1, \ldots, r\}$, define a monomorphism $i_t \colon \mathbb{Z} \hookrightarrow A$ by letting, for every $n \in \mathbb{Z}$,

$$i_t(n) = (\underbrace{0, \ldots, 0, n}_{t}, 0, \ldots, 0; 0) \in A.$$

Then each $\kappa_{j,t} := \varrho_j \circ i_t$, where $j \in \{1, \ldots, s\}$, $t \in \{1, \ldots, r\}$, is a character of $\mathbb{Z}$:

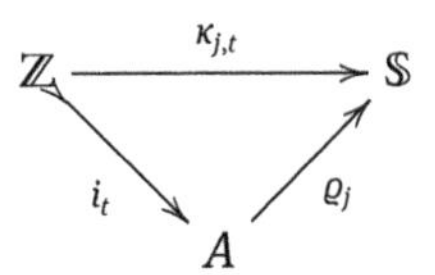

By Proposition 10.2.16, the subset

$$L = U_{\mathbb{Z}}\left(\{\kappa_{j,t}: j \in \{1,\dots,s\}, t \in \{1,\dots,r\}\}; \frac{\pi}{8r}\right) \subseteq \mathbb{Z}$$

is big in $\mathbb{Z}$ and so unbounded from above and from below. Let

$$L^0 = \bigcup_{t=1}^{r} i_t(L) \subseteq A.$$

Then obviously $L^0 \subseteq U_A\left(\varrho_1,\dots,\varrho_s; \frac{\pi}{8r}\right)$, and therefore,

$$L^0_{(4r)} \subseteq U_A\left(\varrho_1,\dots,\varrho_s; \frac{\pi}{2}\right) \subseteq V_{(4)}. \tag{11.15}$$

For every $n \in \mathbb{N}_+$, define $A_n = (-n,n]^r \times F$ and pick $\varepsilon > 0$ such that $\varepsilon < \frac{1}{6k^4}$. Then $\lfloor (\frac{k}{1-k\varepsilon})^2 \rfloor = k^2$, since $k^2 \leq \frac{k^2}{(1-k\varepsilon)^2} \leq \frac{k^2}{1-2k\varepsilon} < \frac{3k^5}{3k^3-1} < k^2+1$. As in Lemma 11.2.4, the sequence $\{A_n\}_{n\in\mathbb{N}_+}$ satisfies the hypotheses of Lemma 11.2.3, and so $|V \cap A_n| > (\frac{1}{k}-\varepsilon)|A_n|$ for sufficiently large $n \in \mathbb{N}_+$. Moreover, we choose $n \in L$. Let

$$G_n = A/(2n\mathbb{Z}^r) \cong \mathbb{Z}(2n)^r \times F$$

and $E = q(V \cap A_n)$, where $q: A \to G_n$ is the canonical projection. Then $q\restriction_{A_n}$ is injective as $(A_n - A_n) \cap \ker q = \{0\}$, so q induces a bijection between A_n and G_n, on the one hand, and between $V \cap A_n$ and E, on the other hand. Thus,

$$|G_n| = |A_n| = (2n)^r|F| \quad \text{and} \quad |E| = |V \cap A_n| > \left(\frac{1}{k} - \varepsilon\right)|A_n|,$$

and so

$$\left(\frac{|G_n|}{|E|}\right)^2 \leq \left(\frac{k}{1-k\varepsilon}\right)^2, \quad \text{hence} \quad \left\lfloor\left(\frac{|G_n|}{|E|}\right)^2\right\rfloor \leq \left\lfloor\left(\frac{k}{1-k\varepsilon}\right)^2\right\rfloor = k^2.$$

To the finite group G_n apply Lemma 11.2.1 to get $\xi_{1,n},\dots,\xi_{m,n} \in G_n^*$, where $m = k^2$, such that

$$U_{G_n}\left(\xi_{1,n},\dots,\xi_{m,n}; \frac{\pi}{2}\right) \subseteq E_{(4)}.$$

For every $j \in \{1,\dots,m\}$, let $\chi_{j,n} = \xi_{j,n} \circ q \in A^*$.

If $a \in A_n \cap U_A(\chi_{1,n},\dots,\chi_{m,n}; \frac{\pi}{2})$, then $q(a) \in U_{G_n}(\xi_{1,n},\dots,\xi_{m,n}; \frac{\pi}{2}) \subseteq E_{(4)}$. Therefore, there exist $b_1, b_2, b_3, b_4 \in V \cap A_n$ and $c = (c_i)_{i=1}^r \in 2n\mathbb{Z}^r$ such that $a = b_1 - b_2 + b_3 - b_4 + c$. Since, for every $i \in \{1,\dots,r\}$, $2n$ divides c_i and $|c_i| \leq 5n$, we conclude that $c_i \in \{0, \pm 2n, \pm 4n\}$. Since $n \in L$, c can be written as a sum of at most $4r$ elements of L^0, so $c \in L^0_{(4r)} \subseteq V_{(4)}$ by (11.15), and consequently $a \in V_{(8)}$. We have proved that

$$A_n \cap U_A\left(\chi_{1,n},\dots,\chi_{m,n}; \frac{\pi}{2}\right) \subseteq V_{(8)}.$$

By Lemma 8.1.3, there exist $\chi_1, \dots, \chi_m \in A^*$ and a subsequence $\{n_l\}_{l\in\mathbb{N}}$ of $\{n\}_{n\in\mathbb{N}_+\cap L}$ such that $\chi_j(a) = \lim_{l\to\infty} \chi_{j,n_l}(a)$ for every $j \in \{1,\dots,m\}$ and for every $a \in A$. Since $A = \bigcup\{A_{n_l} : l > j, n_l \in L\}$ for every $j \in \mathbb{N}_+$, we can conclude, as in the last part of the proof of Lemma 11.2.4, that $U_A\,(\chi_1,\dots,\chi_m; \frac{\pi}{2}) \subseteq V_{(8)}$.

Consider now the general case. Let $g_1, \dots, g_k \in A$ be such that $A = \bigcup_{i=1}^k (g_i + V)$. Suppose that G is a finitely generated subgroup of A containing $g_1, \dots, g_k$. Then $G = \bigcup_{i=1}^k (g_i + (V \cap G))$, and so k translates of $V \cap G$ cover G. By the above argument, there exist $\varphi_{1G}, \dots, \varphi_{mG} \in G^*$, where $m = k^2$, such that

$$U_G\left(\varphi_{1G}, \dots, \varphi_{mG}; \frac{\pi}{2}\right) \subseteq (V \cap G)_{(8)} \subseteq V_{(8)}.$$

By Corollary A.2.5, we can extend each φ_{iG} to a character of A, so that we can assume from now on that $\varphi_{1G}, \dots, \varphi_{mG} \in A^*$, and so

$$G \cap U_A\left(\varphi_{1G}, \dots, \varphi_{mG}; \frac{\pi}{2}\right) = U_G\left(\varphi_{1G}, \dots, \varphi_{mG}; \frac{\pi}{2}\right) \subseteq V_{(8)}. \tag{11.16}$$

Let $\mathcal{G}$ be the family of all finitely generated subgroups G of A containing $g_1, \dots, g_k$, which is a directed set under inclusion. For every $j \in \{1, \dots, m\}$, we get a net $\{\varphi_{jG}\}_{G\in\mathcal{G}}$ in A^*; by Lemma 8.1.3, there exist a subnet $\{\varphi_{jG_\beta}\}_{\beta\in B}$ and $\chi_j \in A^*$ such that

$$\chi_j(a) = \lim_{\beta\in B} \varphi_{jG_\beta}(a) \quad \text{for every } a \in A. \tag{11.17}$$

From (11.16) and (11.17), we conclude, as in the finitely generated case, that

$$U_A\left(\chi_1, \dots, \chi_m; \frac{\pi}{2}\right) \subseteq V_{(8)}. \qquad \square$$

As a corollary of the Følner lemma, we obtain the following *internal* description of the neighborhoods of 0 in the Bohr topology of an abelian group A.

Corollary 11.2.6. *For a subset E of an infinite abelian group A, the following conditions are equivalent:*
(a) *E contains $V_{(8)}$ for some big subset V of A;*
(b) *for every $n \in \mathbb{N}_+$, E contains $V_{(2n)}$ for some big subset V of A;*
(c) *E is a neighborhood of 0 in the Bohr topology of A.*

Proof. (a)⇒(c) follows from Følner lemma 11.2.5, while (c)⇒(b) follows from Corollary 10.2.17 and Proposition 10.2.16, and (b)⇒(a) is obvious. □

The previous and the next corollary should be compared with Remark 10.2.18(b).

Corollary 11.2.7. *For an abelian group G, the Bohr topology $\mathfrak{B}_G = \mathcal{T}_{G^*}$ coincides with the finest precompact group topology $\mathcal{P}_G$.*

Corollary 11.2.8. *For a subgroup H of an abelian group G, the Bohr topology $\mathfrak{B}_{G/H}$ of G/H coincides with the quotient topology of the Bohr topology $\mathfrak{B}_G$.*

Proof. Let $q: G \to G/H$ be the canonical projection. The quotient topology $\overline{\mathfrak{B}}_G$ of the Bohr topology $\mathfrak{B}_G$ is a precompact group topology on G/H (as H is closed in $G^\#$ by Proposition 3.1.9). Hence, $\overline{\mathfrak{B}}_G \le \mathcal{P}_{G/H} = \mathfrak{B}_{G/H}$. On the other hand, $q: G^\# \to (G/H)^\#$ is continuous by Lemma 2.2.12, hence $\mathfrak{B}_{G/H} \le \overline{\mathfrak{B}}_G$ by the properties of the quotient topology. We conclude that $\overline{\mathfrak{B}}_G = \mathfrak{B}_{G/H}$. □

11.3 The Prodanov lemma and independence of characters

In the sequel various linear subspaces of the $\mathbb{C}$-algebra $C^*(G)$ of all bounded complex-valued functions of an abelian group G are used (see Example 5.2.21).

Let G be a topological abelian group. We denote by $\mathfrak{X}(G)$ the $\mathbb{C}$-linear subspace of $C^*(G)$ spanned by all continuous characters of G, and by $\mathfrak{X}_0(G)$ its $\mathbb{C}$-linear subspace spanned by the nontrivial continuous characters of G. Then

$$\mathfrak{X}(G) = \mathbb{C} \cdot 1 \oplus \mathfrak{X}_0(G)$$

(by Corollary 11.3.7) and both $\mathfrak{X}(G)$ and $\mathfrak{X}_0(G)$ are invariant under the action of G on $C^*(G)$ such that, for all $a \in G, f \mapsto f_a$, where $f_a(x) = f(xa)$ for all $x \in G$.

Moreover, let

$$\mathfrak{A}(G) = \{f \in C^*(G): \forall \varepsilon > 0, \exists g \in \mathfrak{X}(G), \|f - g\| \le \varepsilon\},$$

namely, $\mathfrak{A}(G)$ is the closure of $\mathfrak{X}(G)$ in $C^*(G)$ with respect to the uniform convergence topology of $C^*(G)$. Hence, $\mathfrak{A}(G)$ is a $\mathbb{C}$-subalgebra of $C^*(G)$ containing all constants and closed under complex conjugation. Since $\mathfrak{X}(G)$ is contained in $C(G)$ and the uniform limit of a sequence of continuous functions is again continuous, it follows that $\mathfrak{A}(G)$ is a subalgebra of $C(G)$.

Furthermore, let

$$\mathfrak{A}_0(G) = \{f \in C^*(G): \forall \varepsilon > 0, \exists g \in \mathfrak{X}_0(G), \|f - g\| \le \varepsilon\},$$

namely, $\mathfrak{A}_0(G)$ is the closure of $\mathfrak{X}_0(G)$ in $C^*(G)$ with respect to the uniform convergence topology. It is easy to see that $\mathfrak{A}_0(G)$ is a $\mathbb{C}$-linear subspace of $\mathfrak{A}(G)$ (hence of $C^*(G)$ as well). Moreover,

$$\mathfrak{A}(G) = \mathbb{C} \cdot 1 + \mathfrak{A}_0(G),$$

where $\mathbb{C} \cdot 1$ is the one-dimensional $\mathbb{C}$-subalgebra consisting of the constant functions.

Following the spirit of the setting of §B.5.4, we adopt this notation also for an abstract group G, assuming silently that G carries the discrete topology.

11.3.1 The Prodanov lemma and the Følner theorem

The next lemma, due to Prodanov [237], allows us to eliminate the discontinuous characters in uniform approximations of continuous functions via linear combinations of characters. In [99, Lemma 1.4.1] it is proved for abelian groups G that carry a topology τ such that, for every $g \in G$ and $n \in \mathbb{Z}$, the functions $x \mapsto x+g$ and $x \mapsto nx$ are continuous in (G,τ). The fact that this topology is not assumed to be Hausdorff is crucial in the applications of the lemma.

Recall that a subset C of a real or complex vector space is said to be *convex* if, for all $x, y \in C$ and all $t \in [0,1]$, $(1-t)x + ty \in C$.

Lemma 11.3.1 (Prodanov lemma). *Let G be a topological abelian group, U an open set of G, $f: U \to \mathbb{C}$ a continuous function, and M a convex closed set of $\mathbb{C}$. Let $k \in \mathbb{N}_+$, $\chi_1, \dots, \chi_k \in G^*$, and $c_1, \dots, c_k \in \mathbb{C}$ be such that*

$$\sum_{j=1}^{k} c_j\chi_j(x) - f(x) \in M \quad \text{for every } x \in U. \tag{11.18}$$

If $\chi_{m_1}, \dots, \chi_{m_s}$, with $m_1 < \dots < m_s$, $s \in \mathbb{N}$ and $\{m_1, \dots, m_s\} \subseteq \{1, \dots, k\}$, are precisely all continuous characters among $\chi_1, \dots, \chi_k$ (i. e., $\{\chi_{m_1}, \dots, \chi_{m_s}\} = \widehat{G} \cap \{\chi_1, \dots, \chi_k\}$), then $\sum_{i=1}^{s} c_{m_i}\chi_{m_i}(x) - f(x) \in M$ for every $x \in U$.

Proof. Assume that $\chi_k \in G^*$ is not continuous; then it is not continuous at 0. Consequently, there exists a net $\{x_\gamma\}_{\gamma\in A}$ in G such that $x_\gamma \to 0$ and $\{\chi_k(x_\gamma)\}_{\gamma\in A}$ does not converge to 1 in $\mathbb{S}$. Using the compactness of $\mathbb{S}$ and passing to a subnet, we may assume that it converges, i. e., $y_k := \lim_{\gamma\in A} \chi_k(x_\gamma) \neq 1$. Furthermore, exploiting the compactness of $\mathbb{S}$, and passing subsequently to subnets, we can arrange to achieve also the nets $\{\chi_j(x_\gamma)\}_{\gamma\in A}$ to converge for all $j \in \{1, \dots, k-1\}$; let $y_j = \lim_{\gamma\in A} \chi_j(x_\gamma)$. Obviously, $y_j = 1$ when χ_j is continuous because $x_\gamma \to 0$.

For $t \in \mathbb{N}$, $\gamma \in A$, and for every $x \in U$, let

$$z_{t,\gamma}(x) = \sum_{j=1}^{k} c_j\chi_j(x)\chi_j(x_\gamma)^t - f(x + tx_\gamma) = \sum_{j=1}^{k} c_j\chi_j(x + tx_\gamma) - f(x + tx_\gamma).$$

Fix $t \in \mathbb{N}$. Since $x_\gamma \to 0$, we get $x + tx_\gamma \in U$ for every $x \in U$ and for every sufficiently large $\gamma \in A$; thus, by (11.18), $z_{t,\gamma}(x) \in M$ and so, passing to the limit for $\gamma \in A$,

$$z_t(x) := \lim_{\gamma\in A} z_{t,\gamma}(x) = \sum_{j=1}^{k} c_j\chi_j(x)y_j^t - f(x) \in M, \tag{11.19}$$

because f is continuous and M is closed.

Take an arbitrary $n \in \mathbb{N}$. By the convexity of M and (11.19) for $t \in \{0, \dots, n\}$, for every $x \in U$,

$$z^{(n)}(x) := \frac{1}{n+1}\sum_{t=0}^{n} z_t(x) = \frac{1}{n+1}\sum_{t=0}^{n}\left(\sum_{j=1}^{k} c_j\chi_j(x)y_j^t - f(x)\right) \in M.$$

Note that $\sum_{t=0}^{n} y_k^t = \frac{y_k^{n+1}-1}{y_k-1}$ because $y_k \neq 1$. Hence, for every $j \in \{1, \dots, k-1\}$, letting

$$c_{jn} = \frac{c_j}{n+1} \sum_{t=0}^{n} y_j^t,$$

we get, for every $x \in U$,

$$z^{(n)}(x) = \sum_{j=1}^{k-1} c_{jn}\chi_j(x) + \frac{c_k}{1+n}\frac{1-y_k^{n+1}}{1-y_k}\chi_k(x) - f(x) \in M.$$

Now for every $j \in \{1, \dots, k-1\}$,
(1) $|c_{jn}| \leq \frac{|c_j|}{n+1} \sum_{t=0}^{n} |y_j|^t = |c_j|$ (because $|y_j| = 1$), and
(2) if $y_j = 1$, then $c_{jn} = c_j$.

By the boundedness of the sequences $\{c_{jn}\}_{n\in\mathbb{N}}$ for $j \in \{1, \dots, k-1\}$, there exists a subsequence $\{n_m\}_{m\in\mathbb{N}_+}$ of $\{n\}_{n\in\mathbb{N}_+}$ such that all limits $c_j' = \lim_{m\to\infty} c_{jn_m}$ exist. On the other hand, $|y_k| = 1$, so

$$|1 - y_k^{n+1}| \leq 1 + |y_k^{n+1}| \leq 2,$$

and hence

$$\lim_{n\to\infty} \frac{c_k}{1+n}\frac{1-y_k^{n+1}}{1-y_k} = 0.$$

For every $x \in U$, we get

$$\lim_{m\to\infty} z^{(n_m)}(x) = \sum_{j=1}^{k-1} c_j'\chi_j(x) - f(x) \in M; \tag{11.20}$$

moreover, $c_j' = c_j$ for every $j \in \{1, \dots, k-1\}$ such that χ_j is continuous by (2).

The condition in (11.20) is obtained from the hypothesis, removing the discontinuous character χ_k in such a way that the coefficients of the continuous characters remain the same. Iterating this procedure, we can remove all discontinuous characters among $\chi_1, \dots, \chi_k$. □

This lemma allows one to "produce continuity out of nothing" in the process of approximation. Moreover, a careful analysis of the proof shows that one can safely replace the group topology of G with a semitopological group topology such that for every $n \in \mathbb{Z}$, the function $x \mapsto nx$ on G is continuous.

Corollary 11.3.2. *Let G be a topological abelian group, $f: G \to \mathbb{C}$ a continuous function, and $\varepsilon > 0$. If $\| \sum_{j=1}^{k} c_j\chi_j - f\| \leq \varepsilon$ for some $k \in \mathbb{N}_+, \chi_1, \dots, \chi_k \in G^*$, and $c_1, \dots, c_k \in \mathbb{C}$, then also $\| \sum_{i=1}^{s} c_{m_i}\chi_{m_i} - f\| \leq \varepsilon$, where $\{\chi_{m_1}, \dots, \chi_{m_s}\} = \{\chi_1, \dots, \chi_n\} \cap \widehat{G}$, with $m_1 < \dots < m_s$.*

In particular, if $f = \sum_{j=1}^{k} c_j\chi_j$ for some $k \in \mathbb{N}_+$, $\chi_1, \dots, \chi_k \in G^*$, and $c_1, \dots, c_k \in \mathbb{C}$, then also $f = \sum_{i=1}^{s} c_{m_i}\chi_{m_i}$ with $\{\chi_{m_1}, \dots, \chi_{m_s}\} = \{\chi_1, \dots, \chi_n\} \cap \widehat{G}$.

In other words, denoting by G_d the topological abelian group G considered with the discrete topology, $C(G) \cap \mathfrak{X}(G_d)$ coincides with the $\mathbb{C}$-subalgebra $\mathfrak{X}(G)$ of $C^*(G)$ generated by $\widehat{G}$.

Corollary 11.3.3. *For every topological abelian group G, $C(G) \cap \mathfrak{A}(G_d) = \mathfrak{A}(G)$.*

In other words, as far as continuous complex-valued functions are concerned, in the definition of $\mathfrak{A}(G)$ it is irrelevant whether one approximates via (linear combinations of) continuous or discontinuous characters.

Now we give a version for topological abelian groups of the local Stone–Weierstraß theorem (see Corollary B.5.22).

Proposition 11.3.4. *Let G be a topological abelian group and H a subgroup of $\widehat{G}$. If X is a nonempty subset of G and $f \in C^*(X)$, then the following are equivalent:*

(a) *f can be uniformly approximated on X by a linear combination of elements of H with complex coefficients;*
(b) *for every $\varepsilon > 0$, there exist $\delta > 0$ and $\chi_1, \dots, \chi_m \in H$ such that, for every $x, y \in X$, $x - y \in U_G(\chi_1, \dots, \chi_m; \delta)$ yields $|f(x) - f(y)| < \varepsilon$.*

Proof. (a)⇒(b) Let $\varepsilon > 0$. By hypothesis, there exist $c_1, \dots, c_m \in \mathbb{C}$ and $\chi_1, \dots, \chi_m \in H$ such that $\| \sum_{i=1}^{m} c_i\chi_i - f\| < \frac{\varepsilon}{4}$ on X, that is, for every $x \in X$,

$$\left|\sum_{i=1}^{m} c_i\chi_i(x) - f(x)\right| < \frac{\varepsilon}{4}.$$

On the other hand, for every $x, y \in X$,

$$\left|\sum_{i=1}^{m} c_i\chi_i(x) - \sum_{i=1}^{m} c_i\chi_i(y)\right| \leq \sum_{i=1}^{m} |c_i| \cdot |\chi_i(x) - \chi_i(y)|$$

and, for every $i \in \{1, \dots, m\}$,

$$|\chi_i(x - y) - 1| = |\chi_i(x)\chi_i(y)^{-1} - 1| = |\chi_i(x) - \chi_i(y)|.$$

Take

$$\delta = \frac{\varepsilon}{2m \max_{i\in 1,\dots,m} |c_i|}.$$

For $x, y \in X$, $x - y \in U(\chi_1, \dots, \chi_m; \delta)$ implies $|\chi_i(x - y) - 1| \leq |\mathrm{Arg}(\chi_i(x - y))| < \delta$ for every $i \in \{1, \dots, m\}$, so

$$\left|\sum_{i=1}^{m} c_i\chi_i(x) - \sum_{i=1}^{m} c_i\chi_i(y)\right| \leq \sum_{i=1}^{m} |c_i| \cdot |\chi_i(x) - \chi_i(y)| < \frac{\varepsilon}{2};$$

consequently,

$$|f(x)-f(y)| \le \left| f(x) - \sum_{i=1}^{m} c_i\chi_i(x)\right| + \left|\sum_{i=1}^{m} c_i\chi_i(x) - \sum_{i=1}^{m} c_i\chi_i(y)\right| + \left|\sum_{i=1}^{m} c_i\chi_i(y) - f(y)\right| < \varepsilon.$$

(b)⇒(a) Let βX be the Čech–Stone compactification of the discrete space X. If the function $F: X \to \mathbb{C}$ is bounded, there exists a unique continuous extension F^β of F to βX. Let $\mathcal{S}$ be the collection of all complex-valued continuous functions g of βX such that $g = \sum_{j=1}^n c_j \chi_j^\beta$ with $\chi_1, \dots, \chi_n \in H$, $c_1, \dots, c_n \in \mathbb{C}$, and $n \in \mathbb{N}_+$. Then $\mathcal{S}$ is a $\mathbb{C}$-subalgebra of $C(\beta X)$ closed under complex conjugation and containing all constants. To see that $\mathcal{S}$ is a $\mathbb{C}$-subalgebra, it is enough to note that for $\chi, \xi \in H$, $\chi^\beta \xi^\beta = (\chi\xi)^\beta \in \mathcal{S}$. To see that $\mathcal{S}$ is closed under complex conjugation, it suffices to check that $\overline{\chi^\beta} = (\overline{\chi})^\beta \in \mathcal{S}$. Indeed, $\chi\overline{\chi} = 1$ yields $1^\beta = (\chi\overline{\chi})^\beta = \chi^\beta(\overline{\chi})^\beta$, and hence $\overline{\chi^\beta} = (\chi^\beta)^{-1} = \overline{\chi}^\beta$.

Now we see that $\mathcal{S}$ separates the points of βX separated by f^β, to apply the local Stone–Weierstraß theorem (see Corollary B.5.22). Let $x, y \in \beta X$ and $f^\beta(x) \neq f^\beta(y)$. Consider two nets $\{x_i\}_{i\in I}$ and $\{y_i\}_{i \in I}$ in X such that $x_i \to x$ and $y_i \to y$. Since f^β is continuous, $f^\beta(x) = \lim_{i\in I} f(x_i)$ and $f^\beta(y) = \lim_{i \in I} f(y_i)$. Along with $f^\beta(x) \neq f^\beta(y)$, this implies that there exists $\varepsilon > 0$ such that $|f(x_i) - f(y_i)| \ge \varepsilon$ for every sufficiently large $i \in I$. By hypothesis, there exist $\delta > 0$ and $\chi_1, \dots, \chi_m \in H$ such that for $u, v \in X$,

$$\text{if} \quad u - v \in U_G(\chi_1, \dots, \chi_m; \delta) \quad \text{then } |f(u) - f(v)| < \varepsilon. \tag{11.21}$$

Assume that $\chi_j^\beta(x) = \chi_j^\beta(y)$ holds for every $j \in \{1, \dots, m\}$. Then $x_i - y_i \in U_G(\chi_1, \dots, \chi_m; \delta)$ for every sufficiently large $i \in I$. This, together with $|f(x_i) - f(y_i)| \ge \varepsilon$, contradicts (11.21). So, each pair of points of βX separated by f^β is also separated by $\mathcal{S}$.

Since βX is compact, one can apply the local version of the Stone–Weierstraß theorem, that is, Corollary B.5.22, to $\mathcal{S}$ and f^β. So, f^β can be uniformly approximated by elements of $\mathcal{S}$. To conclude, note that if $g = \sum_{j=1}^n c_j\chi_j^\beta$ on βX for some $c_1, \dots, c_n \in \mathbb{C}$ and $\chi_1, \dots, \chi_n \in H$, then $g\restriction_X = \sum_{j=1}^n c_j \chi_j$. □

The reader familiar with uniform spaces will note that item (b) is nothing else but uniform continuity of f with respect to the uniformity on X induced by the uniformity of the whole group G determined by the topology $\mathcal{T}_H$.

The use of the Čech–Stone compactification in the above proof is inspired by Nöbeling and Bauer [25], who proved that if $\mathcal{S}$ is a $\mathbb{C}$-subalgebra of $C^*(X)$ for some nonempty set X, and $\mathcal{S}$ contains the constants and is stable under complex conjugation, then $g \in C^*(X)$ belongs to the closure of $\mathcal{S}$ with respect to the norm topology if and only if for every net $\{x_\alpha\}_{\alpha\in A}$ in X the net $\{g(x_\alpha)\}_{\alpha \in A}$ is convergent whenever the nets $\{f(x_\alpha)\}_{\alpha\in A}$ are convergent for all $f \in \mathcal{S}$.

Theorem 11.3.5 (Følner theorem). *Let G be a topological abelian group. If $k \in \mathbb{N}_+$ and E is a subset of G such that k translates of E cover G, then for every $U \in \mathcal{V}_G(0)$ there exist $\chi_1, \dots, \chi_m \in \widehat{G}$, where $m = k^2$, and $\delta > 0$ such that $U_G(\chi_1, \dots, \chi_m; \delta) \subseteq U - U + E_{(8)}$.*

Proof. We can assume, without loss of generality, that U is open. By Følner lemma 11.2.5, there exist $\varphi_1, \dots, \varphi_m \in G^*$ such that $U_G(\varphi_1, \dots, \varphi_m; \frac{\pi}{2}) \subseteq E_{(8)}$; our aim is to replace these characters by continuous ones by "enlarging" $E_{(8)}$ to $U - U + E_{(8)}$.

It follows from Lemma 3.1.1 that

$$C := \overline{E_{(8)} + U} \subseteq E_{(8)} + U - U. \tag{11.22}$$

Consider the open set $X = U \cup (G \setminus C)$ and the function $f : X \to \mathbb{C}$ defined by

$$f(x) = \begin{cases} 0 & \text{if } x \in U, \\ 1 & \text{if } x \in G \setminus C. \end{cases}$$

Then f is continuous, as $X = U \cup (G \setminus C)$ is partition of X in two clopen sets.

Let $H = \langle \varphi_1, \dots, \varphi_m \rangle$. Take $x, y \in X$ with $x - y \in U_G(\varphi_1, \dots, \varphi_m; \frac{\pi}{2}) \subseteq E_{(8)}$. So, if $y \in U$, then $x \in E_{(8)} + U$, and consequently $x \notin G \setminus \overline{E_{(8)} + U} = G \setminus C$, that is, $x \in U$. In the same way it can be shown that $x \in U$ yields $y \in U$. This gives $f(x) = f(y)$, by the definition of f. By Proposition 11.3.4, one can uniformly approximate f on X by linear combinations of elements of H, namely, one can find $\xi_1, \dots, \xi_t \in H$ and $c_1, \dots, c_t \in \mathbb{C}$ such that, for every $x \in X$,

$$\left| \sum_{j=1}^{t} c_j \xi_j(x) - f(x) \right| \le \frac{1}{3}. \tag{11.23}$$

Since X is open and f is continuous, we can apply Prodanov lemma 11.3.1 to the convex closed set $M = \{z \in \mathbb{C} : |z| \le \frac{1}{3}\}$, and this permits us to assume that all $\xi_1, \dots, \xi_t$ are continuous. Letting $x = 0$ in (11.23), we get $|\sum_{j=1}^{t} c_j| \le \frac{1}{3}$, and consequently,

$$\frac{2}{3} \le \left| \sum_{j=1}^{t} c_j - 1 \right|. \tag{11.24}$$

Let now $\Phi = H \cap \widehat{G}$. By Theorem A.1.1, there exist $\chi_1, \dots, \chi_m \in \Phi$ such that $\Phi = \langle \chi_1, \dots, \chi_m \rangle$. Since for every $j \in \{1, \dots, t\}$ the character ξ_j is continuous and belongs to H, $\xi_j \in \Phi$, so each ξ_j can be written as a product $\xi_j = \chi_1^{s_1(j)} \cdots \chi_m^{s_m(j)}$ for appropriate $s_1(j), \dots, s_m(j) \in \mathbb{Z}$.

Choose $\varepsilon > 0$ with

$$\varepsilon \sum_{j=1}^{t} |c_j| < \frac{1}{3}.$$

By the continuity of $\chi_1, \dots, \chi_m \in \Phi$, there exists $\delta > 0$ such that, for all $j \in \{1, \dots, t\}$,

$$|\xi_j(x) - 1| \le \varepsilon \quad \text{whenever } x \in U_G(\chi_1, \dots, \chi_m; \delta).$$

To prove that $U_G(\chi_1, \dots, \chi_m; \delta) \subseteq U - U + E_{(8)}$, assume for a contradiction that there is some $z \in U_G(\chi_1, \dots, \chi_m; \delta) \setminus (U - U + E_{(8)})$. In view of (11.22), $z \in G \setminus C \subseteq X$. Thus, by the definition of f (which yields $f(z) = 1$), (11.23), (11.24), and $|\xi_j(z) - 1| \leq \varepsilon$ for every $j \in \{1, \dots, t\}$,

$$\frac{2}{3} \leq \left|\sum_{j=1}^{t} c_j - 1\right| \leq \left|\sum_{j=1}^{t} c_j(1 - \xi_j(z))\right| + \left|\sum_{j=1}^{t} c_j\xi_j(z) - f(z)\right| \leq \varepsilon \sum_{j=1}^{t} |c_j| + \frac{1}{3}.$$

These inequalities together give $\frac{1}{3} \leq \varepsilon \sum_{j=1}^{t} |c_j|$, against the choice of ε. □

11.3.2 Independence of characters

Now we apply Prodanov lemma 11.3.1 for an indiscrete abelian group G and $U = G$. Note that this necessarily yields that f is a constant function.

Corollary 11.3.6. *Let G be a topological abelian group, $g \in \mathfrak{A}_0(G)$, and M a convex closed set of $\mathbb{C}$. If $c \in \mathbb{C}$ is such that $g(x) + c \in M$ for every $x \in G$, then $c \in M$. In particular, if $\varepsilon \geq 0$ and $|g(x) - c| \leq \varepsilon$ for every $x \in G$, then $|c| \leq \varepsilon$.*

Proof. Assume first that $g \in \mathfrak{X}_0(G)$. Suppose that for every $x \in G$, $g(x) = \sum_{j=1}^{k} c_j\chi_j(x)$ for some $c_1, \dots, c_k \in \mathbb{C}$ and nonconstant $\chi_1, \dots, \chi_k \in \widehat{G}$. Apply Prodanov lemma 11.3.1 with G indiscrete, $U = G$, and f the constant function $-c$. Since all characters $\chi_1, \dots, \chi_k$ are discontinuous, we conclude that $c \in M$.

Assume now that $g \in \mathfrak{A}_0(G)$ and assume for contradiction that $c \notin M$. Since M is closed, there exists $\varepsilon > 0$ such that $c \notin M + D$, where D is the closed (so compact) ball with center 0 and radius ε. Let $h \in \mathfrak{X}_0(G)$ with $\|g - h\| \leq \varepsilon/2$. Since $M + D$ is still a closed convex set of $\mathbb{C}$ and $h(x) + c \in M + D$ for every $x \in G$, we conclude with the first case of the proof that $c \in M + D$, a contradiction.

For the second assertion, take as M the closed disk of center 0 and radius ε. □

Corollary 11.3.7. *Let G be an abelian group and $\chi_0, \chi_1, \dots, \chi_k \in G^*$ pairwise distinct characters. Then $\chi_0, \chi_1, \dots, \chi_k$ are linearly independent.*

Proof. Let $c_0, c_1, \dots, c_k \in \mathbb{C}$ be such that $\sum_{i=0}^{k} c_i\chi_i(x) = 0$ for every $x \in G$. Then $\sum_{i=1}^{k} c_i\chi_i(x)\chi_0(x)^{-1} + c_0 = 0$ for every $x \in G$. By hypothesis, the function $g\colon G \to \mathbb{C}$, $x \mapsto \sum_{i=1}^{k} c_i\chi_i(x)\chi_0(x)^{-1}$ is in $\mathfrak{X}_0(G)$, so Corollary 11.3.6 with $\varepsilon = 0$ implies $c_0 = 0$. Proceeding by induction, one can prove that $c_i = 0$ for all $i \in \{0, 1, \dots, k\}$. □

Using this corollary, we see that for an abelian group G the characters G^* not only span $\mathfrak{X}(G)$ as a base, but they have a much stronger independence property.

Corollary 11.3.8. *Let G be an abelian group and $\chi_0, \chi_1, \dots, \chi_k \in G^*$ pairwise distinct characters. Then $\|\sum_{j=1}^{k} c_j\chi_j - \chi_0\| \geq 1$ for every $c_1, \dots, c_k \in \mathbb{C}$.*

Proof. Let $\varepsilon = \| \sum_{j=1}^{k} c_j\chi_j - \chi_0 \|$. Then, for every $x \in G$,

$$\left| \sum_{j=1}^{k} c_j\chi_j(x) - \chi_0(x) \right| \leq \varepsilon. \tag{11.25}$$

According to Corollary 11.3.7, $\chi_0, \chi_1, \ldots, \chi_k$ are linearly independent, hence $\varepsilon > 0$.

By our assumption, $\xi_j := \chi_j\chi_0^{-1}$ is nonconstant for every $j \in \{1, \ldots, k\}$. So, $g := \sum_{j=1}^{k} c_j\xi_j \in \mathfrak{X}_0(G)$, and (11.25) yields that, for every $x \in G$,

$$|g(x) - 1| = \left| \sum_{j=1}^{k} c_j\chi_j(x)\chi_0^{-1}(x) - 1 \right| \leq \varepsilon.$$

According to Corollary 11.3.6, $|1| \leq \varepsilon$. □

Corollary 11.3.9. *Let G be an abelian group, H a subgroup of G^*, and let $\chi \in G^*$ be such that there exist $k \in \mathbb{N}_+, \chi_1, \ldots, \chi_k \in H$, and $c_1, \ldots, c_k \in \mathbb{C}$ with*

$$\left\| \sum_{j=1}^{k} c_j\chi_j - \chi \right\| \leq \frac{1}{2}. \tag{11.26}$$

Then there exists $i \in \{1, \ldots, k\}$ such that $\chi = \chi_i$ (in particular, $\chi \in H$).

Proof. We assume without loss of generality that $\chi_1, \ldots, \chi_k$ are pairwise distinct. Assume for a contradiction that $\chi \neq \chi_j$ for all $j \in \{1, \ldots, k\}$. Then Corollary 11.3.8 applied to $\chi, \chi_1, \ldots, \chi_k$ yields $\| \sum_{j=1}^{k} c_j\chi_j - \chi \| \geq 1$, against (11.26). □

As a consequence of Corollary 11.3.9, the continuous characters of $(G, \mathcal{T}_H)$ are precisely the characters of H.

Proposition 11.3.10. *Let G be an abelian group. Then $H = \widehat{(G, \mathcal{T}_H)}$ for every subgroup H of G^*.*

Proof. Obviously, $H \subseteq \widehat{(G, \mathcal{T}_H)}$. Now let $\chi \in \widehat{(G, \mathcal{T}_H)}$. For every fixed $\varepsilon > 0$, the set $O = \{a \in \mathbb{S} : |a - 1| < \varepsilon\}$ is an open neighborhood of 1 in $\mathbb{S}$. Hence, $W = \chi^{-1}(O)$ is $\mathcal{T}_H$-open in G. So, there exist $\chi_1, \ldots, \chi_m \in H$ and $\delta > 0$ such that $U_G(\chi_1, \ldots, \chi_m; \delta) \subseteq W$. If $x - y \in U_G(\chi_1, \ldots, \chi_m; \delta)$, then $\chi(x - y) = \chi(x)\chi(y)^{-1} \in O$, and so $|\chi(x)\chi(y)^{-1} - 1| < \varepsilon$; in particular, $|\chi(x) - \chi(y)| < \varepsilon$. In other words, χ satisfies condition (b) of Proposition 11.3.4, and hence there exist $\chi_1, \ldots, \chi_m \in H$ and $c_1, \ldots, c_m \in \mathbb{C}$ such that $\left\| \sum_{j=1}^{m} c_j\chi_j - \chi \right\| \leq \frac{1}{2}$. By Corollary 11.3.9, $\chi \in H$. □

11.4 Precompact group topologies on abelian groups

Let us recall here that for an abelian group G and a subgroup H of $G^* = \mathrm{Hom}(G, \mathbb{S})$, the group topology $\mathcal{T}_H$ generated by H is the coarsest group topology on G that makes every character from H continuous. We recall its description and properties in the next proposition.

Proposition 11.4.1. *Let G be an abelian group and H a subgroup of G^*. A local base at 0 in $(G, \mathcal{T}_H)$ is given by the sets $U(\chi_1, \dots, \chi_m; \delta)$, where $\chi_1, \dots, \chi_m \in H$ and $\delta > 0$. Moreover, $(G, \mathcal{T}_H)$ is a Hausdorff group if and only if H separates the points of G.*

Now we can characterize the totally bounded group topologies on abelian groups.

Theorem 11.4.2. *Let (G, τ) be a topological abelian group. The following conditions are equivalent:*
(a) *τ is totally bounded;*
(b) *the neighborhoods of 0 in G are big subsets;*
(c) $\tau = \mathcal{T}_{\widehat{(G,\tau)}}$.

Proof. (a)⇒(b) This is the definition of totally bounded topology.

(b)⇒(c) If $H = \widehat{(G,\tau)}$ then $\mathcal{T}_H \le \tau$. Let U, V be open neighborhoods of 0 in (G, τ) such that $V_{(10)} \subseteq U$. Then V is big by hypothesis, and by Følner theorem 11.3.5 there exist $\delta > 0$ and $\chi_1, \dots, \chi_m \in \widehat{G}$ such that $U_G(\chi_1, \dots, \chi_m; \delta) \subseteq V_{(10)} \subseteq U$. Thus, $U \in \mathcal{V}_{\mathcal{T}_H}(0)$, and this proves that $\tau \le \mathcal{T}_H$.

(c)⇒(a) follows from Corollary 10.2.17. □

For the sake of completeness, we give the obvious counterpart for precompactness:

Corollary 11.4.3. *Let (G, τ) be a topological abelian group. The following conditions are equivalent:*
(a) *τ is precompact;*
(b) *τ is Hausdorff and the neighborhoods of 0 in G are big subsets;*
(c) *$H := \widehat{(G,\tau)}$ separates the points of G and $\tau = \mathcal{T}_H$.*

Theorem 11.4.4 allows us to sharpen this property (see Corollary 11.5.3).

Theorem 11.4.4. *For an abelian group G, let*

$$\mathcal{D}(G) = \{H \le G^* : H \text{ separates the points of } G\},$$
$$\mathcal{P} = \{\tau : \tau \text{ precompact group topology on } G\}.$$

Then the following is an order-preserving bijection:

$$T : \mathcal{D}(G) \to \mathcal{P}, \quad H \mapsto \mathcal{T}_H \tag{11.27}$$

(i. e., if $H_1, H_2 \in \mathcal{D}(G)$ then $\mathcal{T}_{H_1} \le \mathcal{T}_{H_2}$ if and only if $H_1 \subseteq H_2$).

Proof. Corollary 11.4.3 yields that $\mathcal{T}_H \in \mathcal{P}$ for every $H \in \mathcal{D}(G)$ and that T is surjective.

By Proposition 11.3.10, $\mathcal{T}_{H_1} = \mathcal{T}_{H_2}$ for $H_1, H_2 \in \mathcal{D}(G)$ yields $H_1 = H_2$. Therefore, T is a bijection. The last statement of the theorem is obvious. □

We proved in Proposition 11.3.10, that for an abelian group G and a subgroup H of G^* the continuous characters of $(G, \mathcal{T}_H)$ are precisely the characters of H. This allows us to prove that $w(G) = \chi(G) = |\widehat{G}|$ for a precompact abelian group G:

Corollary 11.4.5. *If G is an infinite abelian group and H a subgroup of G^* that separates the points of G, then $w(G, \mathcal{T}_H) = \chi(G, \mathcal{T}_H) = |H|$.*

Proof. Let $\kappa = \chi(G, \mathcal{T}_H)$ and note that κ cannot be finite. (Indeed, otherwise $\kappa = 1$ and $\mathcal{T}_H$ must be discrete, so being also precompact, this would imply that G is finite, a contradiction.) According to Exercise 5.4.2, $\kappa \leq w(G, \mathcal{T}_H) \leq |H|$.

We aim to prove that $|H| \leq \kappa$. Pick a base $\mathcal{B}$ of $\mathcal{V}_{\mathcal{T}_H}(0)$ of size κ. We can assume that every $B \in \mathcal{B}$ can be written as $B = U_G(\chi_{1,B}, \dots, \chi_{n_B,B}; \delta_B)$, where $n_B \in \mathbb{N}_+$, $\delta_B > 0$, and $\chi_{i,B} \in H$ for $i \in \{1, \dots, n_B\}$. Then the set

$$X = \{\chi_{i,B} : B \in \mathcal{B}, i \in \{1, \dots, n_B\}\} \subseteq H$$

has size at most κ and $\mathcal{T}_H = \mathcal{T}_X$, by the choice of $\mathcal{B}$. Let $H' = \langle X \rangle$; then $|H'| \leq \kappa$ and $\mathcal{T}_H = \mathcal{T}_X = \mathcal{T}_{H'}$, by Theorem 11.4.4. Since the correspondence (11.27) from Theorem 11.4.4 is bijective, we deduce that $H' = H$. Therefore, $|H| = |H'| \leq \kappa$. □

Corollary 11.4.6. *Let G be an abelian group and H a subgroup of G^* such that $\mathcal{T}_H$ is metrizable. Then H is countable.*

11.5 The Peter–Weyl theorem for compact abelian groups

Let us start with the following important direct consequence of Corollary 11.4.3, which is the Peter–Weyl theorem in the abelian case.

Corollary 11.5.1 (Peter–Weyl theorem). *If K is a compact abelian group, then the group topology on K coincides with $\mathcal{T}_{\widehat{K}}$ and $\widehat{K}$ separates the points of K.*

Corollary 11.5.2. *If K is a compact abelian group, then K is topologically isomorphic to a (closed) subgroup of $\mathbb{T}^{\widehat{K}}$.*

Proof. Since $\widehat{K}$ separates the points of K by Corollary 11.5.1, the diagonal map determined by all characters in $\widehat{K}$ defines a continuous injective homomorphism $\Delta_{\widehat{K}} : K \hookrightarrow \mathbb{T}^{\widehat{K}}$. By the compactness of K and the open mapping theorem (Theorem 8.4.1), $\Delta_{\widehat{K}}$ is the required topological embedding. □

If K is not elementary compact, the power $\mathbb{T}^{\widehat{K}}$ is the smallest possible one with this property. Indeed, if the compact but not elementary compact abelian group K topologically embeds into some power $\mathbb{T}^{\kappa}$, then κ is infinite, so $\kappa = w(\mathbb{T}^{\kappa}) \geq w(K) = |\widehat{K}|$, by Corollary 11.4.5.

As a corollary of Theorem 11.4.4, we obtain the following useful fact that completes Corollary 11.5.1. It is essentially used in the proof of Pontryagin-van Kampen duality theorem 13.4.17.

Corollary 11.5.3. *If (K, τ) is a compact abelian group and H is a subgroup of $\widehat{K}$ that separates the points of K, then $H = \widehat{K}$.*

Proof. By Corollary 11.5.1, $\tau = \mathcal{T}_{\widehat{K}}$. Since $\mathcal{T}_H \leq \mathcal{T}_{\widehat{K}}$ by Theorem 11.4.4 and $\mathcal{T}_H$ is Hausdorff, $\mathcal{T}_H = \mathcal{T}_{\widehat{K}} = \tau$, as τ is compact and due to the open mapping theorem (Theorem 8.4.1). Now again Theorem 11.4.4 yields $H = \widehat{K}$. □

We show now that every compact abelian group is an inverse limit of elementary compact abelian groups (see Definition 9.4.2).

Proposition 11.5.4. *Let K be a compact abelian group and U an open neighborhood of 0 in K. Then there exists a closed subgroup C of K such that $C \subseteq U$ and K/C is an elementary compact abelian group. In particular, K is an inverse limit of elementary compact abelian groups.*

Proof. By Corollary 11.5.1, the topology on K is $\mathcal{T}_{\widehat{K}}$, hence there exists a finite subset F of $\widehat{K}$ such that $C = \bigcap_{\chi \in F} \ker \chi \subseteq U$. Define $g = \prod_{\chi \in F} \chi : K \to \mathbb{T}^F$. Thus, $\ker g = C$ and K/C is topologically isomorphic to the closed subgroup $g(K)$ of $\mathbb{T}^F$ in view of the open mapping theorem (Theorem 8.4.1). So, K/C is elementary compact, by Corollary 9.4.3.

To prove the last statement, let $\{U_i : i \in I\}$ be the family of all open neighborhoods of 0 in K. For every $i \in I$, fix a closed subgroup C_i of K with $C_i \subseteq U_i$ and such that K/C_i is elementary compact. For $i, j \in I$, $K/C_i \cap C_j$ is elementary compact, as $K/C_i \cap C_j$ is topologically isomorphic to a closed subgroup of $K/C_i \times K/C_j$, which is again an elementary compact abelian group. So, enlarging the family $\{C_i : i \in I\}$ with all the finite intersections of its elements, we obtain a family $\{C_j : j \in J\}$ that gives an inverse system $[K/C_j, \nu_{ji}, J]$ of elementary compact abelian groups, where the homomorphisms $\nu_{ji} : K/C_i \to K/C_j$, when $C_i \subseteq C_j$, are simply the canonical projections. Then the inverse limit K' of this inverse system is a compact abelian group together with a continuous homomorphism $f : K \to K'$ induced by the projections $p_i : K \to K/C_i$ with $i \in J$.

Let $x \in K \setminus \{0\}$ and pick on open neighborhood U of 0 in K such that $x \notin U$. By the first part of the proof, there exists $C_i \subseteq U$ for some $i \in J$; hence $x \notin C_i$, and therefore $p_i(x) \neq 0$, so $f(x) \neq 0$ as well. This proves that f is injective. To check the surjectivity of f, take an element $x' = (x_i + C_i)_{i \in J} \in K' \subseteq \prod_{i \in J} K/C_i$. Then the family $\{x_i + C_i : i \in J\}$ of closed cosets in K has the finite intersection property, so has nonempty intersection. For every element x of that intersection, one has $f(x) = x'$. Finally, the continuous isomorphism $f : K \to K'$ must be open, by the compactness of K and the open mapping theorem (Theorem 8.4.1). □

For a topological group G, we say that G has *no small subgroups*, or shortly, G is NSS, if there exists a neighborhood U of 0 such that U contains no nontrivial subgroups of G. For example, $\mathbb{T}$ is NSS.

Corollary 11.5.5. *A compact abelian group K is* NSS *precisely when K is an elementary compact abelian group.*

Proof. If K is NSS, then K must be elementary compact, by Proposition 11.5.4. To prove the inverse implication, note first that $\mathbb{T}$ is NSS. Moreover, the class of NSS abelian groups is stable under taking finite products and subgroups, by Exercise 11.7.4(a). Thus, all powers $\mathbb{T}^n$, as well as their subgroups, are NSS. Since by definition all elementary compact abelian groups can be obtained in this way, we are done. □

11.6 On the structure of compactly generated LCA groups

From now on, all topological groups are Hausdorff; quotients are taken with respect to closed subgroups, and so they are still Hausdorff.

Proposition 11.6.1. *Let G be a compactly generated locally compact abelian group. Then there exists a discrete subgroup H of G such that $H \cong \mathbb{Z}^s$ for some $s \in \mathbb{N}$ and G/H is compact.*

Proof. Suppose first that there exist $g_1, \dots, g_m \in G$ such that $G = \overline{\langle g_1, \dots, g_m \rangle}$. We proceed by induction on $m \in \mathbb{N}_+$. For $m = 1$, apply Theorem 10.2.9: if G is infinite and discrete, take $H = G$, and if G is compact, let $H = \{0\}$. Suppose now that the property holds for $m \geq 1$ and that $G = \overline{\langle g_1, \dots, g_{m+1} \rangle}$. If every $\overline{\langle g_i \rangle}$ is compact, then so is G, and we can take $H = \{0\}$. Suppose without loss of generality that $\langle g_{m+1} \rangle$ is discrete and consider the canonical projection $q: G \to G_1 = G/\langle g_{m+1} \rangle$. Since $G_1 = \overline{\langle q(g_1), \dots, q(g_m) \rangle}$, by inductive hypothesis there exists a discrete subgroup H_1 of G_1 such that $H_1 \cong \mathbb{Z}^n$ and G_1/H_1 is compact. Therefore, $H = q^{-1}(H_1)$ is a closed countable subgroup of G. Thus, H is locally compact and countable, hence discrete by Example 8.1.5(a). Since H is finitely generated, it is isomorphic to $H_2 \times F$, where $H_2 \cong \mathbb{Z}^s$ for some $s \in \mathbb{N}$ and F is a finite abelian group (see Theorem A.1.1). Now G/H is isomorphic to G_1/H_1 and H/H_2 is finite, so G/H_2 is compact thanks to Lemma 8.2.3(b).

Now consider the general case. There exists a compact subset K of G that generates G. By Lemma 8.3.4, we can assume without loss of generality that $K = \overline{U}$, where U is a symmetric neighborhood of 0 in G. We show that there exists a finite subset F of G such that

$$K + K \subseteq \langle F \rangle + K : \tag{11.28}$$

pick a symmetric neighborhood V of 0 in G such that $V + V \subseteq U$; since $K \subseteq \bigcup_{x \in K}(x + V)$, there exists a finite subset F of K such that $K \subseteq \bigcup_{x \in F}(x + V) = F + V$, and so

$$K + K \subseteq F + F + V + V \subseteq \langle F \rangle + U \subseteq \langle F \rangle + K.$$

Since $G = \langle K \rangle$, an easy inductive argument and (11.28) show that

$$G = \langle F \rangle + K.$$

Let $G_1 = \overline{\langle F\rangle}$ and let $q\colon G \to G/G_1$ be the canonical projection. By the equality $G = K + G_1$, the quotient $q(K) = G/G_1$ is compact. By the first part of the proof, there exists a discrete subgroup H of G_1 such that $H \cong \mathbb{Z}^s$ for some $s \in \mathbb{N}$ and G_1/H is compact. By Theorem 3.2.8(b), $(G/H)/(G_1/H) \cong G/G_1$ is compact. Since G_1/H is a compact subgroup of G/H, also G/H is compact, in view of Lemma 8.2.3(b). □

Proposition 11.6.2. *Let G be a compactly generated locally compact abelian group. Then there exists a compact subgroup K of G such that G/K is elementary locally compact.*

Proof. By Proposition 11.6.1, there exists a discrete subgroup $H \cong \mathbb{Z}^s$ of G such that G/H is compact. Consider the canonical projection $q\colon G \to G/H$ and let U be a compact symmetric neighborhood of 0 in G such that $(U + U + U) \cap H = \{0\}$. So, $q(U)$ is a neighborhood of 0 in G/H and, applying Proposition 11.5.4, we find a closed subgroup L of G such that $H \subseteq L$ and the closed subgroup $C = L/H$ of G/H satisfies

$$C \subseteq q(U) \quad \text{and} \quad (G/H)/C \cong G/L \cong \mathbb{T}^t \times F, \tag{11.29}$$

where F is a finite abelian group and $t \in \mathbb{N}$ (i. e., G/L is elementary compact).

The set $K = L \cap U$ is compact, being closed in the compact neighborhood U. Let us see that K is a subgroup of G: if $x, y \in K$, then $x - y \in L$ and $q(x - y) \in C \subseteq q(U)$, thus $q(x - y) = q(u)$ for some $u \in U$; as $q(x - y - u) = 0$ in G/H,

$$x - y - u \in (U + U + U) \cap H = \{0\},$$

and hence $x - y = u \in K$.

Now take $x \in L$; consequently, $q(x) \in C \subseteq q(U)$, and so $q(x) = q(u)$ for some $u \in U$. Clearly, $u \in K$, hence $q(L) = q(K)$. Thus, $L = K + H$, and since also $K \cap H = \{0\}$, the canonical projection $l\colon G \to G/K$ restricted to H is a continuous isomorphism of H onto $l(H) = l(L)$. Let us see now that $l(H)$ is discrete. By Lemma 8.2.2, $l(H)$ is closed. Since H is discrete, $\{0_G\}$ is open in H, so $A = H \setminus \{0_G\}$ is a closed set of H, and hence of G as well. Again by Lemma 8.2.2, $l(A) = l(H) \setminus \{0_{G/K}\}$ is closed in G/K. Hence, $\{0_{G/K}\}$ is open in $l(H)$ and $l(L) = l(H) \cong H \cong \mathbb{Z}^s$ is discrete in G/K.

Observe that (11.29) and Theorem 3.2.8(b) yield the isomorphisms

$$(G/K)/l(L) = (G/K)/(L/K) \cong G/L \cong \mathbb{T}^t \times F.$$

Denote by ϱ the canonical projection $G/K \to G/L$ and recall that $\ker \varrho = l(L) = l(H)$ is a discrete subgroup G/K. Hence, to $\varrho\colon G/K \to G/L$ and the composition $\pi\colon \mathbb{R}^t \to G/L$ of the canonical projection $\mathbb{R}^t \to \mathbb{T}^t$ and the obvious inclusion of $\mathbb{T}^t$ in G/L, apply Lemma 9.1.3[2] to obtain an open continuous homomorphism $f\colon \mathbb{R}^t \to G/K$ such that

2 The reader who is familiar with covering maps may deduce the existence of such a lifting from the facts that ϱ is a covering homomorphism and $\mathbb{R}^t$ is simply connected.

$\varrho \circ f = \pi$:

$$\begin{array}{ccc} & & G/K \\ & {}^{f}\nearrow & \downarrow \varrho \\ \mathbb{R}^t & \xrightarrow[\pi]{} & G/L \end{array}$$

In particular, $N = f(\mathbb{R}^t)$ is an open subgroup of G/K. As $f\colon \mathbb{R}^t \to N$ is open, N is isomorphic to a quotient of $\mathbb{R}^t$, so $N \cong \mathbb{R}^k \times \mathbb{T}^m$, where $k + m \leq t$ by Corollary 9.4.1; in particular, N is elementary locally compact. Since N is divisible, by Corollary A.2.7 there exists a subgroup B of G/K such that $G/K = N \times B$. Since $N \cap B = \{0\}$ and N is open, B is discrete; moreover, B is compactly generated as it is a quotient of G, and so B is finitely generated. Therefore, G/K is elementary locally compact. □

To prove Pontryagin-van Kampen duality theorem 13.4.17, we need the next theorem, generalizing Corollary 11.5.1 and showing that all locally compact abelian groups are MAP.

Theorem 11.6.3. *If G is a locally compact abelian group, then $\widehat{G}$ separates the points of G.*

Proof. Let V be a compact neighborhood of 0 in G. Take $x \in G\setminus\{0\}$. Then $G_1 = \langle V \cup \{x\}\rangle$ is an open (it has nonempty interior) compactly generated subgroup of G. In particular, G_1 is locally compact. By Proposition 11.6.1, there exists a discrete subgroup H of G_1 such that $H \cong \mathbb{Z}^s$ for some $s \in \mathbb{N}$ and G_1/H is compact. Thus, $\bigcap_{n\in\mathbb{N}_+} nH = \{0\}$, and so there exists $n \in \mathbb{N}_+$ such that $x \notin nH$. Since nH is discrete and hence closed in G_1, and H/nH is finite, the quotient $G_2 = G_1/nH$ is compact, by Lemma 8.2.3(b). Consider the canonical projection $q\colon G_1 \to G_2$ and note that $q(x) \neq 0$ in G_2. By Corollary 11.5.1, there exists $\xi \in \widehat{G_2}$ such that $\xi(q(x)) \neq 0$. Consequently, $\chi := \xi \circ q \in \widehat{G_1}$ and $\chi(x) \neq 0$. By Theorem A.2.4, there exists $\widetilde{\chi} \in G^*$ such that $\widetilde{\chi} \restriction_{G_1} = \chi$. Since G_1 is an open subgroup of G, $\widetilde{\chi}$ is continuous, namely, $\widetilde{\chi} \in \widehat{G}$. □

Corollary 11.6.4. *If G is a locally compact abelian group, then every compact subgroup K of G is dually embedded.*

Proof. Clearly, $H = \{\chi \in \widehat{K} : \exists \xi \in \widehat{G},\ \xi\restriction_K = \chi\}$ is a subgroup of $\widehat{K}$. By Theorem 11.6.3, $\widehat{G}$ separates the points of G, so H separates the points of K. Now apply Corollary 11.5.3 to conclude that $H = \widehat{K}$. □

Corollary 11.6.5. *Let G be a nontrivial locally compact abelian group.*
(a) *Then G is connected if and only if $\chi(G) = \mathbb{T}$ for every nontrivial $\chi \in \widehat{G}$.*
(b) *If for every $\chi \in \widehat{G}$ the image $\chi(G)$ is a proper subgroup of $\mathbb{T}$, then G is hereditarily disconnected. The converse implication holds when G is also σ-compact.*

Proof. (a) If G is connected and $\chi \in \widehat{G}$ is nontrivial, then $\chi(G)$ is a nontrivial connected subgroup of $\mathbb{T}$, hence $\chi(G) = \mathbb{T}$ (see Example 6.1.8(b)). If G is not connected, then G

has a proper open subgroup H, by Theorem 8.5.2(b). Pick a nontrivial character ξ of the discrete group G/H such that $\xi(G/H) \neq \mathbb{T}$. Then the composition χ of ξ and the canonical projection $G \to G/H$ gives a nontrivial $\chi \in \widehat{G}$ with $\chi(G) \neq \mathbb{T}$.

(b) Assume that $\chi(G)$ is a proper subgroup of $\mathbb{T}$ for every $\chi \in \widehat{G}$. Since the proper subgroups of $\mathbb{T}$ are hereditarily disconnected (see Example 6.1.8(b)), $\chi(G)$ is hereditarily disconnected for every $\chi \in \widehat{G}$. According to Theorem 11.6.3, the diagonal homomorphism $f: G \to \prod_{\chi\in\widehat{G}} \chi(G)$ is continuous and injective. Since the product is hereditarily disconnected, we deduce that also G is hereditarily disconnected since this property is preserved under taking subgroups and finer topologies.

Now assume that G is σ-compact and hereditarily disconnected. Consider $\chi \in \widehat{G}$ and assume for a contradiction that $\chi(G) = \mathbb{T}$. Then $\chi: G \to \mathbb{T}$ is open by the open mapping theorem (Theorem 8.4.1), so $\mathbb{T}$ is a quotient of G. As hereditary disconnectedness is inherited by quotients of locally compact groups (see Corollary 8.5.9), we conclude that $\mathbb{T}$ must be hereditarily disconnected, a contradiction. □

One cannot remove "σ-compact" in the above corollary.

Example 11.6.6. Let G denote $\mathbb{T}$ equipped with the discrete topology. Then G is hereditarily disconnected, although the identity map provides a character $\chi: G \to \mathbb{T}$ with $\chi(G) = \mathbb{T}$.

Algebraic properties of the dual group $\widehat{G}$ of a compact abelian group G can be described in terms of topological properties of G. We prove in Corollary 11.6.7 that $\widehat{G}$ is torsion precisely when G is hereditarily disconnected; compactness plays an essential role here, and we shall see examples of hereditarily disconnected σ-compact and locally compact abelian groups G such that no continuous character of G is torsion (e. g., $G = \mathbb{Q}_p$).

Corollary 11.6.7. *For a compact abelian group G, the following are equivalent:*
(a) *G is profinite;*
(b) *G is hereditarily disconnected;*
(c) *G is topologically torsion (i. e., $G = G!$);*
(d) *$\widehat{G}$ is torsion;*
(e) *$\chi(G) \neq \mathbb{T}$ for every $\chi \in \widehat{G}$.*

Proof. (a)⇔(b) is Corollary 8.5.7, (b)⇔(e) is Corollary 11.6.5(b), (b)⇒(c) was proved in Exercise 8.7.11.

(b)⇔(d) The image $\chi(G)$ under a continuous character χ of G is a compact, hence closed, subgroup of $\mathbb{T}$. Thus, $\chi(G)$ is a proper subgroup of $\mathbb{T}$ precisely when it is finite. By Corollary 11.6.5(b), G is hereditarily disconnected if and only if $\chi(G)$ is finite for every $\chi \in \widehat{G}$, and this means that the character χ is torsion.

(c)⇒(e) Pick any $\chi \in \widehat{G}$. Since every $x \in G$ is topologically torsion, $\chi(x)$ is a topologically torsion element of $\mathbb{T}$, so $\chi(x) \in \mathbb{T}!$. Hence, $\chi(G) \subseteq \mathbb{T}! \subsetneq \mathbb{T}$, by Exercise 5.4.20. □

Corollary 11.6.8. *For a compact abelian group G and a prime p, the following conditions are equivalent:*
(a) *G is pro-p-finite;*
(b) *G is topologically p-torsion (i. e., $G = G_p$);*
(c) *$\widehat{G}$ is p-torsion;*
(d) *$\chi(G) \subseteq \mathbb{Z}(p^\infty)$ for every $\chi \in \widehat{G}$.*

Proof. (a)⇒(b) This was proved in Exercise 8.7.12.

(b)⇒(d) Pick any $\chi \in \widehat{G}$. Then every $x \in K$ being topologically p-torsion implies that $\chi(x)$ is a topologically p-torsion element of $\mathbb{T}$, so $\chi(x) \in \mathbb{Z}(p^\infty)$ (see Exercise 5.4.15).

(d)⇒(c) By the equivalence (d)⇔(e) in Corollary 11.6.7, $\widehat{G}$ is torsion. Now $\chi(G) \subseteq \mathbb{Z}(p^\infty)$ implies that $\chi(G)$ is a finite p-group, so $\widehat{G}$ is p-torsion.

(c)⇒(a) Since every continuous character of G has as range a finite p-subgroup of $\mathbb{T}$, G has a local base at 0 formed by open subgroups of finite index that is a power of p. Hence, G is pro-p-finite. □

Corollary 11.6.9. *For a locally compact abelian group G, one has $td_p(G) = G_p$ for every prime p.*

Proof. Fix a prime p. In view of Remark 5.3.7, always $td_p(G) \subseteq G_p$. To prove the opposite inclusion, let $x \in G_p$; so $p^n x \to 0$ in G. If $\langle x\rangle$ is finite, then $x \in t_p(G) \subseteq td_p(G)$. If $\langle x\rangle$ is infinite, then $\overline{\langle x\rangle}$ cannot be discrete since $p^n x \to 0$. So, in view of Theorem 10.2.9, $K := \overline{\langle x\rangle}$ is compact. Let $\chi \in \widehat{K}$. Then $\chi(x) \in \mathbb{T}_p = \mathbb{Z}(p^\infty)$ (see Exercise 5.4.15), so $\widehat{K} \cong \mathbb{Z}(p^\infty)$. Since the precompact topology $\mathcal{T}_{\mathbb{Z}(p^\infty)}$ of $\mathbb{Z}$ coincides with $\varpi_{\mathbb{Z}}^p$, we deduce that $x \in td_p(K)$. □

Here is the counterpart of Corollary 11.6.7 for the connected case:

Proposition 11.6.10. *Let G be a topological abelian group.*
(a) *If G is connected, then $\widehat{G}$ is torsion-free.*
(b) *If G is compact, then $\widehat{G}$ is torsion-free if and only if G is connected.*

Proof. (a) For every nonzero $\chi \in \widehat{G}$, the image $\chi(G)$ is a nontrivial connected subgroup of $\mathbb{T}$, so we deduce that $\chi(G) = \mathbb{T}$ (see Example 6.1.8(b)). Hence, $\widehat{G}$ is torsion-free.

(b) If G is compact and not connected, then by Theorem 8.5.2(b) there exists a proper open subgroup N of G. Take any nonzero character ξ of the finite group G/N. Then $m\xi = 0$ for some $m \in \mathbb{N}_+$ (e. g., $m = [G : N]$). Now the composition χ of ξ and the canonical projection $G \to G/N$ satisfies $m\chi = 0$ as well. So, $\chi \in \widehat{G}$ is a nonzero torsion continuous character of G. □

In view of Theorem 11.6.3, every locally compact abelian group (G, τ) is MAP. Then, as a consequence also of Proposition 10.2.13, the group G^+ is simply the group G equipped with the finest totally bounded group topology τ^+ with $\tau^+ \leq \tau$.

According to the following theorem, due to Glicksberg, these two topologies share the same compact sets:

Theorem 11.6.11. *For a locally compact abelian group* (G, τ)*, its Bohr modification* $G^+ = (G, \tau^+)$ *has the same compact sets as* (G, τ).

A proof of this theorem in the discrete case will be given in Theorem 13.4.9.

11.7 Exercises

Exercise 11.7.1. Prove that $G^* \cong G$ for every finite abelian group G.

Exercise 11.7.2. Let G be a torsion abelian group. Apply Proposition 11.1.3(a) to show that G^* is $\mathbb{C}$-linearly independent in $\mathbb{C}^G$.

Hint. Let $\chi_1, \dots, \chi_m$ be pairwise distinct characters of G. There exists a finite subset F of G which separates these characters. Let H be the (finite) subgroup of G generated by F and let ψ_j denote the restriction of χ_j to H for $j \in \{1, \dots, m\}$. According to Proposition 11.1.3(a), $\psi_1, \dots, \psi_m$ are linearly independent. Hence, also $\chi_1, \dots, \chi_m$ are linearly independent.

Exercise 11.7.3. Prove Corollary 11.2.8 using the explicit description of the neighborhoods of 0 in $G^\#$ given in Corollary 11.2.6.

Hint. Since $q: G^\# \to (G/H)^\#$ is continuous, it remains to show that it is also open. To this end, take a neighborhood U of 0 in $G^\#$. Then U contains some $V_{(8)}$, where V is a big subset of G. Since $q(V)$ is big in G/H and $q(V)_{(8)} = q(V_{(8)}) \subseteq q(U)$, we deduce from Corollary 11.2.6 that $q(U)$ is a neighborhood of 0 in $(G/H)^\#$.

Exercise 11.7.4. Prove that:

(a) the class of NSS groups is stable under taking finite product and subgroups; show that no infinite products of nontrivial groups can be NSS;
(b) every NSS group is Hausdorff;
(c) strongly monothetic non-indiscrete groups are NSS;
(d) a monothetic group need not be NSS;
(e) every Hausdorff group topology on $\mathbb{Z}$ is either NSS or admits a coarser Hausdorff linear topology;
(f) every elementary locally compact abelian group is NSS;
(g) if G is an abelian NSS group, $td(G) = t(G)$ and $td_p(G) = t_p(G)$ for $p \in \mathbb{P}$.

Hint. (a), (c) and (g) are immediate.

(b) If G is an NSS group, then the subgroup core(G) is contained in every neighborhood of the identity e_G.

For (d), use the fact that $\mathbb{T}^{\mathbb{N}}$ is monothetic.

For (e) and (f), apply Exercise 3.5.22 and Exercise 9.5.4, respectively.

Exercise 11.7.5. Prove that for a direct product $G = \prod_{i \in I} G_i$ of topological groups, one has $n(G) = \prod_{i \in I} n(G_i)$, $G^+ \cong \prod_{i \in I} G_i^+$, and $bG \cong \prod_{i \in I} bG_i$.

Hint. For finite I, this follows from Exercise 10.4.13 by induction.

In the general case apply Proposition 10.3.5 and the fact that a continuous homomorphism $f: G \to U(n)$ factorizes as $f = f_1 \circ p_J$ through some of the projections $p_J: G \to G_J := \prod_{i \in J} G_i$ for some finite subset J of I and $f_1: G_J \to U(n)$. Indeed, choosing $U \in \mathcal{V}_{U(n)}(e)$ witnessing NSS, there exists $W \in \mathcal{V}_G(e_G)$ with $f(W) \subseteq U$. As W contains a subproduct of the form $A = \prod_{i \in I \setminus J} G_i$ for some finite subset J of I, the subgroup $f(A) \subseteq U$ must be trivial. This gives the desired factorization $f = f_1 \circ p_J$.

By Exercise 10.4.13, $bG_J = \prod_{i \in J} bG_i$, so f_1 can be extended to $\bar{f}_1: \prod_{i \in J} bG_i \to U(n)$. Let $f' = \bar{f}_1 \circ \bar{p}_J$, where $\bar{p}_J: \prod_{i \in I} bG_i \to \prod_{i \in J} bG_i$ is the canonical projection. This provides the factorization $f = f' \circ b$ witnessing the universal property of the Bohr compactification for the inclusion $b: G = \prod_{i \in I} G_i \to \prod_{i \in I} bG_i$.

11.8 Further readings, notes, and comments

It follows from results of Følner [136] obtained by less elementary tools that condition (a) in Corollary 11.2.6 can be replaced by the weaker assumption $V_{(4)} \subseteq E$ (see also the work by Ellis and Keynes [133] or by Cotlar and Ricabarra [63] for further improvements). Nevertheless, the following old problem concerning the group $\mathbb{Z}$ is still open (see Cotlar and Ricabarra [63], Ellis and Keynes [133], Følner [136], Glasner [157], Pestov [227, Question 1025], or Veech [279]):

Question 11.8.1. Does there exist a big subset V of $\mathbb{Z}$ such that $V - V$ is not a neighborhood of 0 in the Bohr topology of G?

Every infinite abelian group G admits a big subset with empty interior with respect to the Bohr topology (see [10]); more precisely, these authors proved that every totally bounded abelian group has a big subset with empty interior.

Using Følner theorem 11.3.5, as well as [98] and [232], Stoyanov [266] proved that the minimal metrizable torsion abelian groups are precompact.

The results from §11.4 can be attributed to Comfort and Ross [60].

Making use of the argument of Proposition 11.5.4 (and Theorem 10.3.3, instead of Corollary 11.5.1), one can prove that for a compact (not necessarily abelian) group K and $U \in \mathcal{V}_K(e_K)$ there exists a closed normal subgroup C of K with $C \subseteq U$ and K/C isomorphic to a subgroup of a finite product $U(n_1) \times \cdots \times U(n_k)$. But since this group is isomorphic to a subgroup of $U(n_1 + \cdots + n_k)$, one can claim that K/C is isomorphic to a subgroup of some $U(m_U)$, $m_U \in \mathbb{N}_+$. This fact calls attention to the closed subgroups of the groups $U(n)$ (known under the term *Lie groups*, defined alternatively as locally Euclidean groups, i. e., groups locally homeomorphic to $\mathbb{R}^n$ for some $n \in \mathbb{N}_+$). We will not enter now in a detailed exposition of the properties of this remarkable class of groups (see, for example, [177]). Let us only mention that the compact Lie groups (defined in this way) are NSS, so these are exactly the compact NSS groups, and from the above argument one can deduce that every compact group is an inverse limit of compact Lie groups. The compact abelian Lie groups are nothing else but the elementary compact abelian groups.

The closed subgroups of compact Lie groups are obviously compact Lie groups. Since a compact Lie group L is locally arcwise connected (being locally homeomorphic to $\mathbb{R}^n$), the subgroup $c(L) = a(L)$ is open, so a finite-index normal subgroup of L.

The notions of topologically torsion element and topologically p-torsion element are widely known, due to their intuitively more natural definition compared to those of quasitorsion and quasi-p-torsion element, introduced by Stoyanov [265] to resolve specific problems on minimal abelian groups (see below). On the other hand, the respective subgroups G_p and $G!$ contain in general too many elements (unlike the respective smaller subgroups $td_p(G)$ and $td(G)$, determined by a condition involving more algebra). The equality $td_p(G) = G_p$ for a compact abelian group G was proved in [99, § 4] (containing also Corollary 11.6.8). Corollary 11.6.9 obviously remains true also for subgroups of locally compact abelian groups (see [81, Proposition 2.13]). Nevertheless, these results do not imply the equality $td(G) = G!$ even for $G = \mathbb{T}$ (see Exercise 5.4.20).

The subgroup $wtd(K)$ of a compact abelian group K is always totally dense (see [265]), hence $wtd(K)$ is a dense totally minimal subgroup of K, by Exercise 8.7.8. As pointed out in §10.5, the minimal abelian groups are precisely the dense essential subgroups of the compact abelian groups. As total density is a transitive property, a subgroup G of a compact abelian group K is totally dense in K if and only if $G \cap wtd(K)$ is totally dense in the much smaller group $wtd(K)$. Hence, to build an embedding of an abelian group G into a compact abelian group K as a dense totally minimal subgroup, it suffices to embed G as a totally dense subgroup in $wtd(K)$. Using this trick the (totally) minimal torsion abelian groups G were described in [98], while their compact completions K (named *exotic tori* there) were described by various equivalent properties (e. g., $wtd(K) = t(K)$, which means that K contains copies of $\mathbb{J}_p$ for no prime p). In particular, a divisible torsion abelian group G admits a minimal topology if and only if $G = (\mathbb{Q}/\mathbb{Z})^n$ for some $n \in \mathbb{N}$.

12 Almost periodic functions and Haar integrals

12.1 Almost periodic functions

12.1.1 The algebra of almost periodic functions

In this chapter, for a topological group G, the Banach space $(C^*(G), \| - \|)$ of bounded complex-valued continuous functions on G with the supremum norm $\| - \|$ will always carry the uniform convergence topology. For $f \in C^*(G)$ and $a \in G$, define the translate $f_a \in C^*(G)$ of f by a letting $f_a(x) = f(xa)$ for all $x \in G$.

The definition of almost periodic functions on $\mathbb{R}$ is due to Bohr [35]; here we use the generalization of the equivalent definition given by Bochner [33].

Definition 12.1.1. For a topological group G, a function $f \in C^*(G)$ is *almost periodic* if $K_f := \overline{\{f_a : a \in G\}}$ is compact in $C^*(G)$.

Denote by $A(G)$ the family of all almost periodic continuous functions on G.

Here we give an equivalent condition to that of Definition 12.1.1.

Lemma 12.1.2. *For a topological group G, a function $f \in C^*(G)$ is almost periodic if and only if every infinite sequence of translates $\{f_{b_m}\}_{m\in\mathbb{N}}$ of f admits a Cauchy subsequence.*

Proof. If f is almost periodic, then $K_f = \overline{\{f_a : a \in G\}}$ is compact by hypothesis, so every sequence in K_f admits a convergent subsequence.

Assume now that every infinite sequence of translates $\{f_{b_n}\}_{n\in\mathbb{N}}$ of f admits a subsequence that is Cauchy in $C^*(G)$. According to Exercise 7.3.8, $C^*(G_d)$ is complete, where G_d denotes the discrete group G. So, each Cauchy sequence of $C^*(G)$ converges uniformly to a bounded function. Since the uniform limit of continuous functions is again continuous, the limit function belongs to $C^*(G)$. This gives the desired convergent subsequence. Hence, K_f is compact and f is almost periodic. □

If G is a compact group, then every continuous function $G \to \mathbb{C}$ is almost periodic. More precisely, the almost periodic continuous functions of a topological group G are related to the Bohr compactification bG of G as follows.

Remark 12.1.3. For a topological group G, every almost periodic continuous function $f: G \to \mathbb{C}$ admits an extension to bG (see [124], and for the proof of this fact in the abelian case see Corollary 12.1.10). In other words, the almost periodic continuous functions of G are precisely the compositions of b_G with complex-valued continuous functions of the compact group bG. Therefore, *G is MAP if and only if the almost periodic continuous functions of G separate the points of G.*

Theorem 12.1.4. *For a topological group G, $A(G)$ is a closed $\mathbb{C}$-subalgebra, stable under complex conjugation, of $C^*(G)$.*

Proof. The closedness of $A(G)$ under complex conjugation is obvious.

https://doi.org/10.1515/9783110654936-012

To check that $A(G)$ is a $\mathbb{C}$-linear subspace of $C^*(G)$, take $f, g \in A(G)$ and $c_1, c_2 \in \mathbb{C}$. Since $h = c_1 f + c_2 g$ is continuous, it remains to see that h is also almost periodic. It suffices to consider the case $c_1 = c_2 = 1$, since $c_1 f$ and $c_2 g$ are obviously almost periodic. Then the closures $K_f = \overline{\{f_a : a \in G\}}$ and $K_g = \overline{\{g_a : a \in G\}}$ are compact. Hence, $K_f + K_g$ is compact as well. Since $(f+g)_a = f_a + g_a \in K_f + K_g$ for every $a \in G$, we conclude that $f + g$ is almost periodic.

Analogously one can prove that if $f, g \in A(G)$, then $fg \in A(G)$ as well.

Next we check that $A(G)$ is closed. Pick $f \in C^*(G)$ and assume that f can be uniformly approximated by almost periodic continuous functions. To prove that $f \in A(G)$, pick a sequence $\{g^{(m)}\}_{m \in \mathbb{N}_+}$ in $A(G)$ such that, for every $m \in \mathbb{N}_+$,

$$\left\| f - g^{(m)} \right\| \leq \frac{1}{2^m}. \tag{12.1}$$

Then, for every sequence $\{f_{a_n}\}_{n \in \mathbb{N}}$ of translates of f, one can inductively define a series of subsequences of $\{a_n\}_{n \in \mathbb{N}}$ as follows.

For the first one, pick a subsequence $\{a_{n_k}\}_{k \in \mathbb{N}}$ of $\{a_n\}_{n \in \mathbb{N}}$ such that the subsequence $\{g^{(1)}_{a_{n_k}}\}_{k \in \mathbb{N}}$ of the sequence $\{g^{(1)}_{a_n}\}_{n \in \mathbb{N}}$ converges in $C^*(G)$ and such that $\|g^{(1)}_{a_{n_k}} - g^{(1)}_{a_{n_l}}\| \leq 1/2$ for all $k, l \in \mathbb{N}$.

Then pick a subsequence $\{a_{n_{k_s}}\}_{s \in \mathbb{N}}$ of $\{a_{n_k}\}_{k \in \mathbb{N}}$ such that the subsequence $\{g^{(2)}_{a_{n_{k_s}}}\}_{s \in \mathbb{N}}$ of the sequence $\{g^{(2)}_{a_{n_k}}\}_{k \in \mathbb{N}}$ converges in $C^*(G)$ and also $\left\| g^{(2)}_{a_{n_{k_s}}} - g^{(2)}_{a_{n_{k_t}}} \right\| \leq 1/2^2$ for all $s, t \in \mathbb{N}$. Continue by taking iterated subsequences of $\{a_n\}_{n \in \mathbb{N}}$ such that at the mth stage the respective translates of the function $g^{(m)}$ obtained by means of the new mth subsequence is uniformly convergent and the members of the sequence of these translates are at pairwise distance $\leq 1/2^m$. Finally, take the diagonal subsequence $\{a_\nu\}_{\nu \in I}$ relative to this infinite series of iterated subsequences, namely, $\{a_\nu\}_{\nu \in I}$ is given by

$$a_1, a_{n_2}, a_{n_{k_3}}, a_{n_{k_{s_4}}}, \ldots$$

Then each m-tail of the sequence $\{a_\nu\}_{\nu \in I}$ is a subsequence of $\{a_n\}_{n \in \mathbb{N}}$ chosen at the mth stage. So, for every $m \in \mathbb{N}_+$, the sequence $\{g^{(m)}_{a_\nu}\}_{\nu \in I}$ converges in $C^*(G)$ and satisfies

$$\left\| g^{(m)}_{a_\nu} - g^{(m)}_{a_\mu} \right\| \leq \frac{1}{2^m} \quad \text{for every } \nu, \mu \geq m. \tag{12.2}$$

By (12.1) and (12.2), for all $m > 2$ and all $\nu, \mu \geq m$,

$$\left\| f_{a_\nu} - f_{a_\mu} \right\| \leq \left\| f_{a_\nu} - g^{(m)}_{a_\nu} \right\| + \left\| g^{(m)}_{a_\nu} - g^{(m)}_{a_\mu} \right\| + \left\| g^{(m)}_{a_\mu} - f_{a_\mu} \right\| \leq \frac{1}{2^{m-2}}.$$

Therefore, $\{f_{a_\nu}\}_{\nu \in I}$ is a Cauchy sequence in $C^*(G)$, and so Lemma 12.1.2 yields that $f \in A(G)$. □

12.1.2 Almost periodic functions and the Bohr compactification of abelian groups

Example 12.1.5. Let $f: \mathbb{R} \to \mathbb{C}$ be a function. One says that $a \in \mathbb{R}$ is a *period* of f if $f(x+a) = f(x)$ for every $x \in \mathbb{R}$ (i. e., $f_a = f$). Clearly, if $a \in \mathbb{R}$ is a period of f, then also ka is a period of f for every $k \in \mathbb{Z}$. More precisely, the periods of f form a subgroup $\Pi(f)$ of $\mathbb{R}$. Call f *periodic* if $\Pi(f) \neq \{0\}$.

It is easy to see that a periodic function $f: \mathbb{R} \to \mathbb{C}$ has period $a \in \mathbb{R} \setminus \{0\}$ if and only if f factorizes through the canonical projection $q: \mathbb{R} \to \mathbb{R}/\langle a \rangle$, that is, there exists a function $f': \mathbb{R}/\langle a \rangle \to \mathbb{C}$ such that $f = f' \circ q$:

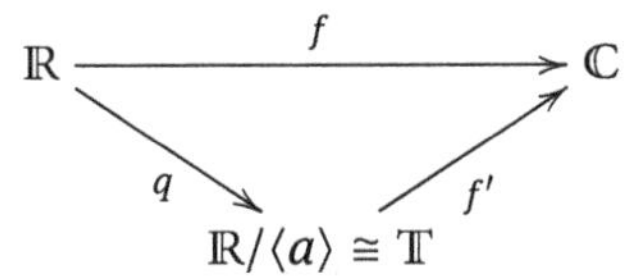

Since $\mathbb{R}/\langle a \rangle \cong \mathbb{T}$ is compact, this explains the great importance of the periodic functions $\mathbb{R} \to \mathbb{C}$, namely, these are the functions that can be factorized through the compact circle group $\mathbb{T}$.

If f is continuous, then also f' is continuous. In this case, since f' is almost periodic by Corollary 12.1.10 below, we can conclude that the periodic continuous function $f: \mathbb{R} \to \mathbb{C}$ is almost periodic.

This example can be generalized to periodic functions on an abelian group (see Exercise 12.3.3).

Definition 12.1.6. Let G be a topological abelian group, $f \in C^*(G)$, and $\varepsilon > 0$. An element $a \in G$ is an *ε-almost period* of f if $\|f - f_a\| \leq \varepsilon$. Moreover, let

$$T(f, \varepsilon) := \{a \in G : a\ \varepsilon\text{-almost period of } f\}.$$

Given an abelian group G and $f \in C^*(G)$, the family $\{T(f,\varepsilon) : \varepsilon > 0\}$ is a filter base of the neighborhoods of 0 in a group topology $\mathcal{T}_f$ on G (see Exercise 12.3.5). Now we use the group topology $\mathcal{T}_f$ to find an equivalent description of almost periodicity of $f \in C^*(G)$.

Proposition 12.1.7. *Let G be a topological abelian group and $f \in C^*(G)$. Then the following conditions are equivalent:*
(a) *f is almost periodic (i. e., $f \in A(G)$);*
(b) *$\mathcal{T}_f$ is totally bounded.*

Proof. Clearly, $\mathcal{T}_f$ is totally bounded if and only if for every $\varepsilon > 0$ the set $T(f,\varepsilon)$ is big, i. e., for every $\varepsilon > 0$ there exist $a_1, \dots, a_n \in G$ such that $G = \bigcup_{j=1}^{n} a_j + T(f,\varepsilon)$.

(a)⇒(b) Arguing for a contradiction, assume that $\mathcal{T}_f$ is not totally bounded. By Lemma 10.2.7, there exists $\varepsilon > 0$ such that $T(f,\varepsilon)$ is small, so there exists a sequence $\{b_n\}_{n\in\mathbb{N}}$ in G such that the sets in $\{b_n + T(f,\varepsilon) : n \in \mathbb{N}\}$ are pairwise disjoint. Since f

is almost periodic, the sequence of translates $\{f_{b_m}\}_{m\in\mathbb{N}}$ admits a Cauchy subsequence, by Lemma 12.1.2. So, there exist $n < m$ in $\mathbb{N}$ such that $\|f_{b_m} - f_{b_n}\| = \|f - f_{b_m - b_n}\| \leq \varepsilon$; this means that $b_m - b_n \in T(f,\varepsilon)$, and hence $(b_m + T(f,\varepsilon)) \cap (b_n + T(f,\varepsilon)) \neq \emptyset$, a contradiction.

(b)⇒(a) By Lemma 12.1.2, it suffices to check that every infinite sequence of translates $\{f_{b_m}\}_{m\in\mathbb{N}}$ of f admits a subsequence that is Cauchy in $C^*(G)$. Assume for a contradiction that some sequence of translates $\{f_{b_m}\}_{m\in\mathbb{N}}$ admits no Cauchy subsequence. This means that for every subsequence $\{f_{b_{m_k}}\}_{k\in\mathbb{N}}$, there exists $\varepsilon > 0$ such that for some subsequence $\{m_{k_s}\}_{s\in\mathbb{N}}$ of $\{m_k\}_{k\in\mathbb{N}}$,

$$\left\| f_{b_{m_{k_s}}} - f_{b_{m_{k_t}}} \right\| = \left\| f - f_{b_{m_{k_s}} - b_{m_{k_t}}} \right\| \geq 2\varepsilon \quad \text{for all } s \neq t. \tag{12.3}$$

By hypothesis $T(f,\varepsilon/2)$ is big, so there exist $a_1,\dots,a_n \in G$ such that $G = \bigcup_{j=1}^n a_j + T(f,\varepsilon/2)$. Then there exists $j \in \{1,\dots,n\}$ such that infinitely many $b_{m_{k_s}}$ are in $a_j + T(f,\varepsilon/2)$. We deduce that for distinct s and t with $b_{m_{k_s}}, b_{m_{k_t}} \in a_j + T(f,\varepsilon/2)$,

$$b_{m_{k_s}} - b_{m_{k_t}} \in T(f,\varepsilon/2) - T(f,\varepsilon/2) \subseteq T(f,\varepsilon).$$

Therefore, $\left\| f - f_{b_{m_{k_s}} - b_{m_{k_t}}} \right\| \leq \varepsilon$, which contradicts (12.3). □

Example 12.1.8. Let G be a topological abelian group and $\chi \in \widehat{G}$. For $\varepsilon > 0$, one has $|\chi(x+a) - \chi(x)| = |\chi(a) - 1|$ for every $x \in G$, hence

$$T(\chi,\varepsilon) = \{a \in G : |\chi(a) - 1| \leq \varepsilon\}.$$

So, there exists $\delta > 0$ such that $U_G(\chi;\delta) \subseteq T(\chi,\varepsilon)$. Since $U_G(\chi;\delta)$ is big by Proposition 10.2.16, also $T(\chi,\varepsilon)$ is big, and Proposition 12.1.7 yields that χ is almost periodic.

Moreover, Theorem 12.1.4 yields that linear combinations with complex coefficients of continuous characters are still almost periodic continuous functions. In other words, $\mathfrak{X}(G) \subseteq A(G)$.

The next theorem reinforces the above inclusion to the more precise equation $A(G) = \mathfrak{A}(G)$ (see [99, Theorem 2.2.2] for more details).

Theorem 12.1.9 (Bohr–von Neumann theorem). *If G is a topological abelian group, then $A(G) = \mathfrak{A}(G)$ (i. e., $f \in C^*(G)$ is almost periodic if and only if f can be uniformly approximated by linear combinations with complex coefficients of continuous characters of G, namely, functions from $\mathfrak{X}(G)$).*

Proof. According to Example 12.1.8, $\mathfrak{X}(G) \subseteq A(G)$. By Theorem 12.1.4, $A(G)$ is closed in $C^*(G)$, and $\mathfrak{A}(G)$ is the closure of $\mathfrak{X}(G)$ in $C^*(G)$, so $\mathfrak{A}(G) \subseteq A(G)$.

To establish the converse inclusion, let $f \in A(G)$ and fix $\varepsilon > 0$. By Proposition 12.1.7, the set $T(f,\varepsilon/10)$ is big. Hence, Følner theorem 11.3.5 applies with $U = E = T(f,\varepsilon/10)$ to give $\chi_1,\dots,\chi_n \in \widehat{G}$ and $\delta > 0$ such that

$$U_G(\chi_1,\dots,\chi_n;\delta) \subseteq T(f,\varepsilon/10)_{(10)} \subseteq T(f,\varepsilon).$$

If $x, y \in G$ satisfy $x - y \in U_G(\chi_1, \dots, \chi_n; \delta)$, then $x - y \in T(f, \varepsilon)$, and so $\|f - f_{x-y}\| \leq \varepsilon$; in particular, $|f(x) - f(y)| \leq \varepsilon$. Then f satisfies condition (b) of Proposition 11.3.4 with $H = \widehat{G}$, and hence $f \in \mathfrak{A}(G)$. □

Corollary 12.1.10. *For a compact abelian group G, every continuous function $G \to \mathbb{C}$ is almost periodic (i. e., $C(G) = A(G) = \mathfrak{A}(G)$).*

Proof. It follows from Corollary 11.5.1 and Theorem B.5.21 that $\mathfrak{X}(G)$ is dense in $C(G)$, so $C(G) = \overline{\mathfrak{X}(G)} = \mathfrak{A}(G) = A(G)$, by Theorem 12.1.9. □

For a compact abelian group G, the above corollary gives $C_0(G) = C(G) = A(G)$. On the other hand, it is easy to prove that if $C_0(G) \cap A(G) \neq \{0\}$ for a topological abelian group, then G is compact. In other words, G is compact if and only if $C_0(G) = A(G)$.

Recall that a map $f: G \to H$ between topological abelian groups is said to be *uniformly continuous* if for every $U \in \mathcal{V}_H(0)$ there exists $V \in \mathcal{V}_G(0)$ such that for $x, y \in G$ with $x - y \in V$ one has $f(x) - f(y) \in U$.

Corollary 12.1.11. *Let (G, τ) be a topological abelian group and $f \in A(G, \tau)$. Then $f: (G, \tau^+) \to \mathbb{C}$ is uniformly continuous.*

Proof. By Theorem 12.1.9, $f: G \to \mathbb{C}$ can be uniformly approximated by functions from $\mathfrak{X}(G)$. On the other hand, $\mathfrak{X}(G, \tau) = \mathfrak{X}(G, \tau^+)$ and these functions are uniformly continuous. Therefore, $f: (G, \tau^+) \to \mathbb{C}$ is uniformly continuous. □

Now we are in position to prove that the almost periodic continuous functions on a topological abelian group G are precisely those that factorize through the Bohr compactification $b_G: G \to bG$.

Theorem 12.1.12. *Let $G = (G, \tau)$ be a topological abelian group. Then a continuous function $f: G \to \mathbb{C}$ is almost periodic if and only if there exists a continuous function $\tilde{f}: bG \to \mathbb{C}$ such that $f = \tilde{f} \circ b_G$:*

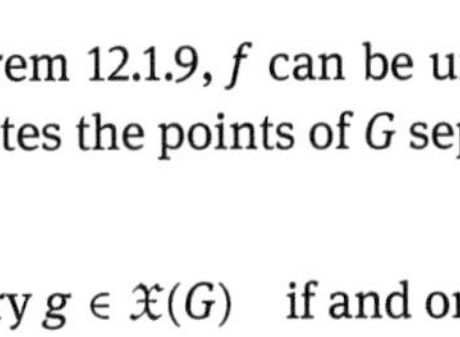

Proof. Assume that there exists a continuous function $\tilde{f}: bG \to \mathbb{C}$ such that $f = \tilde{f} \circ b_G$. Then $\tilde{f}$ is almost periodic by Corollary 12.1.10, and Exercise 12.3.1 implies that f is almost periodic, too.

Now let $f \in A(G)$. By Theorem 12.1.9, f can be uniformly approximated by functions from $\mathfrak{X}(G)$, so $\mathfrak{X}(G)$ separates the points of G separated by f.

Let $x, y \in G$. Then

$$g(x) = g(y) \quad \text{for every } g \in \mathfrak{X}(G) \quad \text{if and only if} \quad b_G(x) = b_G(y).$$

In fact, if $b_G(x) = b_G(y)$, then $\chi(x) = \chi(y)$ for every $\chi \in \widehat{G}$ by Corollary 8.6.3, and so $g(x) = g(y)$ for every $g \in \mathfrak{X}(G)$. Conversely, if $g(x) = g(y)$ for every $g \in \mathfrak{X}(G)$, then $\chi(x) = \chi(y)$ for every $\chi \in \widehat{G}$, and so $b_G(x) = b_G(y)$.

We conclude that $f(x) = f(y)$ whenever $b_G(x) = b_G(y)$. Therefore, recalling that $b_G(G) = G^+$ and bG is the completion of G^+, there exists a function $f'': G^+ \to \mathbb{C}$ such that $f = f'' \circ b_G$.

By Corollary 12.1.11, $f: (G, \tau^+) \to \mathbb{C}$ is uniformly continuous. Since the canonical projection $q: (G, \tau^+) \to G^+$ is an open continuous homomorphism, $f'': b_G(G) \to \mathbb{C}$ is uniformly continuous as well. Now, f'' extends to a uniformly continuous function $\tilde{f}: bG \to \mathbb{C}$ by [134, Theorem 8.3.10], and so $f = \tilde{f} \circ b_G$. □

We recall that for a topological abelian group G, $\mathfrak{A}(G) = \mathfrak{A}_0(G) + \mathbb{C} \cdot 1$, where $\mathbb{C} \cdot 1$ is the one-dimensional subalgebra consisting of the constant functions. We shall see below that $\mathfrak{A}_0(G) \cap \mathbb{C} \cdot 1 = \{0\}$, so $\mathfrak{A}_0(G)$ has codimension 1 in $\mathfrak{A}(G)$.

Lemma 12.1.13. *For every topological abelian group G,*

$$\mathfrak{A}(G) = \mathbb{C} \cdot 1 \oplus \mathfrak{A}_0(G). \tag{12.4}$$

Moreover, if $f \in C^*(G)$ is written as $f(x) = c_f + g_f(x)$, with $c_f \in \mathbb{C}$ and $g_f \in \mathfrak{A}_0(G)$, then $|c_f| \leq \|f\|$. Finally, $c_f \geq 0$ whenever f is real-valued and satisfies $f(x) \geq 0$ for all $x \in G$.

Proof. Assume that $c \cdot 1 = g \in \mathfrak{A}_0(G)$ for some $c \in \mathbb{C}$. Apply Corollary 11.3.6 to $g - c = 0$ and $M = \{0\}$, to get $c = 0$. Hence, $\mathfrak{A}_0(G) \cap \mathbb{C} \cdot 1 = \{0\}$, that is, (12.4) holds.

For $f \in \mathfrak{A}(G)$, the projections $f \mapsto g_f \in \mathfrak{A}_0(G)$ and $f \mapsto c_f \in \mathbb{C} \cdot 1$ related to the factorization in (12.4) can be obtained as follows. By the definition of $\mathfrak{A}(G)$, for every $n \in \mathbb{N}_+$ there exists $h_n \in \mathfrak{X}(G)$ such that $h_n = g_n + c_n$, with $g_n \in \mathfrak{X}_0(G)$ and $c_n \in \mathbb{C}$, and for every $x \in G$,

$$|f(x) - g_n(x) - c_n| \leq \frac{1}{n}. \tag{$*_n$}$$

Applying the triangle inequality to $(*_n)$ and $(*_k)$ one gets, for every $x \in G$,

$$|c_n - c_k + g_n(x) - g_k(x)| \leq \frac{1}{n} + \frac{1}{k}.$$

Corollary 11.3.6, applied to the closed disk M with center 0 and radius $\frac{1}{n} + \frac{1}{k}$, yields that $|c_n - c_k| \leq \frac{1}{n} + \frac{1}{k}$. Hence, $\{c_n\}_{n \in \mathbb{N}}$ is a Cauchy sequence in $\mathbb{C}$. Let $c_f := \lim_{n \to \infty} c_n$. Then $g_f := f - c_f \in \mathfrak{A}_0(G)$ since, according to $(*_n)$,

$$\|f - c_f - g_n\| \leq \|f - c_n - g_n + (c_n - c_f)\| \leq \frac{1}{n} + |c_n - c_f|$$

becomes arbitrarily small when $n \to \infty$.

If $f = 0$, then $c_f = 0$ and there is nothing to prove. Assume $f \neq 0$ and let $\varepsilon = \|f\|$. Then $\|f\| = \|g_f + c_f\| \leq \varepsilon$ yields $|c_f| \leq \varepsilon$, by Corollary 11.3.6.

To prove the last assertion, apply Corollary 11.3.6 to the closed convex subset of $\mathbb{C}$ consisting of all nonnegative real numbers. □

According to Lemma 12.1.13, the projection $\mathfrak{A}(G) \to \mathbb{C}$, $f \mapsto c_f$, is a continuous positive linear functional. We show in the sequel that this is the Haar integral on $\mathfrak{A}(G)$ (see Theorem 12.2.3).

12.2 The Haar integral

Definition 12.2.1. Let G be a topological abelian group and $J(G)$ a translation-invariant $\mathbb{C}$-linear subspace of $C^*(G)$ closed under complex conjugation. A *Haar integral on* $J(G)$ is a nontrivial linear functional $I_{J(G)}: J(G) \to \mathbb{C}$ which is
(i) positive (i. e., $I_{J(G)}(f) \geq 0$ for any real-valued $f \in J(G)$ with $f \geq 0$), and
(ii) invariant (i. e., $I_{J(G)}(f_a) = I_{J(G)}(f)$ for any $f \in J(G)$ and any $a \in G$).

In case $J(G)$ contains all constant functions and $I_{J(G)}(1) = 1$, we call $I_{J(G)}$ a *normed Haar integral on* $J(G)$.

In the sequel we adopt two choices for $J(G)$. In §12.2.1 $J(G) = \mathfrak{A}(G)$ and in §12.2.2 $J(G) = C_0(G)$ for a locally compact abelian group G. Both cases give $\mathfrak{A}(G) = C_0(G) = C(G)$ in case G is compact. To avoid heavy notation, we briefly write I_G instead of $I_{J(G)}$ in both cases, although these two specific choices of $J(G)$ must be kept in mind throughout these two subsections. Following the standard terminology, we will simply say that $I_G = I_{C_0(G)}$ is a *Haar integral on* G in the latter case.

12.2.1 The Haar integral for almost periodic functions on topological abelian groups

We see in Corollary 12.2.4 that there is a *unique* normed Haar integral on $\mathfrak{A}(G)$. To this end, we show first that every Haar integral vanishes on $\mathfrak{A}_0(G)$. Recall that for every $\chi \in G^*$ the inverse χ^{-1} of χ in G^* is the character $\overline{\chi}: G \to \mathbb{S}$, $x \mapsto \overline{\chi(x)}$.

Proposition 12.2.2. *Let G be a topological abelian group and I_G an arbitrary Haar integral on $\mathfrak{A}(G)$. If $\varphi \in \widehat{G}$ is nontrivial, then $I_G(\varphi) = 0$.*

Proof. Let $\varphi \in \widehat{G}$ and $a \in G$ be such that $\varphi(a) \neq 1$. For every $x \in G$, $\varphi_a(x) = \varphi(a)\varphi(x)$, so $I_G(\varphi) = I_G(\varphi_a) = \varphi(a)I_G(\varphi)$, and thus $I_G(\varphi) = 0$. □

Theorem 12.2.3. *For every topological abelian group G, the assignment*

$$I_G: \mathfrak{A}(G) \to \mathbb{C}, \quad f \mapsto c_f,$$

with $c_f \in \mathbb{C}$ defined as in Lemma 12.1.13, gives a normed Haar integral on $\mathfrak{A}(G)$.

Proof. Fix $f \in \mathfrak{A}(G)$. Since I_G is the projection on the first component in (12.4), it is linear. Moreover, it was established in the same lemma that this linear functional is positive. To check the invariance, note that if $f = g_f + c_f$ with $g_f \in \mathfrak{A}_0(G)$, then for every $a \in G$, $(g_f)_a \in \mathfrak{A}_0(G)$ and $f_a = (g_f)_a + c_f$. Hence, $I_G(f_a) = c_f = I_G(f)$. Finally, for $f \equiv 1$, clearly $c_1 = 1$. □

Corollary 12.2.4. *Let G be a topological abelian group. Then the Haar integral I_G defined in Theorem 12.2.3 is the unique normed Haar integral on $\mathfrak{A}(G)$.*

Proof. Let I be a Haar integral on $\mathfrak{A}(G)$. Let $f \in \mathfrak{A}_0(G)$. For every $\varepsilon > 0$, there exists $g \in \mathfrak{X}_0(G)$ such that $\|f - g\| \leq \varepsilon$. Since $I(g) = 0$ by Proposition 12.2.2, we get $|I(f)| \leq \varepsilon$ (see Exercise 12.3.6). Therefore, $I(f) = 0$.

Now assume additionally that I is normed. Then Lemma 12.1.13 guarantees that the functionals I and I_G coincide since they coincide on $\mathbb{C} \cdot 1$ and have $\mathfrak{A}_0(G)$ as kernel. □

According to Corollary 12.1.10, every continuous complex-valued function on a compact abelian group G is almost periodic, namely, $C(G) = A(G) = \mathfrak{A}(G)$. This fact gives a natural way to define the Haar integral on a compact abelian group:

Theorem 12.2.5. *For every compact abelian group G, the assignment $I_G\colon C(G) \to \mathbb{C}$, $f \mapsto c_f$, defines the (unique) normed Haar integral on G.*

12.2.2 The Haar integral on LCA groups

In this section we show that every locally compact abelian group G admits a Haar integral I_G on $C_0(G)$. The following simple property of the Haar integrals will be useful later on.

Lemma 12.2.6. *Let I_G be a Haar integral on a locally compact abelian group G. If $h \in C_0(G)$ is real-valued, $h \geq 0$ on G, and $h(x_0) > 0$ for at least one $x_0 \in G$, then $I_G(h) > 0$.*

Proof. There exists a neighborhood V of 0 in G such that $h(x) \geq a := h(x_0)/2$ for all $x \in x_0 + V$.

There exists $f \in C_0(G)$ with $I_G(f) \neq 0$. Since $f = u + iv$ for some real-valued $u, v \in C_0(G)$, one has either $I_G(u) \neq 0$ or $I_G(v) \neq 0$. So, without loss of generality we may assume that f is real-valued. Setting, for every $x \in G$, $f_+(x) = \max\{f(x), 0\}$ and $f_-(x) = \max\{-f(x), 0\}$ gives functions $f_+, f_- \in C_0(G)$ such that $f_+ \geq 0$ and $f_- \geq 0$ on G and $f = f_+ - f_-$. Thus, either $I_G(f_+) \neq 0$ or $I_G(f_-) \neq 0$. So, we may assume that $f \geq 0$ on G and $I_G(f) \neq 0$. Hence, $I_G(f) > 0$.

Since $f \in C_0(G)$, there exists a compact subset K of G with $f \equiv 0$ on $G \setminus K$. There exists a finite subset F of G such that $K \subseteq F + V$. If $A = \max_{x \in G} f(x)$, then $A > 0$. Moreover, for every $g \in F$ and every $x \in g + V$, one has $h_{x_0 - g}(x) \geq a$, since $x_0 + x - g \in x_0 + V$. Thus, for every $x \in K$, there exists $g \in F$ such that $af(x) \leq Ah_{x_0-g}(x)$. So, for all $x \in G$,

$$f(x) \leq \frac{A}{a} \sum_{g \in F} h_{x_0-g}(x),$$

and therefore

$$0 < I_G(f) \leq \frac{A}{a} |F| \, I_G(h),$$

which shows that $I_G(h) > 0$. □

The following two lemmas are the main steps in the proof of the existence of Haar integrals on locally compact abelian groups.

Lemma 12.2.7. *If G is a discrete abelian group, then G admits a Haar integral.*

Proof. For $f \in C_0(G)$, which has finite support, set $I_G(f) = \sum_{x \in G} f(x)$. One checks easily that I_G is a Haar integral on G. □

Lemma 12.2.8. *If G is a locally compact abelian group and H is a closed subgroup of G such that both H and G/H admit a Haar integral, then also G admits a Haar integral.*

Proof. Let $f \in C_0(G)$. Then $f_y \restriction_H \in C_0(H)$ for every $y \in G$. Let, for every $y \in G$,

$$F(y) = I_H(f_y \restriction_H).$$

Then $F\colon G \to \mathbb{C}$ is a continuous function. Indeed, let $y_0 \in G$; we prove that F is continuous at y_0. Fix $\varepsilon > 0$. There exists a nonempty compact subset K of G such that $f \equiv 0$ on $G \setminus K$. Let U be an arbitrary compact symmetric neighborhood of 0 in G. There exists $h \in C_0(G)$ such that $h\colon G \to [0,1]$ and $h(C) = \{1\}$, where $C = -y_0 + U + K$. Indeed, for $B = G \setminus (C + U)$, the closed subsets C and $\overline{B}$ of G are disjoint, and G is a normal space by Corollary 8.3.3. By Urysohn lemma B.5.2, there exists a continuous function $h\colon G \to [0,1]$ with $h(C) = \{1\}$ and $h(\overline{B}) \subseteq \{0\}$. In particular, the support of h is compact, being contained in $C + U$.

Since f is continuous and $U+K$ is compact, there exists a symmetric neighborhood V of 0 in G such that $V \subseteq U$ and

$$|f(x) - f(y)| \leq \varepsilon \quad \text{for every } x, y \in U + K,\ x - y \in V. \tag{12.5}$$

We show that $|F(y) - F(y_0)| \leq \varepsilon$ for all $y \in y_0 + V$. Given $y \in y_0 + V$ (so, $y - y_0 \in V$), let us first check that, for $x \in G$,

$$|f(x+y) - f(x+y_0)| \leq \varepsilon\, h(x). \tag{12.6}$$

Indeed, if $x \in G$ is such that $f(x+y) = f(x+y_0) = 0$, then (12.6) is obviously true. Assume that either $f(x+y) \neq 0$ or $f(x+y_0) \neq 0$. Then either $x + y \in K$ or $x + y_0 \in K$, so either $x \in -y + K \subseteq -y_0 + V + K$ or $x \in -y_0 + K$. In both cases $x \in -y_0 + U + K$ and $x+y, x+y_0 \in U + K$. Moreover,

$$(x+y) - (x+y_0) = y - y_0 \in V,$$

so (12.5) and $h(x) = 1$ imply $|f(x+y) - f(x+y_0)| \leq \varepsilon = \varepsilon\, h(x)$, i. e., (12.6) holds true.

From (12.6) it follows that $|f_y \restriction_H (x) - f_{y_0} \restriction_H (x)| \leq \varepsilon\, h \restriction_H (x)$ for every $x \in G$, so, as a consequence of Exercise 12.3.7,

$$|F(y) - F(y_0)| \leq I_H(|f_y \restriction_H - f_{y_0} \restriction_H|) \leq \varepsilon\, I_H(h \restriction_H).$$

This proves the continuity of F at y_0.

Next, for any $x, y \in G$ with $x - y \in H$,

$$F(x) = I_H(f_x \restriction_H) = I_H((f_y)_{x-y} \restriction_H) = I_H(f_y \restriction_H) = F(y).$$

Then there exists a continuous function $\widetilde{F}: G/H \to \mathbb{C}$ such that $F = \widetilde{F} \circ p$, where $p: G \to G/H$ is the canonical projection:

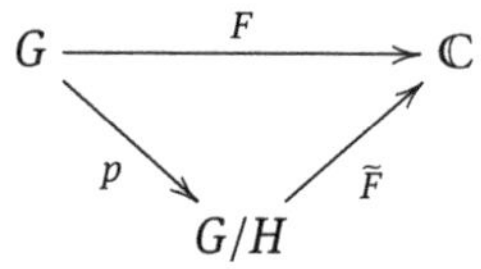

Moreover, $\widetilde{F} \in C_0(G/H)$.

Set, for every $f \in C_0(G)$,

$$I_G(f) := I_{G/H}(\widetilde{F}).$$

We check that I_G is a Haar integral on G.

Indeed, the linearity of I_G follows from that of $I_{G/H}$ and the fact that, for any $\alpha, \beta \in \mathbb{C}$ and any $f_1, f_2 \in C_0(G)$, letting $l = \alpha f_1 + \beta f_2$, we have $L = \alpha F_1 + \beta F_2$, so also $\widetilde{L} = \alpha \widetilde{F}_1 + \beta \widetilde{F}_2$. If $f \in C_0(G)$ is a real-valued function with $f \geq 0$, then $\widetilde{F} \geq 0$, too, so $I_G(f) = I_{G/H}(\widetilde{F}) \geq 0$. To check the invariance, notice that for any $x \in G$, setting $l = f_x$, we have $\widetilde{L} = (\widetilde{F})_{p(x)}$, so

$$I_G(f_x) = I_{G/H}(\widetilde{L}) = I_{G/H}((\widetilde{F})_{p(x)}) = I_{G/H}(\widetilde{F}) = I_G(f).$$

We show that I_G is nontrivial. Take an arbitrary compact neighborhood U of 0 in G. Proceed as in the construction of the function h above, to produce a function $f \in C_0(G)$ such that $f: G \to [0,1]$ and $f(U) = \{1\}$. Then $f \geq 1$ on $U \cap H$, so by Lemma 12.2.6, $F(0_G) = I_H(f \restriction_H) > 0$, which gives $\widetilde{F}(0_{G/H}) > 0$. Moreover, $f \geq 0$ implies $\widetilde{F} \geq 0$ on G/H, so using Lemma 12.2.6 again, $I_G(f) = I_{G/H}(\widetilde{F}) > 0$. Thus, I_G is a Haar integral on G. □

We are ready to prove the existence of Haar integrals on general locally compact abelian groups:

Theorem 12.2.9. *Every locally compact abelian group G admits a Haar integral.*

Proof. If G is compact or discrete, then Theorem 12.2.5 or Lemma 12.2.7 apply, respectively. In case G is compactly generated, G has a discrete subgroup H such that G/H is compact by Proposition 11.6.1. So, both H and G/H admit a Haar integral. It follows from Lemma 12.2.8 that G admits a Haar integral, too.

In the general case, G has an open subgroup H which is compactly generated: just take the subgroup of G generated by an arbitrary compact neighborhood of 0 in G. Such a subgroup H is locally compact (and compactly generated), hence admits a Haar integral by the above argument, while G/H is discrete, so it also admits a Haar integral by Lemma 12.2.7. Finally, Lemma 12.2.8 implies that G admits a Haar integral, too. □

12.2.3 The Haar integral of locally compact groups

Every locally compact group G admits a *right Haar integral* (see [174, Theorem (15.5)]):

Definition 12.2.10. A *right Haar integral* on a locally compact group G is a nontrivial linear functional $I_G\colon C_0(G) \to \mathbb{C}$ which is positive and right invariant (i. e., $I_G(f_a) = I_G(f)$ for any $f \in C_0(G)$ and any $a \in G$).

Moreover, if J is another right Haar integral on G then there exists $c \in \mathbb{R}_{>0}$ such that $I_G = cJ$. Analogously, a locally compact group admits a unique, up to a positive multiplicative constant, *left Haar integral*.

The Haar integral gives the possibility to obtain unitary representations of locally compact groups (see also [174]).

Remark 12.2.11. In the presence of a right Haar integral I_G on a locally compact group G, one can define also a scalar product on $C_0(G)$ making it a pre-Hilbert space, by letting, for $f, g \in C_0(G)$,

$$(f \mid g) = I_G(f\overline{g}).$$

This scalar product is invariant, namely, $(f_a \mid g_a) = (f \mid g)$ for every $a \in G$. Hence, the assignments $f \mapsto f_a$ are unitary operators of the pre-Hilbert space $C_0(G)$. This provides a unitary representation $G \to U(C_0(G))$ similarly as in Definition 10.3.1.

A right Haar integral I_G on a locally compact group G induces – by Riesz representation theorem (see [174, Theorem (11.37)] or [252, Theorem (2.14)]) – a right invariant measure m on the σ-algebra $\mathcal{B}(G)$ of all Borel sets of G such that $I_G(f) = \int f\,dm$ for all $f \in C_0(G)$. Right invariance means that $m(Ba) = m(B)$ holds for all $B \in \mathcal{B}(G)$ and all $a \in G$.

Definition 12.2.12. Let G be a locally compact group. The measure m induced by a right Haar integral on G on $\mathcal{B}(G)$ is called a *right Haar measure*.

A locally compact group G has finite right Haar measure m if and only if G is compact. In such a case m is determined uniquely by the additional condition $m(G) = 1$. Every compact group G admits a unique Haar integral that is right and left invariant, such that its Haar measure satisfies $m(G) = 1$ (see Theorem 12.2.5 in the case of a compact abelian group).

Remark 12.2.13. Alternatively, the *Haar measure* of a compact group G is a function $m\colon \mathcal{B}(G) \to [0,1]$ such that:

(a) m is σ-additive (i. e., $m(\bigcup_{n\in\mathbb{N}} B_n) = \sum_{n\in\mathbb{N}} m(B_n)$ for every family $\{B_n : n \in \mathbb{N}\}$ of pairwise disjoint members of $\mathcal{B}(G)$);
(b) m is left and right invariant (i. e., $m(aB) = m(Ba) = m(B)$ for every $B \in \mathcal{B}(G)$ and every $a \in G$);
(c) $m(G) = 1$.

It easily follows from (b) and (c) that $m(U) > 0$ for every nonempty open set U of G. The Haar measure is unique with the properties (a), (b), and (c).

12.3 Exercises

Exercise 12.3.1. Let G, H be topological groups and $h: G \to H$ a continuous homomorphism. Prove that if $f: H \to \mathbb{C}$ is an almost periodic function, then also $g = f \circ h: G \to \mathbb{C}$ is almost periodic.

Hint. Let $\{a_n\}_{n\in\mathbb{N}}$ be a sequence in G. Letting $b_n = h(a_n)$ for $n \in \mathbb{N}$, the sequence $\{f_{b_n}\}_{n\in\mathbb{N}}$ has a uniformly convergent subsequence $\{f_{b_{n_k}}\}_{k\in\mathbb{N}}$ in $C^*(H)$. Then $\{g_{a_{n_k}}\}_{k\in\mathbb{N}}$ is a convergent subsequence of $\{g_{a_n}\}_{n\in\mathbb{N}}$ in $C^*(G)$. Thus, $g \in A(G)$.

Exercise 12.3.2. For a nonconstant periodic continuous function $f: \mathbb{R} \to \mathbb{C}$, prove that there exists a smallest positive period a of f.

Exercise 12.3.3. Let G be an abelian group. Call $a \in G$ a *period* of a function $f: G \to \mathbb{C}$ if $f(x + a) = f(x)$ for every $x \in G$. Prove that:

(a) the subset $\Pi(f)$ of all periods of f is a subgroup of G and f factorizes through the canonical projection $G \to G/\Pi(f)$;

(b) $\Pi(f)$ is the largest subgroup such that f is constant on each coset of $\Pi(f)$;

(c) if G is a topological group and f is continuous, $\Pi(f)$ is a closed subgroup of G.

Exercise 12.3.4. Let $f: \mathbb{R} \to \mathbb{R}$ be a continuous function. Show that:

(a) f is almost periodic if and only if for every $\varepsilon > 0$ there exists a trigonometric polynomial $P_\varepsilon(x) = \sum_{k=1}^n a_k \cos r_k x + b_k \sin r_k x$, $a_k, b_k, r_k \in \mathbb{R}$ such that $\|f - P_\varepsilon\| \le \varepsilon$;

(b) if f is periodic with period 1, then given $\varepsilon > 0$, there exists a trigonometric polynomial $P_\varepsilon(x) = a_0 + \sum_{k=1}^n a_k \cos 2k\pi x + b_k \sin 2k\pi x$, $a_k, b_k \in \mathbb{R}$ such that $\|f - P_\varepsilon\| \le \varepsilon$.

Hint. (a) Use Theorem 12.1.9 and the fact that a continuous character of $\mathbb{R}$ has the form $\chi: \mathbb{R} \to \mathbb{S}$, $x \mapsto \cos rx + i \sin rx \in \mathbb{S}$ for some $r \in \mathbb{R}$.

(b) Use the fact that f factorizes through the canonical projection $q_0: \mathbb{R} \to \mathbb{T}$, so $f = f' \circ q_0$ and $f': \mathbb{T} \to \mathbb{R}$ is almost periodic, as $\mathbb{T}$ is compact.

Exercise 12.3.5. Let G be a topological abelian group and $f \in C^*(G)$. Prove that $\{T(f, \varepsilon): \varepsilon > 0\}$ is a local base at 0 in a group topology $\mathcal{T}_f$ on G.

Hint. For every $\varepsilon > 0$, $-T(f, \varepsilon) = T(f, \varepsilon)$ and $T(f, \varepsilon/2) + T(f, \varepsilon/2) \subseteq T(f, \varepsilon)$.

Exercise 12.3.6. Let G be a topological abelian group and I_G a Haar integral on a translation-invariant $\mathbb{C}$-linear subspace $J(G)$ of $C^*(G)$ containing all constant functions and closed under complex conjugation which satisfies $I_G(1) = 1$. Prove that if $f \in J(G)$ and $\|f\| \le \varepsilon$, then also $|I_G(f)| \le \varepsilon$.

Hint. First, consider the case when f is real-valued.

Exercise 12.3.7. Let G be a locally compact abelian group and I_G a Haar integral on G. Prove that $|I_G(f)| \leq I_G(|f|)$ for all $f \in C_0(G)$.

Hint. We may assume that $I_G(f) \neq 0$. Choose $\alpha \in \mathbb{S}$ such that $I_G(\alpha f) = |I_G(f)|$. Since $|\alpha f| = |f|$, we may assume that $I_G(f) \geq 0$. So, if $f = u + iv$ with real-valued functions $u, v \in C_0(G)$, $I_G(v) = 0$ and $I_G(f) = I_G(u) \leq I_G(|u|) \leq I_G(|f|)$.

Exercise 12.3.8. For a continuous homomorphism $\phi: G \to H$ of topological abelian groups, define $A(\phi): A(H) \to A(G)$ by $A(\phi)(f) = f \circ \phi$ for $f \in A(H)$. Show that the assignments $G \mapsto A(G)$ and $\phi \mapsto A(\phi)$ define a contravariant functor from the category of abelian topological groups to the category of $\mathbb{C}$-Banach algebras.

12.4 Further readings, notes, and comments

Almost periodic functions on the real line were first introduced in 1923 by Bohr [35], inspired by the work [34] of Bohl on ϵ-periodicity. An equivalent definition was furnished by Bochner [33], which was later generalized to arbitrary topological groups (see [174]). Using Bochner's definition, von Neumann [284] studied almost periodic functions on general groups. A third approach to almost periodic functions using the Bohr compactification was given by Weil [288]. Other classical references for almost periodic functions are [197] and [124].

Theorems 12.2.5, 12.2.9 and Lemma 12.2.8 are [99, Lemma 2.4.3, Theorem 2.4.5, Lemma 2.4.5] (and also [109, Theorem 7.1, Theorem 7.5, Lemma 7.4]), respectively.

Uniform continuity can be defined also for maps between arbitrary topological groups, but we preferred to remain in the abelian case in order to avoid technical complications related to the various uniformities that appear in the nonabelian case. In particular, when G is a compact group, then every continuous function $f: G \to \mathbb{C}$ is uniformly continuous. This property can be extended to pseudocompact groups (see §15.5).

13 The Pontryagin-van Kampen duality

The main aim of this chapter is to introduce in detail the dual group of a topological abelian group and to prove the Pontryagin-van Kampen duality theorem for locally compact abelian groups.

13.1 The dual group

Here we write the circle additively as $\mathbb{T}$ and denote by $q_0: \mathbb{R} \to \mathbb{T} = \mathbb{R}/\mathbb{Z}$ the canonical projection. As $\{(-\frac{1}{3k}, \frac{1}{3k}) : k \in \mathbb{N}_+\}$ is a local base at 0 in $\mathbb{R}$, letting, for every $k \in \mathbb{N}_+$,

$$\Lambda_k := q_0\left(\left(-\frac{1}{3k}, \frac{1}{3k}\right)\right),$$

we obtain a local base $\{\Lambda_k : k \in \mathbb{N}_+\}$ at 0 in $\mathbb{T}$.

For an abelian group G, we let $G^* = \mathrm{Hom}(G, \mathbb{T})$, equipped with the compact topology inherited from $\mathbb{T}^G$ (i. e., the pointwise convergence topology – see Example 8.1.2). This topology coincides with the compact-open topology (see below) if we consider G as a discrete group.

For a subset K of G and a subset U of $\mathbb{T}$, let

$$W_{G^*}(K, U) := \{\chi \in G^* : \chi(K) \subseteq U\}.$$

For any subgroup H of G^*, we abbreviate $H \cap W_{G^*}(K, U)$ as $W_H(K, U)$. When there is no danger of confusion, we shall write only $W(K, U)$ in place of $W_{G^*}(K, U)$.

If G is a topological abelian group, the *dual group* $\widehat{G}$ of all continuous characters of G carries the *compact-open topology* (see Remark 5.2.24): the basic neighborhoods of 0 in $\widehat{G}$ are the sets $W_{\widehat{G}}(K, U)$, where K is a compact subset of G and U is a neighborhood of 0 in $\mathbb{T}$. We show in Theorem 13.1.2(f) that $W_{\widehat{G}}(K, U)$ coincides with $W_{G^*}(K, U)$ when $U \subseteq \Lambda_1$ and K is a neighborhood of 0 in G; therefore, we use mainly the notation $W(K, U)$ when the group is clear from the context.

Let us start with an easy example.

Proposition 13.1.1. *Let G be a topological abelian group.*
(a) *If G is compact, then $\widehat{G}$ is discrete.*
(b) *If G is discrete, then $\widehat{G}$ is compact.*

Proof. If G is compact, then $W_{\widehat{G}}(G, \Lambda_1) = \{0\}$, as Λ_1 contains no subgroup of $\mathbb{T}$ beyond $\{0\}$. If G is discrete, then $\widehat{G} = G^*$ is compact, as explained above. □

Both implications can be inverted as we will see in Corollary 13.4.18.

Now we prove that the dual group of a topological abelian group G is always a topological group (see Theorem 13.1.2(c)), and if G is locally compact, then $\widehat{G}$ is locally

https://doi.org/10.1515/9783110654936-013

compact, too (see Corollary 13.1.3(a)). This is the starting point of the Pontryagin-van Kampen duality theorem.

Theorem 13.1.2. *For a topological abelian group G, the following hold true:*
(a) *if $x \in \mathbb{T}$ and $k \in \mathbb{N}_+$, then $x \in \Lambda_k$ if and only if $x, 2x, \dots, kx \in \Lambda_1$;*
(b) *$\chi \in G^*$ is continuous if and only if $\chi^{-1}(\Lambda_1)$ is a neighborhood of 0 in G;*
(c) *$\widehat{G}$ is a topological group and $\{W_{\widehat{G}}(K, \Lambda_1) : K \subseteq G, K$ compact$\}$ is a local base at 0 in $\widehat{G}$;*
(d) *for every subset A of G and every integer $s > 1$,*

$$W_{\widehat{G}}(A, \Lambda_s) + W_{\widehat{G}}(A, \Lambda_s) \subseteq W_{\widehat{G}}(A, \Lambda_{\lfloor s/2 \rfloor});$$

(e) *if F is a closed set of $\mathbb{T}$, for every subset K of G, $W_{G^*}(K, F)$ is closed in G^* (hence, compact);*
(f) *if U is neighborhood of 0 in G, then:*
  (f_1) *$W_{\widehat{G}}(\overline{U}, V) = W_{G^*}(\overline{U}, V)$ for every neighborhood $V \subseteq \Lambda_1$ of 0 in $\mathbb{T}$;*
  (f_2) *$W(\overline{U}, \Lambda_4)$ has compact closure;*
  (f_3) *if U has compact closure, then $W(\overline{U}, \Lambda_4)$ is a neighborhood of 0 in $\widehat{G}$ with compact closure, so $\widehat{G}$ is locally compact.*

Proof. (a) For $s \in \mathbb{N}_+$, $sx \in \Lambda_1$ if and only if $x \in A_{s,t} := q_0\left(\frac{t}{s}\right) + \Lambda_s$ for some $t \in \{0, \dots, s-1\}$. On the other hand, $A_{s,0} = \Lambda_s$ and $\Lambda_s \cap A_{s+1,t} \neq \emptyset$ if and only if $t = 0$. Hence, if $x \in \Lambda_s$ and $(s+1)x \in \Lambda_1$, then $x \in \Lambda_{s+1}$. Inductively we obtain $sx \in \Lambda_1$ for every $s \in \{1, \dots, k\}$ if and only if $x \in \Lambda_k$.

(b) Suppose that $\chi^{-1}(\Lambda_1) \in \mathcal{V}_G(0)$. We have to see that $\chi^{-1}(\Lambda_k) \in \mathcal{V}_G(0)$ for all $k \in \mathbb{N}_+$. Pick $V \in \mathcal{V}_G(0)$ with $\underbrace{V + \dots + V}_{k} \subseteq \chi^{-1}(\Lambda_1)$. Now $s\chi(y) \in \Lambda_1$ for every $y \in V$ and every $s \in \{1, \dots, k\}$. By item (a), $\chi(y) \in \Lambda_k$ and so $\chi(V) \subseteq \Lambda_k$.

(c) The first assertion follows from Exercise 5.4.12. Moreover, the family $\{W_{\widehat{G}}(K, \Lambda_k) : K \subseteq G, K \text{ compact}, k \in \mathbb{N}_+\}$ is a local base at 0 in $\widehat{G}$. Let $k \in \mathbb{N}_+$ and K a compact subset of G containing $\{0\}$. Define $L = \underbrace{K + \dots + K}_{k}$, which is a compact subset of G, and take $\chi \in W_{\widehat{G}}(L, \Lambda_1)$. For every $x \in K$ and $s \in \{1, \dots, k\}$, $s\chi(x) \in \Lambda_1$, and so $\chi(x) \in \Lambda_k$ by item (a). Hence, $W_{\widehat{G}}(L, \Lambda_1) \subseteq W_{\widehat{G}}(K, \Lambda_k)$. This proves the second assertion.

(d) is obvious.

(e) If $\pi_x : \mathbb{T}^G \to \mathbb{T}$ is the projection defined by the evaluation at x, for $x \in G$, then $W_{G^*}(K, F) = \bigcap_{x \in K} \{\chi \in G^* : \chi(x) \in F\} = \bigcap_{x \in K} (\pi_x^{-1}(F) \cap G^*)$ is closed as each $\pi_x^{-1}(F) \cap G^*$ is closed in G^*.

(f_1) follows immediately from item (b).

(f_2) To prove that the closure of $W_0 = W_{\widehat{G}}(\overline{U}, \Lambda_4)$ is compact it is sufficient to note that $W_0 \subseteq W_1 = W_{\widehat{G}}(\overline{U}, \overline{\Lambda_4})$ and prove that W_1 is compact. Let τ_s denote the subspace topology of W_1 in $\widehat{G}$; we prove that (W_1, τ_s) is compact. Consider on the set W_1 also the weaker topology τ induced from G^* and consequently from $\mathbb{T}^G$. Observe first, that, due to (f_1), $W_1 = W_{\widehat{G}}(\overline{U}, \overline{\Lambda_4}) = W_{G^*}(\overline{U}, \overline{\Lambda_4})$. By item (e), (W_1, τ) is compact. It remains to show that both topologies τ_s and τ coincide on W_1. Since τ_s is finer than τ, it suffices to show that if $\alpha \in W_1$ and K is a compact subset of G, then $(\alpha + W_{\widehat{G}}(K, \Lambda_1)) \cap W_1$ is also a neighborhood of α in (W_1, τ). Since K is compact, there exists a finite subset F of K such that $K \subseteq F + U$. We verify that

$$(\alpha + W_{G^*}(F, \Lambda_2)) \cap W_1 \subseteq (\alpha + W_{\widehat{G}}(K, \Lambda_1)) \cap W_1. \tag{13.1}$$

Let $\xi \in W_{G^*}(F, \Lambda_2)$, so that $\alpha + \xi \in W_1$. As $\alpha \in W_1$ as well, we deduce from item (d) that $\xi(\overline{U}) \subseteq \overline{\Lambda}_2 \subseteq \Lambda_1$, since $\xi = (\alpha + \xi) - \alpha \in W_1 - W_1$. Consequently, $\xi \in \widehat{G}$ by item (b) and $\xi(K) \subseteq \xi(F + U) \subseteq \Lambda_2 + \overline{\Lambda}_2 \subseteq \Lambda_1$. This proves that $\xi \in W_{\widehat{G}}(K, \Lambda_1)$, and so (13.1).

(f_3) follows from (f_2) and the definition of the compact-open topology. □

Corollary 13.1.3. *Let G be a locally compact abelian group. Then:*
(a) *$\widehat{G}$ is locally compact;*
(b) *if G is metrizable, then $\widehat{G}$ is σ-compact;*
(c) *if G is σ-compact, then $\widehat{G}$ is metrizable.*

Proof. (a) follows immediately from Theorem 13.1.2(f_3).

(b) Since G is metrizable, there exists a countable base $\{U_n : n \in \mathbb{N}\}$ of $\mathcal{V}_G(0)$, with $\overline{U}_{n+1} \subseteq U_n$ for every $n \in \mathbb{N}$. By Theorem 13.1.2(f_2), $W(\overline{U}_n, \Lambda_4)$ has compact closure K_n. Let $\chi \in \widehat{G}$. By the continuity of χ, there exists $n \in \mathbb{N}$ such that $\chi(U_n) \subseteq \Lambda_4$, so $\chi(\overline{U}_{n+1}) \subseteq \Lambda_4$, that is, $\chi \in W_{\widehat{G}}(\overline{U}_{n+1}, \Lambda_4)$. In particular, $\chi \in K_{n+1}$. Therefore, $\widehat{G} = \bigcup_{n\in\mathbb{N}} K_n$ is σ-compact.

(c) If G is σ-compact, then G is hemicompact (see Exercise B.7.13), so $G = \bigcup_{n\in\mathbb{N}} K_n$ where each K_n is a compact subset of G and every compact subset K of G is contained in some K_n. Then $W(K, \Lambda_1) \supseteq W(K_n, \Lambda_1)$. Hence, the neighborhoods $W(K_n, \Lambda_1)$ form a countable base of $\mathcal{V}_{\widehat{G}}(0)$ by Theorem 13.1.2(c). By Birkhoff–Kakutani theorem 5.2.17, $\widehat{G}$ is metrizable. □

Local compactness is not needed for the proof of (b), while the proof of (c) needs only hemicompactness, not local compactness (plus σ-compactness).

The proof of Theorem 13.1.2 shows also that, for a topological abelian group G and a neighborhood U of 0 in G, the neighborhood $W(\overline{U}, \Lambda_4)$ of 0 in $\widehat{G}$ carries the same topology in $\widehat{G}$ and G^*; nevertheless, the inclusion map $j : \widehat{G} \hookrightarrow G^*$ need not be an embedding since $W(\overline{U}, \Lambda_4)$ need not be a neighborhood of 0 in $j(\widehat{G})$ equipped with the topology induced by G^*. More precisely, one has:

Corollary 13.1.4. *For a locally compact abelian group G, the following conditions are equivalent:*
(a) *the inclusion map $j : \widehat{G} \hookrightarrow G^*$ is an embedding;*
(b) *$\widehat{G}$ is compact.*

Proof. (a)⇒(b) Assume that $j : \widehat{G} \hookrightarrow G^*$ is an embedding. By Corollary 13.1.3(a), $\widehat{G}$ is locally compact, hence complete by Proposition 8.2.6. By Proposition 7.1.22, $j(\widehat{G}) \cong \widehat{G}$ is closed in the compact group G^*, and consequently $j(\widehat{G})$ is compact. Therefore, $\widehat{G}$ is compact.

(b)⇒(a) Since the compact-open topology of $\widehat{G}$ is finer than the pointwise convergence topology of G^*, the inclusion map $j\colon \widehat{G} \hookrightarrow G^*$ is a continuous injective homomorphism, and j is an embedding by the open mapping theorem (Theorem 8.4.1). □

13.2 Computation of some dual groups

The next lemma is used for the computation of the dual groups in Example 13.2.4.

Lemma 13.2.1. *Every continuous homomorphism* $\chi\colon \mathbb{T} \to \mathbb{T}$ *has the form* $k\, id_{\mathbb{T}}$, *for some* $k \in \mathbb{Z}$.

We give two proofs of this fact. The first, straightforward one, is based on Lemma 9.1.3, the second one on Lemma 4.4.7.

First proof of Lemma 13.2.1. Applying Lemma 9.1.3 to $q := \chi \circ q_0\colon \mathbb{R} \to \mathbb{T}$ and $q_0\colon \mathbb{R} \to \mathbb{T}$, we can find a continuous homomorphism $\eta\colon \mathbb{R} \to \mathbb{R}$ such that $q_0 \circ \eta = q = \chi \circ q_0$, i. e., one has the following commutative diagram:

$$\begin{array}{ccc} \mathbb{R} & \xrightarrow{\ \eta\ } & \mathbb{R} \\ {\scriptstyle q_0}\big\downarrow & \searrow^{q} & \big\downarrow{\scriptstyle q_0} \\ \mathbb{T} & \xrightarrow[\ \chi\ ]{} & \mathbb{T} \end{array} \tag{13.2}$$

As $\chi(q_0(\mathbb{Z})) = \{0\}$, we deduce that $q_0(\eta(\mathbb{Z})) = \{0\}$ as well. Therefore, $\eta(\mathbb{Z}) \subseteq \ker q_0 = \mathbb{Z}$. By Exercise 9.5.8, there exists $\rho \in \mathbb{R}$ such that $\eta(y) = \rho y$ for every $y \in \mathbb{R}$. Thus, $\eta(\mathbb{Z}) \subseteq \mathbb{Z}$ yields $\rho \in \mathbb{Z}$. Hence, $\chi(x) = \rho x$ for every $x \in \mathbb{T}$. □

Second proof of Lemma 13.2.1. For $k \in \mathbb{N}_+$, the endomorphism $\mu_k\colon \mathbb{T} \to \mathbb{T}$, $x \mapsto kx$, is surjective, continuous, and $\ker \mu_k = \mathbb{Z}(k)$. Let now $\chi\colon \mathbb{T} \to \mathbb{T}$ be a nontrivial continuous homomorphism. Then $\ker\chi$ is a proper closed subgroup of $\mathbb{T}$, hence $\ker\chi = \mathbb{Z}(k)$ for some $k \in \mathbb{N}_+$. Let $q\colon \mathbb{T} \to \mathbb{T}/\mathbb{Z}(k)$ be the canonical projection. Since $\chi(\mathbb{T})$ is a connected nontrivial subgroup of $\mathbb{T}$, one has $\chi(\mathbb{T}) = \mathbb{T}$ (see Example 6.1.8(b)). Apply Proposition 3.2.5 with $G = H_1 = H_2 = \mathbb{T}$, $f_2 = \chi$, and $f_1 = \mu_k$. Since $\ker f_1 = \ker f_2 = \mathbb{Z}(k)$, $q_1 = q_2 = q$, and the homomorphism t in Proposition 3.2.5 is the identity map of $\mathbb{T}/\mathbb{Z}(k)$, we obtain this commutative diagram:

$$\begin{array}{ccccc} & & \mathbb{T} & & \\ & {\scriptstyle \mu_k}\swarrow & \big\downarrow{\scriptstyle q} & \searrow{\scriptstyle \chi} & \\ \mathbb{T} & \longleftarrow & \mathbb{T}/\mathbb{Z}(k) & \longrightarrow & \mathbb{T} \\ & & \underset{\eta}{\curvearrowright} & & \end{array} \tag{13.3}$$

Since μ_k and χ are open in view of the open mapping theorem (Theorem 8.4.1), by Proposition 3.2.5 we have that $\eta\colon \mathbb{T} \to \mathbb{T}$ is a topological automorphism, so $\eta = \pm id_{\mathbb{T}}$ according to Lemma 4.4.7. Therefore, $\chi = \pm\mu_k$. □

For a topological abelian group G, the equality $\chi = \pm\xi$ for continuous characters $\chi, \xi: G \to \mathbb{T}$ implies $\ker\chi = \ker\xi$ and $\chi(G) = \xi(G)$. More generally, if $\chi = k\xi$ for some $k \in \mathbb{Z}$, $\ker\xi \subseteq \ker\chi$ and $\chi(G) \subseteq \xi(G)$. Now we see that this implication can be (partially) inverted under appropriate hypotheses.

Corollary 13.2.2. *Let G be a σ-compact locally compact abelian group and $\chi, \xi: G \to \mathbb{T}$ continuous surjective characters of G. Then there exists $m \in \mathbb{Z}$ such that $\xi = m\chi$ if and only if $\ker\chi \subseteq \ker\xi$. If $\ker\chi = \ker\xi$, then $\xi = \pm\chi$.*

Proof. Argue as in the final part of the second proof of Lemma 13.2.1. Since χ and ξ are open by the open mapping theorem (Theorem 8.4.1), Proposition 3.2.5 can be applied with $H_1 = H_2 = \mathbb{T}$ and the diagram (13.3) (with G in place of $\mathbb{T}$ at the top) can be used to conclude. □

Corollary 13.2.3. *Let G be a topological abelian group and $\chi, \xi: G \to \mathbb{T}$ continuous characters of G such that $\ker\xi \subseteq \ker\chi$.*

(a) *If $|\xi(G)| = m$ for some $m \in \mathbb{N}_+$, then $\chi = k\xi$ for some $k \in \mathbb{Z}$; moreover, $\ker\chi = \ker\xi$ if and only if $\chi(G) = \xi(G)$, and in such a case k is coprime to m.*

(b) *If $\ker\chi = \ker\xi = N$ is open and $\chi(G) = \xi(G) = H$, then there exists an automorphism η of H (equipped with the discrete topology), with $\chi = \eta \circ \xi$.*

Proof. (a) If $|\xi(G)| = m$ for some $m \in \mathbb{N}_+$, then $G/\ker\xi \cong \xi(G) = \mathbb{Z}(m) \leq \mathbb{T}$. The hypothesis $\ker\xi \subseteq \ker\chi$ implies that $G/\ker\chi \cong \chi(G) = \mathbb{Z}(n) \leq \mathbb{T}$ with $n \mid m$, as $G/\ker\chi$ is isomorphic to a quotient of $G/\ker\xi \cong \mathbb{Z}(m)$. Therefore, $\chi(G) \subseteq \xi(G)$. Since both $\xi: G \to \xi(G)$ and $\chi: G \to \chi(G)$ are open (as the groups $\xi(G)$ and $\chi(G)$ are discrete), there exists a homomorphism $\eta: \xi(G) \to \chi(G)$ such that $\chi = \eta\circ\xi$, by Proposition 3.2.5. The inclusion $\chi(G) = \mathbb{Z}(n) \subseteq \xi(G) = \mathbb{Z}(m)$ implies that η must be the multiplication by some $k \in \mathbb{Z}$, therefore $\chi = k\xi$. Moreover, $\ker\chi = \ker\xi$ precisely when η is an isomorphism, and so this means $\chi(G) = \xi(G)$. Clearly, $\chi(G) = \xi(G)$ if and only if $k \neq 0$ is coprime to m.

(b) The assertion follows from Proposition 3.2.5. □

Example 13.2.4. Let p be a prime. Then

$$\widehat{\mathbb{Z}(p^\infty)} \cong \mathbb{J}_p, \quad \widehat{\mathbb{J}_p} \cong \mathbb{Z}(p^\infty), \quad \widehat{\mathbb{T}} \cong \mathbb{Z}, \quad \widehat{\mathbb{Z}} \cong \mathbb{T}, \quad \text{and} \quad \widehat{\mathbb{R}} \cong \mathbb{R}.$$

The first isomorphism follows from our definition $\mathbb{J}_p = \operatorname{End}(\mathbb{Z}(p^\infty)) = \operatorname{Hom}(\mathbb{Z}(p^\infty), \mathbb{T}) = \widehat{\mathbb{Z}(p^\infty)}$ and the fact that the topology on $\mathbb{J}_p = \widehat{\mathbb{Z}(p^\infty)}$ described in Example 2.1.5 coincides with the pointwise convergence topology of $\widehat{\mathbb{Z}(p^\infty)} = \mathbb{Z}(p^\infty)^*$.

To verify the isomorphism $\widehat{\mathbb{J}_p} \cong \mathbb{Z}(p^\infty)$, consider first the canonical projection $\eta_n: \mathbb{J}_p \to \mathbb{J}_p/p^n\mathbb{J}_p \cong \mathbb{Z}(p^n) \leq \mathbb{T}$ for $n \in \mathbb{N}_+$. With this identification, we consider $\eta_n \in \mathbb{J}_p^*$. As $\ker\eta_n = p^n\mathbb{J}_p$ is open, $\eta_n \in \widehat{\mathbb{J}_p}$. It is easy to see that under this identification $p\eta_n = \eta_{n-1}$ for every $n \in \mathbb{N}_+$. Therefore, the subgroup H of $\widehat{\mathbb{J}_p}$ generated by the characters η_n is isomorphic to $\mathbb{Z}(p^\infty)$. To see that $H = \widehat{\mathbb{J}}_p$, take any nontrivial continuous character $\chi\colon \mathbb{J}_p \to \mathbb{T}$. Then $N = \ker\chi$ is a proper closed subgroup of $\mathbb{J}_p$. By the open mapping theorem (Theorem 8.4.1), $\chi(G) \cong G/\ker(\chi)$. Moreover, $N \neq \{0\}$ as $\mathbb{J}_p$ is not topologically isomorphic to a subgroup of $\mathbb{T}$. Indeed, $\mathbb{J}_p \not\cong \mathbb{T}$ (since $\mathbb{T}$ is connected, while $\mathbb{J}_p$ is disconnected) and all proper closed subgroups of $\mathbb{T}$ are finite (see Exercise 9.5.7). Thus, $N = p^n\mathbb{J}_p$ for some $n \in \mathbb{N}_+$. Since $N = \ker\eta_n$, Corollary 13.2.3(b) yields $\chi = k\eta_n$ for some $k \in \mathbb{Z}$ coprime to p, so $\chi \in H$. We conclude that $\widehat{\mathbb{J}}_p = H \cong \mathbb{Z}(p^\infty)$.

The isomorphism $g\colon \widehat{\mathbb{Z}} \to \mathbb{T}$ is obtained by setting $g(\chi) = \chi(1)$ for every $\chi\colon \mathbb{Z} \to \mathbb{T}$. It is easy to check that this isomorphism is topological.

According to Lemma 13.2.1, every $\chi \in \widehat{\mathbb{T}}$ has the form $\chi = \mu_k = k\,id_{\mathbb{T}}$ for some $k \in \mathbb{Z}$. This gives a homomorphism $\widehat{\mathbb{T}} \to \mathbb{Z}$, $\mu_k \mapsto k$, which is obviously bijective. Hence, $\widehat{\mathbb{T}} \cong \mathbb{Z}$ since both groups are discrete (see Proposition 13.1.1(a)).

To prove $\widehat{\mathbb{R}} \cong \mathbb{R}$, for every $r \in \mathbb{R}$, consider the map $\rho_r\colon \mathbb{R} \to \mathbb{R}$ defined by $\rho_r(x) = rx$. Then $\chi_r = q_0 \circ \rho_r$ is a continuous character of $\mathbb{R}$. If $r \neq 0$, then $\chi_r \neq 0$, so the homomorphism

$$g\colon \mathbb{R} \to \widehat{\mathbb{R}}, \quad r \mapsto \chi_r,$$

has $\ker g = \{0\}$. To see that g is surjective, consider any nontrivial $\chi \in \widehat{\mathbb{R}}$. Applying Lemma 9.1.3 to $\chi\colon \mathbb{R} \to \mathbb{T}$ and $q_0\colon \mathbb{R} \to \mathbb{T}$, we find a continuous homomorphism $\eta\colon \mathbb{R} \to \mathbb{R}$ such that $q_0 \circ \eta = \chi$:

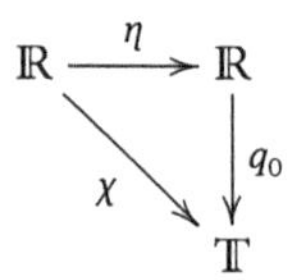

Let $r = \eta(1)$. It is easy to check that $\eta = \rho_r$, and so $\chi = \chi_r$. Then g is an isomorphism. Its continuity follows from the definition of the compact-open topology of $\widehat{\mathbb{R}}$. As $\mathbb{R}$ is σ-compact, g is also open by the open mapping theorem (Theorem 8.4.1).

We need the next result to compute the dual of $\mathbb{Q}_p$.

Proposition 13.2.5. *Let G be a hereditarily disconnected locally compact abelian group. Then $\ker\chi$ is an open subgroup of G for every $\chi \in \widehat{G}$.*

Proof. According to Theorem 8.5.2, G has a local base at 0 formed by open subgroups. Hence, there exists an open subgroup O of G such that $\chi(O) \subseteq \Lambda_1$. Since Λ_1 contains no nontrivial subgroup, $\chi(O) = \{0\}$, and so $O \subseteq \ker\chi$. Therefore, $\ker\chi$ is open. □

Example 13.2.6. Let p be a prime. Then $\widehat{\mathbb{Q}}_p \cong \mathbb{Q}_p$.

Consider the character $\chi_1\colon \mathbb{Q}_p \to \mathbb{T}$ obtained simply by the canonical projection $\mathbb{Q}_p \to \mathbb{Q}_p/\mathbb{J}_p \cong \mathbb{Z}(p^\infty) \subseteq \mathbb{T}$. As $\mathbb{J}_p$ is open in $\mathbb{Q}_p$, $\chi_1 \in \widehat{\mathbb{Q}}_p$. For every $\xi \in \mathbb{Q}_p$, consider the map $\mu_\xi\colon \mathbb{Q}_p \to \mathbb{Q}_p$, $x \mapsto \xi x$. Then its composition $\chi_\xi = \chi_1 \circ \mu_\xi$ with χ_1 gives a continuous character of $\mathbb{Q}_p$. If $\xi \neq 0$, then $\chi_\xi \neq 0$, so the homomorphism

$$g\colon \mathbb{Q}_p \to \widehat{\mathbb{Q}}_p, \quad \xi \mapsto \chi_\xi,$$

has $\ker g = \{0\}$. To see that g is surjective consider any nontrivial $\chi \in \widehat{\mathbb{Q}_p}$. By Proposition 13.2.5, $N = \ker\chi$ is an open subgroup of $\mathbb{Q}_p$, and so $N = p^m\mathbb{J}_p$ for some $m \in \mathbb{Z}$. Let χ' be defined by $\chi'(x) = \chi(p^{-m}x)$ for all $x \in \mathbb{Q}_p$. Then $\ker\chi' = \mathbb{J}_p = \ker\chi_1$. On the other hand, $\chi_1(\mathbb{Q}_p) = \chi'(\mathbb{Q}_p) = \mathbb{Z}(p^\infty)$. By Corollary 13.2.3(b), there exists an automorphism η of $\mathbb{Z}(p^\infty)$ such that $\chi' = \eta \circ \chi_1$. Moreover, there exists $\xi \in \mathbb{J}_p$ such that $\eta(x) = \xi x$ for every $x \in \mathbb{Z}(p^\infty)$. Since all three homomorphisms $\chi_1\colon \mathbb{Q}_p \to \mathbb{Z}(p^\infty)$, $\chi'\colon \mathbb{Q}_p \to \mathbb{Z}(p^\infty)$, and $\eta\colon \mathbb{Z}(p^\infty) \to \mathbb{Z}(p^\infty)$ are $\mathbb{J}_p$-module homomorphisms, we deduce that $\chi'(x) = \chi_1(\xi x)$ for all $x \in \mathbb{Q}_p$. Consequently, $\chi(x) = \chi'(p^m\xi x)$ for all $x \in \mathbb{Q}_p$. In other words, $\chi = \chi_{p^m\xi} = g(p^m\xi)$. Therefore, $g\colon \mathbb{Q}_p \to \widehat{\mathbb{Q}_p}$ is an isomorphism.

To check its continuity, note first that every compact subset of $\mathbb{Q}_p$ is contained in some of the compact open subgroups $p^{-m}\mathbb{J}_p$ for $m \in \mathbb{N}$. Then the basic neighborhood $U_m := W(p^{-m}\mathbb{J}_p, \Lambda_1)$ of 0 in $\widehat{\mathbb{Q}_p}$ coincides with the set of all $\chi \in \widehat{\mathbb{Q}_p}$ that vanish on $p^{-m}\mathbb{J}_p$ (as Λ_1 contains no nontrivial subgroups). Hence, for every $m \in \mathbb{N}$, $g^{-1}(U_m)$ is open as it contains the open subgroup $p^m\mathbb{J}_p$ of $\mathbb{Q}_p$. This proves the continuity of g. As $\mathbb{Q}_p$ is σ-compact, since $\mathbb{Q}_p = \bigcup_{n\in\mathbb{N}_+} p^{-n}\mathbb{J}_p$, g is also open by the open mapping theorem (Theorem 8.4.1).

13.3 Some general properties of the dual group

13.3.1 The dual of direct products and direct sums

We start proving that the dual group of a finite product of topological abelian groups is the product of their dual groups.

Lemma 13.3.1. *If G, H are topological abelian groups, then $\widehat{G \times H} \cong \widehat{G} \times \widehat{H}$.*

Proof. The isomorphism $\Phi\colon \widehat{G}\times\widehat{H} \to \widehat{G \times H}$, defined by $\Phi(\chi_1,\chi_2)(x_1,x_2) = \chi_1(x_1) + \chi_2(x_2)$ for every $(\chi_1,\chi_2) \in \widehat{G} \times \widehat{H}$ and $(x_1,x_2) \in G \times H$, is a particular case of (A.3) when the target group is $\mathbb{T}$ and only continuous homomorphisms are taken into account.

Now we show that Φ is continuous. Let $W(K,U)$ be an open neighborhood of 0 in $\widehat{G\times H}$ (namely, K is a compact subset of $G \times H$ and U is an open neighborhood of 0 in $\mathbb{T}$). Since the projections π_G and π_H of $G \times H$ onto G and H, respectively, are continuous, $K_G := \pi_G(K)$ and $K_H := \pi_H(K)$ are compact in G and H, respectively. Taking an open neighborhood V of 0 in $\mathbb{T}$ with $V + V \subseteq U$, since $K \subseteq K_G \times K_H$, it follows that $\Phi(W(K_G,V) \times W(K_H,V)) \subseteq W(K,U)$.

It remains to prove that Φ is open. Consider two open neighborhoods $W(K_G,U_G)$ of 0 in $\widehat{G}$ and $W(K_H,U_H)$ of 0 in $\widehat{H}$, where $K_G \subseteq G$ and $K_H \subseteq H$ are compact and U_G, U_H are open neighborhoods of 0 in $\mathbb{T}$. Then $K := (K_G \cup \{0\}) \times (K_H \cup \{0\})$ is a compact subset of $G \times H$ and $U := U_G \cap U_H$ is an open neighborhood of 0 in $\mathbb{T}$ with $W(K,U) \subseteq \Phi(W(K_G,U_G) \times W(K_H,U_H))$. □

By Examples 13.2.4 and 13.2.6, $\mathbb{T}$, $\mathbb{Z}$, $\mathbb{Z}(p^\infty)$, $\mathbb{J}_p$, $\mathbb{R}$, and $\mathbb{Q}_p$ satisfy $\widehat{\widehat{G}} \cong G$, namely, they satisfy the Pontryagin-van Kampen duality theorem (a more precise form is given in Definition 13.4.4, but we prefer to anticipate this weaker one). Using Lemma 13.3.1, this property extends to all finite direct products of these groups.

As a matter of fact, all finite groups (see Example 13.3.3(a)), as well as the groups $\mathbb{R}$ and $\mathbb{Q}_p$ (see Examples 13.2.4 and 13.2.6) satisfy the even stronger condition $\widehat{G} \cong G$ (which immediately implies $\widehat{\widehat{G}} \cong G$ in view of the obvious fact that if $G \cong H$, then also $\widehat{G} \cong \widehat{H}$). This motivates the following notion:

Definition 13.3.2. Call a topological abelian group G *selfdual* if $\widehat{G} \cong G$.

Example 13.3.3. (a) Any finite abelian group F is selfdual. Indeed, $\widehat{F} = F^*$ and we prove that $F^* \cong F$. Recall that F has the form $F \cong \mathbb{Z}(n_1) \times \cdots \times \mathbb{Z}(n_m)$ for suitable $n_1, \dots, n_m \in \mathbb{N}_+$, by Theorem A.1.1. So, applying Lemma 13.3.1, we are left with the proof of the isomorphism $\mathbb{Z}(n)^* \cong \mathbb{Z}(n)$ for every $n \in \mathbb{N}_+$. The elements x of $\mathbb{T}$ satisfying $nx = 0$ are precisely those of the unique cyclic subgroup $\mathbb{Z}(n)$ of order n of $\mathbb{T}$, so $\mathbb{Z}(n)^* \cong \mathrm{Hom}(\mathbb{Z}(n), \mathbb{Z}(n)) \cong \mathbb{Z}(n)$.

(b) As already mentioned, $\mathbb{R}$ and $\mathbb{Q}_p$, by Examples 13.2.4 and 13.2.6, and all finite abelian groups by item (a) are selfdual. By Lemma 13.3.1, finite direct products of selfdual groups are selfdual, hence all groups of the form $F \times \mathbb{R}^m \times \mathbb{Q}_p^n$, where F is a finite abelian group and $m, n \in \mathbb{N}$, are selfdual.

Example 13.3.3 and Lemma 13.3.1 provide a large supply of topological abelian groups G satisfying $\widehat{\widehat{G}} \cong G$:

Proposition 13.3.4. *Let P_1, P_2, P_3 be finite sets of primes, $m, n, k, k_p \in \mathbb{N}$ with $p \in P_3$, $n_p \in \mathbb{N}_+$ with $p \in P_1$ and $m_p \in \mathbb{N}_+$ with $p \in P_2$. Every group of the form*

$$G = \mathbb{T}^n \times \mathbb{Z}^m \times \mathbb{R}^k \times F \times \prod_{p\in P_1} \mathbb{Z}(p^\infty)^{n_p} \times \prod_{p\in P_2} \mathbb{J}_p^{m_p} \times \prod_{p\in P_3} \mathbb{Q}_p^{k_p},$$

where F is a finite abelian group, satisfies $\widehat{\widehat{G}} \cong G$. Such a group G is selfdual if and only if $n = m$, $P_1 = P_2$, and $n_p = m_p$ for all $p \in P_1 = P_2$. In particular, $\widehat{\widehat{G}} \cong G$ holds true for all elementary locally compact abelian groups G.

Proof. Example 13.2.4 gives that $\widehat{\mathbb{Z}} \cong \mathbb{T}$ and $\widehat{\mathbb{T}} \cong \mathbb{Z}$, hence $\mathbb{Z} \cong \widehat{\widehat{\mathbb{Z}}}$ and $\mathbb{T} \cong \widehat{\widehat{\mathbb{T}}}$. Analogously, $\widehat{\mathbb{Z}(p^\infty)} \cong \mathbb{J}_p$ and $\widehat{\mathbb{J}}_p \cong \mathbb{Z}(p^\infty)$ yield $\mathbb{Z}(p^\infty) \cong \widehat{\widehat{\mathbb{Z}(p^\infty)}}$ and $\mathbb{J}_p \cong \widehat{\widehat{\mathbb{J}}}_p$. So, $H \in \{\mathbb{T}, \mathbb{Z}, \mathbb{Z}(p^\infty), \mathbb{J}_p\}$ satisfies $\widehat{\widehat{H}} \cong H$. Moreover, $\widehat{\mathbb{R}} \cong \mathbb{R}$, $\widehat{\mathbb{Q}}_p \cong \mathbb{Q}_p$, and $\widehat{F} \cong F$ were already checked (see Examples 13.2.4, 13.2.6, and 13.3.3(a)). The conclusions follow from Lemma 13.3.1. □

The problem of characterizing all selfdual locally compact abelian groups is still open (see [140, 141]).

Theorem 13.3.5. *Let $\{D_i : i \in I\}$ be a family of discrete abelian groups and let $\{G_i : i \in I\}$ be a family of compact abelian groups. Then*

$$\widehat{\bigoplus_{i\in I} D_i} \cong \prod_{i\in I} \widehat{D_i} \quad \text{and} \quad \widehat{\prod_{i\in I} G_i} \cong \bigoplus_{i\in I} \widehat{G_i}. \tag{13.4}$$

Proof. Let $D = \bigoplus_{i\in I} D_i$, let $\chi : D \to \mathbb{T}$ be a character and, for $i \in I$, let $\chi_i : D_i \to \mathbb{T}$ be its restriction to D_i. The isomorphism $\Phi : \widehat{D} \to \prod_{i\in I} \widehat{D_i}$ in (13.4) is defined by $\chi \mapsto (\chi_i)_{i\in I}$. To see that Φ is continuous, pick a prebasic neighborhood $V = W_{D_{i_0}}(F_0, \Lambda_1) \times \prod_{i\in I\setminus\{i_0\}} \widehat{D_i} \in \mathcal{V}_{\prod_{i\in I}\widehat{D_i}}(0)$, with $i_0 \in I$ and F_0 a finite subset of D_{i_0}. Let $F = F_0 \oplus \bigoplus_{i\in I\setminus\{i_0\}}\{0\}$; then $\Phi(W_D(F, \Lambda_1)) \subseteq V$. Finally, Φ is open, by Corollary 8.4.2 and Proposition 13.1.1(b).

Let $G = \prod_{i\in I} G_i$, pick a continuous character $\chi : G \to \mathbb{T}$ and $V \in \mathcal{V}_G(0)$ with $\chi(V) \subseteq \Lambda_1$. There exists a finite subset F of I such that $V \supseteq B := \prod_{i\in F}\{0\} \times \prod_{i\in I\setminus F} G_i$. Since $\chi(B) \subseteq \Lambda_1$ is a subgroup of $\mathbb{T}$, $\chi(B) = \{0\}$. Hence, χ factorizes through the projection $p : G \to \prod_{i\in F} G_i \cong G/B$; so there exists a character $\chi' : \prod_{i\in F} G_i \to \mathbb{T}$ such that $\chi = \chi' \circ p$. Obviously, $\chi' \in \bigoplus_{i\in F} \widehat{G_i} \subseteq \bigoplus_{i\in I} \widehat{G_i}$. Then the assignment $\chi \mapsto \chi'$ induces the second isomorphism in (13.4). Since both groups are discrete by Proposition 13.1.1(a), this is a topological isomorphism. □

In order to extend the isomorphisms (13.4) to the general case of topological abelian groups, one has to consider a specific topology on the direct sum (see [18, 14.11] or [182]).

Example 13.3.6. Using the isomorphism $\mathbb{Q}/\mathbb{Z} \cong \bigoplus_{p\in\mathbb{P}} \mathbb{Z}(p^\infty)$, where $\mathbb{Q}/\mathbb{Z}$ is discrete, Example 13.2.4 and Theorem 13.3.5, we obtain $\widehat{\mathbb{Q}/\mathbb{Z}} \cong \prod_{p\in\mathbb{P}} \mathbb{J}_p$.

13.3.2 Extending the duality functor to homomorphisms

Let G, H be topological abelian groups. If $f : G \to H$ is a continuous homomorphism, then the function $\widehat{f} : \widehat{H} \to \widehat{G}$, $\chi \mapsto \chi \circ f$, is well-defined since $\chi \circ f$ is continuous. Clearly, a subgroup H of a topological abelian group G is dually embedded if and only if the map $\widehat{i} : \widehat{G} \to \widehat{H}$ is surjective, where $i : H \to G$ is the inclusion map. Open subgroups H are dually embedded (for a continuous character $\chi : H \to \mathbb{T}$, any extension $\chi' : G \to \mathbb{T}$ of χ is continuous, and such an extension χ' exists by Theorem A.2.4.

Lemma 13.3.7. *If G, H are topological abelian groups and $f : G \to H$ is a continuous homomorphism, then $\widehat{f} : \widehat{H} \to \widehat{G}$ is a continuous homomorphism as well.*

(a) *If $f(G)$ is dense in H, then $\widehat{f}$ is bijective.*
(b) *If f is an embedding and $f(G)$ is dually embedded in H, then $\widehat{f}$ is surjective.*
(c) *If f is a surjective homomorphism such that every compact subset of H is covered by some compact subset of G, then $\widehat{f}$ is an embedding.*
(d) *If f is a quotient homomorphism and G is locally compact, then $\widehat{f}$ is an embedding.*
(e) *If $H = G$ and $f = id_G$, then $\widehat{f} = id_{\widehat{G}}$.*
(f) *If $g : H \to L$ is a continuous homomorphism, then $\widehat{g \circ f} = \widehat{f} \circ \widehat{g}$.*

Proof. Clearly, $\widehat{f}$ is a homomorphism. Assume that K is a compact subset of G. Then $f(K)$ is a compact subset of H, so $W := W_{\widehat{H}}(f(K), \Lambda_1)$ is a neighborhood of 0 in $\widehat{H}$ and $\widehat{f}(W) \subseteq W_{\widehat{G}}(K, \Lambda_1)$. This proves the continuity of $\widehat{f}$.

The proofs of items (e) and (f) are immediate.

(a) If $\widehat{f}(\chi) = 0$ for $\chi \in \widehat{H}$, then $\chi \circ f = 0$. By the density of $f(G)$ in H, this yields $\chi = 0$. By Exercise 7.3.14, dense subgroups are dually embedded, so $\widehat{f}$ is surjective.

(b) Let $\chi \in \widehat{G}$. Then $\chi': f(G) \to \mathbb{T}$, defined by $\chi'(f(x)) = \chi(x)$ for $x \in G$, is continuous since f is an embedding. Since $f(G)$ is dually embedded in H, χ' can be extended to a continuous character ξ of H. This yields $\widehat{f}(\xi) = \chi$.

(c) Assume that L is a compact subset of H. By hypothesis, there exists a compact subset K of G such that $f(K) = L$. Therefore, we get $\widehat{f}\,(W_{\widehat{H}}(L, \Lambda_1)) = \widehat{f}(\widehat{H}) \cap W_{\widehat{G}}(K, \Lambda_1)$. This proves that $\widehat{f}$ is an embedding.

(d) follows from (c) and Lemma 8.2.5. □

Corollary 13.3.8. *Let G be a locally compact abelian group, H a subgroup of G, and $i: H \to G$ the canonical inclusion of H in G. Then:*
(a) *$\widehat{i}: \widehat{G} \to \widehat{H}$ is surjective if H is dense or open, or compact;*
(b) *$\widehat{i}$ is injective if and only if H is dense in G.*

Proof. (a) If H is compact, apply Corollary 11.6.4, otherwise Lemma 13.3.7(b).

(b) If H is dense, then $\widehat{i}$ is injective by Lemma 13.3.7(a). Conversely, assume that $\overline{H}$ is a proper subgroup of G and let $q: G \to G/\overline{H}$ be the canonical projection. By Theorem 11.6.3, there exists a nonzero $\chi \in \widehat{G/\overline{H}}$. Then $\xi = \chi \circ q \in \widehat{G}$ is nonzero and satisfies $\xi(H) = \{0\}$, i. e., $\widehat{i}(\xi) = 0$. Then $\widehat{i}$ is not injective. □

If $\mathcal{H}$ denotes the category of all Hausdorff topological abelian groups, the *Pontryagin–van Kampen duality functor*, defined by

$$\widehat{\ }: \mathcal{H} \to \mathcal{H}, \quad G \mapsto \widehat{G} \quad \text{and} \quad f \mapsto \widehat{f},$$

for objects G and morphisms $f: G \to H$ in $\mathcal{H}$, is a contravariant representable functor (see Lemma 13.3.7). In particular, the following property holds.

Corollary 13.3.9. *If G, H are topological abelian groups and $f: G \to H$ is a topological isomorphism, then $\widehat{f}: \widehat{H} \to \widehat{G}$ is a topological isomorphism as well.*

The next corollary is valid for arbitrary locally compact abelian groups, but that stronger form will be proved later on. The case of discrete abelian groups considered in this corollary follows immediately from Corollary 13.3.8(a).

Corollary 13.3.10. *If G is a discrete abelian group and H a subgroup of G, then $|\widehat{H}| \leq |\widehat{G}|$.*

We use this corollary to compute the size of $\widehat{G} = G^*$ for a discrete abelian group G:

Theorem 13.3.11 (Kakutani theorem). *If G is an infinite discrete abelian group, then $|\widehat{G}| = 2^{|G|}$.*

Proof. The inequality $|\widehat{G}| \leq 2^{|G|}$ is obvious since $\widehat{G}$ is contained in the Cartesian power $\mathbb{T}^G$ which has cardinality $2^{|G|}$. It remains to prove the inequality $|\widehat{G}| \geq 2^{|G|}$. We consider several cases using each time the inequality $|\widehat{G}| \geq |\widehat{H}|$ from Corollary 13.3.10 for an appropriate subgroup H of G with $|\widehat{H}| \geq 2^{|G|}$.

CASE 1. If $r(G)$ is finite, G is countable by Lemma A.4.14, so we have to check that $|\widehat{G}| \geq \mathfrak{c}$. According to Lemma A.4.14, $G \cong G_0 \oplus F \oplus \bigoplus_{i=1}^m \mathbb{Z}(p_i^\infty)$, where the primes $p_1, \dots, p_m$ are not necessarily distinct, F is a finite abelian group, and G_0 is a subgroup of $\mathbb{Q}^n$ for $n = r_0(G)$. Since G is infinite, either $n > 0$ or $m > 0$. In the first case, G contains a subgroup $H \cong \mathbb{Z}$, in the latter case, G contains a subgroup $H \cong \mathbb{Z}(p_i^\infty)$ for some $i \in \{1, \dots, m\}$. In both cases, $|\widehat{H}| = \mathfrak{c}$, as $|\widehat{\mathbb{Z}}| = |\mathbb{T}| = \mathfrak{c}$ and $|\widehat{\mathbb{Z}(p_i^\infty)}| = |\mathbb{J}_p| = \mathfrak{c}$, respectively. This proves the desired inequality $|\widehat{G}| \geq \mathfrak{c}$.

CASE 2. If $r(G)$ is infinite, according to Lemma A.4.14, G has a subgroup H such that $H = \bigoplus_{i\in I} C_i$, $|I| = |G|$ and each C_i is cyclic. Then $\widehat{H} = \prod_{i\in I} \widehat{C}_i$ by Theorem 13.3.5, and so $|\widehat{H}| = 2^{|I|} = 2^{|G|}$ since each $\widehat{C}_i$ is either finite or has size $\mathfrak{c}$. This proves the desired inequality $|\widehat{G}| \geq 2^{|G|}$ in this case. □

From Corollary 11.4.5 and Theorem 13.3.11, we obtain:

Corollary 13.3.12. *For every infinite abelian group G, $w(G^\#) = 2^{|G|}$.*

Remark 13.3.13. As we shall see in the sequel, every compact abelian group K has the form $K = \widehat{G}$ for some discrete abelian group G. Moreover, G can be taken to be $\widehat{K}$. Applying Kakutani theorem 13.3.11 to $K = \widehat{G}$, we obtain $|K| = |\widehat{G}| = 2^{|G|}$, while $w(K) = |\widehat{K}| = |G|$ according to Corollary 11.4.5. Combining these equalities yields $|K| = 2^{w(K)}$.

This property can be established for *arbitrary compact groups*. Since the inequality $|K| \leq 2^{w(K)}$ holds true for every Hausdorff group, it remains to use the deeply nontrivial fact, due to the joint efforts of Hagler–Gerlits–Efimov (see [256] for a direct proof), that a compact group K contains a copy of the Cantor cube $\{0,1\}^{w(K)}$ having size $2^{w(K)}$. The compactness plays a relevant role in this embedding theorem. Indeed, there are precompact groups that contain no copy of $\{0,1\}^{\aleph_0}$ (e. g., all groups of the form $G^\#$ with G an abelian group, as they contain no nontrivial convergent sequences in view of Glicksberg theorem 11.6.11, whereas $\{0,1\}^{\aleph_0}$ contains nontrivial convergent sequences).

We conclude this section studying the dual group $\widehat{\mathbb{Q}}$.

Example 13.3.14. Let K denote the compact group $\widehat{\mathbb{Q}}$. Then:
(a) K contains a closed subgroup H isomorphic to $\widehat{\mathbb{Q}/\mathbb{Z}}$ such that $K/H \cong \mathbb{T}$;
(b) $\widehat{K} \cong \mathbb{Q}$.

To verify (a), consider the continuous character $\rho: K \to \widehat{\mathbb{Z}} \cong \mathbb{T}$ obtained by the restriction to $\mathbb{Z}$ of every $\chi \in K$ (i. e., $\rho = \widehat{j}$, where $j: \mathbb{Z} \hookrightarrow \mathbb{Q}$). Then ρ is surjective by Lemma 13.3.7(b) and $H := \ker\rho = \{\chi \in K : \chi(\mathbb{Z}) = \{0\}\}$ is a closed subgroup of K with

$K/H \cong \rho(K) \cong \mathbb{T}$. To see that $H \cong \widehat{\mathbb{Q}/\mathbb{Z}}$, note that the characters of $\mathbb{Q}/\mathbb{Z}$ correspond precisely to those characters of $\mathbb{Q}$ that vanish on $\mathbb{Z}$, and these are precisely the elements of H.

To prove (b), note that K is divisible and torsion-free, by Exercise 13.7.3. Hence, every $r \in \mathbb{Q} \setminus \{0\}$ defines a continuous automorphism λ_r of K by setting $\lambda_r(\chi) = r\chi$ for every $\chi \in K$ (see Example 8.7.4). Then the composition $\rho \circ \lambda_r : K \to \widehat{\mathbb{Z}} \cong \mathbb{T}$ defines a character $\eta_r \in \widehat{K}$ with $\ker \eta_r = r^{-1}H$. For the sake of completeness, let $\eta_0 = 0$. We prove now that $\widehat{K} = \{\eta_r : r \in \mathbb{Q}\} \cong \mathbb{Q}$.

Indeed, let $\eta \in \widehat{K}$ be nonzero. Then $\eta(K)$ is a nonzero closed divisible subgroup of $\mathbb{T}$, hence $\eta(K) = \mathbb{T}$. By Exercise 13.7.4, H is a totally disconnected compact group. On the other hand, $N = \ker \eta$ is a proper closed subgroup of K such that $N + H \neq K$, as $\eta(H)$ is a proper closed subgroup of $\mathbb{T}$, by Corollary 11.6.5(b). Hence, $\eta(H)$ is finite, say of order $m \in \mathbb{N}_+$. Then $N + H$ contains N as a finite-index subgroup, more precisely $[N+H : N] = [H : N \cap H] = m$, and so $mH \subseteq N$. Consider the character $\eta_{m^{-1}}$ of K having $\ker \eta_{m^{-1}} = mH \subseteq N$. By Corollary 13.2.2, there exists $k \in \mathbb{Z}$ such that $\eta = k\eta_{m^{-1}} = \eta_r$, where $r = km^{-1} \in \mathbb{Q}$.

13.4 The natural transformation ω

Let G be a topological abelian group. Define $\omega_G : G \to \widehat{\widehat{G}}$, by $\omega_G(x)(\chi) = \chi(x)$ for every $x \in G$ and for every $\chi \in \widehat{G}$.

Proposition 13.4.1. *If G is a topological abelian group, then $\omega_G(x) \in \widehat{\widehat{G}}$ for every $x \in G$ and ω_G is a homomorphism. The restriction of ω_G to every compact subset of G is continuous. In particular, if G is locally compact or metrizable, then ω_G is continuous.*

Proof. For every $\chi, \psi \in \widehat{G}$ and $x \in G$,

$$\omega_G(x)(\chi + \psi) = (\chi + \psi)(x) = \chi(x) + \psi(x) = \omega_G(x)(\chi) + \omega_G(x)(\psi).$$

If U is an open neighborhood of 0 in $\mathbb{T}$, then $\omega_G(x)(W(\{x\}, U)) \subseteq U$. This proves that $\omega_G(x)$ is a continuous character of $\widehat{G}$, that is, $\omega_G(x) \in \widehat{\widehat{G}}$.

Moreover, ω_G is a homomorphism, since for every $x, y \in G$ and every $\chi \in \widehat{G}$,

$$\omega_G(x + y)(\chi) = \chi(x + y) = \chi(x) + \chi(y) = \omega_G(x)(\chi) + \omega_G(y)(\chi).$$

Fix a compact subset C of G and $x \in C$. We are going to show that $\omega_G \restriction_C$ is continuous in x. Let K be a compact subset of $\widehat{G}$. Since $C - C$ is still compact, $U := W_{\widehat{G}}(C - C, \Lambda_2)$ is a nonempty open set of $\widehat{G}$, so there is a finite subset F of K such that $K \subseteq F + U$. Fix $V \in \mathcal{V}_G(0)$ such that $\chi(V) \subseteq \Lambda_2$ for all $\chi \in F$. Our aim is to prove that

$$\omega_G \restriction_C ((x + V) \cap C) \in \omega_G(x) + W_{\widehat{\widehat{G}}}(K, \Lambda_1). \tag{13.5}$$

Since K is an arbitrary compact subset of $\widehat{G}$, this shows that $\omega_G \restriction_C$ is continuous. To prove (13.5), pick $y \in (x + V) \cap C$ and $\chi \in K$. Then there are $\chi_1 \in F$, $\chi_2 \in U$ and $v \in V$ such that $\chi = \chi_1 + \chi_2$ and $y = x + v \in C$. As $v \in C - C$ and $\chi_2 \in U$, we have $\chi_2(v) \in \Lambda_2$. Moreover, $\chi_1(v) \in \Lambda_2$ as $v \in V$ and $\chi_1 \in F$, so

$$\omega_G \restriction_C (y)(\chi) - \omega_G \restriction_C (x)(\chi) = \chi(y) - \chi(x) = \chi(v) = \chi_1(v) + \chi_2(v) \in \Lambda_2 + \Lambda_2 = \Lambda_1.$$

In other words, $\omega_G \restriction_C (y) - \omega_G \restriction_C (x) \in W_{\widehat{G}}(K, \Lambda_1)$. This proves (13.5).

Since continuity is a local property and every point in a locally compact abelian group G has a compact neighborhood, we obtain that ω_G is continuous in this case.

Finally, let G be a metrizable group. Since the continuity of ω_G can be shown by considering the images of convergent sequences in G (which are compact), also in this case ω_G is continuous. □

Remark 13.4.2. For a topological abelian group G, ω_G is injective if and only if $\widehat{G}$ separates the points of G. Hence, by Theorem 11.6.3, ω_G is injective for every locally compact abelian group. Moreover, $\omega_G(G)$ is a subgroup of $\widehat{\widehat{G}}$ that separates the points of $\widehat{G}$.

Let us see that local compactness is essential in Proposition 13.4.1.

Example 13.4.3. For an infinite discrete abelian group G the map $\omega_{G^\#}: G^\# \to \widehat{\widehat{G^\#}} = \widehat{G^*}$ is not continuous. Indeed, $\widehat{G^\#} = G^*$, since the groups $G^\#$ and G have the same dual group, namely, $\mathrm{Hom}(G, \mathbb{T})$. Furthermore, the only compact subsets of $G^\#$ are the finite ones, according to Glicksberg theorem 11.6.11 (see also Theorem 13.4.9 for a proof in the discrete case, relevant in this example). Hence, $\widehat{G^\#}$ has the same compact topology as G^*, namely, the pointwise convergence topology, so $\widehat{\widehat{G^\#}} = \widehat{G^*}$ is discrete, by Proposition 13.1.1(a). The canonical mapping $\omega_{G^\#}: G^\# \to \widehat{\widehat{G^\#}}$ is not continuous, since $G^\#$ is an infinite precompact group by Corollary 10.2.17, so $\{0_{G^\#}\}$ is not open, but it is the inverse image under $\omega_{G^\#}$ of the open set $\{0_{\widehat{\widehat{G^\#}}}\}$.

Here we adopt a more precise approach to the Pontryagin-van Kampen duality theorem, by asking ω_G to be a topological isomorphism:

Definition 13.4.4. A topological abelian group G is said to *satisfy the Pontryagin-van Kampen duality theorem*, or shortly, to be *reflexive*, if ω_G is a topological isomorphism.

Lemma 13.4.5. *If, for $i \in \{1, \dots, n\}$, the topological abelian groups G_i are reflexive, then also $G = \prod_{i=1}^n G_i$ is reflexive.*

Proof. To obtain a topological isomorphism $j: \prod_{i=1}^n \widehat{\widehat{G}}_i \to \widehat{\widehat{G}}$, apply Lemma 13.3.1 twice. Then the product $\pi: G \to \prod_{i=1}^n \widehat{\widehat{G}}_i$ of the topological isomorphisms $\omega_{G_i}: G_i \to \widehat{\widehat{G}}_i$ composed with j gives precisely ω_G. □

Let $\mathcal{L}$ be the full subcategory of $\mathcal{H}$ having as objects all locally compact abelian groups. According to Corollary 13.1.3(a), the functor $\widehat{\ }: \mathcal{H} \to \mathcal{H}$ sends $\mathcal{L}$ to itself, i. e., defines a functor $\widehat{\ }: \mathcal{L} \to \mathcal{L}$. Pontryagin-van Kampen duality theorem 13.4.17 states that ω is a natural equivalence from $1_{\mathcal{L}}$ to $\widehat{\widehat{\ }}: \mathcal{L} \to \mathcal{L}$, where $\widehat{\widehat{\ }} = \widehat{\ } \cdot \widehat{\ }$. We start by proving that ω is a natural transformation.

Proposition 13.4.6. *ω is a natural transformation from $1_{\mathcal{L}}$ to $\widehat{\widehat{\ }}: \mathcal{L} \to \mathcal{L}$.*

Proof. By Proposition 13.4.1, ω_G is continuous for every $G \in \mathcal{L}$. Moreover, for every continuous homomorphism $f: G \to H$ of locally compact abelian groups, the following diagram commutes:

$$\begin{array}{ccc} G & \xrightarrow{f} & H \\ {\scriptstyle \omega_G}\downarrow & & \downarrow{\scriptstyle \omega_H} \\ \widehat{\widehat{G}} & \xrightarrow[\widehat{\widehat{f}}]{} & \widehat{\widehat{H}} \end{array}$$

In fact, if $x \in G$ and $\xi \in \widehat{H}$, then $\omega_H(f(x))(\xi) = \xi(f(x))$. On the other hand,

$$(\widehat{\widehat{f}}(\omega_G(x)))(\xi) = (\omega_G(x) \circ \widehat{f})(\xi) = \omega_G(x)(\widehat{f}(\xi)) = \omega_G(x)(\xi \circ f) = \xi(f(x)).$$

Hence, $\omega_H(f(x)) = \widehat{\widehat{f}}(\omega_G(x))$ for every $x \in G$. □

13.4.1 The compact or discrete case

Now we can prove the Pontryagin-van Kampen duality theorem in the case when the topological abelian group G is either compact or discrete.

Theorem 13.4.7. *If the topological abelian group G is either compact or discrete, then ω_G is a topological isomorphism.*

Proof. If G is discrete, then $\widehat{G}$ separates the points of G by Corollary A.2.6, and if G is compact, then $\widehat{G}$ separates the points of G by Corollary 11.5.1. Therefore, ω_G is injective by Remark 13.4.2.

If G is discrete, then $\widehat{G}$ is compact by Proposition 13.1.1(b) and the characters from $\omega_G(G)$ separate the points of $\widehat{G}$, in view of Remark 13.4.2. Hence, $\omega_G(G) = \widehat{\widehat{G}}$ by Corollary 11.5.3. Since $\widehat{\widehat{G}}$ is discrete, ω_G is a topological isomorphism.

Let now G be compact. Then ω_G is a continuous injective homomorphism by Proposition 13.4.1 and Remark 13.4.2. Suppose by contradiction that $\omega_G(G)$ is a proper subgroup of $\widehat{\widehat{G}}$. By the compactness of G, $\omega_G(G)$ is compact, hence closed in the compact group $\widehat{\widehat{G}}$ (see Proposition 13.1.1). By Corollary 11.5.1 applied to $\widehat{\widehat{G}}/\omega_G(G)$, there exists $\xi \in \widehat{\widehat{\widehat{G}}} \setminus \{0\}$ such that $\xi(\omega_G(G)) = \{0\}$. Since $\widehat{G}$ is discrete by Proposition 13.1.1(a),

$\omega_{\widehat{G}}$ is a topological isomorphism by the first part of the proof, and so there exists $\chi \in \widehat{G}$ such that $\omega_{\widehat{G}}(\chi) = \xi$. So, for $x \in G$,

$$0 = \xi(\omega_G(x)) = \omega_{\widehat{G}}(\chi)(\omega_G(x)) = \omega_G(x)(\chi) = \chi(x).$$

Hence, $\chi \equiv 0$, and so also $\xi \equiv 0$, a contradiction. Therefore, ω_G is surjective.

Finally, ω_G is open by the open mapping theorem (Theorem 8.4.1). □

Our next step is to prove the Pontryagin-van Kampen duality theorem for elementary locally compact abelian groups.

Theorem 13.4.8. *If G is an elementary locally compact abelian group, then ω_G is a topological isomorphism.*

Proof. According to Lemma 13.4.5 and Theorem 13.4.7, it suffices to prove that $\omega_{\mathbb{R}}$ is a topological isomorphism. The mapping $g\colon \mathbb{R} \to \widehat{\mathbb{R}}$, $r \mapsto \chi_r$, where $\chi_r(x) = q_0(rx) \in \mathbb{T}$ for $x \in \mathbb{R}$, was shown to be a topological isomorphism in Example 13.2.4. For $x, r \in \mathbb{R}$, we have $\omega_{\mathbb{R}}(x)(g(r)) = g(r)(x) = \chi_r(x) = \chi_x(r)$, so $\omega_{\mathbb{R}}(x) \circ g = \chi_x = g(x)$. Since $\omega_{\mathbb{R}}(x) \circ g = \widehat{g}(\omega_{\mathbb{R}}(x)) = \widehat{g} \circ \omega_{\mathbb{R}}(x)$, this implies $(g^{-1} \circ \widehat{g} \circ \omega_{\mathbb{R}})(x) = x$, so $g^{-1} \circ \widehat{g} \circ \omega_{\mathbb{R}} = id_{\mathbb{R}}$. Hence, $\omega_{\mathbb{R}}$ is a topological isomorphism, since g^{-1} and $\widehat{g}$ are topological isomorphisms. □

Here is a second argument proving that $\omega_{\mathbb{R}}$ is a topological isomorphism, using in a crucial way the ($\mathbb{Z}$-)bilinear map

$$\lambda\colon \mathbb{R} \times \mathbb{R} \to \mathbb{T}, \quad (x, y) \mapsto \lambda(x, y) := q_0(xy).$$

For every $y \in \mathbb{R}$, the map $\chi_y := \lambda(-, y)\colon \mathbb{R} \to \mathbb{T}$, $x \mapsto \lambda(x, y)$, is an element of $\widehat{\mathbb{R}}$. Hence, the second copy $\{0\} \times \mathbb{R}$ of $\mathbb{R}$ in $\mathbb{R} \times \mathbb{R}$ can be identified with $\widehat{\mathbb{R}}$, since by Example 13.2.4 every continuous character of $\mathbb{R}$ has this form. On the other hand, every element $x \in \mathbb{R}$ gives a continuous character $\mathbb{R} \to \mathbb{T}$, $y \mapsto \lambda(x, y)$, so can be considered as the element $\omega_{\mathbb{R}}(x)$ of $\widehat{\widehat{\mathbb{R}}}$. We have seen in Example 13.2.4 that every $\xi \in \widehat{\widehat{\mathbb{R}}}$ has this form, that is, $\omega_{\mathbb{R}}$ is surjective. Since $\omega_{\mathbb{R}}$ is continuous by Proposition 13.4.1, we conclude that $\omega_{\mathbb{R}}$ is a topological isomorphism by the open mapping theorem (Theorem 8.4.1) and Theorem 11.6.3.

The following is an application of Theorem 13.4.8 to the Bohr topology.

Theorem 13.4.9. *For G an abelian group, $G^\#$ has no infinite compact subsets. In particular, $G^\#$ admits no nontrivial convergent sequence.*

Proof. The assertion is obviously true when G is finite, so we suppose from now on that G is infinite.

Suppose that C is an infinite compact subset of $G^\#$ and pick a countably infinite subset X of C. Then the subgroup $H = \langle X \rangle$ of G is countably infinite and closed in $G^\#$ by Proposition 3.1.9. Therefore, $C_1 = C \cap H$ is compact, as a closed set of C, and C_1 is countably infinite, as H is countable and $X \subseteq C_1$ is infinite. Since the countable subgroup H of G carries its Bohr topology, we shall work from now on in $H^\#$.

According to Theorem B.5.11, C_1 has a nontrivial convergent sequence $\{x_n\}_{n\in\mathbb{N}}$. Since it is contained in H along with its limit, it is not restrictive to assume that it is a null sequence in $H^\#$. This means that $\{\chi(x_n)\}_{n\in\mathbb{N}}$ converges to 0 in $\mathbb{T}$ for all $\chi \in \widehat{H^\#}$. Algebraically, $\widehat{H^\#}$ coincides with the dual group K of the discrete group H. By Proposition 13.1.1(b), K is compact and

$$\omega_H(x_n)(\chi) = \chi(x_n) \to 0 \text{ in } \mathbb{T} \quad \text{for every } \chi \in K. \tag{13.6}$$

Hence, letting for every $n \in \mathbb{N}$, $F_n = \{\chi \in K: \forall m \geq n, \omega_H(x_m)(\chi) \in \overline{\Lambda_4}\}$, we get an increasing chain $F_0 \subseteq F_1 \subseteq \cdots \subseteq F_n \subseteq \cdots$ of closed sets of K with $K = \bigcup_{n\in\mathbb{N}} F_n$. Since K is compact, from Theorem B.5.20 we deduce that some F_{n_0} must have nonempty interior, i. e., there exist $\xi \in K$ and $U \in \mathcal{V}_K(0)$ with $\xi + U \subseteq F_{n_0}$. Hence, $\omega_H(x_m)(\xi + U) \subseteq \overline{\Lambda_4}$ for all $m \geq n_0$. From (13.6) we deduce that there exists $n_1 \in \mathbb{N}$ such that $\omega_H(x_m)(\xi) \in \overline{\Lambda_4}$ for all $m \geq n_1$. Therefore, $\omega_H(x_m)(U) \subseteq \overline{\Lambda_2}$ for all $m \geq n_2 = \max\{n_0, n_1\}$.

From (13.6) we deduce that, for every $\chi \in K$, there exists $n_\chi \in \mathbb{N}$ such that $\chi(x_n) \in \Lambda_2$ for all $n \geq n_\chi$. By the compactness of $K = \bigcup_{\chi\in K}(\chi + U)$, there exist $\chi_1, \dots, \chi_s \in K$ such that $K = \bigcup_{i=1}^{s}(\chi_i + U)$. Let $n_3 = \max\{n_{\chi_1}, \dots, n_{\chi_s}\}$ and $n_4 = \max\{n_2, n_3\}$. Then $\omega_H(x_m)(K) \subseteq \Lambda_1$ for all $m \geq n_4$. As Λ_1 contains no nontrivial subgroups, $\omega_H(x_m) = 0$ for all $m \geq n_4$. This entails $x_m = 0$ for all $m \geq n_4$, since ω_H is an isomorphism in view of Theorem 13.4.8. Thus, $\{x_n\}_{n\in\mathbb{N}}$ is trivial, a contradiction. □

13.4.2 Exactness of the duality functor

For a topological abelian group G and a subset X of G, the *annihilator* of X in $\widehat{G}$ is

$$A_{\widehat{G}}(X) := \{\chi \in \widehat{G}: \chi(X) = \{0\}\},$$

and for a subset Y of $\widehat{G}$ the *annihilator* of Y in G is

$$A_G(Y) := \{x \in G: \chi(x) = 0 \text{ for every } \chi \in Y\}.$$

When no confusion is possible, we omit the subscripts and write $A(X)$ and $A(Y)$.

Lemma 13.4.10. *Let G be a topological abelian group and M a subset of G. Then:*
(a) *$A_{\widehat{G}}(M)$ is a closed subgroup of $\widehat{G}$;*
(b) *the correspondence $M \mapsto A_{\widehat{G}}(M)$ is monotone decreasing;*
(c) *$A_{\widehat{G}}(M) = A_{\widehat{G}}(\overline{\langle M\rangle})$;*
(d) *if $0 \in M \subseteq G$ and $0 \in M' \subseteq G$, then $A_{\widehat{G}}(M + M') = A_{\widehat{G}}(M) \cap A_{\widehat{G}}(M')$;*
(e) *$A_{\widehat{G}}(mG) = \widehat{G}[m]$ for every $m \in \mathbb{N}_+$.*

Proof. (a) Assume that $\{\chi_\alpha\}_{\alpha\in A}$ is a net in $A_{\widehat{G}}(M)$ converging to χ in $\widehat{G}$. Since the net converges also in the (coarser) pointwise-convergence topology, $\chi_\alpha(x) \to \chi(x)$ for every

$x \in M$. As $\chi_\alpha(x) = 0$ for all $\alpha \in A$, we deduce that $\chi(x) = 0$ for all $x \in M$, i. e., $\chi \in A_{\widehat{G}}(M)$, hence this set is closed.

(b) is obvious.

(c) Use (b) and the fact that when $\chi \in \widehat{G}$ vanishes on M, it also vanishes on $\overline{\langle M\rangle}$.

(d) The inclusion $A_{\widehat{G}}(M+M') \subseteq A_{\widehat{G}}(M) \cap A_{\widehat{G}}(M')$ easily follows from (b) and $M \subseteq M+M' \supseteq M'$. The other inclusion is obvious.

(e) For $\xi \in \widehat{G}$, $\xi \in \widehat{G}[m]$ if and only if $\xi(mG) = \{0\}$, namely, $\xi \in A_{\widehat{G}}(mG)$. □

Next we show a first use of the annihilator in a direct consequence of Theorem 11.6.3.

Proposition 13.4.11. *Every closed subgroup of a locally compact abelian group G is dually closed.*

Proof. Let H be a closed subgroup of G and let $a \in G \setminus H$. By Theorem 11.6.3, G/H is MAP, that is, $\widehat{G/H}$ separates the points of G/H, and so there exists $\xi \in \widehat{G/H}$ such that $\xi(q(a)) \neq 0$, where $q\colon G \to G/H$ is the canonical projection. Now $\chi = \xi \circ q \in A_{\widehat{G}}(H)$ and $\chi(a) \neq 0$. □

Call a continuous homomorphism $f\colon G \to H$ of topological groups *proper* if $f\colon G \to f(G)$ is open, when $f(G)$ carries the topology inherited from H. A continuous surjective homomorphism is proper if and only if it is open, while an injective continuous homomorphism is proper if and only if it is an embedding.

A short sequence $0 \to G_1 \xrightarrow{f} G \xrightarrow{h} G_2 \to 0$ in $\mathcal{L}$, where f and h are continuous homomorphisms, is *exact* if f is injective, h is surjective and $\operatorname{im} f = \ker h$; we call it *proper* if f and h are proper.

Proposition 13.4.12. *Let G be a locally compact abelian group, H a closed subgroup of G, $i\colon H \to G$ the canonical inclusion, and $q\colon G \to G/H$ the canonical projection. Then the sequence*

$$0 \to \widehat{G/H} \xrightarrow{\widehat{q}} \widehat{G} \xrightarrow{\widehat{i}} \widehat{H}$$

is exact, $\widehat{q}$ is proper, and $\operatorname{im}\widehat{q} = A_{\widehat{G}}(H)$. If H is open or compact, $\widehat{i}$ is open and surjective.

Proof. According to Lemma 13.3.7(d), $\widehat{q}$ is an embedding, so it is proper. We have that $\widehat{i} \circ \widehat{q} = \widehat{q \circ i} = 0$, hence $\operatorname{im}\widehat{q} \subseteq \ker\widehat{i}$. If $\xi \in \ker\widehat{i} = A_{\widehat{G}}(H)$, then $\xi(H) = \{0\}$. So, there exists $\xi_1 \in \widehat{G/H}$ such that $\xi = \xi_1 \circ q$ (i. e., $\xi = \widehat{q}(\xi_1)$), and we can conclude that $\ker\widehat{i} = \operatorname{im}\widehat{q} = A_{\widehat{G}}(H)$. Thus, the sequence is exact.

If H is open or compact, Corollary 13.3.8(a) implies that $\widehat{i}$ is surjective. It remains to show that $\widehat{i}$ is open. If H is compact, then $\widehat{H}$ is discrete by Proposition 13.1.1(a), so $\widehat{i}$ is obviously open. If H is open, let K be a compact neighborhood of 0 in G such that $K \subseteq H$. Then $W = W_{\widehat{G}}(K, \overline{\Lambda}_4)$ is a compact neighborhood of 0 in $\widehat{G}$. Since $\widehat{i}$ is continuous and surjective, $V = \widehat{i}(W) = W_{\widehat{H}}(K, \overline{\Lambda}_4)$ is a compact neighborhood of 0 in $\widehat{H}$. Now $M = \langle W\rangle$ and $M_1 = \langle V\rangle$ are compactly generated open subgroups of $\widehat{G}$ and $\widehat{H}$, respectively, and $\widehat{i}(M) = M_1$. Since M is σ-compact by Lemma 8.3.2, we apply the open mapping theorem (Theorem 8.4.1) to $\widehat{i}\restriction_M: M \to M_1$, and so $\widehat{i}$ is open. □

The next corollary of Proposition 13.4.12 says that the duality functor preserves proper exactness for some sequences.

Corollary 13.4.13. *If the sequence $0 \to G_1 \xrightarrow{f} G \xrightarrow{h} G_2 \to 0$ in $\mathcal{L}$ is proper exact, with G_1 compact or G_2 discrete (i. e., G_1 open), then $0 \to \widehat{G_2} \xrightarrow{\widehat{h}} \widehat{G} \xrightarrow{\widehat{f}} \widehat{G_1} \to 0$ is proper exact with the same property (i. e., $\widehat{G_1}$ is discrete or $\widehat{G_2}$ is compact).*

Corollary 13.4.14. *For a locally compact abelian group G and a closed subgroup H of G, $\widehat{G/H} \cong A_{\widehat{G}}(H)$. If H is open or compact, then $\widehat{H} \cong \widehat{G}/A_{\widehat{G}}(H)$.*

Corollary 13.4.15. *Let G be a locally compact abelian group and H a closed subgroup of G. If H is open (respectively, compact), then $A_{\widehat{G}}(H)$ is compact (respectively, open).*

Proof. If H is open, then G/H is discrete by Lemma 3.2.10(a) and so $\widehat{G/H} \cong A_{\widehat{G}}(H)$ is compact by Corollary 13.4.14 and Proposition 13.1.1(b). If H is compact, then $\widehat{H} \cong \widehat{G}/A_{\widehat{G}}(H)$ is discrete by Corollary 13.4.14 and Proposition 13.1.1(a), so $A_{\widehat{G}}(H)$ is open by Lemma 3.2.10(a). □

For the duality functor, f is an epimorphism in $\mathcal{L}$ if and only if $\widehat{f}$ is a monomorphism:

Corollary 13.4.16. *Let $f: G \to H$ be a continuous homomorphism of locally compact abelian groups. Then the following conditions are equivalent:*
(a) *$f(G)$ is dense in H;*
(b) *$\widehat{f}$ is injective (i. e., $\widehat{f}$ is a monomorphism in the category $\mathcal{L}$);*
(c) *if $g \circ f = h \circ f$ for a pair of morphisms $g, h: H \to L$ in $\mathcal{L}$, then $g = h$ (i. e., f is an epimorphism in the category $\mathcal{L}$);*
(d) *if $g \circ f = 0$ for a morphism $g: H \to L$ in $\mathcal{L}$, then $g = 0$.*

Proof. (a)⇒(b) If $f(G)$ is dense in H then $\widehat{f}$ is injective by Lemma 13.3.7(a). As far as the second assertion is concerned, it is easy to check the monomorphisms in $\mathcal{L}$ are precisely the continuous injective homomorphisms (see Exercise 13.7.5).

(b)⇒(c) Assume that $g \neq h$. Then the subgroup $N = \{x \in H: g(x) = h(x)\}$ is a proper closed subgroup of H containing $f(G)$. By Proposition 13.4.11, there exists a nonzero $\chi \in A_{\widehat{H}}(N)$. Since $f(G) \subseteq N$, one has $\chi \circ f = \widehat{f}(\chi) = 0$. This contradicts the injectivity of $\widehat{f}$.

(c)⇒(d) Apply (c) for the pair of morphisms g and 0.

(d)⇒(a) Assume that $f(G)$ is not dense in H. Then $\overline{f(G)}$ is a proper closed subgroup of H. Let $a \in H \setminus \overline{f(G)}$. By Proposition 13.4.11, there exists $\chi \in A_{\widehat{H}}(\overline{f(G)})$ such that $\chi(a) \neq 0$. In particular, $\chi \neq 0$. On the other hand, our choice of χ entails $\chi \circ f = 0$, which leads to $\chi = 0$, according to our hypothesis applied to $g = \chi$. This contradiction proves that $f(G)$ is dense in H. □

13.4.3 Proof of the Pontryagin-van Kampen duality theorem

Now we can prove the Pontryagin-van Kampen duality theorem, namely, that ω is a natural equivalence from $1_{\mathcal{L}}$ to $\widehat{\widehat{\ }}: \mathcal{L} \to \mathcal{L}$.

Theorem 13.4.17. *If G is a locally compact abelian group, then $\omega_G\colon G \to \widehat{\widehat{G}}$ is a topological isomorphism.*

Proof. We know by Proposition 13.4.6 that ω is a natural transformation from $1_{\mathcal{L}}$ to $\widehat{\widehat{\ }}: \mathcal{L} \to \mathcal{L}$. Our plan is to chase the given locally compact abelian group G into an appropriately chosen proper exact sequence $0 \to G_1 \xrightarrow{f} G \xrightarrow{h} G_2 \to 0$ in $\mathcal{L}$, with G_1 compact or G_2 discrete, such that both G_1 and G_2 satisfy the Pontryagin-van Kampen duality theorem. By Corollary 13.4.13, the sequences

$$0 \to \widehat{G_2} \xrightarrow{\widehat{h}} \widehat{G} \xrightarrow{\widehat{f}} \widehat{G_2} \to 0 \quad \text{and} \quad 0 \to \widehat{\widehat{G_1}} \xrightarrow{\widehat{\widehat{f}}} \widehat{\widehat{G}} \xrightarrow{\widehat{\widehat{h}}} \widehat{\widehat{G_2}} \to 0$$

are proper exact. By Proposition 13.4.6, the following diagram commutes:

$$\begin{array}{ccccccccc}
0 & \longrightarrow & G_1 & \xrightarrow{f} & G & \xrightarrow{h} & G_2 & \longrightarrow & 0 \\
 & & \downarrow{\scriptstyle \omega_{G_1}} & & \downarrow{\scriptstyle \omega_G} & & \downarrow{\scriptstyle \omega_{G_2}} & & \\
0 & \longrightarrow & \widehat{\widehat{G_1}} & \xrightarrow[\widehat{\widehat{f}}]{} & \widehat{\widehat{G}} & \xrightarrow[\widehat{\widehat{h}}]{} & \widehat{\widehat{G_2}} & \longrightarrow & 0
\end{array}$$

According to Remark 13.4.2 and Proposition 13.4.1, ω_{G_1}, ω_G, and ω_{G_2} are injective continuous homomorphisms. Moreover, ω_{G_1} and ω_{G_2} are surjective by our choice of G_1 and G_2. Then ω_G must be surjective too, by Lemma A.5.3.

If G is compactly generated, by Proposition 11.6.2 we can choose G_1 compact and G_2 elementary locally compact. Then G_1 and G_2 satisfy the Pontryagin-van Kampen duality theorem by Theorems 13.4.7 and 13.4.8, so ω_G is a continuous isomorphism by Lemma A.5.3. Since G is σ-compact by Lemma 8.3.2, we conclude with the open mapping theorem (Theorem 8.4.1) that ω_G is a topological isomorphism.

In the general case, we can take a compactly generated open subgroup G_1 of G. This produces a proper exact sequence $0 \to G_1 \xrightarrow{f} G \xrightarrow{h} G_2 \to 0$ with G_1 compactly generated and $G_2 \cong G/G_1$ discrete. By the previous case, ω_{G_1} is a topological isomorphism and ω_{G_2} is an isomorphism thanks to Theorem 13.4.7. Therefore, ω_G is a continuous isomorphism by Lemma A.5.3. Since f is an embedding with open image, to show that ω_G is open, it is enough to verify that $\omega_G \circ f = \widehat{\widehat{f}} \circ \omega_{G_1}$ (the equality due to Proposition 13.4.6) is open. As ω_{G_1} is a topological isomorphism, it is enough to prove that $\widehat{\widehat{f}}\colon \widehat{\widehat{G_1}} \to \widehat{\widehat{G}}$ is open. Since $\widehat{G_2}$ is compact, Corollary 13.4.13 applied to $0 \to \widehat{G_2} \xrightarrow{\widehat{h}} \widehat{G} \xrightarrow{\widehat{f}} \widehat{G_1} \to 0$ yields that the homomorphism $\widehat{\widehat{f}}$ is an embedding. Further, $\operatorname{im} \widehat{\widehat{f}} = \ker \widehat{\widehat{h}} = A_{\widehat{\widehat{G}}}(\operatorname{im} \widehat{h})$ by Proposition 13.4.12. As $\operatorname{im} \widehat{h}$ is compact, we conclude that $A_{\widehat{\widehat{G}}}(\operatorname{im} \widehat{h})$ is open by Corollary 13.4.15. So, $\widehat{\widehat{f}}$ is open. □

We give a first application of Theorem 13.4.8, with Proposition 13.1.1:

Corollary 13.4.18. *A locally compact abelian group G is compact (respectively, discrete) if and only if $\widehat{G}$ is discrete (respectively, compact).*

Proof. If $\widehat{G}$ is compact (respectively, discrete), Proposition 13.1.1 implies that $\widehat{\widehat{G}}$ is discrete (respectively, compact). So, the assertions follow from Theorem 13.4.17. □

Finally, let us discuss the uniqueness of the Pontryagin-van Kampen duality. For topological abelian groups G, H denote by $\mathrm{Chom}(G, H)$ the group of all continuous homomorphisms $G \to H$ equipped with the compact-open topology. It was pointed out already by Pontryagin that the group $\mathbb{T}$ is the unique locally compact abelian group L with the property $\mathrm{Chom}(\mathrm{Chom}(\mathbb{T}, L), L) \cong \mathbb{T}$ (note that this is much weaker than asking $\mathrm{Chom}(-, L)$ to define a duality of $\mathcal{L}$).

Much later Roeder proved:

Theorem 13.4.19 (Roeder theorem [250]). *The Pontryagin-van Kampen duality functor is the unique functorial duality of $\mathcal{L}$, i. e., the unique involutive duality of $\mathcal{L}$.*

See §16.2.1 for a more rigorous formulation and further results in the realm of topological modules.

13.5 Further properties of the annihilators

Our last aim concerning the annihilators is to prove that they define an inclusion-inverting bijection between the family of all closed subgroups of a locally compact abelian group G and the family of all closed subgroups of its dual $\widehat{G}$. We use the fact that one can identify G and $\widehat{\widehat{G}}$ by the topological isomorphism ω_G.

Lemma 13.5.1. *If G is a locally compact abelian group and H a closed subgroup of G, then $H = A_G(A_{\widehat{G}}(H)) = \omega_G^{-1}(A_{\widehat{\widehat{G}}}(A_{\widehat{G}}(H)))$.*

Proof. Clearly, $H \subseteq A_G(A_{\widehat{G}}(H))$, the equality $H = A_G(A_{\widehat{G}}(H))$ follows immediately from Proposition 13.4.11. The second equality follows from the first and Exercise 13.7.14. □

Remark 13.5.2. Let G be a locally compact abelian group.

(a) Let H a subgroup of G. The equality $H = A_G(A_{\widehat{G}}(H))$ holds if and only if H is a closed subgroup of G by Lemma 13.5.1, Lemma 13.4.10(a), and by Pontryagin-van Kampen duality theorem 13.4.17. In particular, one can deduce from this equivalence and Lemma 13.4.10(c) that $A_G(A_{\widehat{G}}(H)) = \overline{H}$ for an arbitrary subgroup H of G.

(b) According to Lemma 13.4.10(e), $\widehat{G}[m] = A_{\widehat{G}}(mG)$ for every $m \in \mathbb{N}_+$, hence Lemma 13.5.1 and item (a) give $A_G(\widehat{G}[m]) = A_G(A_{\widehat{G}}(mG)) = \overline{mG}$.

(c) If K is a compact subgroup of $\widehat{G}$, then $A_G(K)$ is open in G. Indeed, $A_{\widehat{\widehat{G}}}(K)$ is open in $\widehat{\widehat{G}}$ by Corollary 13.4.15, and $A_G(K) = \omega_G^{-1}(A_{\widehat{\widehat{G}}}(K))$ by Exercise 13.7.14.

For a locally compact abelian group G, let $\mathcal{S}(G)$ denote the complete lattice of all closed subgroups of G.

Corollary 13.5.3. *Let G be a locally compact abelian group. The pair of maps*

$$A_{\widehat{G}}: \mathcal{S}(G) \to \mathcal{S}(\widehat{G}) \quad \text{and} \quad A_G: \mathcal{S}(\widehat{G}) \to \mathcal{S}(G)$$

define a complete lattice antiisomorphism. For every family $\{H_i : i \in I\}$ in $\mathcal{S}(G)$,

$$A_{\widehat{G}}\Big(\sum_{i\in I} H_i\Big) = \bigcap_{i\in I} A_{\widehat{G}}(H_i) \quad \text{and} \quad A_{\widehat{G}}\Big(\bigcap_{i\in I} H_i\Big) = \overline{\sum_{i\in I} A_{\widehat{G}}(H_i)}. \tag{13.7}$$

Analogously, for every family $\{L_i : i \in I\}$ in $\mathcal{S}(\widehat{G})$,

$$A_G\Big(\sum_{i\in I} L_i\Big) = \bigcap_{i\in I} A_G(L_i) \quad \text{and} \quad A_G\Big(\bigcap_{i\in I} L_i\Big) = \overline{\sum_{i\in I} A_G(L_i)}. \tag{13.8}$$

Proof. The map $A_{\widehat{G}}: \mathcal{S}(G) \to \mathcal{S}(\widehat{G})$ is monotone decreasing by Lemma 13.4.10(b). By Lemma 13.5.1, the maps $A_{\widehat{G}}: \mathcal{S}(G) \to \mathcal{S}(\widehat{G})$ and $A_G: \mathcal{S}(\widehat{G}) \to \mathcal{S}(G)$ are inverse to each other, so they are bijective and A_G is monotone decreasing as well.

To prove (13.7), note that the inclusion $A_{\widehat{G}}(\sum_{i\in I} H_i) \subseteq \bigcap_{i\in I} A_{\widehat{G}}(H_i)$ follows from the monotonicity of $A_{\widehat{G}}$, while the opposite inclusion is immediate from the definition of the annihilator. Analogously, one can prove that $A_G(\sum_{i\in I} L_i) = \bigcap_{i\in I} A_G(L_i)$ in (13.8). As the subgroups of the second equalities in (13.7) and (13.8) are closed, it is sufficient to prove that their annihilators in G, respectively in $\widehat{G}$, are the same. Remark 13.5.2(a) gives $A_G(\overline{\sum_{i\in I} A_{\widehat{G}}(H_i)}) = A_G(\sum_{i\in I} A_{\widehat{G}}(H_i)) = \bigcap_{i\in I} A_G(A_{\widehat{G}}(H_i)) = \bigcap_{i\in I} H_i = A_G(A_{\widehat{G}}(\bigcap H_i))$. The remaining equality in (13.8) can be shown analogously. □

Corollary 13.5.4. *Let G be a discrete abelian group. If H is a pure subgroup of G, then $A_{\widehat{G}}(H)$ is a pure subgroup of $\widehat{G}$.*

Proof. Put for brevity $K = \widehat{G}$ and pick an $m \in \mathbb{N}_+$. The purity of H gives $mH = mG \cap H$, so taking the annihilators and making use of Corollary 13.5.3, as $A_K(mG) = K[m]$ by Lemma 13.4.10(e) and Lemma 13.5.1, we get $A_K(mH) = A_K(mG) + A_K(H) = K[m] + A_K(H)$. As $A_K(mH) = \{\chi \in K : m\chi \in A_K(H)\}$ by Example A.4.9(b), this shows that $A_K(H)$ is a pure subgroup of K. □

We can extend the isomorphism in Corollary 13.4.14 to the general case:

Proposition 13.5.5. *Let G be a locally compact abelian group and H a closed subgroup of G. Then $A_{\widehat{G}}(H) \cong \widehat{G/H}$ and $\widehat{G}/A_{\widehat{G}}(H) \cong \widehat{H}$. More precisely, $\widehat{G}/A_{\widehat{G}}(H) \to \widehat{H}, \chi + A_{\widehat{G}}(H) \mapsto \chi\restriction_H$, is a topological isomorphism; in particular, this shows that H is dually embedded in G.*

Proof. The first assertion follows from Corollary 13.4.14. We apply this result to $\widehat{G}/A_{\widehat{G}}(H)$ in order to deduce that

$$A_{\widehat{\widehat{G}}}(A_{\widehat{G}}(H)) \to \widehat{\widehat{G}/A_{\widehat{G}}(H)}, \quad \eta \mapsto (\chi + A_{\widehat{G}}(H) \mapsto \eta(\chi)),$$

is a topological isomorphism. According to Lemma 13.5.1, $\omega_G(H) = A_{\widehat{\widehat{G}}}(A_{\widehat{G}}(H))$. Composing $\omega_G \restriction_H$ with the above topological isomorphism gives that

$$\phi: H \to \widehat{\widehat{G}/A_{\widehat{G}}(H)}, \quad h \mapsto (\chi + A_{\widehat{G}}(H) \mapsto \omega_G(h)(\chi) = \chi(h)),$$

is a topological isomorphism. Hence, also $\widehat{\phi}$ is a topological isomorphism by Corollary 13.3.9. Composing with the topological isomorphism $\omega_{\widehat{G}/A_{\widehat{G}}(H)}$ (see Pontryagin-van Kampen duality theorem 13.4.17), one obtains the topological isomorphism

$$\rho: \widehat{G}/A_{\widehat{G}}(H) \to \widehat{H}, \quad \chi + A_{\widehat{G}}(H) \mapsto \widehat{\phi} \circ \omega_{\widehat{G}/A_{\widehat{G}}(H)}(\chi + A_{\widehat{G}}(H)).$$

For $h \in H$,

$$\widehat{\phi} \circ \omega_{\widehat{G}/A_{\widehat{G}}(H)}(\chi + A_{\widehat{G}}(H))(h) = \omega_{\widehat{G}/A_{\widehat{G}}(H)}(\chi + A_{\widehat{G}}(H))(\phi(h)) = \chi(h).$$

This shows that $\rho(\chi + A_{\widehat{G}}(H)) = \chi \restriction_H$, as desired. □

13.6 Duality for precompact abelian groups

Another duality can be obtained for precompact abelian groups as follows. First of all, for every topological abelian group G, the dual $\widehat{G}$ of G equipped with the topology of the pointwise convergence instead of the finer compact-open topology will be denoted by $\widehat{G}_{pw}$. Then $\widehat{\widehat{G}_{pw}}$ is a subgroup of $\widehat{\widehat{G}}$. Moreover, for every $x \in G$ obviously $\omega_G(x)(W(\{x\}, \Lambda_1)) \subseteq \Lambda_1$, i. e., $\omega_G(x) \in \widehat{\widehat{G}_{pw}}$. Therefore, the image of ω_G is actually contained in $\widehat{\widehat{G}_{pw}}$ for every topological abelian group G. This allows us to define the homomorphism $\gamma_G: G \to \widehat{(\widehat{G}_{pw})}_{pw}$ by simply putting $\gamma_G(x) = \omega_G(x)$ for every $x \in G$. Hence, disregarding the topology of the second dual group (and the larger codomain), this map coincides with ω_G.

The next lemma follows directly from the definitions.

Lemma 13.6.1. *For a topological abelian group G, $\chi_1, \dots, \chi_n \in \widehat{G}$, $\delta > 0$, and the neighborhood $U = \{z \in \mathbb{S}: |\mathrm{Arg}(z)| < \delta\}$ of 0 in $\mathbb{T} \cong \mathbb{S}$, we have $U_G(\chi_1, \dots, \chi_n; \delta) = \gamma_G^{-1}\left(W_{\widehat{\widehat{G}}}(\{\chi_1, \dots, \chi_n\}, U)\right)$.*

Immediately we can give the promised duality for precompact abelian groups.

Theorem 13.6.2. *The assignment $G \mapsto \widehat{G}_{pw}$ defines a duality in the category of precompact abelian groups, more precisely $\gamma_G: G \to \widehat{(\widehat{G}_{pw})}_{pw}$ is a topological isomorphism for every precompact abelian group G.*

Proof. By the definition of $\widehat{G}_{pw}$, its topology coincides with $\mathcal{T}_G$. This proves that γ_G is surjective in view of Proposition 11.3.10. The injectivity of γ_G follows from Remark 13.4.2 and the precompactness of G. The continuous characters of G separate the points of G by Proposition 3.1.20(a). That γ_G is a homeomorphism follows from Lemma 13.6.1 and the fact that a basic neighborhood of 0 in $\widehat{(\widehat{G}_{pw})}_{pw}$ has the form $W_{\widehat{\widehat{G}}}(\{\chi_1, \dots, \chi_n\}, U)$ for some $\chi_1, \dots, \chi_n \in \widehat{G}$ and a neighborhood U of 0 in $\mathbb{T} \cong \mathbb{S}$. □

13.7 Exercises

Exercise 13.7.1. For $\chi, \xi \in \mathbb{Z}^*$ with $\ker\chi = \ker\xi = \{0\}$, prove that $\mathcal{T}_\chi = \mathcal{T}_\xi$ if and only if $\xi = \chi^{\pm 1}$.

Hint. Use the completion and that $\pm id_{\mathbb{T}}$ are the only topological automorphisms of $\mathbb{T}$.

Exercise 13.7.2. For an infinite abelian group G, prove that $d(G^\#) = |G|$, $\chi(G^\#) = w(G^\#) = 2^{|G|}$, while $\psi(G^\#) = \log|G|$.

Hint. To prove the first equality, use the fact that every subgroup of $G^\#$ is closed. To prove the second chain of equalities, apply Corollary 11.4.5 and use the fact that $|G^*| = 2^{|G|}$ (see Theorem 13.3.11). For the third one, use the fact that if $|G| \leq 2^\kappa$ for some infinite cardinal κ, then G is isomorphic to a subgroup of $\mathbb{T}^\kappa$.

Exercise 13.7.3. Let G be a discrete abelian group, p a prime, and $\chi \in \widehat{G}$. Prove that:
(a) $\chi \in p\widehat{G}$ if and only if $\chi(G[p]) = \{0\}$;
(b) $p\chi = 0$ in $\widehat{G}$ if and only if $\chi(pG) = \{0\}$.

Conclude that:
(c) a discrete abelian group G is divisible (respectively, torsion-free) if and only if $\widehat{G}$ is torsion-free (respectively, divisible);
(d) the groups $\widehat{\mathbb{Q}}$ and $\widehat{\mathbb{Q}}_p$ are torsion-free and divisible.

Exercise 13.7.4. Let G be a discrete abelian torsion group. Show that $\widehat{G}$ is a hereditarily disconnected compact group.

Hint. For every $x \in G$, the neighborhood $W_{\widehat{G}}(\langle x\rangle, \Lambda_1)$ is an open subgroup of $\widehat{G}$.

Exercise 13.7.5. Prove that the monomorphisms in $\mathcal{L}$ are precisely the continuous injective homomorphisms.

Hint. Clearly, all continuous injective homomorphisms between locally compact abelian groups are monomorphisms in $\mathcal{L}$. If $f: G \to H$ is a noninjective continuous homomorphism of locally compact abelian groups, pick $a \in \ker f \setminus \{0\}$ and consider the homomorphism $g: \mathbb{Z} \to G$ defined by $g(1) = a$. Obviously, $f \circ g = 0 = f \circ 0$, yet $g \neq 0$, where 0 denotes the zero morphism $\mathbb{Z} \to G$.

Exercise 13.7.6. Prove that $b_G: G \to bG$ is a bimorphism in $\mathcal{L}$ for every locally compact abelian group G. Deduce that a locally compact abelian group G is:
(a) compact if and only if every bimorphism $G \to H$ in $\mathcal{L}$ is an isomorphism;
(b) discrete if and only if every bimorphism $H \to G$ in $\mathcal{L}$ is an isomorphism.

Deduce from (a) and (b) an alternative proof of Proposition 13.1.1.

Hint. Apply Corollary 13.4.16 and Exercise 13.7.5.

Exercise 13.7.7. Let H be a subgroup of $\mathbb{R}^n$. Prove that every $\chi \in \widehat{H}$ extends to a continuous character of $\mathbb{R}^n$.

Hint. Applying Theorem 7.1.18, extend $\chi \in \widehat{H}$ to a continuous character of $\overline{H}$ and apply Theorem 9.2.2.

Exercise 13.7.8. Prove without recourse to the Pontryagin-van Kampen duality theorem that a discrete abelian group G satisfies $\widehat{\widehat{G}} \cong G$ whenever:
(a) G is divisible;
(b) G is free;
(c) G is of finite exponent;
(d) G is torsion and every primary component of G is of finite exponent.

Hint. (a) Use Example 13.2.4, Example 13.3.14(b), and the fact that every divisible abelian group is a direct sum of copies of $\mathbb{Q}$ and of $\mathbb{Z}(p^\infty)$, with p prime (see Theorem A.2.17).

(c), (d) Use that fact that every abelian group of finite exponent is a direct sum of cyclic subgroups (i. e., Prüfer theorem A.1.4).

Exercise 13.7.9. Prove that, for every locally compact abelian group G, the group $G\times\widehat{G}$ is selfdual.

Exercise 13.7.10. Give an example of:
(a) a selfdual locally compact abelian group G such that some quotients and some closed subgroups of G are not selfdual;
(b) a selfdual locally compact abelian group G such that all nontrivial quotients and all proper nontrivial closed subgroups of G are not selfdual.

Exercise 13.7.11. Let G be a reflexive group. Show that also $\widehat{G}$ is reflexive.

Exercise 13.7.12. Show that a compact group K admits a continuous homomorphism $f\colon \mathbb{R} \to K$ with $\overline{f(\mathbb{R})} = K$ if and only if K is connected, abelian, and $w(K) \leq \mathfrak{c}$.

Hint. The necessity follows from the connectedness of $f(\mathbb{R})$ and the fact that $f(\mathbb{R})$ is separable, so $w(K) = w(f(\mathbb{R})) \leq \mathfrak{c}$. For the sufficiency, note that if G is connected and $w(K) \leq \mathfrak{c}$, then $\widehat{K}$ is a torsion-free abelian group of size $\leq \mathfrak{c}$. Hence, there exists an injective homomorphism $j\colon \widehat{K} \to \mathbb{R}$. Then composing $\widehat{j}\colon \widehat{\mathbb{R}} \to \widehat{\widehat{K}}$ with the topological isomorphisms $\mathbb{R} \cong \widehat{\mathbb{R}}$ and ω_K^{-1} we obtain the desired continuous homomorphism $f\colon \mathbb{R} \to K$. By Corollary 13.4.16, it has dense image since j is a monomorphism in $\mathcal{L}$.

Exercise 13.7.13. Show that $a(K)$ is dense in $c(K)$ for every compact abelian group K.

Hint. It suffices to see that $a(K)$ is dense in K when K is a connected compact abelian group. Let $H = \overline{a(K)}$ and note that H contains the union U of all connected metrizable closed subgroups of K, by Exercise 13.7.12. In view of $\langle U\rangle \leq H$, it suffices to show that $\langle U\rangle$ is dense in K. To this end consider the torsion-free dual $X = \widehat{K}$ and pick for every non-zero $x \in X$ a subgroup Y_x of X maximal with the property $Y_x \cap \langle x\rangle = \{0\}$. Then X/Y_x is a rank-one torsion-free abelian group, so its dual $K_x = \widehat{X/Y_x} = A(Y_x)$ is a connected metrizable closed subgroup of K, so $K_x \subseteq U$. Since $\bigcap_{x\in X\setminus\{0\}} Y_x = \{0\}$ and $Y_x = A(K_x)$, we get $A(\sum_{x\in X\setminus\{0\}} K_x) = \bigcap_{x\in X\setminus\{0\}} A(K_x) = \{0\}$, so $\overline{\sum_{x\in X\setminus\{0\}} K_x} = K$. Since $\sum_{x\in X\setminus\{0\}} K_x \subseteq \langle U\rangle$, this yields that $\langle U\rangle$ is dense in K, and we are done.

Exercise 13.7.14. Let G be a locally compact abelian group and Y a subset of $\widehat{G}$. Prove that $A_{\widehat{\widehat{G}}}(Y) = \omega_G(A_G(Y))$.

Exercise 13.7.15. Deduce from Glicksberg theorem 11.6.11 that for locally compact abelian groups G, H, a group homomorphism $f: G \to H$ is continuous if and only if $\chi \circ f: G \to \mathbb{T}$ is continuous for every $\chi \in \widehat{H}$.

13.8 Further readings, notes, and comments

Here we briefly explain how the Pontryagin-van Kampen duality theorem was extended to some non-locally compact topological abelian groups and the relation to the duality theory of locally convex spaces.

The dual space V' of a normed vector space V over the field $K \in \{\mathbb{R}, \mathbb{C}\}$ is the set of all continuous linear forms $V \to K$. With pointwise defined operations and endowed with the strong topology (a neighborhood base at 0 is given by suitably defined polars of bounded subsets), it is again a normed space, which allows one to iterate this process. A normed space V is called *reflexive* if the canonical map $V \to V''$, $x \mapsto (f \mapsto f(x))$ is a topological isomorphism. Many famous and important Banach spaces, e. g., the sequence spaces ℓ^p for $p \in \{1, \infty\}$, are not reflexive in this sense. Surprisingly, Smith [262] showed that every Banach space satisfies the Pontryagin-van Kampen duality theorem. Although the dual space of a real normed vector space is canonically isomorphic to its dual group (when considered as a topological abelian group), the dual objects are endowed with different topologies: the compact-open topology is strictly coarser than the strong topology in case V has infinite dimension.

Now we discuss a property of reflexive groups that is not sufficient for reflexivity but much easier to deal with. Vilenkin [282] introduced the notions of quasi-convexity and local quasi-convexity as follows. A subset A of a topological abelian group G is called *quasi-convex* if for every $x \in G \setminus A$ there exists $\chi \in \widehat{G}$ such that $\chi(A) \subseteq \mathbb{T}_+$ and $\chi(x) \notin \mathbb{T}_+$ where $\mathbb{T}_+ = \{t + \mathbb{Z}: |t| \leq \frac{1}{4}\}$. This notion generalizes convexity in locally convex vector spaces due to the description of closed convex sets given by the Hahn–Banach theorem. A topological abelian group is *locally quasi-convex* if it has a neighborhood base at 0 consisting of quasi-convex sets. The dual of every topological abelian group is locally quasi-convex, so reflexive groups are locally quasi-convex, while locally quasi-convex Hausdorff groups are MAP.

A topological vector space is locally quasi-convex if and only if it is locally convex (see [18]). The class of locally quasi-convex groups is closed under taking arbitrary products and subgroups, but it fails to be closed under taking Hausdorff quotients (actually, every Hausdorff group is a quotient of a locally quasi-convex Hausdorff group). Since every reflexive group is locally quasi-convex, this property is necessary but not sufficient – as we shall see soon – for reflexivity.

While the local quasi-convexity of a topological abelian group G implies that the canonical map ω_G is open with respect to its image, the injectivity of ω_G is equivalent to G being MAP. However, the continuity and surjectivity of ω_G are not as well classified. While the continuity is somehow related to k-space properties of the group G, the surjectivity requires some forms of completeness (e. g., a metrizable reflexive group must be complete). However, the group $\mathsf{Z} = \mathrm{L}^2_{\mathbb{Z}}([0,1])$ of all square-integrable almost everywhere integer valued functions on $[0,1]$ is a closed subgroup of the Hilbert space $\mathrm{L}^2([0,1])$, so complete and metrizable. Nevertheless, the dual homomorphism $\widehat{i}$ of the inclusion map $i\colon \mathsf{Z} \to \mathrm{L}^2([0,1])$ is a topological isomorphism (see [11]), which implies that $\mathsf{Z} = \mathrm{L}^2_{\mathbb{Z}}([0,1])$ is not reflexive. The group Z is locally quasi-convex (being a subgroup of the Hilbert space $\mathrm{L}^2([0,1])$) and ω_{Z} is continuous by Proposition 13.4.1. So, ω_{Z} is an embedding with closed image, which is not surjective.

In general, the dual group of a metrizable abelian group G is not only σ-compact (see Corollary 13.1.3(b)), but also hemicompact which means that $\widehat{G}$ can be written as a countable union of compact sets such that every compact subset of $\widehat{G}$ is contained in one of these. Further, $\widehat{G}$ is a k-space (see [11, 44]). Due to Proposition 13.4.1, $\omega_{\widehat{G}}$ is continuous. Since $\widehat{G}$ is locally quasi-convex, the canonical mapping $\omega_{\widehat{G}}$ is an embedding. For example, the group $\mathsf{Z} = L^2_{\mathbb{Z}}([0,1])$ is metrizable but not reflexive. Its character group is topologically isomorphic to the reflexive character group of $\mathrm{L}^2([0,1])$, hence reflexive. So, in this particular case $\omega_{\widehat{\mathsf{Z}}}$ is not only an embedding, but a topological isomorphism.

However, it is an open question whether the dual group of a metrizable abelian group G is reflexive in general (see [11, (5.23)]). Only the surjectivity of $\omega_{\widehat{G}}$ remains to be proved or disproved. Since the dual group of a metrizable abelian group is a hemicompact k-space (and by Corollary 13.1.3 also vice versa), the above question can be equivalently reformulated: let H be a topological abelian group which is a hemicompact k-space. Is $\widehat{H}$ reflexive?

Also in this situation, examples which suggest a positive answer are available: for every compact space K, the free abelian topological group $A(K)$ is a hemicompact k-space, but $A(K)$ is reflexive if and only if K is hereditarily disconnected (see [11, 220, 226]), while its character group is reflexive for all compact spaces (see [11, 151]).

It was shown by Kaplan [182] that arbitrary products of reflexive groups are reflexive. Leptin [196] gave an example of a closed subgroup C of a product P of discrete groups such that ω_C is not continuous but open and bijective. (Observe that P is reflexive due to Kaplan's result.) The same failure of reflexivity arises with quotient groups: every infinite dimensional Banach space E has a discrete free subgroup D which is not dually embedded and with E/D minimally almost periodic (see [18]).

As described above, closed subgroups and Hausdorff quotient groups of Banach spaces are in general not reflexive. The situation changes when we consider locally convex nuclear vector spaces instead of Banach spaces. The intersection of these two classes of topological vector spaces consists of the finite-dimensional vector spaces. Although the definition of a locally convex nuclear vector space is a bit technical, many famous function spaces, such as the space of harmonic functions or that of holomorphic functions (with the compact-open topology), are nuclear vector spaces. It was shown by T. and Y. Kōmura [186] that every real locally convex nuclear space can be embedded into a product of the sequence space Σ of rapidly decreasing sequences

$$\Sigma = \left\{(x_n)_{n\in\mathbb{N}_+} \in \mathbb{R}^{\mathbb{N}_+} : \sup\{n^r|x_n| : n \in \mathbb{N}_+\} < \infty\ \forall\, r \in \mathbb{N}\right\}$$

when Σ is endowed with the vector space topology induced by the family $\{p_r : r \in \mathbb{N}\}$ of norms where $p_r : \Sigma \to \mathbb{R}_+$, $(x_n)_{n\in\mathbb{N}_+} \mapsto \sup\{n^r|x_n| : n \in \mathbb{N}_+\}$. Locally convex nuclear vector spaces share many common properties with locally compact abelian groups. This motivated Banaszczyk to introduce the class of *nuclear groups* containing the class of all locally convex nuclear vector spaces and all locally compact abelian groups (see [18]). This class of groups consists of Hausdorff groups and is closed under taking subgroups, Hausdorff quotient groups, arbitrary products, and countable sums. Banaszczyk proved that every nuclear group is locally quasi-convex. Since every Hausdorff quotient group of a nuclear group is again nuclear, hence a locally quasi-convex Hausdorff group, it is MAP (as mentioned above). This easily implies that every closed subgroup of a nuclear group is dually closed. Moreover, every subgroup of a nuclear group is dually embedded. Nuclear groups have very strong duality properties, as the following examples shall demonstrate: every complete metrizable nuclear group is reflexive (see [18]) and the canonical mapping ω_G of every complete nuclear group G is surjective (see [11]). However, Leptin's example C, which is a subgroup of a product of discrete groups and hence nuclear, provides an example of a nonreflexive nuclear group. There exists noncompact precompact groups (which are, of course, nuclear), which are reflexive (see [154, 5]). This shows that completeness is sufficient for the surjectivity of ω_G in the class of nuclear groups, but it is not necessary.

Characterizations of reflexive groups were proposed by Venkatamaran [280] and Kye [193], but they contained flaws. These gaps were removed in the recent paper [170] of Hernández. While a vector space V that is topologically isomorphic to its bidual V'' need not be reflexive (in the vector space sense), the question whether every topological group G which is topologically isomorphic to $\widehat{\widehat{G}}$ is reflexive is still open. Further references to this topic can be found also in [44, 151, 172, 45].

According to Glicksberg theorem 11.6.11, the compact sets of a locally compact group (G,τ) coincide with the compact sets of the Bohr modification (G,τ^+) of (G,τ). In [19] it was shown that a nuclear group (G,τ) and its Bohr modification (G,τ^+) share the same compact, countably compact, and pseudocompact subsets.

We do not discuss here noncommutative versions of duality for locally compact groups. The difficulties arise already in the compact case: there is no appropriate (or at least, comfortable) structure on the set of irreducible unitary representations of a nonabelian compact group. The reader is referred to [174] for a historical panorama of this trend (Tanaka–Kreĭn duality, etc.). In the locally compact case, one should see the pioneering paper [47] of H. Chu, as well as the monograph [175] of Heyer (see also [176]).

The reader can find the last achievements in this field in the survey [153] of Galindo, Hernández, and Wu (see also [84, 171]).

The well-known Mackey–Arens theorem says that a Hausdorff locally convex vector space has a finest compatible topology. A locally convex vector space that carries this finest compatible topology is called a Mackey space. Varopoulos [278] studied the analogue question for locally precompact groups. In 1999, this was further generalized for locally quasi-convex groups by Chasco, Martín Peinador, and Tarieladze [46]: for such a group (G,τ), one can define similarly the notion of a compatible (locally quasi-convex) group topology η by asking (G,η) to have the same continuous characters as (G,τ), i. e., $\widehat{(G,\tau)} = \widehat{(G,\eta)}$ as abstract groups. Denoting by $\mathcal{C}(G,\tau)$ the poset of all compatible with τ topologies on G, ordered by inclusion, one can immediately see that $\mathcal{C}(G,\tau)$ has a bottom element, namely, the weak topology τ^+. The question of whether $\mathcal{C}(G,\tau)$ has also a top element (named Mackey topology) already posed in 1999 remained open for almost twenty years. The survey [14] presents recent (negative) answers to this problem (see [12, 13, 150]) and many other related results.

14 Applications of the duality theorem

14.1 The structure of compact abelian groups

First, we give a characterization of the connected compact abelian groups.

Proposition 14.1.1. *For a compact abelian group K, the following are equivalent:*
(a) *K is connected;*
(b) *K is divisible;*
(c) *$\widehat{K}$ is torsion-free.*

Proof. (a)⇔(c) is Proposition 11.6.10.

(b)⇔(c) For $m \in \mathbb{N}_+$, since $A_{\widehat{K}}(mK) = \widehat{K}[m]$ by Lemma 13.4.10(e), $mK = K$ if and only if $\widehat{K}[m] = \{0\}$. So, K is divisible precisely when $\widehat{K}[m] = \{0\}$ for every $m \in \mathbb{N}_+$, namely, $\widehat{K}$ is torsion-free. □

Remark 14.1.2. Proposition 14.1.1 fails in the noncompact case: (b)⇏(a) is witnessed by $\mathbb{Q}_p \cong \widehat{\mathbb{Q}}_p$ divisible (see Example 13.2.6), but not connected; while (c)⇏(b) is witnessed by Exercise 14.5.1.

On the other hand, one can prove that if G is a divisible locally compact abelian group, then $\widehat{G}$ is torsion-free following the proof of Proposition 14.1.1, while a connected locally compact abelian group G is of the form $\mathbb{R}^n \times K$, where $n \in \mathbb{N}$ and K is a connected compact abelian group by Corollary 14.2.11(a) below, and so G is divisible since K is divisible by Proposition 14.1.1.

As an application of the Pontryagin-van Kampen duality theorem, we describe the structure of some classes of compact abelian groups, such as the monothetic or the bounded ones.

Theorem 14.1.3. *Let K be a compact abelian group. Then K is monothetic if and only if $\widehat{K}$ admits an injective homomorphism into $\mathbb{T}$.*

Proof. The group K is monothetic if and only if there exists a homomorphism $\mathbb{Z} \to K$ with dense image. According to Corollary 13.4.16, this is equivalent to the existence of an injective homomorphism $\widehat{K} \to \mathbb{T}$. □

Corollary 14.1.4. *Let K be a compact abelian group.*
(a) *If K is connected, then K is monothetic if and only if $w(K) \leq \mathfrak{c}$.*
(b) *If K is hereditarily disconnected, then K is monothetic if and only if K is isomorphic to a quotient group of $\prod_{p\in\mathbb{P}} \mathbb{J}_p$.*

Proof. By Proposition 13.1.1(a), $G = \widehat{K}$ is discrete.

(a) By Proposition 14.1.1, G is torsion-free, so G admits an injective homomorphism into $\mathbb{T}$ precisely when $r_0(G) \leq \mathfrak{c}$ by Example A.2.16(b); equivalently, $|G| \leq \mathfrak{c}$. It remains to recall that $w(K) = |G|$ by Corollary 11.4.5. To conclude, apply Theorem 14.1.3.

https://doi.org/10.1515/9783110654936-014

(b) By Corollary 11.6.7, G is torsion. Since $t(\mathbb{T}) = \mathbb{Q}/\mathbb{Z}$, the torsion abelian group G admits an injective homomorphism into $\mathbb{T}$ if and only if G admits an injective homomorphism into $\mathbb{Q}/\mathbb{Z}$. This is equivalent to being K isomorphic to a quotient group of $\prod_{p\in\mathbb{P}} \mathbb{J}_p \cong \widehat{\mathbb{Q}/\mathbb{Z}}$, by Proposition 13.4.12 and Theorem 13.4.17. □

Now we provide more information on the set M_K of topological generators of a compact abelian group K.

Theorem 14.1.5. *Let K be a compact abelian group. Then:*
(a) *$M_K = \{x \in K : \omega_K(x) \text{ is injective}\}$;*
(b) *M_K is the intersection of at most $|\widehat{K}|$ open sets;*
(c) *if K is connected and metrizable, M_K is a dense G_δ-set of K.*

Proof. (a) Let $x \in M_K$ and $\chi \in \widehat{K}$. If $\omega_K(x)(\chi) = \chi(x) = 0$, then $\chi(K) = \{0\}$, and so $\chi \equiv 0$ (by assumption). This shows that $\ker \omega_K(x) = \{0\}$ and hence that $\omega_K(x)$ is injective. Conversely, assume that $x \in K$ is not a topological generator of K, i. e., $\overline{\langle x\rangle} \neq K$. By Proposition 13.4.11, there exists a nontrivial continuous character $\chi \in A_{\widehat{K}}(\overline{\langle x\rangle})$. This character satisfies $0 = \chi(x) = \omega_K(x)(\chi)$. Since χ was not the trivial character, this yields that $\omega_K(x)$ is not injective.

(b) For $\chi \in \widehat{K} \setminus \{0\}$, the set $U_\chi = K \setminus \ker\chi$ is open. In view of (a), $M_K = \bigcap_{\chi\in\widehat{K}\setminus\{0\}} U_\chi$, so M_K is an intersection of at most $|\widehat{K}|$ open sets.

(c) Since K is compact and metrizable, $\widehat{K}$ is countable by Corollary 13.1.3(b). Moreover, as K is connected, $\ker\chi$ has empty interior for every $\chi \in \widehat{K} \setminus \{0\}$, i. e., the open set $U_\chi = K \setminus \ker\chi$ is dense. Theorem B.5.20, applied to the compact group K, yields that $M_K = \bigcap_{\chi\in\widehat{K}\setminus\{0\}} U_\chi$ is a dense G_δ-set. □

Now we describe the torsion compact abelian groups.

Theorem 14.1.6. *Every torsion compact abelian group K is bounded, and there exist $m_1, \dots, m_n \in \mathbb{N}_+$ and cardinals $\alpha_1, \dots, \alpha_n$ such that $K \cong \prod_{i=1}^n \mathbb{Z}(m_i)^{\alpha_i}$.*

Proof. Since $K = \bigcup_{n\in\mathbb{N}_+} K[n!]$ is a union of closed subgroups, we conclude, with Theorem B.5.20, that $K[n!]$ is open for some $n \in \mathbb{N}_+$, so must have finite index, by the compactness of K. This yields $mK = \{0\}$ for some $m \in \mathbb{N}_+$, so also $m\widehat{K} = \{0\}$. By Prüfer theorem A.1.4, there exist $m_1, \dots, m_n \in \mathbb{N}_+$ and cardinals $\alpha_1, \dots, \alpha_n$ such that $\widehat{K} \cong \prod_{i=1}^n \mathbb{Z}(m_i)^{(\alpha_i)}$. Then by Theorems 13.4.7 and 13.3.5, as well as Example 13.3.3(a), $K \cong \widehat{\widehat{K}} \cong \prod_{i=1}^n \mathbb{Z}(m_i)^{\alpha_i}$. □

Next we compute the density character of a compact abelian group K as a function of its weight $w(K) = |\widehat{K}|$ (see Corollary 11.4.5 for this equality). More precisely, given the already known inequality $w(K) \leq 2^{d(K)}$ from Lemma 5.1.5, valid for all topological groups, now we see that the density character $d(K)$ has the smallest possible value (with respect to $w(K)$).

Proposition 14.1.7. *For K an infinite compact abelian group, $d(K) = \log w(K)$.*

Proof. Let $\kappa = \min\{\beta: w(K) \leq 2^\beta\}$. Then the inequality $w(K) \leq 2^{d(K)}$ (see Lemma 5.1.5) implies $\kappa \leq d(K)$.

Since $r_p(\mathbb{T}^\kappa) = r_0(\mathbb{T}^\kappa) = 2^\kappa$, the inequality $|\widehat{K}| = w(K) \leq 2^\kappa$, given by Corollary 11.4.5, and the divisibility of $\mathbb{T}^\kappa$ ensure that there exists an injective homomorphism $j: \widehat{K} \to \mathbb{T}^\kappa$. Therefore, the continuous homomorphism $\widehat{j}: \mathbb{Z}^{(\kappa)} \cong \widehat{\mathbb{T}^\kappa} \to \widehat{\widehat{K}} \cong K$ has dense image by Corollary 13.4.16. This proves $d(K) \leq \kappa$. □

Now we consider the case of connected compact abelian groups.

Proposition 14.1.8. *For a connected compact abelian group K, the subgroup $t(K)$ is dense in K if and only if $\widehat{K}$ is reduced. Consequently, every connected compact abelian group K has the form $K \cong K_1 \times \widehat{\mathbb{Q}}^\alpha$ for some cardinal α, where $K_1 = \overline{t(K)}$.*

Proof. Since $\widehat{K}$ is discrete by Proposition 13.1.1(a) and torsion-free by Proposition 14.1.1, $\widehat{K}$ is reduced if and only if $\bigcap_{m\in\mathbb{N}_+} m\widehat{K} = \{0\}$, by Proposition A.4.6(e). Since $A_K(m\widehat{K}) = K[m]$ by Lemma 13.4.10(e), so Corollary 13.5.3 implies

$$A_K\Big(\bigcap_{m\in\mathbb{N}_+} m\widehat{K}\Big) = \overline{\sum_{m\in\mathbb{N}_+} K[m]} = \overline{\bigcup_{m\in\mathbb{N}_+} K[m]} = \overline{t(K)}.$$

Therefore, the equality $\bigcap_{m\in\mathbb{N}_+} m\widehat{K} = \{0\}$ is equivalent to the density of $t(K)$ in K.

To prove the second assertion, consider the torsion-free and discrete dual $\widehat{K}$ and its decomposition $\widehat{K} = \mathrm{div}(\widehat{K}) \times R$, where R is a reduced subgroup of $\widehat{K}$, given by Theorem A.4.3. Since $\widehat{K}$ is torsion-free, there exists a cardinal α such that $\mathrm{div}(\widehat{K}) \cong \mathbb{Q}^{(\alpha)}$, by Theorem A.2.17. Therefore, $\widehat{\mathrm{div}(\widehat{K})} \cong \widehat{\mathbb{Q}}^\alpha$ in view of Theorem 13.3.5. On the other hand, by the first part of the proof, the connected compact abelian group $K_1 = \widehat{R}$ has dense torsion part $t(K_1)$. Since $K \cong \widehat{\widehat{K}} \cong \widehat{\mathbb{Q}}^\alpha \times K_1$ by Theorems 13.4.7 and 13.3.5, and $\widehat{\mathbb{Q}}^\alpha$ is torsion-free by Exercise 13.7.3(d), the torsion subgroup of $\widehat{\widehat{K}}$ coincides with $t(K_1)$, so its closure gives K_1. □

We conclude with a result on the structure of hereditarily disconnected compact abelian groups.

Theorem 14.1.9. *A compact abelian group K is hereditarily disconnected if and only if $K = \prod_{p\in\mathbb{P}} K_p$, where each K_p is a closed topological $\mathbb{J}_p$-module. The closed subgroups M of K are of the form $M = \prod_{p\in\mathbb{P}} M_p$ where M_p is a closed subgroup of K_p for every $p \in \mathbb{P}$.*

Proof. By Corollary 11.6.5(b), K is hereditarily disconnected if and only if $\widehat{K}$ is torsion, which is equivalent to $\widehat{K} = \bigoplus_{p\in\mathbb{P}} t_p(\widehat{K})$. According to Corollary 11.6.8, the group $X_p := \widehat{t_p(\widehat{K})}$ is topologically p-torsion, and so $X_p \subseteq K_p$. Since $K \cong \prod_{p\in\mathbb{P}} X_p$ by Theorems 13.3.5 and 13.4.7, and as $(X_q)_p = \{0\}$ for every $q \in \mathbb{P} \setminus \{p\}$ by Corollary 11.6.9 and Exercise 5.4.14(d), we conclude that for every prime p necessarily $K_p \cong X_p$; in particular, K_p is closed as X_p is compact by Proposition 13.1.1(b). Thus, $K \cong \prod_{p\in\mathbb{P}} K_p$, where each K_p is a $\mathbb{J}_p$-module by Remark 5.3.7.

The last assertion follows from the first one applied to M and the fact that $M_p = M \cap K_p$ (see Exercise 5.4.14(a)), so M_p is a closed subgroup of K_p. □

By Corollary 8.5.7, K in Theorem 14.1.9 is equivalently profinite. Each K_p is a pro-p-group in view of Corollary 11.6.8.

Remark 14.1.10. For an arbitrary compact abelian group K, the sum of the family of all closed hereditarily disconnected subgroups of K coincides with $td(K)$ and plays a prominent role (recall that $td(N) = N$ for all closed hereditarily disconnected subgroups N of K, by Exercise 8.7.11). Since $K_p = td_p(K) \subseteq td(K)$ for all primes p (by Corollary 11.6.9), $td(K)$ contains the sum $wtd(K)$ of all subgroups K_p (see Exercise 5.4.14(d)) and it is, like the subgroups K_p, functorial, so $td(\prod_{i\in I} K_i) = \prod_{i\in I} td(K_i)$, $td(H) = H \cap td(K)$, and $td(K/H) = (td(K) + H)/H$ for a closed subgroup H of K (see Exercises 14.5.3 and 5.4.14).

14.2 The structure of LCA groups

14.2.1 The subgroup of compact elements of an LCA group

Definition 14.2.1. For a topological group G, let $B(G)$ be the union of all compact subgroups of G. Moreover, G is *compactly covered* if $G = B(G)$.

Clearly, if G is compact, then $G = B(G)$, that is, G is compactly covered. On the other hand, a locally compact group G satisfies $G = B(G)$ precisely when G contains no subgroups isomorphic to $\mathbb{Z}$, according to Theorem 10.2.9.

Example 14.2.2. (a) For every $n \in \mathbb{N}$, $B(\mathbb{R}^n) = \{0\}$.
(b) If G is a proper subgroup of $\mathbb{T}$ properly containing $t(\mathbb{T})$, then $B(G) = t(\mathbb{T}) \neq G$. In particular, $B(G)$ is a proper dense subgroup of G. This shows that $B(G)$ need not be closed.

Remark 14.2.3. Let G be a topological group.
(a) For every closed subgroup H of G, $B(H) = B(G) \cap H$.
(b) If $f\colon G \to H$ is a continuous homomorphism, then $f(B(G)) \subseteq B(H)$.
(c) If $G = \prod_{i\in I} G_i$ for a family $\{G_i : i \in I\}$ of topological groups, then $B(G) = \prod_{i\in I} B(G_i)$.

The following lemma is a direct consequence of the definition.

Lemma 14.2.4. *Let G be a topological abelian group. Then $B(G)$ is a subgroup of G and $t(G) \subseteq B(G)$.*

In the nonabelian case, $B(G)$ may fail to be a subgroup of the topological group G. Indeed, $B(G) = t(G)$ in a discrete group G, so it suffices to take a discrete group G such that $t(G)$ is not a subgroup of G (e. g., $G = \mathbb{Z} \rtimes \mathbb{Z}(2)$).

Our aim is to prove in Corollary 14.3.10 that the conclusion of the following proposition characterizes the hereditarily disconnected locally compact abelian groups.

Proposition 14.2.5. *For a locally compact abelian group G and every prime p, the subgroup G_p is contained in $B(G)$. If G is hereditarily disconnected, then $B(G)$ is open and G_p is closed in G.*

Proof. Pick an element $x \in G_p$. If x is torsion, then $x \in B(G)$. Otherwise, the infinite cyclic subgroup $\langle x\rangle$ is nondiscrete, as $p^n x \to 0$. By Theorem 10.2.9, $\langle x\rangle$ is contained in $B(G)$.

If G is hereditarily disconnected, then G has a compact open subgroup K, by van Dantzig theorem 8.5.1. Since $K \subseteq B(G)$, this entails that $B(G)$ is open. Moreover, by Theorem 14.1.9, K_p is compact and hence $K_p = K \cap G_p$ (see Exercise 5.4.14(a)) is a compact open subgroup of G_p, so G_p is locally compact, in particular closed in G by Proposition 8.2.6. □

The locally compact abelian groups G with $B(G) = G$ and $c(G) = \{0\}$ are known also under the name *periodic* locally compact abelian groups:

Definition 14.2.6 ([169]). A locally compact abelian group G is *periodic* if G is hereditarily disconnected and compactly covered.

Van Dantzig theorem 8.5.1 implies that a locally compact abelian group G is periodic if and only if G has a hereditarily disconnected compact open subgroup K such that G/K is torsion and (necessarily) discrete. Therefore, for such a group G, the dual $\widehat{G}$ is periodic as well, since $A(K) \cong \widehat{G/K}$ is a hereditarily disconnected compact open subgroup of $\widehat{G}$ and $\widehat{G}/A(K) \cong \widehat{K}$ is torsion (see also Corollary 14.2.16).

Remark 14.2.7. Any two compact open subgroups K, K_1 of a (necessarily) locally compact abelian group are *commensurable*, namely, the indices $[K : K\cap K_1]$ and $[K_1 : K\cap K_1]$ are finite.

The next theorem, due to Braconnier (see [41]), describes the structure of the periodic locally compact abelian groups via local direct products.

Theorem 14.2.8. *For a periodic locally compact abelian group G, the subgroup G_p is closed for every prime p and $G \cong \prod^{\mathrm{loc}}_{p\in\mathbb{P}}(G_p, K_p)$, where $K = \prod_{p\in\mathbb{P}} K_p$ is a compact open subgroup of G.*

Proof. By van Dantzig theorem 8.5.1, G has a compact open subgroup K, which splits in a direct product $K = \prod_{p\in\mathbb{P}} K_p$, as in Theorem 14.1.9.

According to Proposition 14.2.5, the subgroup G_p is closed. Moreover, the openness of K in G yields that $G_p = \{x \in G : \exists n \in \mathbb{N},\ p^n x \in K_p\}$. In particular, G_p/K_p is a p-torsion group.

By Exercise 5.4.14(d) and Corollary 11.6.9, $wtd(G) = \bigoplus_{p\in\mathbb{P}} G_p$.

Let us see that $G = K + wtd(G)$. Indeed, fix $g \in G$. Since the group G/K is compactly covered and discrete, it is torsion. So, there exists $m = p_1^{k_1}\cdots p_s^{k_s} \in \mathbb{N}_+$ such that $mg \in K$, where $p_1,\dots,p_s$ are pairwise distinct primes, $k_1,\dots,k_s \in \mathbb{N}_+$, and $s \in \mathbb{N}_+$. Let $P =$

$\{p_1, \ldots, p_s\}$ and $M = \prod_{p \in \mathbb{P}\setminus P} K_p$; we prove that

$$g \in M + \sum_{p \in P} G_p \subseteq K + wtd(G).$$

Write $mg = h' + h''$, where $h' \in \prod_{p \in P} K_p$ and $h'' \in M$. Since $M = mM$ by Remark 5.3.7, we can write $h'' = mz$, with $z \in M$. Then $m(g - z) = h' \in \prod_{p \in P} K_p$. Letting $g_1 = g - z$, it suffices to show that $g_1 \in \sum_{p \in P} G_p$. Observe that in case $s = 1$ this means that $h' \in K_{p_1}$ and then $mg_1 = p_1^{k_1} g_1 = h' \in K_{p_1}$, so $g_1 \in G_{p_1}$.

Let $m_i = m/p_i^{k_i}$ for every $i \in \{1, \ldots, s\}$, so the greatest common divisor of $m_1, \ldots, m_s$ is 1, and hence $1 = \sum_{i=1}^s u_i m_i$ for suitable $u_1, \ldots, u_s \in \mathbb{Z}$. Therefore, $g_1 = \sum_{i=1}^s u_i m_i g_1$. The above discussed case $s = 1$ applies to show that $m_i g_1 \in G_{p_i}$ for every $i \in \{1, \ldots, s\}$, and this proves that $g_1 \in \sum_{p \in P} G_p$. Consequently, $g = z + g_1 \in M + \sum_{p \in P} G_p$, as required. □

14.2.2 The structure theory of LCA groups

Using the full power of the Pontryagin-van Kampen duality theorem one can prove the following structure theorem.

Theorem 14.2.9. *Let G be a compactly generated locally compact abelian group. Then $G \cong \mathbb{R}^n \times \mathbb{Z}^m \times K$, where $n, m \in \mathbb{N}$ and K is a compact abelian group.*

Proof. According to Proposition 11.6.2, there exists a compact subgroup K of G such that G/K is an elementary locally compact abelian group. Taking a bigger compact subgroup, one can get the quotient G/K to be of the form $\mathbb{R}^n \times \mathbb{Z}^m$ for some $n, m \in \mathbb{N}$. Now, by Proposition 13.5.5 and Theorem 13.3.5, $A(K) \cong \widehat{G/K} \cong \mathbb{R}^n \times \mathbb{T}^m$ is an open subgroup of $\widehat{G}$, as $A(K) = W(K, \Lambda_1)$ (by the definition of the compact-open topology). Since $A(K)$ is divisible, $\widehat{G} \cong \mathbb{R}^n \times \mathbb{T}^m \times D$ by Corollary A.2.7, and $D \cong \widehat{G}/A(K) \cong \widehat{K}$ (again by Proposition 13.5.5). Taking the duals, Pontryagin-van Kampen duality theorem 13.4.17 gives $G \cong \widehat{\widehat{G}} \cong \mathbb{R}^n \times \mathbb{Z}^m \times K$. □

Making sharper use of the annihilators, we describe in Theorem 14.2.18 the structure of locally compact abelian groups. Theorem 14.2.9 can be obtained as a corollary, but we need Theorem 14.2.9 for the proof of Theorem 14.2.18.

We first propose some consequences of Theorem 14.2.9.

Corollary 14.2.10. *If G is a locally compact abelian group, then $B(G)$ is closed.*

Proof. Let $x \in \overline{B(G)}$ and let H be a compactly generated open subgroup of G with $x \in H$. By Theorem 14.2.9, $H \cong \mathbb{R}^n \times \mathbb{Z}^m \times K$ for some compact subgroup K of G and $n, m \in \mathbb{N}$. Hence, $B(H) = K$ is closed in H.

Next we show that $\overline{B(G)} \cap H \subseteq \overline{B(G) \cap H}$. Indeed, pick $y \in \overline{B(G)} \cap H$ and a neighborhood W of y in G. Since $y \in H$ and H is open, we can assume that $W \subseteq H$. So, $W \cap (B(G) \cap H) = W \cap B(G) \neq \emptyset$. Hence, $y \in \overline{B(G) \cap H}$. This gives, also by Remark 14.2.3(a),

$x \in \overline{B(G)} \cap H \subseteq \overline{B(G) \cap H} = \overline{B(H)} = B(H) \subseteq B(G)$. So, $x \in B(G)$, and this proves that $B(G)$ is closed. □

Corollary 14.2.11. *Let G be a connected locally compact abelian group. Then:*
(a) *G is compactly generated, so $G \cong \mathbb{R}^n \times C$ for some connected compact abelian group C and $n \in \mathbb{N}$;*
(b) *$C = B(G)$ is the largest compact subgroup of G (so fully invariant) and n is the largest possible dimension of an affine subgroup V (i. e., $V \cong \mathbb{R}^m$ for some $m \in \mathbb{N}$) of G, in particular n is uniquely determined by G.*

Proof. (a) If U is a compact neighborhood of 0 in G, then U generates an open subgroup H of G that is obviously compactly generated. Since G is connected, $H = G$. From Theorem 14.2.9 we deduce that $G \cong \mathbb{R}^n \times C$ for some compact group C that is necessarily connected, as G is connected.

(b) Let $p: G \cong \mathbb{R}^n \times C \to \mathbb{R}^n$ be the canonical projection. By Remark 14.2.3, $p(B(G)) \subseteq B(\mathbb{R}^n) = \{0\}$, according to Example 14.2.2. Therefore, $B(G) \subseteq C$, and hence $C = B(G)$ is the largest compact subgroup of G.

Now assume that $V \cong \mathbb{R}^m$ is an affine (so, necessarily closed) subgroup of G. Since the kernel C of the projection p is compact, p is a closed map, by Lemma 8.2.2. Hence, $p(V)$ is a closed subgroup of $\mathbb{R}^n$. Since $C \cap V = \{0\}$, the restriction $p \restriction_V: V \to p(V)$ is a continuous isomorphism between locally compact abelian groups.

Since V is σ-compact, the open mapping theorem (Theorem 8.4.1) implies that $p \restriction_V$ is open. So, $p(V) \cong V \cong \mathbb{R}^m$ is a connected closed subgroup of $\mathbb{R}^n$. This implies $m \leq n$, by Theorem 9.2.2. □

Remark 14.2.12. (a) It follows from the above corollary that for an arbitrary locally compact abelian group G there is a highest dimension n for the affine subgroups $V \cong \mathbb{R}^n$ of G (it suffices to note that each V is necessarily contained in the closed subgroup $c(G)$, so Corollary 14.2.11 applies to $c(G)$).
(b) It follows from Exercise 13.7.13 and the above corollary that $a(G)$ is dense in $c(G)$ for every locally compact abelian group G.

Theorem 14.2.13. *If G is a locally compact abelian group, then*

$$c(G) = A_G(B(\widehat{G})) \quad \text{and} \quad B(G) = A_G(c(\widehat{G})).$$

Proof. Let $\{K_i : i \in I\}$ be the family of all compact subgroups of $\widehat{G}$. By definition, $B(\widehat{G}) = \sum_{i\in I} K_i$. By Remark 13.5.2(c), $\{A_G(K_i) : i \in I\}$ is precisely the family of all open subgroups of G, so Theorem 8.5.2(b) yields $\bigcap_{i\in I} A_G(K_i) = c(G)$. To conclude, Corollary 13.5.3 and Lemma 13.4.10(c) give

$$c(G) = \bigcap_{i\in I} A_G(K_i) = A_G\Big(\sum_{i\in I} K_i\Big) = A_G(B(\widehat{G})).$$

This equality and Theorem 13.4.17 imply that

$$c(\widehat{G}) = A_{\widehat{G}}(B(\widehat{\widehat{G}})) = A_{\widehat{G}}(\omega_G(B(G))) = A_{\widehat{G}}(B(G)).$$

Hence, as $B(G)$ is closed by Corollary 14.2.10, Remark 13.5.2(a) implies that $B(G) = A_G(A_{\widehat{G}}(B(G))) = A_G(c(\widehat{G}))$, as desired. □

Proposition 14.2.14. *Let G be a locally compact abelian group. Then $B(G)+c(G)$ is open and $B(G) \cap c(G)$ is the maximal connected compact subgroup of G.*

Proof. Let H be a compactly generated open subgroup of G. By Theorem 14.2.9, $H = R \oplus D \oplus K$ topologically, where K is a compact subgroup of G, $R \cong \mathbb{R}^n$, $D \cong \mathbb{Z}^m$ and $n, m \in \mathbb{N}$. Clearly, $R \oplus K$ is an open subgroup of H, hence of G as well. On the other hand, $R \subseteq c(G)$ and $K \subseteq B(G)$. So, $R \oplus K$ is an open subgroup of G contained in $c(G) + B(G)$. Therefore, $c(G) + B(G)$ is open.

By Corollary 14.2.11, $c(G) = S \oplus C$, where $S \cong \mathbb{R}^m$ and $C = B(c(G))$ is a connected compact subgroup of $c(G)$. As $C = B(G) \cap c(G)$ (see Remark 14.2.3(a)) obviously contains all connected compact subgroups of G, we are done. □

Corollary 14.2.15. *For K a compact abelian group, $c(K) = A(t(\widehat{K}))$.*

Proof. Now $\widehat{K}$ is discrete by Proposition 13.1.1(a), hence $B(\widehat{K}) = t(\widehat{K})$, so Theorem 14.2.13 applies. □

From Theorem 14.2.13 and Corollary 14.2.10, we obtain also the following characterization of hereditarily disconnected locally compact abelian groups as those locally compact abelian groups with compactly covered dual group, or vice versa, a characterization of compactly covered locally compact abelian groups as those locally compact abelian groups with hereditarily disconnected dual group.

Corollary 14.2.16. *A locally compact abelian group G is hereditarily disconnected (respectively, compactly covered) if and only if $\widehat{G}$ is compactly covered (respectively, hereditarily disconnected). In particular, G is periodic if and only if $\widehat{G}$ is periodic.*

The next is the last step before proving Theorem 14.2.18.

Lemma 14.2.17. *Let G be a locally compact abelian group and let K, L be closed subgroups of G such that algebraically G is the direct sum of the subgroups K and L and K is compactly generated. Then the topology on G coincides with the product topology of $K \oplus L$.*

Proof. Let C be a compact subset of K that generates K and let U be a compact neighborhood of 0 in G. Then $H = \langle C + U \rangle$ is a compactly generated open subgroup of G containing K and algebraically $H = K \oplus (L \cap H)$, as $K \subseteq H$.

Since the inclusion maps of K and $L \cap H$ in H are continuous, the identity isomorphism $id_H: K \oplus (L \cap H) \to H$ is continuous when $K \oplus (L \cap H)$ is endowed with the product topology. As K and $L \cap H$ (as closed subgroups of H) are σ-compact, $id_H: K \oplus (L \cap H) \to H$

is open by the open mapping theorem (Theorem 8.4.1). Therefore, also the identity map $id_G : K \oplus L \to G$ is a topological isomorphism, as $L \cap H$ is open in L. □

Theorem 14.2.18. *If G is a locally compact abelian group, then $G \cong \mathbb{R}^n \times G_0$, where $n \in \mathbb{N}$ and G_0 is a closed subgroup of G containing a compact open subgroup.*

Proof. By Corollary 14.2.11(a), there exist connected compact subgroups C of G and K of $\widehat{G}$ and closed subgroups $R \cong \mathbb{R}^n$ of G and $S \cong \mathbb{R}^m$ of $\widehat{G}$ such that $c(G) = R \oplus C$ and $c(\widehat{G}) = S \oplus K$ topologically. We verify that

$$R + A_G(S) = G \quad \text{and} \quad S + A_{\widehat{G}}(R) = \widehat{G}. \tag{14.1}$$

Observe first that $S + A_{\widehat{G}}(R)$ is an open subgroup of $\widehat{G}$. Indeed, $R \subseteq c(G)$ and hence Theorem 14.2.13 implies $B(\widehat{G}) = A_{\widehat{G}}(c(G)) \subseteq A_{\widehat{G}}(R)$. So, $c(\widehat{G}) + A_{\widehat{G}}(R) \supseteq c(\widehat{G}) + B(\widehat{G})$, which is an open subgroup according to Proposition 14.2.14. Analogously, $R + A_G(S)$ is an open subgroup of G. By Example 14.2.2 and Remark 14.2.3, $\{0\} = B(R) = B(G) \cap R$. By Corollary 13.5.3 and Theorem 14.2.13, $\widehat{G} = c(\widehat{G}) + A_{\widehat{G}}(R)$, as the latter subgroup is open and hence closed. Analogously, $G = c(G) + A_G(S)$. By Theorem 14.2.13 and the compactness of C,

$$C \subseteq B(G) = A_G(c(\widehat{G})) \subseteq A_G(S),$$

where the last inclusion follows from $S \subseteq c(\widehat{G})$, by Lemma 13.4.10(b). Consequently,

$$R + A_G(S) \supseteq R + C + A_G(S) = c(G) + A_G(S) = G,$$

and this proves the first equality in (14.1). The second equality is proved similarly.

By Corollary 13.5.3, (14.1) yields $A_{\widehat{G}}(R) \cap S = \{0\}$ and $A_G(S) \cap R = \{0\}$. Consequently,

$$R \oplus A_G(S) = G \quad \text{and} \quad S \oplus A_{\widehat{G}}(R) = \widehat{G}.$$

By Lemma 14.2.17, G and $\widehat{G}$ are endowed with the respective product topology, since R and S are compactly generated. By Remark 14.2.12 and by the choice of R, this implies that $A_G(S)$ contains no copies of $\mathbb{R}$.

It remains to prove that $A_G(S)$ admits a compact open subgroup. Let H be a compactly generated open subgroup of $A_G(S)$; as $A_G(S)$ contains no copies of $\mathbb{R}$, Theorem 14.2.9 implies that $H \cong \mathbb{Z}^k \times H_1$, where H_1 is a compact subgroup of H and $k \in \mathbb{N}$. Since H_1 is open in H, it is open also in $A_G(S)$. □

This is the strongest structure theorem concerning locally compact abelian groups. The affine subgroup $\mathbb{R}^n$ is not uniquely determined (one can see that in $\mathbb{R} \times \mathbb{T}$), but its dimension n is. According to [9, p. 595], the group G_0 is unique up to topological isomorphism. If $n = 0$, we say that G is *line-free*.

According to Corollary 11.4.5, $w(K) = |\widehat{K}| = w(\widehat{K})$ for a compact abelian group K. Now we extend this equality between weights.

Corollary 14.2.19. *If G is a locally compact abelian group, then $w(G) = w(\widehat{G})$.*

Proof. Since this equality is obviously true for finite groups, we assume that G is infinite. By Theorem 14.2.18, $G \cong \mathbb{R}^n \times G_0$, where G_0 is a closed subgroup of G containing a compact open subgroup K. Then $\widehat{G} \cong \mathbb{R}^n \times \widehat{G_0}$. So, it suffices to prove that $w(G_0) = w(\widehat{G_0})$ in view of Theorem 5.1.15.

As $A_{\widehat{G_0}}(K) \cong \widehat{G_0/K}$ by Proposition 13.5.5 and it is a compact open subgroup of $\widehat{G_0}$ (by Remark 13.5.2(c)), $\widehat{G_0}/A_{\widehat{G_0}}(K) \cong \widehat{K}$ is discrete. Next we use the fact that, for an open subgroup H_0 of a topological group H, one has $w(H) = w(H_0) \cdot |H/H_0|$. Applying this to the pairs $(A_{\widehat{G_0}}(K), \widehat{G_0})$ and (K, G_0), we get

$$w(\widehat{G_0}) = w(A_{\widehat{G_0}}(K)) \cdot |\widehat{K}| = |G_0/K| \cdot w(K) = w(G_0),$$

since $w(K) = |\widehat{K}|$ and $w(A_{\widehat{G_0}}(K)) = |G_0/K|$, by Corollary 11.4.5. □

As another consequence of Theorem 14.2.18, one obtains:

Corollary 14.2.20. *Every locally compact abelian group is topologically isomorphic to a closed subgroup of a group of the form $\mathbb{R}^n \times D \times C$, where $n \in \mathbb{N}$, D is a discrete divisible abelian group, and C is a compact abelian group.*

Proof. Let $G \cong \mathbb{R}^n \times G_0$ with n, G_0 and K as in Theorem 14.2.18. By Corollary 11.5.2, there exist a cardinal κ and an embedding $j: K \to \mathbb{T}^\kappa$. Since $\mathbb{T}^\kappa$ is divisible, one can extend j to a homomorphism $j_1: G_0 \to \mathbb{T}^\kappa$, which is continuous by the continuity of j and by the openness of K in G_0.

Since G_0/K is discrete, there exists an injective homomorphism $j_2: G_0/K \to D$ with D a discrete divisible abelian group. Then the diagonal map $f = (j_1, j_2 \circ \pi): G_0 \to \mathbb{T}^\kappa \times D$, where $\pi: G_0 \to G_0/K$ is the canonical projection, is injective and continuous. Since K is compact, the restriction of f to K is an embedding, by the open mapping theorem (Theorem 8.4.1). Since K is open in G_0, this yields that $f: G_0 \to \mathbb{T}^\kappa \times D$ is an embedding. This provides an embedding v of $G \cong \mathbb{R}^n \times G_0$ into the group $\mathbb{R}^n \times \mathbb{T}^\kappa \times D$. The image $v(G) \cong G$ is a closed subgroup of $\mathbb{R}^n \times \mathbb{T}^\kappa \times D$ since locally compact groups are complete (see Propositions 8.2.6 and 7.1.22). □

14.3 Topological features of LCA groups

14.3.1 Dimension of locally compact groups

In the sequel we discuss the dimension of a locally compact abelian group (see [78, 254, 255] for further information on dimension theory for topological groups). There are three major dimension functions for a topological space X: the covering dimension $\dim X$, the small inductive dimension $\operatorname{ind} X$, and the large inductive dimension $\operatorname{Ind} X$ (see Definition 14.3.1). We are not going to define the covering dimension here, since according to a well-known theorem of Pasynkov [225], in the realm of locally compact groups X,

$$\dim X = \operatorname{Ind} X = \operatorname{ind} X. \tag{14.2}$$

The small inductive dimension ind was introduced by Urysohn and Menger. The intuitive idea behind it is that a nonempty topological space X has $\operatorname{ind} X = 0$ precisely when it is zero-dimensional in the sense of Definition B.6.6(b); the dimension function $\operatorname{Ind} X$ is introduced similarly. Here are the formal definitions:

Definition 14.3.1. Let X be a topological space.
(i) Put $\operatorname{ind} X = -1$ for $X = \emptyset$, and for $n \in \mathbb{N}_+$ put $\operatorname{ind} X \leq n$ if X has a base of open sets U, such that $\operatorname{ind} \operatorname{Fr}(U) < n$.
(ii) Put $\operatorname{Ind} X = -1$ for $X = \emptyset$, and for $n \in \mathbb{N}$ put $\operatorname{Ind} X \leq n$ if every closed set F of X has a base of open sets U such that $\operatorname{ind} \operatorname{Fr}(U) < n$.

Then $\operatorname{ind} X = n$ for some $n \in \mathbb{N}$, if $\operatorname{ind} X \leq n$ and $\operatorname{ind} X \nleq n-1$; if $\operatorname{ind} X \nleq n$ for all $n \in \mathbb{N}$, we put $\operatorname{ind} X = \infty$. Analogously for $\operatorname{Ind} X$.

Remark 14.3.2. For a topological space X and $n \in \mathbb{N}$, note that $\operatorname{ind} X \leq n$ is a local property: if every point $x \in X$ has a neighborhood U with $\operatorname{ind} U \leq n$ then $\operatorname{ind} X \leq n$.

Obviously, $\operatorname{ind} X \leq \operatorname{Ind} X$ for every T_1-space X. Moreover, $\dim X \leq \operatorname{Ind} X$ for a normal space X, and furthermore $\operatorname{ind} X \leq \operatorname{Ind} X = \dim X$ for every metrizable space X, according to a theorem of Katětov. While P. S. Alexandrov established that every compact Hausdorff space X satisfies $\dim X \leq \operatorname{ind} X \leq \operatorname{Ind} X$ (see [134]). In particular, (14.2) holds when X is a compact metric space. Moreover, according to a theorem of Urysohn (see [134]), the equality (14.2) is available for all separable metric spaces, in particular for all spaces considered in the next example.

Example 14.3.3. (a) It is easy to see that $\operatorname{ind} \mathbb{R} = 1$ and $\operatorname{ind} \mathbb{R}^n \leq n$ for every $n \in \mathbb{N}_+$. By Remark 14.3.2, $\operatorname{ind} \mathbb{T}^n \leq n$ for every $n \in \mathbb{N}_+$, since $\mathbb{T}^n$ and $\mathbb{R}^n$ are locally homeomorphic.
(b) The highly nontrivial equality $\dim \mathbb{R}^n = n$ is due to Lebesgue who introduced the covering dimension. Since $\mathbb{T}^n$ and $\mathbb{R}^n$ are locally homeomorphic, we deduce that $\dim \mathbb{T}^n = n$.
(c) The Menger–Nöbeling theorem from 1932 (see [134]) states that if X is a separable compact metric space with $\dim X = n$, then it embeds as a subspace in $\mathbb{R}^{2n+1}$.

Since all locally compact groups satisfy (14.2), from now on we use only $\dim G$ for such groups. Recall that for such groups zero-dimensionality is equivalent to hereditary disconnectedness (see Vedenissov theorem B.6.10).

The following classical additivity result for the covering dimension turns out to be crucial for the dimension theory of locally compact groups. We refer to [217] for a proof. Here, G/H denotes the quotient space, that need not be a group if the subgroup H is not normal.

Theorem 14.3.4. *Let G be a locally compact group and H a closed subgroup of G. Then $\dim G = \dim G/H + \dim H$. In particular, if K is another locally compact group, then $\dim(G \times K) = \dim G + \dim K$.*

Theorem 14.3.4 implies that the dimension function is monotone decreasing with respect to taking quotients and closed subgroups. In items (a) and (c) of the next remark, we discuss two extremal values of the dimension.

Remark 14.3.5. (a) By van Dantzig theorem 8.5.1, hereditarily disconnected locally compact groups are zero-dimensional and have a local base of compact open subgroups (local compactness plays a relevant role here, since $\dim \mathbb{Q}/\mathbb{Z} = 0$, yet $\mathbb{Q}/\mathbb{Z}$ does not carry a linear topology, see Example 6.1.8).

(b) Theorem 14.3.4 yields $\dim G = \dim c(G)$ for a locally compact group G, as $\dim G/c(G) = 0$. Thus, the dimension theory of locally compact groups is worth studying in connected groups.

(c) In Theorem 14.3.4, the locally compact group G is not necessarily finite-dimensional, so it implies that $\dim G = \infty$ precisely when (at least) one of G/H and H, where H is a closed subgroup of G, is infinite-dimensional.
According to [51, Theorem 1] (see also [51, Remarks (ii)]), a locally compact group G is homeomorphic to $\mathbb{R}^n \times K \times D$, where K is a compact subgroup of G and D is a discrete space. Since dimension is invariant under homeomorphisms, this (in conjunction with (b)) shows that $\dim G = \dim(\mathbb{R}^n \times K \times D) = \dim(\mathbb{R}^n \times K) = n + \dim K$. Hence, a locally compact group is infinite-dimensional if and only if it contains an infinite-dimensional compact subgroup. This is why we mainly concentrate on the compact case.

(d) In order to better exploit Theorem 14.3.4 also in the case of infinite-dimensional connected compact groups K, only for such groups K we put $\dim K := w(K)$ (e. g., $\dim \mathbb{T}^\kappa = \kappa$ for an arbitrary infinite cardinal κ).
For every compact connected group K, the group $K/Z(K)$ is center-free and $K/Z(K) \cong K'/(Z(K) \cap K') \cong \prod_{i\in I} L_i$, where I is either finite (when K' is a Lie group), or infinite with $|I| = w(K)$ and each L_i is a simple connected compact Lie group. Moreover, $K = K'c(Z(K))$.

(e) With K and $\{L_i : i \in I\}$ as in (d), denote by $\widetilde{L}_i$ the covering group of L_i. This is a connected compact Lie group with finite center $Z(\widetilde{L}_i)$ such that $\widetilde{L}_i/Z(\widetilde{L}_i) \cong L_i$. Moreover, for $L = \prod_{i\in I} \widetilde{L}_i$ there exists a closed totally disconnected subgroup N of $c(Z(K)) \times Z(L)$, such that $K \cong (c(Z(K)) \times Z(L))/N$ and $K' \cong L/L \cap N$.

Theorem 14.3.6. *Let K be a compact abelian group. Then $\dim K = r_0(\widehat{K})$ and:*

(a) *for every hereditarily disconnected closed subgroup H of K such that $K/H \cong \mathbb{T}^\kappa$, one has $\kappa = \dim K = \dim K/H$;*

(b) *there exists a subgroup H of K as in (a), so $r_0(K) \geq \max\{\mathfrak{c}, 2^{\dim K}\} = \max\{\mathfrak{c}, 2^{r_0(\widehat{K})}\}$.*

Proof. For H as in (a), one has $\dim H = 0$, by Remark 14.3.5(a). So, Theorem 14.3.4 applied to K and H gives $\dim K = \dim K/H = \dim \mathbb{T}^\kappa = \kappa$, where the last equality comes from Remark 14.3.5(d).

Next we put $G = \widehat{K}$ and $\rho = r_0(G)$, and we prove (b) and $\rho = \kappa$. Fix a free subgroup $F \cong \mathbb{Z}^{(\rho)}$ of G such that $Y = G/F$ is torsion. By Lemma 13.4.10(a), $H = A_K(F)$ is a closed subgroup of K such that, in view of Corollary 13.4.14, $K/H \cong \widehat{F} \cong \mathbb{T}^\rho$. By Corollary 11.6.7, $H \cong \widehat{Y}$ is hereditarily disconnected. This proves the first assertion in (b). By the conclusion of item (a), $\rho = \dim K = \kappa$.

The second assertion in (b) follows from $r_0(K) \geq r_0(\mathbb{T}^\rho) \geq \max\{\mathfrak{c}, 2^\rho\}$. □

The next corollary is covered by the more general Theorem 14.3.8, yet we prefer to give a direct proof in this case:

Corollary 14.3.7. *A nontrivial finite-dimensional connected compact abelian group K is metrizable, $|K| = \mathfrak{c}$, and $r_p(K) \leq \dim K$ for every prime p.*

Proof. Let $n = \dim K$ and $G = \widehat{K}$. Then G is torsion-free by Proposition 11.6.10(b) with $r_0(G) = n$ by Theorem 14.3.6. Hence, $D(G) \cong \mathbb{Q}^n$ by Lemma A.2.14, and in particular G is countable. Thus, K is metrizable by Corollary 13.1.3(c). Moreover, $|K| = 2^{|G|} = \mathfrak{c}$, by Theorem 13.3.11.

Since $A_G(K[p]) = pG$, by Lemmas 13.4.10(e) and 13.5.1, we deduce from Corollary 13.4.14 that $\widehat{K[p]} \cong G/pG$. As G is a torsion-free abelian group of rank n, $r_p(G/pG) \leq n$. Indeed, if $g_1, \dots, g_{n+1} \in G$, then there exist $k_1, \dots, k_{n+1} \in \mathbb{Z}$ not all 0 such that $\sum_{j=1}^{n+1} k_j g_j = 0$. We may assume that p does not divide all k_j. Then $0 = \sum_{j=1}^{n+1} k_j g_j \in pG$, and so $g_1 + pG, \dots, g_{n+1} + pG$ are dependent in G/pG. Hence, $r_p(G/pG) \leq n$ and each G/pG is finite. By Example 13.3.3(a), $K[p] \cong \widehat{K[p]} \cong G/pG$ is finite as well, and we conclude that $r_p(K) = r_p(K[p]) = r_p(G/pG) \leq n$. □

Theorem 14.3.8. *Let K be a nontrivial connected compact abelian group with $\sigma = \dim K$.*

(a) *There exists a continuous surjective homomorphism $q\colon \widehat{\mathbb{Q}}^\sigma \to K$ with $N = \ker q$ hereditarily disconnected and $N_p \cong \mathbb{J}_p^{\gamma_p}$ for cardinals $\gamma_p \leq \sigma$ for every prime p, consequently, $r_0(K) = |K| = 2^{w(K)}$.*
(b) *For every q as in (a), $\gamma_p = r_p(K)$, if $\gamma_p < \infty$, otherwise $2^{\gamma_p} = r_p(K)$.*

Proof. (a) Since K is connected, $G = \widehat{K}$ is discrete and torsion-free by Proposition 13.1.1(a) and Proposition 11.6.10(a), and $\beta := |G| = w(K)$ according to Corollary 11.4.5. By Theorem 14.3.6, $r_0(G) = \sigma$, so G contains a subgroup $F \cong \mathbb{Z}^{(\sigma)}$ with G/F torsion and $\beta = \max\{\sigma, \omega\}$. The divisible hull of G is $D(G) \cong \mathbb{Q}^{(\sigma)}$, by Lemma A.2.14. Put

$$L = \widehat{D(G)} \cong \widehat{\mathbb{Q}}^\sigma,$$

so $w(L) = |\mathbb{Q}^{(\sigma)}| = \beta$ and L is a compact torsion-free abelian group by Proposition 13.1.1(b) and Exercise 13.7.3(c). We identify in the sequel $\widehat{G}$ with K in view of

Pontryagin-van Kampen duality theorem 13.4.17. Let us see that the dual homomorphism $q: L \to K = \widehat{G}$ of the inclusion $G \to D(G)$, which is obviously continuous and surjective, is the desired surjective homomorphism. Indeed, the group $D(G)/G$ is torsion and divisible, hence $D(G)/G \cong \bigoplus_{p\in\mathbb{P}} \mathbb{Z}(p^\infty)^{(\gamma_p)}$, where

$$\gamma_p = r_p(D(G)/G) \le r_p(D(G)/F) = \sigma \le \beta \tag{14.3}$$

by Proposition A.4.13. Further, $N := \ker q \cong \widehat{D(G)/G}$ is hereditarily disconnected by Corollary 11.6.7, and $N \cong \prod_{p\in\mathbb{P}} N_p$ by Theorems 13.3.5 and 14.1.9, where $N_p \cong \mathbb{J}_p^{\gamma_p}$ for every prime p.

Since L is torsion-free, $r_0(L) = |L| = 2^{w(L)} = 2^\beta$, by Remark 13.3.13. By Theorem 14.3.6, $r_0(K) \ge 2^\beta$, and K is a quotient of L, so we get $2^\beta = r_0(L) \ge r_0(K) \ge 2^\beta$. Hence, $r_0(K) = 2^\beta = |K|$, by Theorem 13.3.11.

(b) For L, q, and N as in (a), put $Y_p = \{y \in L: py \in N_p\}$ for every prime p. We prove first that

$$K[p] = q(Y_p) \cong Y_p/N_p \cong N_p/pN_p,$$

so $r_p(K) = r_p(N_p/pN_p)$. Indeed, the inclusion $q(Y_p) \subseteq K[p]$ is trivial. To check the opposite inclusion, pick $x \in K[p]$. Then $x = q(y)$ for some $y \in L$ with $py \in N = \prod_{p\in\mathbb{P}} N_p$. Let

$$M = \prod_{r\in\mathbb{P}\setminus\{p\}} N_r, \quad \text{so that } N = N_p \times M.$$

Then we can write $py = (y_r)_{r\in\mathbb{P}} \in N$ as $py = y_p + z$, where $y_p \in N_p$ and $z = (z_r)_{r\in\mathbb{P}\setminus\{p\}} \in M$. Since N_r is p-divisible for all primes $r \neq p$ by Theorem 14.1.9 and Remark 5.3.7, M is p-divisible as well. So, $z = pu$ for some $u \in M$ and $py = pu + y_p$. This means that $y_p = p(y-u)$, so $y - u \in Y_p$. As $q(u) = 0$, we have $x = q(y) = q(y-u) \in q(Y_p)$. This proves that $K[p] = q(Y_p)$.

Our next aim is to compute $\ker q \restriction_{Y_p} = Y_p \cap N$. Clearly, $Y_p \cap N \supseteq N_p$. On the other hand, $p(Y_p \cap N) \subseteq N_p$ and $pN \cap N_p = pN_p$, since N_p is a direct summand of N. Hence, $p(Y_p \cap N) \le pN \cap N_p = pN_p$. As L is torsion-free, we deduce that $Y_p \cap N \subseteq N_p$, so $\ker q \restriction_{Y_p} = Y_p \cap N = N_p$.

Combining the equalities $\ker q \restriction_{Y_p} = N_p$ and $K[p] = q(Y_p)$, we obtain that $K[p] \cong Y_p/N_p$ (since q is continuous and Y_p is compact, Frobenius theorem 3.2.3 and the open mapping theorem (Theorem 8.4.1) imply that this isomorphism is topological). Now it only remains to note that $pY_p = N_p$, as L is divisible; moreover, the injective continuous homomorphism $\mu_p: L \to L$, $x \mapsto px$, induces a topological isomorphism $j: Y_p \to N_p$, which induces in turn a topological isomorphism $Y_p/N_p \cong pY_p/pN_p = N_p/pN_p \cong \mathbb{Z}(p)^{\gamma_p}$. Hence,

$$r_p(K) = r_p(N_p/pN_p) = r_p(\mathbb{Z}(p)^{\gamma_p}).$$

Therefore, either $r_p(K) = \gamma_p < \infty$, or $r_p(K) = 2^{\gamma_p}$ when $\gamma_p \ge \omega$. □

In the sequel we discuss properties of the subgroup G_p, first in case the group G is connected and compact, and then for G hereditarily disconnected and locally compact abelian.

Theorem 14.3.9. *For a connected compact abelian group K, the subgroup K_p is dense and proper for every prime p.*

Proof. Let us start with $K = \widehat{\mathbb{Q}}$, which has $\dim K = r_0(\mathbb{Q}) = 1$ in view of Theorem 14.3.6. Since K is divisible, for every prime p the group K_p is divisible by Remark 5.3.7; so, $H := \overline{K_p}$ is divisible as well (see Exercise 3.5.18). There exists a continuous surjective homomorphism $q: K \to \mathbb{T}$, by Example 13.3.14(a). By Exercise 14.5.4(b), $q(K_p) = \mathbb{T}_p \neq \{0\}$, hence $H \neq \{0\}$ is a nontrivial connected compact abelian group by Proposition 14.1.1, and consequently $\dim H > 0$. By Theorem 14.3.4, $\dim K/H = 0$, as $\dim K = 1$ and $\dim H > 0$. Since the quotient K/H is connected, we deduce that $H = K$, i. e., K_p is dense in K.

Now let $K = \widehat{\mathbb{Q}}^\sigma$ for a cardinal σ. According to Exercise 5.4.14 and Corollary 11.6.9, $K_p = ((\widehat{\mathbb{Q}})_p)^\sigma$, hence K_p is dense in K as a consequence of the fact above that $(\widehat{\mathbb{Q}})_p$ is dense in $\widehat{\mathbb{Q}}$.

In the general case, if $\sigma = \dim K$, there exists a continuous surjective homomorphism $q: \widehat{\mathbb{Q}}^\sigma \to K$, by Theorem 14.3.8. Since $L := ((\widehat{\mathbb{Q}})_p)^\sigma$ is dense in $\widehat{\mathbb{Q}}^\sigma$ and $q(L) \subseteq K_p$ by Exercise 5.4.14, we conclude that K_p is dense in K. By Exercise 5.4.14(d) and Corollary 11.6.9, the sum of all K_p is direct, hence the density of each K_p yields properness. □

The above property does not hold in the noncompact case: for example, $\mathbb{R}_p = \{0\}$ for every prime p.

Corollary 14.3.10. *A locally compact abelian group G is hereditarily disconnected if and only if G is line-free and all subgroups G_p are closed.*

Proof. We can assume without loss of generality that G contains no copies of $\mathbb{R}$, so that G has a compact open subgroup K by Theorem 14.2.18.

If G is hereditarily disconnected, then each G_p is closed by Proposition 14.2.5. If G is not hereditarily disconnected, then K is not hereditarily disconnected either. Hence, $c(K)_p$ is a proper dense subgroup of $c(K)$ by Theorem 14.3.9. As $c(K)_p = G_p \cap c(K)$, and $c(K)_p$ is not closed in $c(K)$, we deduce that G_p is not closed in G. □

The compact group $\widehat{\mathbb{Q}}$ is closely related to the Adele ring $\mathbf{A}_{\mathbb{Q}}$ of the field $\mathbb{Q}$, the subring of $\mathbb{R} \times \prod_{p\in\mathbb{P}} \mathbb{Q}_p$ defined by $\mathbf{A}_{\mathbb{Q}} := \mathbb{R} \times \prod^{\mathrm{loc}}_{p\in\mathbb{P}}(\mathbb{Q}_p, \mathbb{J}_p)$ (more details can be found in [94, 117, 205, 289]).

Theorem 14.3.11. *The diagonal subgroup $Q := \{(r, (r)_{p\in\mathbb{P}}) \in \mathbf{A}_{\mathbb{Q}} : r \in \mathbb{Q}\}$ of $\mathbf{A}_{\mathbb{Q}}$ is discrete and $\mathbf{A}_{\mathbb{Q}}/Q \cong \widehat{\mathbb{Q}}$ is compact.*

Proof. Let $K = \mathbf{A}_{\mathbb{Q}}/Q$ and let $q\colon \mathbf{A}_{\mathbb{Q}} \to K$ be the canonical projection. We show first that K is compact.

To this end, consider the closed subgroup $N = \{0\} \times \prod^{\mathrm{loc}}_{p\in\mathbb{P}}(\mathbb{Q}_p, \mathbb{J}_p)$ of $\mathbf{A}_{\mathbb{Q}}$, its compact subgroup $H = \{0\} \times \prod_{p\in\mathbb{P}} \mathbb{J}_p$ and $u = (0, (1_p)_{p\in\mathbb{P}}) \in H$. Then the subgroup $\mathbb{Q} \cdot u = \{(0, (r)_{p\in\mathbb{P}}) : r \in \mathbb{Q}\}$ is contained in N, since for $r = n/m \in \mathbb{Q}$ one has $r \cdot 1_p \in \mathbb{J}_p$ for all primes $p \nmid m$. Moreover, a standard proof using the Chinese remainder theorem shows that $\langle u\rangle$ is dense in H. This implies that

$$\mathbb{Q} \cdot u + H = N \quad \text{and} \quad \mathbb{Q} \cdot u \text{ is dense in } N.$$

Indeed, for $x \in N$ there exists $m \in \mathbb{N}_+$ with $mx \in H$. As mH is open and $\langle u\rangle$ is dense in H, there exists $k \in \mathbb{Z}$ such that $ku \in mx + mH$, i. e., $mx \in ku + mH$. Pick a $\xi \in H$ with $mx = ku + m\xi$, then $x = (k/m)u + \xi \in \mathbb{Q} \cdot u + H$. This proves the equality $N = \mathbb{Q} \cdot u + H$, which yields the density of $\mathbb{Q} \cdot u$ in N, as $\langle u\rangle$ is dense in H.

Consider the open neighborhood $U = ((-1/2, 1/2) \times \{0\}) + H$ of 0 in $\mathbf{A}_{\mathbb{Q}}$. Since $\mathbb{Q} \cap \mathbb{J}_p = \left\{\frac{k}{m} : k \in \mathbb{Z}, p \nmid m\right\}$ for every prime p, we get $U \cap Q = \{0\}$, so the subgroup Q is discrete, hence closed in $\mathbf{A}_{\mathbb{Q}}$. Thus, $L := Q + H$ is closed in $\mathbf{A}_{\mathbb{Q}}$, by Lemma 8.2.1(a). Since $\mathbb{Q} \times \{0\} + \mathbb{Q} \cdot u = (\mathbb{Q} \times \{0\}) + Q$, one has

$$N \subseteq (\mathbb{Q} \times \{0\}) + N = (\mathbb{Q} \times \{0\}) + \mathbb{Q} \cdot u + H = (\mathbb{Q} \times \{0\}) + Q + H = (\mathbb{Q} \times \{0\}) + L.$$

Therefore, $\mathbf{A}_{\mathbb{Q}} = (\mathbb{R} \times \{0\}) + N = (\mathbb{R} \times \{0\}) + L$. As $Q \cap H = \{0\}$, $q(L) = q(H) \cong H$, and so $q(L)$ is a compact subgroup of K; moreover,

$$K/q(L) \cong (\mathbf{A}_{\mathbb{Q}}/Q)/(L/Q) \cong \mathbf{A}_{\mathbb{Q}}/L \cong (\mathbb{R} \times \{0\}) + L/L \cong \mathbb{R} \times \{0\}/(\mathbb{R} \times \{0\}) \cap L \cong \mathbb{R}/\mathbb{Z},$$

where the isomorphism $\mathbb{R}\times\{0\}/(\mathbb{R}\times\{0\})\cap L \to \mathbf{A}_{\mathbb{Q}}/L \cong (\mathbb{R}\times\{0\})+L/L$ is topological. Indeed, it is continuous, by Theorem 3.2.8(c), and it is open since the domain is compact. This proves that $K/q(L)$ is compact. Therefore, K is compact, by Lemma 8.2.3(b).

It remains to see that $K \cong \widehat{\mathbb{Q}}$. By using Pontryagin-van Kampen duality theorem 13.4.17, it is enough to check that $\widehat{K} \cong \mathbb{Q}$.

Next we prove that K is connected. To this end, we note that $(\mathbb{R}\times\{0\})+Q$ is a dense subgroup of $\mathbf{A}_{\mathbb{Q}}$ because it contains the dense subgroup $\mathbb{Q} \cdot u$ of N. Since $q((\mathbb{R} \times \{0\}) + Q) = q(\mathbb{R}\times\{0\})$, this is a dense connected subgroup of K, so K is connected as well. By Proposition 11.6.10(a), $\widehat{K}$ is torsion-free.

Moreover, since Q is discrete, Theorem 14.3.4 gives $\dim K = \dim \mathbf{A}_{\mathbb{Q}} = \dim c(\mathbf{A}_{\mathbb{Q}}) = \dim \mathbb{R} = 1$. Hence, $r_0(\widehat{K}) = 1$, by Theorem 14.3.6.

Finally, to show that $K = \mathbf{A}_{\mathbb{Q}}/Q$ is torsion-free, we consider the following algebraic isomorphisms: the subgroup $B := (Q + N)/Q \cong N$ of K is torsion-free and $K/B = (\mathbf{A}_{\mathbb{Q}}/Q)/(Q + N)/Q \cong ((\mathbb{R} \times \{0\}) + N)/((\mathbb{Q} \times \{0\}) + N) \cong \mathbb{R}/\mathbb{Q}$ is torsion-free as well. By Exercise 13.7.3(c), $\widehat{K}$ is divisible.

Since we have seen that the discrete abelian group $\widehat{K}$ is torsion-free, divisible and of rank 1, we deduce that $\widehat{K} \cong \mathbb{Q}$. □

Example 14.3.12. This theorem allows us to consider the quotient $K = \mathbf{A}_{\mathbb{Q}}/Q$ as the dual of $\mathbb{Q}$. According to Remark 8.5.11(b), $q(a(\mathbf{A}_{\mathbb{Q}})) = a(K)$. As $a(\mathbf{A}_{\mathbb{Q}}) = c(\mathbf{A}_{\mathbb{Q}}) = \mathbb{R}\times\{0\}$ and $q \restriction_{\mathbb{R}\times\{0\}}$ is injective, this entails that $a(K) = q(\mathbb{R}\times\{0\})$ is a continuous isomorphic image of $\mathbb{R}$. For the surjective continuous homomorphism $q\colon \mathbf{A}_{\mathbb{Q}} \to K$, one has $q(c(\mathbf{A}_{\mathbb{Q}})) \neq c(K) = K$. Indeed, otherwise $\mathbb{R} \to K$, $x \mapsto q(x,(0))$, would be a continuous isomorphism and by the open mapping theorem (Theorem 8.4.1) a topological isomorphism. (Compare this with Corollary 8.5.10.)

14.3.2 The Halmos problem: the algebraic structure of compact abelian groups

Halmos [163] noticed that the compact group $\widehat{\mathbb{Q}}$, being divisible and torsion-free with $|\widehat{\mathbb{Q}}| = \mathfrak{c}$, is algebraically isomorphic to $\mathbb{R} = \mathbb{Q}^{(\mathfrak{c})}$, and deduced that one can endow the reals with a compact group topology. He posed the problem to determine all abelian groups that support a compact group topology, in other words, to describe the algebraic structure of all compact abelian groups.

In the attempt to give a solution to the Halmos problem, a new relevant class of abelian groups, namely, that of algebraically compact groups, was introduced by Kaplansky as those abelian groups that are summands of abelian groups admitting a compact group topology.

Definition 14.3.13. An abelian group G is *algebraically compact* if it is a direct summand of every abelian group containing it as a pure subgroup.

These groups are also named *pure-injective* because of the following equivalent form, due to Maranda, that we are not going to use here:

Fact 14.3.14. *An abelian group G is algebraically compact if and only if for every abelian group H, any homomorphism from a pure subgroup of H to G can be extended to a homomorphism $H \to G$.*

Remark 14.3.15. We recall some properties of algebraically compact abelian groups (even if not all are used in the sequel):
(a) ([174, Theorem (25.21)]) compact abelian groups are algebraically compact ;
(b) ([223]) a reduced abelian group is algebraically compact if and only if it is Hausdorff and complete in the $\mathbb{Z}$-adic topology ;
(c) ([138]) every abelian group can be embedded as a pure subgroup in an algebraically compact group;
(d) ([137, Theorem 3]) every linearly compact group is algebraically compact.

Remark 14.3.16. The connected component $c(K)$ of a compact abelian group K is divisible by Corollary 14.1.1 and $K/c(K)$ is hereditarily disconnected, so profinite. Therefore, $K \cong c(K)\times K/c(K)$ algebraically, and $\operatorname{div}(K) = c(K)$ admits a connected compact group topology, while the reduced profinite part $R = K/c(K)$ admits a hereditarily disconnected compact group topology. Since hereditarily disconnected compact groups are profinite (see Remark 8.5.5), we deduce that the abelian groups admitting hereditarily disconnected compact group topologies are actually residually finite.

Since, for an abelian group G, Theorem A.4.3 implies that $G \cong \operatorname{div}(G) \times R$ where R is a reduced subgroup of G, if we equip $\operatorname{div}(G)$ with a compact (necessarily connected) group topology and R with a compact (necessarily hereditarily disconnected) group topology, we can take the compact product topology on G.

In this way we are left with these two cases. We start with the second.

Theorem 14.3.17. *A reduced abelian group K admits a compact group topology if and only if there exist cardinals $\{\sigma_p : p \in \mathbb{P}\}$ and $\{\alpha_{n,p} : p \in \mathbb{P}, n \in \mathbb{N}\}$ such that*

$$K \text{ is algebraically isomorphic to } \prod_{p\in\mathbb{P}} \left(\mathbb{J}_p^{\sigma_p} \times \prod_{n\in\mathbb{N}_+} \mathbb{Z}(p^n)^{\alpha_{n,p}} \right). \tag{14.4}$$

Proof. Starting with the necessity of (14.4), assume that K is a reduced abelian group that admits a compact, necessarily hereditarily disconnected, group topology. Then $K \cong \prod_{p\in\mathbb{P}} K_p$, where each K_p is a $\mathbb{J}_p$-module, by Theorem 14.1.9. First, we note that $\bigcap_{n\in\mathbb{N}} p^n K_p = \{0\}$ for all primes p. Indeed, the discrete dual group $\widehat{K_p}$ is a p-group, so $\widehat{K_p} = \sum_{n\in\mathbb{N}} \widehat{K_p}[p^n]$, and we obtain $\{0\} = \bigcap_{n\in\mathbb{N}} A_{K_p}(\widehat{K_p}[p^n]) = \bigcap_{n\in\mathbb{N}} p^n K_p$ from Corollary 13.5.3, Lemma 13.5.1, and Lemma 13.4.10(e).

On the other hand, K_p is q-divisible for all primes $q \neq p$, so one has $K_p = \bigcap_{m\in\mathbb{N}_+,(m,p)=1} mK$. Therefore, the subgroups $K_p = \{x \in K : p^n x \to 0\}$ are determined also by a purely algebraic condition. Hence, one can "localize" the problem by studying the algebraic structure of the hereditarily disconnected compact abelian groups of the form K_p.

According to Theorem A.4.11, the dual $X = \widehat{K_p}$ has a basic subgroup B. Then $B = \bigoplus_{n\in\mathbb{N}_+} \mathbb{Z}(p^n)^{(\alpha_{n,p})}$, for cardinals $\alpha_{n,p}$ with $n \in \mathbb{N}_+$, B is a pure subgroup of X and $X/B \cong \mathbb{Z}(p^\infty)^{(\sigma_p)}$, with a suitable cardinal σ_p, is divisible. Identifying $\widehat{X}$ with K_p in view of Theorem 13.4.7, we obtain a closed subgroup $N = A_{K_p}(B)$ of K_p such that, in view of Proposition 13.5.5, Example 13.2.4, and Theorem 13.3.5,

$$N \cong \widehat{X/B} \cong \mathbb{J}_p^{\sigma_p} \quad \text{and} \quad K_p/N \cong \widehat{B} \cong \prod_{n\in\mathbb{N}_+} \mathbb{Z}(p^n)^{\alpha_{n,p}}.$$

According to Corollary 13.5.4, the purity of B implies that N is a pure subgroup of K_p. Since $N \cong \mathbb{J}_p^{\sigma_p}$ is compact, we deduce from Remark 14.3.15(a) that N is algebraically compact, hence *algebraically* splits in K_p. In other words,

$$K_p \text{ is algebraically isomorphic to } \mathbb{J}_p^{\sigma_p} \times \prod_{n\in\mathbb{N}_+} \mathbb{Z}(p^n)^{\alpha_{n,p}}. \tag{14.5}$$

Summarizing, we showed that if a reduced abelian group K admits a compact group topology, then $K = \prod_{p\in\mathbb{P}} K_p$ and the subgroups K_p (which are determined by a purely algebraic condition regardless of the specific compact group topology on K) satisfy (14.5). This proves the necessity of (14.4)

On the other hand, if an abelian group K satisfies (14.4) for appropriate cardinals σ_p and $\alpha_{n,p}$, then we can first identify (algebraically) K with the product $\prod_{p\in\mathbb{P}}(\mathbb{J}_p^{\sigma_p} \times \prod_{n\in\mathbb{N}_+} \mathbb{Z}(p^n)^{\alpha_{n,p}})$. Since each $K_p = \mathbb{J}_p^{\sigma_p} \times \prod_{n\in\mathbb{N}_+} \mathbb{Z}(p^n)^{\alpha_{n,p}}$ admits an obvious compact group topology, taking the product topology on $K = \prod_{p\in\mathbb{P}} K_p$, we obtain also a compact group topology on K. □

The splitting (14.5) need not be topological in general (see Exercise 14.5.7).

In the next theorem we describe the algebraic structure of the connected (so, divisible) compact abelian groups. Since the trivial group $G = \{0\}$ is both divisible and connected in any topology, we rule out this trivial case in the next theorem.

Theorem 14.3.18. *A divisible abelian group $G \neq \{0\}$ admits a compact group topology if and only if there exist cardinals β and $\{\gamma_p : p \in \mathbb{P}\}$ with*

$$\beta \geq \omega,\ r_0(G) = 2^\beta \quad \text{and} \quad \forall p \in \mathbb{P} \quad \gamma_p \leq \beta \quad \text{and} \quad r_p(G) = \begin{cases} \gamma_p & \text{if } \gamma_p < \infty, \\ 2^{\gamma_p} & \text{otherwise.} \end{cases} \tag{14.6}$$

Furthermore, G admits a finite-dimensional compact group topology if and only if $r_0(G) = \mathfrak{c}$ and all $\{r_p(G) : p \in \mathbb{P}\}$ are bounded by some $m \in \mathbb{N}$. In this case $\dim G \geq \sup_{p\in\mathbb{P}} r_p(G)$ when G carries such a topology.

Proof. Suppose first that G is a divisible abelian group which admits a compact group topology that is necessarily connected. Put $\sigma = \dim G$ and $\beta = \max\{\sigma, \omega\}$. Then, $|G| = 2^\beta$; indeed, if σ is finite, this is a consequence of Corollary 14.3.7; in case σ is infinite, $|G| = 2^{|\widehat{G}|} = 2^{w(G)} = 2^\sigma$ by Theorem 13.3.11, Corollary 14.2.19, and Remark 14.3.5(d). So, Theorem 14.3.8 gives $r_0(G) = 2^\beta$ and $r_p(G)$ for appropriate cardinals $\{\gamma_p : p \in \mathbb{P}\}$ as in (14.6).

To verify the sufficiency, let G be a divisible abelian group satisfying (14.6) for cardinals β and $\{\gamma_p : p \in \mathbb{P}\}$ as stated there.

Put $\rho = \sup_{p\in\mathbb{P}} r_p(G)$. In case $\rho < \infty$ and $2^\beta = \mathfrak{c}$, put $\sigma = \rho$, otherwise let $\sigma = \beta$ (so $\mathfrak{c}^\sigma = \mathfrak{c}^\beta = 2^\beta$ in both cases). Let $K = \widehat{\mathbb{Q}}^\sigma$. Dualizing the short exact sequence $0 \to \mathbb{Z} \to \mathbb{Q} \to \mathbb{Q}/\mathbb{Z} \to 0$, one gets a closed subgroup $H \cong \prod_{p\in\mathbb{P}} \mathbb{J}_p$ of $\widehat{\mathbb{Q}}$ with $\widehat{\mathbb{Q}}/H \cong \mathbb{T}$. Therefore, K contains a closed subgroup $L \cong H^\sigma \cong \prod_{p\in\mathbb{P}} \mathbb{J}_p^\sigma$, such that $K/L \cong \mathbb{T}^\sigma$. Moreover, L obviously contains a closed subgroup $N = \prod_{p\in\mathbb{P}} N_p$, with $N_p \cong \mathbb{J}_p^{\gamma_p}$, $\gamma_p \leq \sigma$ for every prime p. Since $\dim N = 0$, Theorem 14.3.4 gives $\dim K/N = \sigma$. This yields $w(K/N) = \beta$ in case σ is infinite (so $\sigma = \beta$), since in this case $\sigma = \dim K/N = w(K/N)$. Otherwise, if σ is finite, K/N is metrizable by Corollary 14.3.7, so $w(K/N) = \omega$, so $2^{w(K/N)} = \mathfrak{c} = 2^\beta$, by the definition of σ. Hence, $2^{w(K/N)} = 2^\beta$ in both cases. Now Theorem 14.3.8, applied to the nontrivial connected compact group K/N, yields $r_0(K/N) = 2^{w(K/N)} = 2^\beta$, $r_p(K/N) = \gamma_p = r_p(G)$ in case $r_p(G)$ is finite; otherwise, $r_p(K/N) = 2^{\gamma_p} = r_p(G)$ when $r_p(G)$ is infinite. Since both groups G and K/N are divisible and have the same free-rank and the same p-rank for every prime p, these groups are algebraically isomorphic in virtue of Theorem A.2.17. This isomorphism of abstract abelian groups produces a connected compact group topology on G.

The necessity of the condition $|G| = \mathfrak{c}$ and $\rho \leq \dim G < \infty$ follows from Corollary 14.3.7, when G admits a finite-dimensional connected compact group topology (see also (14.3)). On the other hand, if G satisfies the condition $|G| = \mathfrak{c}$ and $\rho < \infty$, then choosing *any finite* σ with $\rho \leq \sigma < \infty$ the above argument produces a connected compact group topology on G with $\dim G = \dim K = \sigma$. □

Theorem 14.3.18 shows that the existence of a compact group topology on an abelian group may bring relevant restraints on the algebraic structure of the group. One may try to relax compactness to a weaker compactness-like property:

(a) In the case of precompactness, one has an immediate answer: every abelian group admits a precompact group topology (for example, the Bohr topology). Yet some nonabelian groups admit no precompact group topology, e. g., the infinite permutation groups $S(X)$ (see Exercise 10.4.2), or the group $\mathrm{SL}_2(\mathbb{C})$ (see [285]), just to mention two examples.
(b) In the case of pseudocompactness, the problem has been completely resolved (see [101] for a partial solution), reported in Chapter 15, where some information about the countable compactness can be found as well.

14.3.3 The Bohr topology of abelian groups

For a discrete abelian group G, the Bohr compactification bG is simply the completion of $G^{\#}$ (see the proof of Theorem 10.2.15). The Pontryagin-van Kampen duality can be used to easily build the *Bohr compactification* bG in the more general case of a locally compact abelian group G. Indeed, next we see that the Bohr compactification of any locally compact abelian group G can be obtained as $\widehat{\widehat{G}_d}$, where $\widehat{G}_d$ denotes the group $\widehat{G}$ equipped with the discrete topology. This result can be seen as a consequence of [4, Theorem 2] characterizing the Bohr compactification of a locally compact abelian group by means of dual groups. For a comment on the nonabelian case, see [82, 153].

Theorem 14.3.19. *Let G be a locally compact abelian group, $\widehat{G}_d$ its dual equipped with the discrete topology, and $i\colon \widehat{G}_d \to \widehat{G}$ the continuous identity map. The Bohr compactification $b_G\colon G \to bG$ of G coincides, up to isomorphism, with the composition $\widehat{\imath} \circ \omega_G$ of $\omega_G\colon G \to \widehat{\widehat{G}}$ and $\widehat{\imath}\colon \widehat{\widehat{G}} \to \widehat{\widehat{G}_d}$.*

Proof. To see that $\phi := \widehat{\imath} \circ \omega_G\colon G \to \widehat{\widehat{G}_d}$ is the Bohr compactification of G, pick a continuous homomorphism $f\colon G \to K$ to a compact group K. We can assume without loss of generality that $f(G)$ is dense in K. Then K is abelian, by Theorem 3.1.22. The dual homomorphism $\widehat{f}\colon \widehat{K} \to \widehat{G}$, as well as $g := i^{-1} \circ \widehat{f}\colon \widehat{K} \to \widehat{G}_d$, are continuous since $\widehat{K}$ is discrete, by Proposition 13.1.1(a). Then $\widehat{g}\ :\ \widehat{\widehat{G}_d} \to \widehat{\widehat{K}}$ is a continuous homomorphism, by Lemma 13.3.7. Moreover, as $\widehat{f} = i \circ g$, we deduce that $\widehat{g} \circ \widehat{i} = \widehat{\widehat{f}}$ by Lemma 13.3.7(f). Composing with ω_G, we get

$$\omega_K \circ f = \widehat{\widehat{f}} \circ \omega_G = \widehat{g} \circ \widehat{i} \circ \omega_G = \widehat{g} \circ \phi.$$

Since ω_K is a topological isomorphism by Pontryagin-van Kampen duality theorem 13.4.17, $f = \omega_K^{-1} \circ \widehat{g} \circ \phi$. Then $f' := \omega_K^{-1} \circ \widehat{g}$ with $f' \circ \phi = f$ witnesses the universal property of the Bohr compactification of G, so $\phi = b_G: G \to \widehat{\widehat{G}_d}$ is the Bohr compactification of G. □

From a rapid look at the cardinal invariants (of topological nature, introduced so far in this book) of the group $G^\#$, for an abelian group G, one can see that they depend only on $|G|$ and no other features of the group G (see Theorem 13.3.12 and Exercise 13.7.2). This motivated E. van Douwen to pose in [277] the following challenging problem (see also [153]): *if G and H are abelian groups of the same size, must $G^\#, H^\#$ be homeomorphic*? The first negative solution was obtained by Kunen [190]. Following his notation, we write $\mathbb{V}_m^\kappa := \mathbb{Z}(m)^{(\kappa)}$ for a cardinal κ, and $m \in \mathbb{N}_+$.

Theorem 14.3.20 ([190]). *For primes $p \neq q$, $(\mathbb{V}_p^\omega)^\#$ and $(\mathbb{V}_q^\omega)^\#$ are not homeomorphic.*

Independently, and by using a simpler argument, Watson and the second named author proved in [116] that $(\mathbb{V}_2^\kappa)^\#$ and $(\mathbb{V}_3^\kappa)^\#$ are not homeomorphic whenever $\kappa > 2^{2^{\mathfrak{c}}}$. Motivated by these results, we give the following:

Definition 14.3.21. A pair G, H of infinite abelian groups is:
(i) *Bohr-homeomorphic* if $G^\#$ and $H^\#$ are homeomorphic;
(ii) *weakly Bohr-homeomorphic* if $G^\#$ can be homeomorphically embedded into $H^\#$, and $H^\#$ can also be homeomorphically embedded into $G^\#$.

Obviously, Bohr-homeomorphic abelian groups are weakly Bohr-homeomorphic, the status of the converse implication is unclear so far (see Question 14.3.30(ii)). Clearly, two finite groups G, H satisfy the condition in (ii) if and only if $|G| = |H|$. Moreover, van Douwen's problem has obviously a positive answer for finite groups. This is why in this definition, as well as in the sequel, we consider only pairs of infinite abelian groups.

The notion of weak Bohr-homeomorphism provides a more flexible tool for studying the Bohr topology than the more *rigid* notion of Bohr-homeomorphism, e. g., $\mathbb{V}_p^\omega$ and $\mathbb{V}_q^\omega$ are not even weakly Bohr-homeomorphic for distinct primes p and q (see [190]).

Remark 14.3.22. If G is an abelian group such that $G^\#$ homeomorphically embeds into $H^\#$ and H is a bounded abelian group, then also G must be a bounded abelian group (see [156, Theorem 5.1]). So, boundedness is invariant under weak Bohr-homeomorphisms, i. e., if G, H are weakly Bohr-homeomorphic abelian groups and

G is bounded, then H must be bounded as well. Hence, the study of (weak) Bohr-homeomorphisms can be carried out separately for bounded and for unbounded abelian groups.

Starting with the class of bounded abelian groups, recall that every bounded abelian group has the form $\prod_{i=1}^n \mathbb{V}_{m_i}^{\kappa_i}$ for certain integers $m_i \in \mathbb{N}_+$ and cardinals κ_i, by Prüfer theorem A.1.4. For this reason, the study of the Bohr topology of bounded abelian groups can be focused on the groups $\mathbb{V}_m^{\kappa}$.

Definition 14.3.23 ([65, 156]). For a bounded abelian group G, the *essential order* $\mathrm{eo}(G)$ of G is the smallest $m \in \mathbb{N}_+$ such that mG is finite.

Example 14.3.24. A bounded abelian p-group $G = \mathbb{Z}(p)^{(\alpha_1)} \oplus \cdots \oplus \mathbb{Z}(p^n)^{(\alpha_n)}$, for $p \in \mathbb{P}$, has $\mathrm{eo}(G) = 1$ in case G is finite and $\mathrm{eo}(G) = p^k$ for some $k \in \{1, \dots, n\}$, when α_k is infinite, but all α_i, with $k < i \leq n$, are finite, provided $k < n$. Therefore, $r_p(G) = \omega$ for a countably infinite bounded abelian group G and a prime p, if and only if $p \mid \mathrm{eo}(G)$.

Definition 14.3.25. Two infinite abelian groups G, H are:
(i) ([165]) *almost isomorphic* if G and H have isomorphic finite index subgroups;
(ii) ([65]) *weakly isomorphic* if a finite-index subgroup of G is isomorphic to a subgroup of H and a finite-index subgroup of H is isomorphic to a subgroup of G.

The first of these notions was motivated by the following result of Hart and Kunen:

Theorem 14.3.26 ([165]). *Almost isomorphic abelian groups are Bohr-homeomorphic.*

It is unclear whether the implication in the above theorem can be inverted for bounded abelian groups:

Question 14.3.27 ([190]). Are Bohr-homeomorphic bounded abelian groups almost isomorphic?

By [65], a pair of infinite bounded abelian groups G, H is weakly isomorphic if and only if

$$|mG| = |mH| \quad \text{whenever } m \in \mathbb{N} \text{ and } |mG| \cdot |mH| \geq \omega. \tag{14.7}$$

By Theorem 14.3.26, weakly isomorphic bounded abelian groups are weakly Bohr-homeomorphic. On the other hand, if $G^\#$ embeds into $H^\#$ for bounded abelian groups G, H, then $\mathrm{eo}(G) \mid \mathrm{eo}(H)$ and $r_p(G) \leq r_p(H)$ whenever $r_p(G)$ is infinite, according to [65, Theorem 1.16]. Hence, weakly Bohr-homeomorphic bounded abelian groups satisfy the following simple algebraic conditions:

$$\mathrm{eo}(G) = \mathrm{eo}(H) \quad \text{and} \quad r_p(G) = r_p(H) \quad \text{for all } p \in \mathbb{P} \text{ with } r_p(G) \cdot r_p(H) \geq \omega. \tag{B}$$

The above arguments show that the following implications hold for a pair of infinite bounded abelian groups:

$$\text{weakly isomorphic} \Rightarrow \text{weakly Bohr-homeomorphic} \Rightarrow (B). \qquad (14.8)$$

Now we discuss the opposite implications. By Example 14.3.24, if $\mathrm{eo}(G) = \mathrm{eo}(H)$ for countable bounded abelian groups G, H, then they satisfy (B) and they are weakly isomorphic (see Exercise 14.5.14(a)). By Exercise 14.5.14(b), for bounded abelian groups of square-free essential order, (B) implies almost isomorphism, hence Bohr-homeomorphism, by Theorem 14.3.26. In summary:

Theorem 14.3.28 ([65, 156]). *For a pair G, H of infinite bounded abelian groups that are either countable or have square-free essential order, all three properties in* (14.8) *are equivalent.*

Therefore, the essential order is an invariant that alone allows for a complete classification, up to (weak) Bohr-homeomorphism, of all countable bounded abelian groups.

The situation changes completely even for the simplest *uncountable* bounded abelian groups of essential order 4. Indeed, $G = \mathbb{V}_4^{\omega_1}$ and $H = \mathbb{V}_2^{\omega_1} \times \mathbb{V}_4^{\omega}$ are not weakly isomorphic because $\omega_1 = |2G| > |2H| = \omega$. However, we do not know whether these groups are weakly Bohr-homeomorphic. More generally:

Question 14.3.29. Let p be a prime, $\kappa \geq \omega$ a cardinal, and $k > 1$ an integer.
(i) Are $\mathbb{V}_{p^k}^{\kappa}$ and $\mathbb{V}_p^{\kappa} \times \mathbb{V}_{p^k}^{\omega}$ weakly Bohr-homeomorphic? Can this depend on p?
(ii) If $(\mathbb{V}_{p^k}^{\kappa})^{\#}$ can be homeomorphically embedded into $(\mathbb{V}_{p^{k-1}}^{\kappa} \times \mathbb{V}_{p^k}^{\lambda})^{\#}$ for an infinite cardinal λ, is then necessarily $\lambda \geq \kappa$?

If the answer to Question 14.3.29(i) were positive for all cardinals $\kappa \geq \omega$, primes p, and integers $k > 1$, then two bounded abelian groups G, H would be weakly Bohr-homeomorphic if and only if (B) holds. Item (ii) is an equivalent form of the strongest negative answer to (i).

Question 14.3.30 ([190]). (i) Are $\mathbb{V}_4^{\omega}$ and $\mathbb{V}_2^{\omega} \times \mathbb{V}_4^{\omega}$ Bohr-homeomorphic?
(ii) Are weakly Bohr-homeomorphic bounded groups *always* Bohr-homeomorphic?

Question 14.3.31. Suppose that G, H are infinite bounded abelian groups such that $G^{\#}$ homeomorphically embeds into $H^{\#}$. Does there exist a subgroup G' of G of finite index that algebraically embeds into H?

A positive answer to this question would imply, in particular, that weak Bohr-homeomorphism coincides with weak isomorphism for infinite bounded abelian groups. Hence, a positive answer to this question would imply a positive answer to Question 14.3.29(ii).

Now we leave the "bounded world" and turn to the class of unbounded abelian groups. The implication of Theorem 14.3.26 and the first implication in (14.8) cannot be inverted: $\mathbb{Q}$ and $\mathbb{Q}/\mathbb{Z} \times \mathbb{Z}$ are Bohr-homeomorphic (see [53]), and yet these groups are not even weakly isomorphic.

The question on whether the pairs $\mathbb{Z}, \mathbb{Z}^2$ and $\mathbb{Z}, \mathbb{Q}$ are Bohr-homeomorphic is raised in [56, 153]. Let us consider here the version for weak Bohr-homeomorphisms:

Question 14.3.32. (i) Are $\mathbb{Z}$ and $\mathbb{Q}$ weakly Bohr-homeomorphic?
(ii) Are $\mathbb{Z}$ and $\mathbb{Q}/\mathbb{Z}$ weakly Bohr-homeomorphic?

A positive answer to Question 14.3.32(i) would yield that all torsion-free abelian groups of a fixed finite free-rank are pairwise weakly Bohr-homeomorphic.

14.4 Precompact group topologies determined by sequences

14.4.1 Characterized subgroups of $\mathbb{T}$

Large and lacunary sets (mainly in $\mathbb{Z}$ or elsewhere) are largely studied in number theory, harmonic analysis, and dynamical systems (see [63, 133, 152, 153, 157, 227]).

Let us consider a specific problem. For a strictly increasing sequence $A = \{a_n\}_{n\in\mathbb{N}}$ of integers, the interest in the distribution of the multiples $\{a_n\alpha : n \in \mathbb{N}\}$ of a nontorsion element α of $\mathbb{T} = \mathbb{R}/\mathbb{Z}$ has roots in ergodic theory (e. g., Sturmian sequences and Hartman sets – see [290]) and furthermore in number theory, according to the Weyl theorem: a sequence $\{x_n\}_{n\in\mathbb{N}}$ in $\mathbb{T}$ is *uniformly distributed* if for all $[a, b] \subseteq \mathbb{T}$ (with $a, b \in [0, 1)$),

$$\frac{|\{j \in \{0, \dots, n\} : x_j \in [a, b]\}|}{n} \longrightarrow |a - b|;$$

by the Weyl theorem, the set $\mathcal{W}_A := \{\beta \in \mathbb{T} : \{a_n\beta\}_{n\in\mathbb{N}} \text{ is uniformly distributed}\}$ has Lebesgue measure 1 (e. g., see [189]).

Definition 14.4.1. For a sequence $A = \{a_n\}_{n\in\mathbb{N}}$ of integers, the subgroup

$$t_A(\mathbb{T}) := \{\alpha \in \mathbb{T} : a_n\alpha \to 0\}$$

of $\mathbb{T}$ is called *topologically A-torsion subgroup*. We say also that the sequence A *characterizes* a subgroup H of $\mathbb{T}$, or that H is *characterized* by A, if $H = t_A(\mathbb{T})$; we briefly call H a *characterized* subgroup if $H = t_A(\mathbb{T})$ for some sequence $A = \{a_n\}_{n\in\mathbb{N}}$ of integers.

With respect to the above definition, note that we can always assume that the sequence A is in $\mathbb{N}_+$.

Topologically A-torsion subgroups were introduced in [81, 99], while the term characterized subgroup was first coined in [32].

Example 14.4.2. (a) If H is a finite subgroup of $\mathbb{T}$, then H is cyclic and it is straightforward to verify that H is characterized.

(b) A sequence A in $\mathbb{Z}$ characterizes $\mathbb{T}$ precisely when A is eventually zero. Indeed, in case $A = \{a_n\}_{n\in\mathbb{N}}$ characterizes $\mathbb{T}$, then $a_n \to 0$ in $\mathbb{Z}^\#$, and so A is eventually zero by Theorem 13.4.9. The converse implication is obvious.

Example 14.4.3. It was observed by Armacost [8] that, for $A = \{p^n\}_{n\in\mathbb{N}}$, where p is a prime, $t_A(\mathbb{T}) = t_p(\mathbb{T}) = \mathbb{Z}(p^\infty)$.

Armacost posed the question of describing the subgroup $t_A(\mathbb{T})$ for the sequence $A = \{n!\}_{n\in\mathbb{N}_+}$. This was done in [36], [99, § 4.4.2], and [66].

In [66, 89, 99] an analogous more general problem was considered of describing $t_A(\mathbb{T})$ for an increasing sequences $A = \{a_n\}_{n\in\mathbb{N}}$ of integers with $a_0 = 1$ and $a_n \mid a_{n+1}$ for every $n \in \mathbb{N}$.

For a sequence $A = \{a_n\}_{n\in\mathbb{N}}$ of integers, the subgroup $t_A(\mathbb{T})$ is a Borel set, since

$$t_A(\mathbb{T}) = \bigcap_{N\geq 2} \bigcup_{m\in\mathbb{N}} \bigcap_{n\geq m} \left\{x \in \mathbb{T}: \|a_n x\| \leq \frac{1}{N}\right\};$$

actually, this proves that $t_A(\mathbb{T})$ is an $F_{\sigma\delta}$-set.

Remark 14.4.4. Inspired by a construction of Aaronson–Nadkarni [1], Biró [31] showed that the F_σ-subgroups of $\mathbb{T}$ need not be characterized. His proof is based on the crucial point that the characterized subgroups of $\mathbb{T}$ are *Polishable*, that is, they admit a finer Polish group topology; this topology is unique by a result of Solecki [263].

We state now Biró's result; first recall that a nonempty subset K of $\mathbb{T}$ is a *Kronecker set* if K is compact and, for every continuous function $f: K \to \mathbb{T}$ and every $\delta > 0$, there exists $n \in \mathbb{Z}$ such that $\max_{k\in K} \|f(k) - nk\| < \delta$ (i. e., f can be uniformly approximated by characters of $\mathbb{T}$).

Theorem 14.4.5 (Biró theorem [31]). *Let K be an uncountable Kronecker set in $\mathbb{T}$. Then the subgroup $\langle K\rangle$ is not Polishable. In particular, $\langle K\rangle$ cannot be characterized.*

Since $\langle K\rangle$ is obviously F_σ, this provides an example of a noncharacterized F_σ-subgroup of $\mathbb{T}$, thereby answering a question of the second named author (see also [91], where some special classes of F_σ-subgroups were shown to be characterized).

Being a Borel set, $t_A(\mathbb{T})$ is measurable and either countable or has size $\mathfrak{c}$ (see [184, Theorem 13.6]).

Egglestone [127] proved that, for a sequence of positive integers $A = \{a_n\}_{n\in\mathbb{N}}$, the asymptotic behavior of the sequence of ratios $\{q_n\}_{n\in\mathbb{N}}$, with $q_n = \frac{a_{n+1}}{a_n}$ for every $n \in \mathbb{N}$, may have an impact on the size of the subgroup $t_A(\mathbb{T})$ in terms of the following remarkable "dichotomy":

Theorem 14.4.6 ([127]). *Let $A = \{a_n\}_{n\in\mathbb{N}}$ be an increasing sequence in $\mathbb{N}_+$.*

(a) *If $\lim_{n\to\infty} \frac{a_{n+1}}{a_n} = +\infty$, then $|t_A(\mathbb{T})| = \mathfrak{c}$.*

(b) *If $\{\frac{a_{n+1}}{a_n}\}_{n\in\mathbb{N}}$ is bounded, then $t_A(\mathbb{T})$ is countable.*

If A is sequence of integers and $t_A(\mathbb{T}) \neq \mathbb{T}$, then $t_A(\mathbb{T})$ has Haar measure zero, as every proper subgroup of $\mathbb{T}$ has infinite index. A measure zero subgroup H of $\mathbb{T}$ of size $\mathfrak{c}$ that is not even contained in any proper characterized subgroup $t_A(\mathbb{T})$ of $\mathbb{T}$ was built in [21] (under the assumption of Martin axiom) and later in [166, 167] (in ZFC).

Theorem 14.4.7 (Borel theorem [36]). *Every countable subgroup of $\mathbb{T}$ is characterized.*

Three proofs of Borel theorem 14.4.7 were given in [32].

Unaware of Borel's result, Larcher [194], and later Kraaikamp and Liardet [188], proved that the cyclic subgroups of $\mathbb{T}$ are characterized. In this specific case they explicitly described characterizing sequences of a subgroup $\langle q_0(\alpha)\rangle$ of $\mathbb{T}$, with $\alpha \in \mathbb{R}\setminus\mathbb{Q}$, in terms of the continued fraction expansion of α.

In particular, as an illustration we propose the following example (related to Example 5.3.3(c) and Exercise 14.5.16), which answers [81, Question 3.11].

Example 14.4.8 ([22]). Let ϕ be the golden ratio, that is, $\phi = \frac{1+\sqrt{5}}{2}$. Then

$$\langle q_0(\phi)\rangle = t_F(\mathbb{T}), \tag{14.9}$$

where $F = \{f_n\}_{n\in\mathbb{N}}$ is the Fibonacci sequence.

This is a particular case of the more general form of sequences $A = \{a_n\}_{n\in\mathbb{N}}$ in $\mathbb{Z}$ considered in [81], namely, those which satisfy

$$a_n \mid a_{n+1} - a_{n-1} \quad \text{for all } n \in \mathbb{N}_+. \tag{14.10}$$

Inspired by (14.9) and, more generally, by the class of sequences satisfying (14.10), the paper [23] describes the algebraic structure of the subgroup $t_A(\mathbb{T})$ when the sequence $A = \{a_n\}_{n\in\mathbb{N}}$ verifies a linear recurrence relation of order $\leq k$, that is,

$$a_n = u_n^{(1)} a_{n-1} + u_n^{(2)} a_{n-2} + \cdots + u_n^{(k)} a_{n-k}$$

for every $n > k$ with $u_n^{(i)} \in \mathbb{Z}$ for $i \in \{1, \ldots, k\}$.

14.4.2 Characterized subgroups of topological abelian groups

In [32] the authors conjectured that Borel theorem 14.4.7 of characterizability of the countable subgroups of $\mathbb{T}$ can be extended to compact abelian groups in place of $\mathbb{T}$, without providing any precise formulation. There is a natural way to extend the definition of $t_A(\mathbb{T})$ to an arbitrary topological abelian group G by letting, for a sequence $A = \{a_n\}_{n\in\mathbb{N}}$ of integers,

$$t_A(G) := \{x \in G \colon a_n x \to 0 \text{ in } G\}.$$

Prominent examples are provided by the sequence of integers $\{p^n\}_{n\in\mathbb{N}}$, with p a prime, and $\{n!\}_{n\in\mathbb{N}_+}$, defining respectively the topologically p-torsion and the topologically torsion elements – see Definition 5.3.4. For locally compact abelian groups G, one can easily reduce the computation of $t_A(G)$ to that of $t_A(\mathbb{T})$ (see [81]), and independently on their relevance in other questions, the subgroups $t_A(G)$ turned out to be of no help in the characterization of countable subgroups of compact abelian groups other than $\mathbb{T}$. Indeed, a much weaker condition, turned out to determine the circle group $\mathbb{T}$ in the class of all locally compact abelian groups:

Theorem 14.4.9 ([67]). *In a locally compact abelian group G, every cyclic subgroup of G is an intersection of subgroups of the form $t_A(G)$, where A is a sequence of integers, if and only if $G \cong \mathbb{T}$.*

One can remove the "abelian restraint" in the above theorem, recalling that in the nonabelian case $t_A(G)$ is just a subset of G, not a subgroup in general (see [66]).

Theorem 14.4.9 suggested to use in [95] a different approach to the problem, replacing the sequences of integers (that can be seen as characters of $\mathbb{T}$) by a sequence $A = (a_n)_{n\in\mathbb{N}}$ in the dual group $\widehat{G}$ of a locally compact abelian group G.

Definition 14.4.10 ([95]). Let G be a topological abelian group and $A = \{a_n\}_{n\in\mathbb{N}}$ a sequence in $\widehat{G}$. Define

$$s_A(G) := \{x \in G: a_n(x) \to 0 \text{ in } \mathbb{T}\}.$$

A subgroup H of G is called *characterized* if there exists a sequence A in $\widehat{G}$ such that $H = s_A(G)$. We say that H is characterized by A and that A characterizes H.

If $G = \mathbb{T}$, then we can identify $\widehat{\mathbb{T}} = \mathbb{Z}$, so $t_A(\mathbb{T}) = s_A(\mathbb{T})$ for a sequence A in $\mathbb{Z}$.

For a topological abelian group G,

$$s_A(G) = \bigcap_{N\geq 2}\bigcup_{m\in\mathbb{N}}\bigcap_{n\geq m}\left\{x \in G: \|a_n(x)\| \leq \frac{1}{N}\right\}.$$

So, if K is a compact abelian group, then:

(1) $s_A(K)$ is a Borel set (actually, an $F_{\sigma\delta}$-set), and so either $s_A(K)$ is countable or $|s_A(K)| \geq \mathfrak{c}$;
(2) $\mu(s_A(K)) = 0$ if A is faithfully indexed (see [57, Lemma 3.10], and see [242] for locally compact abelian groups).

For G an abelian group and a subgroup H of G, let

$$\mathfrak{g}_G(H) := \bigcap\{s_A(G): A \in \widehat{G}^{\mathbb{N}},\ H \leq s_A(G)\};$$

we say that the subgroup H of G is $\mathfrak{g}$-*closed* if $H = \mathfrak{g}_G(H)$.

It is proved in [95] that a topological abelian group G is maximally almost periodic if and only if every cyclic subgroup of G is $\mathfrak{g}$-closed. The following positive answer to [95, Problem 5.1] was given independently and simultaneously in three papers.

Theorem 14.4.11 ([28, 91, 199]). *All countable subgroups of a compact abelian group are $\mathfrak{g}$-closed.*

Moreover, Borel theorem 14.4.7 was extended independently in two papers:

Theorem 14.4.12 ([28, 91]). *If K is a metrizable compact abelian group, then every countable subgroup of K is characterized.*

As observed above, every characterized subgroup of a metrizable compact abelian group K is $F_{\sigma\delta}$. Inspired by Biró theorem 14.4.5, Gabriyelyan proved in [146] that if K is an uncountable Kronecker set of an infinite metrizable compact abelian group X, then $\langle K\rangle$ is not Polishable. In particular, $\langle K\rangle$ cannot be characterized.

Moreover, Gabriyelyan [146] showed that every characterized subgroup H of a compact metrizable abelian group K is Polishable (and the finer Polish group topology is also locally quasi-convex). For further results in this direction, see [90, 147, 145, 219].

14.4.3 *TB*-sequences

Another motivation for the study of characterized subgroups $t_A(\mathbb{T})$ comes from the fact that they lead to the description of precompact group topologies on $\mathbb{Z}$ that make the sequence of integers $A = \{a_n\}_{n\in\mathbb{N}}$ converge to 0 in $\mathbb{Z}$ (see Corollary 14.4.16).

In other words, we discuss a counterpart of the notion of T-sequences, defined with respect to group topologies induced by characters, i. e., precompact group topologies. Recall from Theorem 11.4.2 and Corollary 11.4.3 that τ is a totally bounded group topology on an abelian group G precisely when $\tau = \mathcal{T}_H$ for some subgroup H of $\widehat{G}$, and moreover τ is precompact if and only if H separates the points of G.

Definition 14.4.13 ([21, 23]). A sequence $A = \{a_n\}_{n\in\mathbb{N}}$ in an abelian group G is a *TB-sequence* if there exists a precompact group topology τ on G such that $a_n \to 0$ in τ.

Clearly, every *TB*-sequence is a T-sequence (see Proposition 14.4.18 for a T-sequence in $\mathbb{Z}$ that is not a *TB*-sequence). The advantage of *TB*-sequences over T-sequences is in the easier way of determining sufficient conditions for a sequence to be a *TB*-sequence (see [21, 23]).

Let us start by an easy to prove general fact:

Fact 14.4.14 ([21]). *A sequence $A = \{a_n\}_{n\in\mathbb{N}}$ in a totally bounded abelian group G converges to 0 in G if and only if $\chi(a_n) \to 0$ in $\mathbb{T}$ for every $\chi \in \widehat{G}$.*

In the case of $G = \mathbb{Z}$, the characters of G are simply elements of $\mathbb{T}$, i. e., a totally bounded group topology on $\mathbb{Z}$ has the form $\mathcal{T}_H$ for some subgroup H of $\mathbb{T}$. Thus, Fact 14.4.14 for $G = \mathbb{Z}$ can be reformulated as follows: for a subgroup H of $\mathbb{T}$, a sequence $A = \{a_n\}_{n\in\mathbb{N}}$ in $\mathbb{Z}$ converges to 0 in $\mathcal{T}_H$ if and only if $a_n x \to 0$ for every $x \in H$ (i. e., simply $H \subseteq t_A(\mathbb{T})$). This was generalized in [95] for any abelian group:

Theorem 14.4.15 ([95]). *Let $A = \{a_n\}_{n\in\mathbb{N}}$ be a nontrivial sequence in an abelian group G. Then:*
(a) *given a subgroup H of $\widehat{G}$, $a_n \to 0$ in $\mathcal{T}_H$ if and only if $H \subseteq s_A(\widehat{G})$;*
(b) *$\mathcal{T}_{s_A(\widehat{G})}$ is the finest totally bounded group topology on G with $a_n \to 0$ in $\mathcal{T}_{s_A(\widehat{G})}$;*
(c) *A is a TB-sequence if and only if $s_A(\widehat{G})$ is dense in $\widehat{G}$.*

Corollary 14.4.16. *A sequence $A = \{a_n\}_{n\in\mathbb{N}}$ in $\mathbb{Z}$ is a TB-sequence if and only if the subgroup $t_A(\mathbb{T})$ of $\mathbb{T}$ is infinite.*

Example 14.4.17. If $A = \{a_n\}_{n\in\mathbb{N}}$ is a sequence in $\mathbb{Z}$ with $\lim_{n\to\infty} \frac{a_{n+1}}{a_n} = +\infty$, then A is a *TB*-sequence (see [127]). On the other hand, there exists a *TB*-sequence $A = \{a_n\}_{n\in\mathbb{N}}$ in $\mathbb{Z}$ with $\lim_{n\to\infty} \frac{a_{n+1}}{a_n} = 1$ (see [23]).

Here is an example of a T-sequence in $\mathbb{Z}$ that is not a *TB*-sequence.

Proposition 14.4.18 ([240]). *For every TB-sequence $A = \{a_n\}_{n\in\mathbb{N}}$ in $\mathbb{Z}$ such that $t_A(\mathbb{T})$ is countable, there exists a sequence $\{c_n\}_{n\in\mathbb{N}}$ in $\mathbb{Z}$ such that the sequence $\{q_n\}_{n\in\mathbb{N}}$, defined by $q_{2n} = c_n$ and $q_{2n-1} = a_n$, is a T-sequence but not a TB-sequence.*

Proof. Let $\{z_1, \dots, z_n, \dots\}$ be an enumeration of $t_A(\mathbb{T})$.

According to Lemma 5.3.12, there exists a sequence $B = \{b_n\}_{n\in\mathbb{N}}$ in $\mathbb{Z}$ such that for every choice of the sequence $\{e_n\}_{n\in\mathbb{N}}$, where $e_n \in \{0,1\}$, the sequence $\{q_n\}_{n\in\mathbb{N}}$ defined by $q_{2n} = b_n + e_n$ and $q_{2n-1} = a_n$, is a T-sequence. Now we define the sequence $\{q_n\}_{n\in\mathbb{N}}$ with $q_{2m-1} = a_m$ and $q_{2m} = b_m$ when m is not a prime power. Let $p_1, \dots, p_n, \dots$ be all prime numbers enumerated one-to-one. Now fix k and define $e_k \in \{0,1\}$ depending on $\lim_{n\to\infty} b_{p_k^n} z_k$ as follows:
(1) if $\lim_{n\to\infty} b_{p_k^n} z_k = 0$, let $e_k = 1$;
(2) if $\lim_{n\to\infty} b_{p_k^n} z_k \neq 0$ (in particular, if the limit does not exists) let $e_k = 0$.

Now let $q_{2p_k^n} = b_{p_k^n} + e_k$ for $n \in \mathbb{N}$. To see that $Q = \{q_n\}_{n\in\mathbb{N}}$ is not a *TB*-sequence, assume that $\chi\colon \mathbb{Z} \to \mathbb{T}$ is a character such that $\chi(q_n) \to 0$ in $\mathbb{T}$. Then $x = \chi(1) \in \mathbb{T}$ satisfies $q_n x \to 0$, so $x \in t_Q(\mathbb{T}) \subseteq t_A(\mathbb{T})$. Hence, there exists $k \in \mathbb{N}$ with $x = z_k$. In particular, $q_{2p_k^n} z_k \to 0$.

Assume first that $\lim_{n\to\infty} b_{p_k^n} z_k = 0$. Then $e_k = 1$ and $0 = \lim_{n\to\infty} q_{2p_k^n} z_k = z_k$ implies $z_k = 0$. Conversely, if $\lim_{n\to\infty} b_{p_k^n} z_k \neq 0$, then $e_k = 0$, and hence one obtains the contradiction $0 = \lim_{n\to\infty} q_{2p_k^n} z_k \neq 0$. This proves that every character $\chi\colon \mathbb{Z} \to \mathbb{T}$ such that $\chi(q_n) \to 0$ in $\mathbb{T}$ is trivial. In particular, Q not a *TB*-sequence. □

The above proof gives more. Since $q_n \to 0$ in $\tau_{\{q_n\}}$ (the topology $\tau_{\{q_n\}}$ was introduced in §5.3), it shows that every $\tau_{\{q_n\}}$-continuous character of $\mathbb{Z}$ is trivial, i. e., $\widehat{(\mathbb{Z}, \tau_{\{q_n\}})} = \{0\}$. Therefore, $(\mathbb{Z}, \tau_{\{q_n\}})$ is a minimally almost periodic group.

14.5 Exercises

Exercise 14.5.1. Let p be a prime, $G = \mathbb{J}_p^{\mathbb{N}}$, and $K = pG$. Equip G with the topology that makes the compact abelian group K, equipped with the product topology, open in G. Prove that G is a torsion-free periodic locally compact abelian group, yet $\widehat{G}$ is neither divisible nor torsion-free.

Hint. Since $D = G/K \cong \mathbb{Z}(p)^{\mathbb{N}}$ is discrete, the dual $\widehat{G}$ has a compact open subgroup $\widehat{D} \cong \mathbb{Z}(p)^{\mathfrak{c}}$ with $\widehat{G}/\widehat{D} \cong \widehat{K} \cong \mathbb{Z}(p^\infty)^{(\mathbb{N})}$ countable. Hence, $\widehat{G}$ is torsion and $p\widehat{G}$, being isomorphic to a quotient of $\widehat{G}/\widehat{D}$, is countable, thus $\widehat{G} \neq p\widehat{G}$ is not divisible.

Exercise 14.5.2. Prove that:
(a) a compact abelian group G is separable if and only if $|\widehat{G}| \leq \mathfrak{c}$, if and only if G is isomorphic to a quotient of $(\widehat{\mathbb{Q}} \times \prod_{p\in\mathbb{P}} \mathbb{J}_p)^{\mathfrak{c}}$;
(b) a product of at most $\mathfrak{c}$ separable compact abelian groups is still separable.

Exercise 14.5.3. Let K be a compact abelian group. Prove that:
(a) $td(K) = \omega_K^{-1}(\mathrm{Hom}(\widehat{K}, \mathbb{Q}/\mathbb{Z}))$;
(b) if $f\colon K \to G$ is a continuous surjective homomorphism onto a compact group G, then $f(td(K)) = td(G)$;
(c) $td(\widehat{\mathbb{Q}})$ contains a subgroup $H \cong \prod_{p\in\mathbb{P}} \mathbb{J}_p$ such that $td(\widehat{\mathbb{Q}}/H) \cong td(\mathbb{T}) \cong \mathbb{Q}/\mathbb{Z}$.

Hint. (a) Fix $x \in td(K)$ and $\chi \in \widehat{K}$. If $\langle x\rangle$ is finite, then $\chi(x) \in \mathbb{Q}/\mathbb{Z}$. In case $\langle x\rangle$ is an infinite cyclic group endowed with a nondiscrete linear topology, the continuity of χ implies that an open (nontrivial) subgroup is mapped to Λ_1 and hence to 0. This yields that $\chi(G)$ is finite and hence $\chi(x) \in \mathbb{Q}/\mathbb{Z}$.

On the other hand, if $x \in \omega_K^{-1}(\mathrm{Hom}(\widehat{K}, \mathbb{Q}/\mathbb{Z}))$, then for every $\chi \in \widehat{K}$ the subgroup $\langle\chi(x)\rangle$ of $\mathbb{T}$ is finite. Since $\langle x\rangle$ is isomorphic to a subgroup of $\prod_{\chi\in\widehat{K}}\langle\chi(x)\rangle$ and the latter group has a linear topology, we deduce that $\langle x\rangle$ has a linear topology as well, so $x \in td(G)$.

(b) Since $\widehat{f}$ is injective and $\mathbb{Q}/\mathbb{Z}$ is divisible, the following holds: for every $\eta \in \mathrm{Hom}(\widehat{G}, \mathbb{Q}/\mathbb{Z})$, there exists $\eta' \in \mathrm{Hom}(\widehat{K}, \mathbb{Q}/\mathbb{Z})$ such that $\eta' \circ \widehat{f} = \eta$. Combined with (a) this yields that for every $g \in td(G)$ exists $x \in td(K)$ such that $\omega_G(g) = \omega_K(x) \circ \widehat{f} = \widehat{\widehat{f}}(\omega_K(x)) = \omega_G(f(x))$. Since ω_G is injective, one obtains $g \in f(td(K))$.

Exercise 14.5.4. Let G be a discrete abelian group and $K = \widehat{G}$. Prove that:
(a) $K_p = \mathrm{Hom}(G, \mathbb{Z}(p^\infty))$ equipped with the topology induced from the product topology of $\mathbb{Z}(p^\infty)^G$, where $\mathbb{Z}(p^\infty)$ carries the topology induced by $\mathbb{T}$;
(b) deduce that if $f\colon K \to L$ is a surjective homomorphism of compact abelian groups, then $f(K_p) = L_p$ and $f(wtd(K)) = wtd(L)$.

Exercise 14.5.5. Deduce Theorem 14.2.9 from Theorem 14.2.18.

Hint. Let G be a compactly generated locally compact abelian group and let C be a compact subset of G generating G. By Theorem 14.2.18, we can write $G = \mathbb{R}^n \times G_0$, where G_0 is a closed subgroup of G containing a compact open subgroup K. Since the quotient group $G_0/K \cong G/(\mathbb{R}^n \times K)$ is discrete, the image of C in $G/(\mathbb{R}^n \times K)$ is finite. Since G is generated by C, this yields that $G/(\mathbb{R}^n \times K)$ is finitely generated, so isomorphic to $\mathbb{Z}^d \times F$ for some finite abelian group F and $d \in \mathbb{N}$. By taking a suitable compact subgroup K_1 of G containing K, we can assume that $G/(\mathbb{R}^n \times K) \cong \mathbb{Z}^d$. Since the group $\mathbb{Z}^d$ is free, the group G splits as $G = \mathbb{R}^n \times K \times \mathbb{Z}^d$.

Exercise 14.5.6. Show that:

(a) divisible abelian groups and bounded abelian groups are algebraically compact;

(b) summands and products of algebraically compact groups are algebraically compact.

Exercise 14.5.7. Let p be a prime, $G = \prod_{n\in\mathbb{N}_+} \mathbb{Z}(p^n)$, and $X = t(G)$. Show that:

(a) $B = \bigoplus_{n\in\mathbb{N}_+} \mathbb{Z}(p^n)$ is a basic subgroup of X and $X/B \cong \mathbb{Z}(p^\infty)^{(\mathfrak{c})}$, so the short exact sequence $0 \to B \to X \to X/B \to 0$ does not split;

(b) the subgroup $N = A(B)$ of $K = \widehat{X}$ is pure and satisfies $N \cong \mathbb{J}_p^{\mathfrak{c}}$, so it splits algebraically, yet N does not split topologically.

Hint. (a) It is easy check that B is a pure subgroup of X. To show that X/B is divisible it is enough to check that $X = pX + B$. Since X is reduced, X cannot contain a subgroup isomorphic to X/B, so the subgroup B does not split.

(b) The subgroup N splits algebraically by Remark 14.3.15(a). The rest follows from the exactness of the duality functor (see also the proof of Theorem 14.3.17). For an alternative argument to see that the subgroup N does not split topologically, assume that $K \cong N \times G$. Then obviously $t(K) = \bigcup_{n\in\mathbb{N}} K[p^n]$, being contained in G, is not dense. On the other hand, $A(t(K)) = \bigcap_{n\in\mathbb{N}} A(K[p^n]) = \bigcap_{n\in\mathbb{N}} p^n X = \{0\}$ by Remark 13.5.2(b) and Corollary 13.5.3, so $\overline{t(K)} = K$ by Remark 13.5.2(a), a contradiction.

Exercise 14.5.8. Let K be a connected compact abelian group. Show that the following conditions are equivalent:

(a) for some prime p the group $\mathbb{J}_p^{\mathbb{N}}$ embeds into K;

(b) for all primes p the group $\mathbb{J}_p^{\mathbb{N}}$ embeds into K;

(c) $\dim K = \infty$.

Exercise 14.5.9. Show that a nontrivial connected compact abelian group K admits $2^{|K|}$ involutive discontinuous automorphisms.

Hint. By Theorem 14.3.8, K is a divisible abelian group with $r_0(K) = |K|$, there is an algebraic isomorphism $K \cong t(K) \times \mathbb{Q}^{(|K|)}$. Since $\mathrm{Aut}(\mathbb{Q}^{(|K|)})$ is isomorphic to a subgroup of $\mathrm{Aut}(K)$ and contains $2^{|K|}$ involutions, so does $\mathrm{Aut}(K)$. Using Pontryagin–van Kampen duality conclude that at most $|K| = 2^{\widehat{K}} \geq \mathrm{Aut}(\widehat{K})$ of them are continuous. So, the remaining $2^{|K|}$ involutive automorphisms are discontinuous.

Exercise 14.5.10. Show that a nontrivial connected compact abelian group K admits an extension G, such that $[G : K] = 2$ and the compact topology of K cannot be extended to a group topology of G.

Hint. By Exercise 14.5.9, there exists a discontinuous automorphism f of K of order 2. Arguing as in Example 4.4.8, show that the group $G = K \rtimes \langle f\rangle$ has the desired property.

Exercise 14.5.11. Show that for every locally compact abelian group G the subgroup $wtd(G)$ is dense in $B(G)$.

Hint. Since $B(G)$ is a union of compact subgroups, it is enough to consider the case when G itself is compact. Apply Theorem 14.3.9 to deduce that $N = \overline{wtd(G)}$ contains $c(G)$. Let $q\colon G \to G/c(G)$ be the canonical projection. Since $G/c(G)$ is hereditarily disconnected, $G/c(G) \cong \prod_{p\in\mathbb{P}}(G/c(G))_p$, so $wtd(G/c(G)) = \bigoplus_{p\in\mathbb{P}}(G/c(G))_p$ is dense in $G/c(G)$. Since $q(wtd(G)) = wtd(G/c(G))$ by Exercise 14.5.4, we deduce that $q(N)$ is dense in $G/c(G) = \overline{wtd(G/c(G))}$. As $q(N)$ is compact and contains $\ker q = c(G)$, this implies $N = K$.

Exercise 14.5.12. Let K be a compact abelian group and p a prime. Show that $K[p] \cong \mathbb{Z}(p)^{r_p(\widehat{K}/p\widehat{K})}$. In particular, $r_p(K) = r_p(\widehat{K}/p\widehat{K})$ if these cardinals are finite, otherwise $r_p(K) = 2^{r_p(\widehat{K}/p\widehat{K})}$.

Exercise 14.5.13. Let G be a discrete abelian group. Show that:
(a) $w(bG) = \chi(bG) = \psi(bG) = 2^{|G|}$ in case G is infinite;
(b) bG is metrizable if and only if G is finite;
(c) bG is connected if and only if G is divisible;
(d) bG is hereditarily disconnected if and only if G is bounded torsion;
(e) bG is torsion-free if and only if G is torsion-free, in such a case

$$bG \cong \widehat{\mathbb{Q}}^{2^{|G|}} \times \prod_{p\in\mathbb{P}} \mathbb{Z}_p^{\beta_p},$$

where $\beta_p = r_p(G/pG)$ if $r_p(G/pG) < \infty$, otherwise $\beta_p = 2^{r_p(G/pG)}$;
(f) describe bG for the following discrete abelian groups G: $\mathbb{Z}$, $\mathbb{Q}$, $\mathbb{Z}(p^\infty)$, $\mathbb{Q}/\mathbb{Z}$, $\mathbb{Z}(p)^{(\mathbb{N})}$, and $\mathbb{Z}(p^m)^{(\kappa)}$, with $p \in \mathbb{P}$, $\kappa \geq \omega$, and $m \in \mathbb{N}_+$, making use of Theorem 14.3.19.

Hint. (a) Use the fact that $bG = \widetilde{G^\#}$ and apply Corollary 11.4.5.

(b) follows from (a).

(c), (d) Use the fact that the topological properties (connectedness or hereditary disconnectedness) depend only on the algebraic properties of $\widehat{G}$. Therefore, bG is connected if and only if $\widehat{G}$ is torsion-free if and only if G is divisible, while bG is hereditarily connected if and only if $\widehat{G}$ is torsion if and only if G is bounded torsion.

(e) If bG is torsion-free, then G is torsion-free as a subgroup of bG. If G is torsion-free, then $\widehat{G}$ is connected, hence divisible. Therefore, bG, as a dual of the discrete group $\widehat{G}_d$, is torsion-free.

(f) It is easy to see that $b\mathbb{Z} \cong \widehat{\mathbb{Q}}^{\mathfrak{c}} \times \prod_{p\in\mathbb{P}} \mathbb{J}_p$, $b\mathbb{Q} = \widehat{\mathbb{Q}}^{\mathfrak{c}}$, $b(\mathbb{Z}(p^m)^{(\mathbb{N})}) = \mathbb{Z}(p^m)^{\mathfrak{c}}$, and $b(\mathbb{Z}(p)^{(\kappa)}) = \mathbb{Z}(p)^{2^\kappa}$.

For $G = \mathbb{Z}(p^\infty)$, note that $\widehat{G} = \mathbb{J}_p$. For the (discrete) group $\mathbb{J}_p$, the subgroup $\mathbb{Z}$ is dense in the natural topology of $\mathbb{J}_p$, hence $\mathbb{Z} + n\mathbb{J}_p = \mathbb{J}_p$ for every $n \in \mathbb{N}$. This means that the quotient group $\mathbb{J}_p/\mathbb{Z}$ is divisible. A more careful look at this group shows that $\mathbb{J}_p/\mathbb{Z} \cong \mathbb{Q}^{(\mathfrak{c})} \oplus \bigoplus_{q\in\mathbb{P}\setminus\{p\}} \mathbb{Z}(q^\infty)$. Therefore, bG has a (large) closed torsion-free subgroup $N \cong \widehat{\mathbb{Q}}^{\mathfrak{c}} \times \prod_{q\in\mathbb{P}\setminus\{p\}} \mathbb{J}_q$ such that $bG/N \cong \mathbb{T}$.

Show that for $G = \mathbb{Q}/\mathbb{Z}$, the dual group $\widehat{G} = \prod_{p\in\mathbb{P}} \mathbb{J}_p$ contains a dense pure infinite cyclic subgroup. Then an argument similar to the previous one shows that $b(\mathbb{Q}/\mathbb{Z})$ contains a closed subgroup $N \cong \widehat{\mathbb{Q}}^{\mathfrak{c}}$ such that $b(\mathbb{Q}/\mathbb{Z})/N \cong \mathbb{T}$.

Exercise 14.5.14. Let G, H be bounded abelian groups. Prove that:
(a) if G, H are countable with $\mathrm{eo}(G) = \mathrm{eo}(H)$, then they are weakly isomorphic;
(b) if (B) holds and $\mathrm{eo}(G) = \mathrm{eo}(H)$ is square-free, then G, H are almost isomorphic.

Exercise 14.5.15. Give examples of (countable) bounded abelian groups G, H that are weakly isomorphic, but not almost isomorphic.

Hint. Try $\mathbb{V}_4^\omega$ and $\mathbb{V}_2^\omega \times \mathbb{V}_4^\omega$.

Exercise 14.5.16. (a) Prove that the Fibonacci sequence is a *TB*-sequence.

(b) Let $\{a_n\}_{n\in\mathbb{N}_+}$ be a sequence of positive integers. Define the sequence $\{u_n\}_{n\in\mathbb{N}}$ by letting $u_0 = 1$, $u_1 = a_1$ and $u_n = a_n u_{n-1} + u_{n-2}$ for any $n > 1$. Prove that $\{u_n\}_{n\in\mathbb{N}}$ is a *TB*-sequence.

Hint. (a) See Example 5.3.3(c).

(b) Apply Exercise 5.4.18 to the irrational number α determined by the continued fraction $[a_1, a_2, a_3, \ldots]$.

14.6 Further readings, notes, and comments

An application of the structure theorem for locally compact abelian groups reduces the characterization of arcwise connected or locally connected locally compact abelian groups to compact abelian groups with the corresponding properties. It was shown by Pontryagin [228] that a compact abelian group is locally connected if and only if it is the character group of a discrete abelian L-group. A discrete abelian group D is called an *L-group* if every finite subset F of D is contained in a finitely generated subgroup H of D such that D/H is torsion-free. Free abelian groups are L-groups and countable torsion-free L-groups are free. Hence, the countable products of tori are the only examples of connected and locally connected compact metrizable groups.

A compact abelian group K is arcwise connected if and only if the discrete group $D = \widehat{K}$ is a Whitehead group, which means that $\operatorname{Ext}(D, \mathbb{Z}) = \{0\}$. Obviously, free abelian groups are Whitehead groups, while Whitehead groups are always torsion-free. Whitehead asked whether every Whitehead group is free. Shelah proved that this question is undecidable under the axioms of Zermelo–Fraenkel and the axiom of choice (see [259] or [128] or [177, p. 654]). This question is, of course, equivalent to: is every arcwise connected compact abelian group a product of tori? The answer is affirmative in the metrizable case, since countable Whitehead groups are free (see [138]).

As we mentioned in §8.8, arcwise connected compact (abelian) groups are locally (arcwise) connected, so Whitehead groups are L-groups. The discrete Specker group $\mathbb{Z}^\mathbb{N}$ is an L-group but not a Whitehead group (see [123, 138]), so its dual group is a connected and locally connected compact abelian group which is not arcwise connected.

The survey papers [68, 87, 147] offer more information on characterized subgroups of $\mathbb{T}$ (the first one, for this direction see also [67]), on characterized subgroups of arbitrary topological abelian groups (the second and third, whereas the latter offers also a discussion of the nonabelian counterpart of the characterized subgroups of a compact abelian group). For the quite recent variation of *statistically* characterized subgroups of the circle, using statistical convergence in place of the usual one, see [37, 83].

For the connection of characterized subgroups of $\mathbb{T}$ (and of $\mathbb{R}$) with thin sets from harmonic analysis, such as Arbault sets and Dirichlet sets, see [20], where Dirichlet sets of $\mathbb{T}$ are studied, in connection to the Erdös–Kunen–Mauldin theorem from [135] (see also the improvement by Eliaš [130]). Moreover, in [24] the problem of the inclusion of a characterized subgroup in another is resolved under suitable hypotheses. This problem is connected to Arbault sets (see Eliaš [129]).

J. Pelant conjectured in 1996 that $\mathbb{V}_2^\omega$ and $\mathbb{V}_3^\omega$ are not Bohr-homeomorphic, he had a proof with a gap he could not fill.

Many nice properties of $\mathbb{Z}^\#$ can be found in [191]. For a fast growing sequence $\{a_n\}_{n\in\mathbb{N}}$ in $\mathbb{Z}^\#$, the range is a closed discrete set of $\mathbb{Z}^\#$ (see [153] for further properties of the lacunary sets in $\mathbb{Z}^\#$), whereas for a sequence $\{a_n\}_{n\in\mathbb{N}}$ determined by a polynomial $P(x) \in \mathbb{Z}[x]$, that is, $a_n = P(n)$ for every $n \in \mathbb{N}$, the range has no isolated points (see [191, Theorem 5.4]). Moreover, the range $P(\mathbb{Z})$ is closed when $P(x) = x^k$ is a monomial. For quadratic polynomials $P(x) = ax^2 + bx + c$ with $a, b, c, \in \mathbb{Z} \setminus \{0\}$, the situation is already more complicated: the range $P(\mathbb{Z})$ is closed if and only if there is at most one prime that divides a, but does not divide b (see [191, Theorem 5.6]). This leaves open the general question from [191]:

Problem 14.6.1 ([102, Problem 954]). *Characterize the polynomials $P(x) \in \mathbb{Z}[x]$ such that $P(\mathbb{Z})$ is closed in $\mathbb{Z}^\#$.*

15 Pseudocompact groups

This chapter deals with pseudocompact groups, which will be supposed to be Hausdorff. The first section gives fairly general properties of pseudocompact and countably compact topological spaces.

15.1 General properties of countably compact and pseudocompact spaces

Lemma 15.1.1. (a) *Closed subspaces and continuous images of countably compact spaces are countably compact.*
(b) *Continuous images of pseudocompact spaces are pseudocompact.*
(c) *Countable compactness implies pseudocompactness.*

Proof. (a) The first assertion is obvious. To prove the second, assume that X is a countably compact space and $f: X \to Y$ is a continuous surjective function. Let $\mathcal{U} = \{U_n : n \in \mathbb{N}\}$ be an open cover of Y, so $\mathcal{V} = \{f^{-1}(U_n) : n \in \mathbb{N}\}$ is an open cover of X; by hypothesis, there exists a finite subcover $\{f^{-1}(U_n) : n \in \{1, \dots, k\}\}$ of $\mathcal{V}$, and, clearly, $\{U_n : n \in \{1, \dots, k\}\}$ is a finite subcover of $\mathcal{U}$.

(b) This is obvious.

(c) Assume that X is a countably compact space and consider a continuous function $f: X \to \mathbb{R}$. For every $n \in \mathbb{N}$, let $U_n = \{x \in X : |f(x)| < n\}$. Then $\{U_n : n \in \mathbb{N}\}$ is a countable open cover of X, which has a finite subcover since X is countably compact. Therefore, f is bounded. □

Pseudocompact T_4-spaces are countably compact (see Exercise 15.4.3). In particular, pseudocompact metric spaces are countably compact, hence compact.

There exist regular spaces without nonconstant real-valued functions, they are vacuously pseudocompact. This is why pseudocompactness is treated only in Tichonov spaces. Here comes a criterion for pseudocompactness of such spaces. Recall that a family $\{A_i : i \in I\}$ of nonempty subsets of a topological space X is *locally finite* if for every $x \in X$ there exists an open neighborhood U of x in X such that $U \cap A_i \neq \emptyset$ for finitely many $i \in I$.

Theorem 15.1.2. *For a Tichonov space X, the following are equivalent:*
(a) *X is pseudocompact;*
(b) *every locally finite family of nonempty open sets of X is finite;*
(c) *$\bigcap_{n\in\mathbb{N}} V_n \neq \emptyset$ for every chain $\{V_n : n \in \mathbb{N}\}$ of nonempty open sets of X with*

$$\overline{V}_{n+1} \subseteq V_n \quad \textit{for every } n \in \mathbb{N}. \tag{15.1}$$

https://doi.org/10.1515/9783110654936-015

Proof. (a)⇒(b) Assume that $\{V_n : n \in \mathbb{N}_+\}$ is an infinite locally finite family of nonempty open sets of X. For every $n \in \mathbb{N}_+$, fix a point $x_n \in V_n$; since X is Tichonov, there exists a continuous function $f_n : X \to [0,1]$ such that f_n vanishes on $X \setminus V_n$ and $f_n(x_n) = 1$. Define a function $f : X \to \mathbb{R}$ by letting $f(x) = \sum_{n\in\mathbb{N}_+} nf_n(x)$ for every $x \in X$. Since $\{V_n : n \in \mathbb{N}_+\}$ is locally finite and each f_n is continuous, f is continuous as well. Obviously, f is unbounded, as $f(x_n) = n$ for every $n \in \mathbb{N}_+$. This contradicts the pseudocompactness of X.

(b)⇒(c) is obvious.

(c)⇒(a) Assume that $f : X \to \mathbb{R}$ is an unbounded continuous function. Then for every $n \in \mathbb{N}$ the open set $V_n = f^{-1}(\mathbb{R} \setminus [-n, n])$ is nonempty, and obviously (15.1) and $\bigcap_{n\in\mathbb{N}} V_n = \emptyset$ hold, a contradiction. □

Proposition 15.1.3. *If X is a dense pseudocompact subspace of a Tichonov space Y, then X is G_δ-dense in Y.*

Proof. Let O be a nonempty G_δ-set of Y. Then there exist $y \in O$ and open sets U_n of Y such that $O = \bigcap_{n\in\mathbb{N}} U_n$. By the regularity of Y, we can find, for each $n \in \mathbb{N}$, an open set V_n of Y such that $\overline{V}_n \subseteq V_{n-1}$ for every $n \in \mathbb{N}_+$ and $y \in V_n \subseteq \overline{V_n} \subseteq U_n$ for every $n \in \mathbb{N}$. Hence, $y \in O' = \bigcap_{n\in\mathbb{N}} V_n \subseteq O$ and O' is a G_δ-set in Y.

For every $n \in \mathbb{N}$, let $A_n = X \cap V_n$, which is an open set of X with $\overline{A_n}^Y = \overline{V_n}^Y$, in view of the density of X in Y and Lemma B.1.19. So, for every $n \in \mathbb{N}_+$, $\overline{A_n}^X = X \cap \overline{A_n}^Y = X \cap \overline{V_n}^Y \subseteq X \cap V_{n-1} = A_{n-1}$, and hence $\bigcap_{n\in\mathbb{N}} \overline{A_n}^X = \bigcap_{n\in\mathbb{N}} A_n \neq \emptyset$, by Theorem 15.1.2. Therefore, $X \cap O \supseteq X \cap O' = X \cap \bigcap_{n\in\mathbb{N}} V_n = \bigcap_{n\in\mathbb{N}} A_n \neq \emptyset$. □

If the topological space Y is only regular, one can prove that, if X is a dense countably compact subspace of Y, then X is G_δ-dense in Y (see Exercise 15.4.2).

A consequence of Proposition 15.1.3 is the next useful criterion.

Corollary 15.1.4. *A Tichonov space X is pseudocompact if and only if X is G_δ-dense in βX.*

Proof. Assume that X is not pseudocompact and arrange for an unbounded continuous function $f : X \to \mathbb{R}$ with $f(x) \geq 1$ for all $x \in X$. Then $g := 1/f : X \to (0,1]$ has $\inf\{g(x) : x \in X\} = 0$, so its continuous extension $\bar{g} : \beta X \to [0,1]$ has $0 \in \bar{g}(\beta X)$. Thus, $\bar{g}^{-1}(0) \neq \emptyset$ is a G_δ-set of βX that does not meet X, namely, X is not G_δ-dense in βX.

The converse implication follows from Proposition 15.1.3. □

Theorem 15.1.5. *Every pseudocompact Tichonov space X is a Baire space.*

Proof. Any compactification Y of X is a Baire space, by Theorem B.5.20. Since X is G_δ-dense in Y by Proposition 15.1.3, Exercise 15.4.1 applies. □

15.2 The Comfort–Ross criterion for pseudocompact groups

15.2.1 The Comfort–Ross criterion and first applications

The following characterization of pseudocompact groups, due to Comfort and Ross [61], makes them a very convenient class to work with.

Theorem 15.2.1 (Comfort–Ross criterion). *A topological group G is pseudocompact if and only if G is precompact and G_δ-dense in its (compact) completion.*

Proof. Assume that the topological group G is not precompact. By Lemma 10.2.7, there are a symmetric neighborhood $V \in \mathcal{V}_G(e_G)$ and a sequence $\{g_n\}_{n\in\mathbb{N}}$ in G such that $g_n V \cap g_m V = \emptyset$ whenever $m \neq n$ in $\mathbb{N}$. Pick a symmetric open $W \in \mathcal{V}_G(e_G)$ with $WW \subseteq V$. Then the family $\{g_n W : n \in \mathbb{N}\}$ of nonempty open sets is locally finite. This contradicts the pseudocompactness of G, by Theorem 15.1.2. For the second assertion, apply Proposition 15.1.3. So, every pseudocompact group is a dense – and by Proposition 15.1.3 even a G_δ-dense – subgroup of its compact completion.

Conversely, let G be precompact and G_δ-dense in its completion K. Assume for contradiction that $\mathcal{F} = \{x_n U_n : n \in \mathbb{N}\}$ is an infinite locally finite family in G, where $U_n \in \mathcal{V}_G(e_G)$ is open and $x_n \in G$ for all $n \in \mathbb{N}$. Write $U_n = G \cap W_n$ for an appropriate open $W_n \in \mathcal{V}_K(e_K)$, for $n \in \mathbb{N}$. Then $x_n W_n \cap G = x_n U_n$ for all $n \in \mathbb{N}$ and $W = \bigcap_{n\in\mathbb{N}} W_n$ is a G_δ-set of K. For every $n \in \mathbb{N}$, let $V_n = \bigcup_{k>n} x_k U_k$ and $V_n^* = \bigcup_{k>n} x_k W_k$; note that $V_n^* \cap G = V_n$. The local finiteness of $\mathcal{F}$ implies $\overline{V_n}^G = \bigcup_{k>n} \overline{x_k U_k}^G$, hence

$$\bigcap_{n\in\mathbb{N}} \overline{V_n}^G = \emptyset, \tag{15.2}$$

since every $z \in G$ has an open neighborhood O that does not meet $\overline{x_k U_k}^G$ for all $k \geq n$ and sufficiently large $n \in \mathbb{N}$, hence $z \notin \overline{V_n}^G$. Since K is compact, there exists an accumulation point x of the sequence $\{x_k\}_{k\in\mathbb{N}}$ in K. By the G_δ-density of G in K, there exists $g \in xW \cap G$. Then $g = xy$ for some $y \in W$. For a fixed $n \in \mathbb{N}$, since $g = xy$ is an accumulation point of the sequence $\{x_k y\}_{k\in\mathbb{N}}$ in K and $x_k y \in x_k W \subseteq x_k W_k \subseteq V_n^*$ for every $k > n$, we deduce that

$$g \in \overline{V_n^*}^K \cap G = \overline{V_n^* \cap G}^G = \overline{V_n}^G.$$

Then $g \in \bigcap_{n\in\mathbb{N}} \overline{V_n}^G$, which obviously contradicts (15.2). So, the family $\mathcal{F}$ must be finite, and G is pseudocompact by Theorem 15.1.2. □

Corollary 15.2.2. *If G is a G_δ-dense subgroup of a pseudocompact group H, then G is pseudocompact.*

Proof. Since G_δ-density is transitive, G is G_δ-dense in the completion of H. So, Comfort–Ross criterion 15.2.1 applies. □

Corollary 15.2.3. *Direct products of pseudocompact groups are pseudocompact.*

Proof. If $\{G_i : i \in I\}$ is a family of pseudocompact groups, then their completions $\{K_i : i \in I\}$ are compact and the completion of $G = \prod_{i\in I} G_i$ is $K = \prod_{i\in I} K_i$. Then K is compact and G_i is G_δ-dense in K_i for each $i \in I$, so G is G_δ-dense in K. Now Comfort–Ross criterion 15.2.1 applies. □

Corollary 15.2.4. *Every countably compact group is precompact. In particular, a countably compact group is compact precisely when it is complete.*

Proof. This follows from Comfort–Ross criterion 15.2.1, since every countably compact group is pseudocompact by Lemma 15.1.1(c) and Theorem 10.2.6. □

The following example shows that not all countably compact groups are compact. For an example of a pseudocompact group that is not countably compact, see Exercise 15.4.6.

Example 15.2.5. The Σ-product $\Sigma\mathbb{T}^{\mathbb{R}} = \bigcup\{\mathbb{T}^A : A \subseteq \mathbb{R}, |A| \leq \omega\}$ (here $\mathbb{T}^A$ is considered as a subgroup of $\mathbb{T}^{\mathbb{R}}$) is countably compact and noncompact.

Corollary 15.2.4 and Example 15.2.5 allow us to distinguish countable compactness from compactness. In particular, countably compact groups need not be complete. Nevertheless, countably compact groups are sequentially complete:

Remark 15.2.6. To see that a countably compact group G is sequentially complete, consider its completion $\widetilde{G}$. If G is sequentially closed in $\widetilde{G}$, then G is sequentially complete. Assume that G is not sequentially closed in $\widetilde{G}$. Then there exist $x \in \widetilde{G} \setminus G$ and a sequence $\{x_n\}_{n\in\mathbb{N}}$ in G such that $x_n \to x$. So, $\{x_n\}_{n\in\mathbb{N}}$ is a Cauchy sequence of G with no accumulation points in G, against the countable compactness of G (see Proposition B.5.8).

15.2.2 The G_δ-refinement and its relation to pseudocompact groups

Now we present an alternative way to prove that a pseudocompact group is G_δ-dense in its (compact) completion, by showing that the G_δ-refinement of a compact group is linear.

Notation 15.2.7. Let (X, τ) be a topological space. The family of all nonempty G_δ-sets of (X, τ) is a base of a finer topology τ_δ on X, called *G_δ-refinement* of τ. The space (X, τ) is a *P-space* if $\tau_\delta = \tau$.

The following elementary properties can be proved straightforward.

Lemma 15.2.8. (a) *The G_δ-refinement of a topological space is a P-space.*
(b) *The G_δ-refinement of a topological group is a topological group.*

(c) *If* $f: (X, \tau) \to (Y, \sigma)$ *is a continuous map (respectively, an embedding of topological spaces), then so is* $f_\delta: (X, \tau_\delta) \to (Y, \sigma)$, $x \mapsto f(x)$.
(d) *A subset* Y *of a topological space* (X, τ) *is* G_δ*-dense if and only if it is dense in the topology* τ_δ.

For a topological group G, we define

$$\Lambda(G) := \{H \leq G : H \text{ closed } G_\delta\text{-set}\} \quad \text{and} \quad \Lambda_\trianglelefteq(G) := \{N \trianglelefteq G : N \text{ closed } G_\delta\text{-set}\}.$$

Lemma 15.2.9. *Let* (G, τ) *be a topological group. Then* $\Lambda(G)$ *is a neighborhood base at* e_G *of* (G, τ_δ). *If* G *is precompact, then* $\Lambda_\trianglelefteq(G)$ *is a neighborhood base at* e_G *of* (G, τ_δ); *in particular, the* G_δ*-refinement is linear.*

Proof. It is clear that every $H \in \Lambda(G)$ is a neighborhood of e_G in τ_δ. Conversely, let $O = \bigcap_{n\in\mathbb{N}} U_n$, where each U_n is an open neighborhood of e_G in τ. We can build inductively a chain $\{W_n : n \in \mathbb{N}\}$ of symmetric neighborhoods of e_G in τ such that, for every $n \in \mathbb{N}$, $W_n \subseteq U_n$ and $\overline{W_{n+1}W_{n+1}} \subseteq W_n$. Then $H := \bigcap_{n\in\mathbb{N}} W_n \subseteq O$ is a closed G_δ-subgroup of G, as desired. In case G is compact, we can additionally assume that $y^{-1}W_{n+1}y \subseteq W_n$ for all $n \in \mathbb{N}$ and $y \in G$. In this case, H is a closed normal G_δ-subgroup of G.

Finally, assume that G is precompact and let $\widetilde{G}$ be the compact completion. There exist a sequence $\{\widetilde{U}_n\}_{n\in\mathbb{N}}$ of open neighborhoods of $e_{\widetilde{G}}$ such that $\widetilde{U}_n \cap G = U_n$ and $\widetilde{N} \in \Lambda_\trianglelefteq(\widetilde{G})$ contained in $\bigcap_{n\in\mathbb{N}} \widetilde{U}_n$. Then $N := \widetilde{N} \cap G \in \Lambda_\trianglelefteq(G)$ has the desired properties. □

Lemma 15.2.10. *Let* (G, τ) *be a topological group and* H *a subgroup of* G. *Then:*
(a) *if* $H \in \Lambda(G)$ *and* $L \in \Lambda(H)$, *then* $L \in \Lambda(G)$;
(b) H *is* G_δ*-dense in* G *if and only if* $HL = G$ *for every* $L \in \Lambda(G)$;
(c) *if* H *is* G_δ*-dense in* G *and* $N \in \Lambda(G)$, *then* $N \cap H$ *is* G_δ*-dense in* N; *if additionally,* $N \in \Lambda_\trianglelefteq(G)$, *then* $H/H \cap N \cong HN/N = G/N$.

Proof. (a) The subgroup L of G is τ-closed and $L \in (\tau\restriction_H)_\delta \subseteq \tau_\delta$, since $H \in \tau_\delta$.

(b) The subgroup H is G_δ-dense in G if and only if H is dense in (G, τ_δ). This is equivalent to $HL = G$ for all $L \in \Lambda(G)$, by Lemma 15.2.9.

(c) The first assertion follows from Lemma B.1.19 applied to the topology τ_δ, since $N \in \tau_\delta$. Assume now that $N \in \Lambda_\trianglelefteq(G)$. Since $H \cap N$ is in particular dense in N, the second assertion is a consequence of Theorem 3.2.9 and (b). □

Remark 15.2.11. Let K be a compact group. Obviously, $\psi(K/N) \leq \omega$ for a closed normal subgroup N of K implies $N \in \Lambda_\trianglelefteq(K)$. On the other hand, if $N \in \Lambda_\trianglelefteq(K)$, then $N = \bigcap_{n\in\mathbb{N}} O_n$ with $O_n \in \mathcal{V}(e_K)$ open for all $n \in \mathbb{N}$. By Lemma 8.2.1, for all $n \in \mathbb{N}$, there exists $U_n \in \mathcal{V}(e_K)$ with $NU_n \subseteq O_n$. Obviously, $N = \bigcap_{n\in\mathbb{N}} NU_n$, hence $\{NU_n : n \in \mathbb{N}\}$ witnesses $\psi(K/N) \leq \omega$. Since compact groups of countable pseudocharacter are metrizable by Theorem 8.2.7 and Birkhoff–Kakutani theorem 5.2.17, we deduce that $N \in \Lambda_\trianglelefteq(K)$ precisely when K/N is metrizable.

In conclusion, an alternative way to establish the G_δ-density of a pseudocompact group G in $K = \widetilde{G}$ is the following. By Lemma 15.2.10(b), this is equivalent to proving that if $N \in \Lambda_{\trianglelefteq}(K)$ then $GN = K$, namely, the canonical projection $q: K \to K/N$ satisfies $q(G) = K/N$. By Remark 15.2.11, K/N is metrizable, so its dense subgroup $q(G)$ is metrizable as well. On the other hand, $q(G)$ is pseudocompact (as a continuous image of G). Hence, $q(G)$ is compact, and by its density in K/N we conclude that $q(G) = K/N$.

15.3 C-embedded subsets and Moscow spaces

15.3.1 C- and C^*-embedded subgroups and subspaces

Let us recall that a subset Y of a topological space X is *C-embedded* (respectively, *C^*-embedded*) if every continuous function $f: Y \to \mathbb{R}$ (respectively, $f: Y \to [0,1]$) extends to X. According to Tietze theorem B.5.3, a closed subset of a normal space is C-embedded. Every C-embedded subspace Y of a topological space X is also C^*-embedded. Indeed, let $f: Y \to [0,1]$ be a continuous function. Then there is a continuous extension $g: X \to \mathbb{R}$ of f. The function $\bar{f} = \max\{\min\{g, 1\}, 0\}$ is continuous, extends f, and has values in $[0,1]$.

Let us see that C^*-embeddedness is equivalent to C-embeddedness for a pseudocompact G_δ-dense subset.

Lemma 15.3.1. *Let X be a topological space and Y a pseudocompact G_δ-dense subset of X. Then Y is C-embedded in X if and only if Y is C^*-embedded in X.*

Proof. By the above remark, it is sufficient to prove that Y is C-embedded if it is C^*-embedded. Pick a continuous function $f: Y \to \mathbb{R}$. Since Y is pseudocompact, f is bounded, and so we can assume without loss of generality that $f(Y) \subseteq (0,1)$. There exists a continuous extension $\bar{f}: X \to [0,1]$ of f. Since Y is G_δ-dense in X, $f(Y)$ is G_δ-dense in $\bar{f}(X)$, thus $\bar{f}(X) = f(Y) \subseteq (0,1)$ as $f(Y)$ is pseudocompact by Lemma 15.1.1(b), and so compact. Then $\bar{f}: X \to \mathbb{R}$ is a continuous extension of f, and therefore Y is C-embedded in X. □

Claim 15.3.2. *If X is a Tichonov space and O is a nonempty G_δ-set of X with $x \in O$, then there exists a nonempty zero-set Z of X such that $x \in Z \subseteq O$.*

Proof. Assume that $O \neq \emptyset$ is a G_δ-set of X, and let $O = \bigcap_{n\in\mathbb{N}_+} U_n$ for open sets U_n of X. Pick $x \in O$ and build a zero-set $x \in Z \subseteq O$ as follows. As X is a Tichonov space, for every $n \in \mathbb{N}_+$ there exists a continuous function $f_n: X \to [0,1]$ such that $f_n(x) = 0$ and $f_n(X \setminus U_n) = \{1\}$. Put

$$f = \sum_{n\in\mathbb{N}_+} \frac{1}{2^n} f_n,$$

which is a continuous function $f: X \to [0,1]$ and obviously $x \in Z := f^{-1}(0)$. If $z \in Z$, then $f(z) = 0$; thus, $f_n(z) = 0$, and so $z \in U_n$, for every $n \in \mathbb{N}_+$. Hence, $z \in O$. This proves that $Z \subseteq O$. □

There is a well-known fact, due to Gillman and Jerison, that connects C-embeddedness of a dense subspace of a Tichonov space to its G_δ-density:

Theorem 15.3.3. *If a dense subspace Y of a Tichonov space X is C-embedded in X, then Y is G_δ-dense in X.*

Proof. Assume that Y is dense but not G_δ-dense in X. We build a continuous real-valued function on Y that cannot be extended to X.

Let O be a nonempty G_δ-set contained in $X \setminus Y$ and let $x \in O$. By Claim 15.3.2, there exists a zero-set Z of X such that $x \in Z \subseteq O$. Let the continuous function $f: X \to \mathbb{R}$ witness $Z = f^{-1}(0)$. Then $0 \notin f(Y)$, so the function $g := 1/f: Y \to \mathbb{R}$ is continuous. Since Y is dense in X, g cannot be extended to a continuous function $\bar{g}: X \to \mathbb{R}$, as the equality $f(y)g(y) = 1$, valid for every $y \in Y$, cannot be extended to $f(y)\bar{g}(y) = 1$ on $X \setminus Y$. □

Corollary 15.1.4 ensures the pseudocompactness of a topological group G which is G_δ-dense in βG. In Corollary 15.3.7 we see that $K = \beta G$ in case G is a G_δ-dense subgroup of a compact group K. This amounts to check that every continuous function $f: G \to [0,1]$ extends to K, in other words, that G is C^*-embedded in K.

Unfortunately, the implication in Theorem 15.3.3 goes in the "wrong direction". Indeed, we need to deduce C^*-embeddedness from G_δ-density and this is apparently not always possible in the framework of topological spaces. But fortunately, this is the case when G is a G_δ-dense subgroup of a compact group K.

The argument we give here is inspired by the original proof of the Comfort–Ross criterion with due modifications and simplifications.

We make use of the following sharpening of the property from Lemma 15.2.9:

Lemma 15.3.4. *In a compact group K, for every nonempty open set U of K, there exists $N \in \Lambda_{\trianglelefteq}(K)$ such that $\overline{U} = \overline{U}N$.*

Proof. According to [251, Theorem 1.6], $\overline{U}$ is a Baire set, hence [164, Theorem 64 G] applies. □

In other words, if $e_K \in U$, then $\overline{U}$ not only contains a subgroup $N \in \Lambda_{\trianglelefteq}(K)$, but it *is actually a union of cosets of N.*

Remark 15.3.5. If in Lemma 15.3.4 the compact group K is metrizable, then one can take $N = \{e_K\}$ to achieve that $\overline{U} = \overline{U}N$. Also when $K = \prod_{i\in I} M_i$, where each M_i is a metrizable compact group, the statement of Lemma 15.3.4 is obviously true for $U = W \times \prod_{i\in I\setminus J} M_i$, a basic open set of K, with $W \subseteq \prod_{j\in J} M_j$ open and J a finite subset of I: indeed, take $N = \prod_{j\in J}\{e_K\} \times \prod_{i\in I\setminus J} M_i$.

The following lemma, combined with Theorem 15.3.3, shows that a precompact group is C-embedded in a maximal possible subgroup of its completion.

Lemma 15.3.6. *Let G be a precompact group with completion (K,τ). Then G is C-embedded in $H := \overline{G}^{\tau_\delta}$, the closure of G in (K,τ_δ).*

Proof. Fix a continuous function $f: G \to \mathbb{R}$ and a countable base of open sets $\mathcal{B} = \{\Delta_n : n \in \mathbb{N}\}$ of the topology on $\mathbb{R}$. For every $n \in \mathbb{N}$, let $U_n = f^{-1}(\Delta_n)$ and find an open set W_n of K such that $U_n = G \cap W_n$. By Lemma 15.3.4, for $n \in \mathbb{N}$ there exists $N_n \in \Lambda_{\trianglelefteq}(K)$ such that

$$\overline{W_n N_n} = \overline{W_n} \tag{15.3}$$

(here and in the sequel all closures are taken in K). Let $N = \bigcap_{n\in\mathbb{N}} N_n \in \Lambda_{\trianglelefteq}(K)$. Let us see first that $\overline{W_n N} = \overline{W_n}$ for every $n \in \mathbb{N}$: the inclusion $\overline{W_n} \subseteq \overline{W_n N}$ is obvious, while $\overline{W_n N} \subseteq \overline{W_n N_n} = \overline{W_n}$ comes by (15.3).

We check that, for $x, y \in G$,

$$\text{if} \quad xN = yN \quad \text{then} \quad f(x) = f(y). \tag{15.4}$$

Assume for a contradiction that $f(x) \neq f(y)$ in $\mathbb{R}$. Then there exist $n, m \in \mathbb{N}$ such that $\overline{\Delta_n} \cap \overline{\Delta_m} = \emptyset$ and $f(x) \in \Delta_n$, while $f(y) \in \Delta_m$, that is, $x \in U_n$ and $y \in U_m$. From $xN = yN$, we deduce that $y \in xN \subseteq \overline{W_n N} = \overline{W_n}$, and so $y \in \overline{W_m} \cap \overline{W_n}$. Moreover, $\overline{W_n} = \overline{U_n}$ by Lemma B.1.19. Therefore, $f(y) \in f(\overline{U_n} \cap G) \subseteq \overline{f(U_n)} \subseteq \overline{\Delta_n}$. On the other hand, $f(y) \in \Delta_m \subseteq \overline{\Delta_m}$, and so $f(y) \in \overline{\Delta_m} \cap \overline{\Delta_n}$, against $\overline{\Delta_n} \cap \overline{\Delta_m} = \emptyset$.

By (15.4), there exists a correctly defined function $f': G/(N \cap G) \to \mathbb{R}$ such that $f = f' \circ q$, where $q: G \to G/(N \cap G)$ is the canonical projection. Since $f = f' \circ q$ is continuous, the standard property of the quotient topology of $G/(N \cap G)$ guarantees that f' is continuous as well.

$$\begin{array}{ccc} G & \xrightarrow{\quad f \quad} & \mathbb{R} \\ & {\scriptstyle q}\searrow \quad \nearrow{\scriptstyle f'} & \\ & G/(N \cap G) & \end{array} \tag{15.5}$$

Since G is G_δ-dense in H and $N \cap H \in \Lambda_{\trianglelefteq}(H)$, Lemma 15.2.10(c) yields that $G/(G \cap N)$ is topologically isomorphic to $H/(H \cap N)$. So, the composition $H \to H/(H \cap N) \to G/(G \cap N) \xrightarrow{f'} \mathbb{R}$ is a continuous extension of f. □

The following corollary reinforces one of the implications of Comfort–Ross criterion 15.2.1.

Corollary 15.3.7. *If G is a G_δ-dense subgroup of a compact group K, then $K = \beta G$ and G is pseudocompact.*

Proof. In view of Lemma 15.3.6, G is C-embedded in K, and hence $K = \beta G$. Since G is G_δ-dense in $K = \beta G$, G is pseudocompact by Corollary 15.1.4. □

Corollary 15.3.8. *If G is a pseudocompact group, then $bG = \widetilde{G} = \beta G$.*

Proof. Since G is pseudocompact, G is G_δ-dense in $\widetilde{G}$ by Theorem 15.2.1. By Corollary 15.3.7, $\widetilde{G} = \beta G$. That $bG = \widetilde{G}$ follows from Theorem 10.2.15, since G is precompact by Theorem 15.2.1. □

15.3.2 $\mathbb{R}$-factorizable groups and Moscow spaces

We discuss here two remarkable notions in view of their utility for the study of topological groups, even if we are not using them.

The above proofs of Lemma 15.3.6 and Corollary 15.3.7 show also the following interesting property.

Proposition 15.3.9. *If G is a pseudocompact group and $f: G \to \mathbb{R}$ is a continuous function, then there exist a second countable group M, a continuous homomorphism $h: G \to M$, and a continuous function $f': M \to \mathbb{R}$ such that $f = f' \circ h$.*

The factorization property in Proposition 15.3.9 appeared implicitly in many papers and monographs, starting with Pontryagin [228]. Topological groups with this property were called $\mathbb{R}$-*factorizable* by Tkachenko who carried out a deep study of this remarkable class of groups [272, 273]. In particular, he proved that *precompact groups are* $\mathbb{R}$-*factorizable* (see [7] where further results can be found).

A Tichonov space X is said to be a *Moscow space* if for every open set U in X and $x \in \overline{U}$ there exists a G_δ-set P of X such that $x \in P \subseteq \overline{U}$ (hence, the closure of an open set is a union of closures of G_δ-sets). It was proved by Tkachenko and Uspenskij (see [7]) that a Tichonov space X is a Moscow space if and only if every dense subset Y of X is C-embedded in its G_δ-closure in X.

Lemma 15.3.4 shows that compact groups are Moscow spaces. Actually, all precompact groups are Moscow spaces (see [7]).

15.3.3 Submetrizable pseudocompact groups are compact

To prove the statement in the subsection title, we need the following:

Lemma 15.3.10. *If X is a pseudocompact Tichonov space, then every singleton $\{x\}$ in X that is a G_δ-set in X is also a G_δ-set in βX.*

Proof. According to Claim 15.3.2, if a singleton $\{x\}$ in a Tichonov space X is a G_δ-set, then $\{x\}$ is functionally closed, i. e., there exists a continuous function $f: X \to [0,1]$ such that $f^{-1}(0) = \{x\}$. Let us see that $\{x\}$ is functionally closed in βX as well. Indeed, let $\bar{f}: \beta X \to [0,1]$ be the continuous extension of f. Assume that $y \in \bar{f}^{-1}(0) \setminus X$, and pick a continuous function $g: \beta X \to [0,1]$ with $g(x) = 1$ and $g(y) = 0$. Then, for $h = g + \bar{f}$, one has $h > 0$ on X, while $h(y) = 0$. On the other hand, $h(X) \subseteq (0, 2]$ is compact, in view of the pseudocompactness of X. Hence, $h(\beta X) = h(\overline{X}) \subseteq \overline{h(X)} = h(X) \subseteq (0, 2]$, a contradiction (in view of $h(y) = 0$). □

Corollary 15.3.11. *A pseudocompact submetrizable group G is compact.*

Proof. By Theorem 15.2.1, G is precompact. Let K be the compact completion of G. If G admits a coarser metrizable group topology, then the singleton $\{e_G\}$ is a G_δ-set of G. Since $K = \beta G$ by Corollary 15.3.7, Lemma 15.3.10 yields that $\{e_G\}$ is a G_δ-set of $K = \beta G$ as well. Therefore, K is metrizable, so G is metrizable as well. Since G is G_δ-dense in K by Theorem 15.2.1, we get that $G = K$ is compact. □

Corollary 15.3.12. *A metrizable compact group admits no strictly finer pseudocompact group topology.*

Proof. If (G, τ) is a metrizable compact group and $\tau' \geq \tau$ is a finer pseudocompact group topology on G, then τ' is submetrizable, hence compact by Corollary 15.3.11. Since compact groups are minimal, this entails $\tau' = \tau$. □

This corollary is a part of a more precise theorem due to Comfort and Robertson [58], treating the question of when a compact abelian group G admits a strictly finer pseudocompact group topology:

Theorem 15.3.13 ([58]). *A compact abelian group G admits a strictly finer pseudocompact group topology if and only if G is not metrizable.*

15.4 Exercises

Exercise 15.4.1. Let Y be a Baire space and X be a G_δ-dense subspace of Y. Prove that X is a Baire space as well.

Exercise 15.4.2. Prove that if X is a dense countably compact subspace of a regular space Y, then X is G_δ-dense in Y.

Hint. Proceed as in the proof of Lemma 15.1.3.

Exercise 15.4.3. Prove that a pseudocompact T_4-space X is countably compact.

Hint. Assume for a contradiction that X is not countably compact. Then there exists a sequence $A = \{x_n\}_{n\in\mathbb{N}}$ in X with no accumulation point; in particular, A is discrete and closed. Let $f: A \to \mathbb{R}$ be defined by $f(x_n) = n$ for every $n \in \mathbb{N}$. By Tietze theorem B.5.3, there exists a continuous extension of f to X, which is clearly unbounded.

Exercise 15.4.4. Prove that a group having a dense pseudocompact subgroup is necessarily pseudocompact.

Exercise 15.4.5. Show that a countably infinite Hausdorff group cannot be pseudocompact.

Hint. If G is a countably infinite Hausdorff group and it is also pseudocompact, then it is discrete, being a Baire space in view of Theorem 15.1.5. By Theorem 15.2.1, pseudocompact groups are precompact, so G must be finite, a contradiction.

Exercise 15.4.6. Give examples of pseudocompact groups having closed nonpseudocompact subgroups. Deduce from Lemma 15.1.1 that there exist pseudocompact groups that are not countably compact.

Hint. Let $K = \mathbb{T}^{\mathbb{R}}$ and $G_0 = \Sigma\mathbb{T}^{\mathbb{R}}$; moreover, let $Q = (\mathbb{Q}/\mathbb{Z})^{\mathbb{R}} \cap \Delta_K \cong \mathbb{Q}/\mathbb{Z}$, where Δ_K denotes the diagonal. By Example 15.2.5 and Exercise 15.4.4, $G = G_0 + Q$ is pseudocompact. Yet $Q = G \cap \Delta_K$ is a closed subgroup of G that is not pseudocompact, in view of Exercise 15.4.5.

Exercise 15.4.7. Show that for an infinite abelian group G, the group $G^{\#}$ cannot be pseudocompact.

Hint. Show first that G admits a subgroup H such that G/H is countably infinite. Then use the fact that H is a closed subgroup of $G^{\#}$ and $(G/H)^{\#}$ coincides with the quotient $G^{\#}/H$, so must be pseudocompact. This contradicts Exercise 15.4.5.

15.5 Further readings, notes, and comments

The fact that metrizable pseudocompact groups are compact triggered another question that was aggressively attacked by Comfort and his coauthors. Namely, the necessity in Theorem 15.3.13 can be interpreted by saying that a metrizable pseudocompact group (G, τ) is *r-extremal*, in the sense that it admits no strictly finer pseudocompact group topology. The question of whether this implication can be inverted, i. e., whether the r-extremal pseudocompact groups are necessarily metrizable turned out to be rather hard (see [59, 88] for some partial results). The final affirmative solution came only twentyfive years later in [62]. A generalization of this problem for α-pseudocompact groups was resolved in [155].

The structure of groups admitting pseudocompact group topologies, as well as many other features of pseudocompact groups, is discussed in [101]. The final solution of the problem of the description of the algebraic structure of pseudocompact groups was given by Shakhmatov and the second named author (unpublished). An alternative proof of the necessity part of this characterization was given later in [85] as an application of the new cardinal invariant (*divisible weight*) introduced there. This cardinal invariant motivated the introduction in [107] of a similar one (*divisible rank*) that turned out to have relevant applications (along with the divisible weight) for the solution of various problems (see [106, 107, 108, 257]).

As a relevant consequence of Corollary 15.3.7, one can deduce that all real-valued continuous functions on a pseudocompact group are uniformly continuous.

16 Topological rings, fields, and modules

16.1 Topological rings and fields

Definition 16.1.1. A topology τ on a ring A is a *ring topology* if the maps $f: A \times A \to A$, $(x, y) \mapsto x - y$, and $m: A \times A \to A$, $(x, y) \mapsto xy$, are continuous when $A \times A$ carries the product topology. A *topological ring* is a pair (A, τ) of a ring A and a ring topology τ on A.

For every ring A, the discrete and indiscrete topologies on A are ring topologies. Nontrivial examples of topological rings are provided by the fields $\mathbb{R}$ and $\mathbb{C}$.

Example 16.1.2. For every prime p, the group $\mathbb{J}_p$ of p-adic integers carries also a ring structure and its compact group topology is also a ring topology.

Other examples of ring topologies are given in Example 16.1.7.

Obviously, a topology τ on a ring A is a ring topology if and only if $(A, +, \tau)$ is a topological group and the map $m: A \times A \to A$ is continuous. We shall exploit this fact and in particular that, for $a \in A$, the filter $\mathcal{V}_\tau(a)$ coincides with $a + \mathcal{V}_\tau(0)$.

The following theorem is a counterpart of Theorem 2.1.10.

Theorem 16.1.3. *Let A be a ring and $\mathcal{V}(0_A)$ the filter of all neighborhoods of 0_A in some ring topology τ on A. Then (gt1), (gt2), and the following conditions hold:*
(rt1) *for every $U \in \mathcal{V}(0_A)$ and $a \in A$, there is $V \in \mathcal{V}(0_A)$ with $Va \cup aV \subseteq U$;*
(rt2) *for every $U \in \mathcal{V}(0_A)$, there exists $V \in \mathcal{V}(0_A)$ with $VV \subseteq U$.*

Conversely, if $\mathcal{V}$ is a filter on A satisfying (gt1), (gt2), (rt1), and (rt2), then there exists a unique ring topology τ on A such that $\mathcal{V} = \mathcal{V}_\tau(0_A)$.

Proof. Since $(A, +, \tau)$ is a topological group, (gt1) and (gt2) hold by Theorem 2.1.10. To prove (rt2), it suffices to apply the definition of continuity of the multiplication m at $(0_A, 0_A) \in A \times A$. Analogously, for (rt1) use the continuity of the multiplication m at $(0_A, a) \in A \times A$ and $(a, 0_A) \in A \times A$.

Let $\mathcal{V}$ be a filter on A satisfying all conditions (gt1), (gt2), (rt1), and (rt2). By Theorem 2.1.10, there exists a group topology τ on $(A, +)$ such that $\mathcal{V} = \mathcal{V}_\tau(0_A)$. It remains to check that τ is a ring topology, i. e., the multiplication $m: A \times A \to A$ is continuous at every pair $(a, b) \in A \times A$. Pick a neighborhood of $ab \in A$; it is not restrictive to take it of the form $ab + U$, with $U \in \mathcal{V}$. Next, choose $V \in \mathcal{V}$ such that $V + V + V \subseteq U$ and pick $W \in \mathcal{V}$ with $WW \subseteq V$, $aW \subseteq V$, and $Wb \subseteq V$. Then

$$m((a + W) \times (b + W)) = ab + aW + Wb + WW \subseteq ab + V + V + V \subseteq ab + U.$$

This proves the continuity of the multiplication $m: A \times A \to A$ at (a, b). □

https://doi.org/10.1515/9783110654936-016

16.1.1 Examples and general properties of topological rings

Remark 16.1.4. Let $\mathcal{V} = \{J_i : i \in I\}$ be a filter base consisting of two-sided ideals of a ring A. Since $\mathcal{V}$ satisfies (gt1), (gt2), (rt1), and (rt2), by Theorem 16.1.3 it generates a ring topology on A having as a local base at $a \in A$ the family of cosets $\{a + J_i : i \in I\}$. Ring topologies of this type are called *linear ring topologies.*

Let (A, τ) be a topological ring and let $\mathfrak{A}$ be a two-sided ideal of A. The quotient ring $A/\mathfrak{A}$, equipped with the quotient topology of the underlying abelian group $(A/\mathfrak{A}, +)$, is a topological ring that we call the *quotient ring.*

If (A, τ) is a topological ring, the closure of a two-sided (respectively, left, right) ideal of A is again a two-sided (respectively, left, right) ideal of A. In particular, $\mathfrak{A} = \operatorname{core}(A)$ is a closed two-sided ideal. As we already know, the quotient ring $\mathfrak{h}A = A/\mathfrak{A}$ is Hausdorff and shares many of the properties of the topological ring (A, τ). This is why we consider exclusively Hausdorff rings.

Definition 16.1.5. A Hausdorff ring (A, τ) is called *complete* if it is complete as a topological group.

If (A, τ) is a Hausdorff ring, the completion $\widetilde{A}$ of the topological group $(A, +, \tau)$ carries a natural ring structure, obtained by the extension of the ring operation of A to $\widetilde{A}$ by continuity using the fact that the product of two Cauchy nets of A is a Cauchy net (see Exercise 16.3.1). In this way, $\widetilde{A}$ becomes a topological ring.

As far as connectedness is concerned, one has the following easy to prove fact:

Theorem 16.1.6. *The connected component of a topological ring is a two-sided ideal. Hence, every topological ring that is a division ring is either connected or hereditarily disconnected.*

Let us see some basic examples of linear ring topologies.

Example 16.1.7. Let A be a ring and $\mathfrak{A}$ a two-sided ideal of A. Then $\{\mathfrak{A}^n : n \in \mathbb{N}\}$ is a filter base of a ring topology, named the *$\mathfrak{A}$-adic topology.*

(a) The p-adic topology of the ring $\mathbb{J}_p$ coincides also with the $p\mathbb{J}_p$-adic topology of the ring $\mathbb{J}_p$, generated by the ideal $p\mathbb{J}_p$.
(b) Let k be a field and $A = k[x]$ the polynomial ring over k. Take $\mathfrak{A} = xA = (x)$; then the $\mathfrak{A}$-adic topology has as basic neighborhoods of 0 the principal ideals $x^n A$.
(c) The completion $\widetilde{A}$ of the ring $A = k[x]$, equipped with the (x)-adic topology is the ring $k[[x]]$ of formal power series over k (elements of $k[[x]]$ are the formal power series of the form $\sum_{n=0}^{\infty} a_n x^n$, with $a_n \in k$ for all n). The topology of the completion $\widetilde{A}$ coincides with the $x\widetilde{A}$-adic topology of $\widetilde{A}$.
(d) Let k be a field, $n \in \mathbb{N}_+$, and $A = k[x_1, \dots, x_n]$ the ring of polynomials of n variables over k. Take $\mathfrak{A} = x_1 A + \cdots + x_n A$; then the $\mathfrak{A}$-adic topology has as basic neighborhoods of 0 the ideals $(\mathfrak{A}^m)$, $m \in \mathbb{N}$, where the power $\mathfrak{A}^m$ consists of all polynomials having no terms of degree less than m.

(e) For every $n \in \mathbb{N}_+$, the completion $\widetilde{A}$ of the ring $A = k[x_1, \dots, x_n]$, equipped with the $(x_1, \dots, x_n)$-adic topology, is the ring $k[[x_1, \dots, x_n]]$ of formal power series of n variables over k. The topology of $\widetilde{A}$ coincides with the $(x_1\widetilde{A}+\dots+x_n\widetilde{A})$-adic topology. Moreover, $\widetilde{A}$ is compact precisely when k is finite.

16.1.2 Topological fields

Definition 16.1.8. A topology τ on a field K is said to be a *field topology* if (K, τ) is a topological ring and the map $\iota: K \setminus \{0\} \to K \setminus \{0\}$, $x \mapsto x^{-1}$, is continuous. A *topological field* is a pair (K, τ) of a field K and a field topology τ on K.

The next example provides instances of infinite locally compact fields.

Example 16.1.9. (a) Clearly, $\mathbb{R}$ and $\mathbb{C}$ are (connected) locally compact fields.
(b) For every prime p, the field $\mathbb{Q}_p$ is a locally compact field.

The example in item (a) gives all connected locally compact fields:

Theorem 16.1.10 (Pontryagin [228]). *The only connected locally compact fields are $\mathbb{R}$ and $\mathbb{C}$.*

According to Theorem 16.1.6, a topological field that is not connected is necessarily hereditarily disconnected. In the sequel we discuss the hereditarily disconnected locally compact fields.

A subset B of a topological ring A is *bounded* if for every $U \in \mathcal{V}(0)$ there exists a $V \in \mathcal{V}(0)$ such that $VB \subseteq U$ and $BV \subseteq U$. If $A = K$ is a topological field, a subset $B \subseteq K$ containing 0 is *retrobounded* if $(K \setminus B)^{-1}$ is bounded. If every neighborhood of 0 in K is retrobounded, K is said to be *locally retrobounded*.

Example 16.1.11. An *absolute value* on a field K is a norm $x \mapsto |x|$ of the additive group $(K, +)$ such that $|xy| = |x| \cdot |y|$ for $x, y \in K$. The metric topology τ generated by this norm makes (K, τ) a retrobounded topological field, where the bounded sets are precisely those which are bounded for the metric (see Exercise 16.3.5). The usual absolute values in $\mathbb{R}$ and $\mathbb{C}$, as well as the p-adic norm $| - |_p$ in $\mathbb{Q}_p$, are absolute values in this sense.

Theorem 16.1.12. *If K is a locally retrobounded field, then:*
(a) ([216, Theorems 2 and 6]) *every finite dimensional K-vector space admits a unique topological K-vector space topology, namely, the product topology;*
(b) ([287, Theorem 13.9]) *the completion $\widetilde{K}$ is a locally retrobounded field.*

Since locally compact fields are locally retrobounded (see [287]), item (a) of this theorem applies for locally compact fields. Item (b) provides examples of topological fields that are not locally retrobounded (see Exercise 16.3.6).

The locally compact fields from Example 16.1.9 have characteristic 0. Kowalsky [187] proved that a hereditarily disconnected locally compact field K of characteristic 0 is necessarily a finite extension of $\mathbb{Q}_p$, for some prime p. By Theorem 16.1.12(a), we can assume that K carries the Tichonov topology, so $K \cong \mathbb{Q}_p^d$ as a $\mathbb{Q}_p$-vector space, where $d = [K : \mathbb{Q}_p]$.

Here comes an example of a locally compact field of positive characteristic.

Example 16.1.13. Take the compact ring $k[[x]]$, where k is a finite field, and its field of fractions $K = k((x))$, consisting of formal Laurent power series of the form $\sum_{n=n_0}^{\infty} a_n x^n$, $n_0 \in \mathbb{Z}$ and $a_n = 0$ for all $n < n_0$. By declaring the subring $k[[x]]$ of K open, with its compact topology, one obtains a locally compact field topology on K having the same characteristic as k. Moreover, since $K \cong k^{(\mathbb{N})} \times k^{\mathbb{N}}$ as topological groups, where $k^{\mathbb{N}} \cong k[[x]]$ is open and compact while the first summand is discrete, one has $K \cong \widehat{K}$, i. e., K is selfdual.

The topology in the above example is generated by an appropriate absolute value of K (e. g., letting $|f| = 2^{-n_0}$ for a nonzero $f = \sum_{n=n_0}^{\infty} a_n x^n \in K$ with $a_{n_0} \neq 0$, and $|0| = 0$) and the same applies to the finite extensions of $\mathbb{Q}_p$ mentioned above. More generally, every locally compact field K admits an absolute value and carries the metric topology induced by this absolute value (see [180, 289]). An infinite locally compact field K of finite characteristic p is necessarily of the form described in Example 16.1.13 for some finite field k of characteristic p (see [289, Theorem 8]).

Since $\mathbb{R}$, $\mathbb{Q}_p$, and $\mathbb{Z}/p\mathbb{Z}((x))$ are selfdual and this property is preserved by taking finite products, we deduce that all locally compact fields are selfdual.

16.2 Topological modules

Definition 16.2.1. Let A be a topological ring. A (left) A-module M is a *(left) topological A-module* if M is a topological group and the module multiplication $A \times M \to M$ is continuous when $A \times M$ carries the product topology.

A proof in the line of the proof of Theorem 16.1.3 shows that:

Theorem 16.2.2. *If A is a topological ring, then a left A-module M equipped with a group topology τ is a topological A-module if and only if its filter $\mathcal{V}(0_M)$ of neighborhoods satisfies, beyond (gt1) and (gt2), also these three properties:*

(mt1) *for every $W \in \mathcal{V}(0_M)$, there exist $U \in \mathcal{V}(0_A)$ and $V \in \mathcal{V}(0_M)$ with $UV \subseteq W$;*

(mt2) *for every $U \in \mathcal{V}(0_M)$ and $x \in M$, there exists $V \in \mathcal{V}(0_A)$ with $Vx \subseteq U$;*

(mt3) *for every $U \in \mathcal{V}(0_M)$ and $a \in A$, there exists $V \in \mathcal{V}(0_M)$ with $aV \subseteq U$.*

Conversely, if $\mathcal{V}$ is a filter on M satisfying (gt1), (gt2), (mt1), (mt2), and (mt3), then there exists a unique topological A-module topology τ on M such that $\mathcal{V} = \mathcal{V}_\tau(0_M)$.

Remark 16.2.3. Combining (mt1) and (mt2), one can show that for every $U \in \mathcal{V}(0_M)$ and $x \in M$ there exists $V \in \mathcal{V}(0_A)$ and $W \in \mathcal{V}_M(x)$ such that $VW \subseteq U$. Using this, one can show that for every $U \in \mathcal{V}(0_M)$ and every compact set $K \in M$ there exists $V \in \mathcal{V}(0_A)$ such that $VK \subseteq U$.

16.2.1 Uniqueness of the Pontryagin-van Kampen duality

In this subsection R will be a locally compact commutative unitary ring and $\mathcal{L}_R$ the category of locally compact topological unitary R-modules and continuous R-module homomorphisms. The commutativity of R allows us to avoid distinguishing between left and right R-modules and gives the possibility to define, for a morphism $f: M \to N$ in $\mathcal{L}_R$ and $r \in R$, the morphism $rf: M \to N$ by $(rf)(x) = rf(x)$ for $x \in M$. For the discrete ring $\mathbb{Z}$, $\mathcal{L}_\mathbb{Z}$ is the whole class of locally compact abelian groups. However, for the discrete ring $\mathbb{Q}$, $\mathcal{L}_\mathbb{Q}$ consists only of the torsion-free divisible locally compact abelian groups (for more examples, see Exercise 16.3.8).

The Pontryagin-van Kampen dual $\widehat{M}$ of $M \in \mathcal{L}_R$ has a natural structure of an R-module (with $(r\chi)(x) = \chi(rx)$ for $\chi \in \widehat{M}$ and $r \in R$) and $\widehat{M} \in \mathcal{L}_R$ (see Exercise 16.3.10). For a morphism $f: M \to N$ in $\mathcal{L}_R$ and $r \in R$, one has $\widehat{rf} = r\widehat{f}$.

Roeder theorem 13.4.19 on the uniqueness of the Pontryagin-van Kampen duality functor was rediscovered independently by Prodanov [235] several years later in the much more general setting of the category $\mathcal{L}_R$, as follows.

Definition 16.2.4. A *functorial duality* $^\#: \mathcal{L}_R \to \mathcal{L}_R$ is an involutive duality of $\mathcal{L}_R$ such that $(rf)^\# = rf^\#$ for each morphism $f: M \to N$ in $\mathcal{L}_R$ and $r \in R$.

The above observation shows that the restriction of the Pontryagin-van Kampen duality functor on $\mathcal{L}_R$ is a functorial duality, so *there is always a functorial duality on* $\mathcal{L}_R$, that we shall briefly call Pontryagin-van Kampen functor (duality) of $\mathcal{L}_R$.

Theorem 16.2.5 (Stoyanov [268]). *If R is a compact commutative ring, then the Pontryagin-van Kampen functor is the only functorial duality on $\mathcal{L}_R$.*

Surprisingly, the case of a discrete ring R turned out to be more complicated. From now on R is discrete.

Let us recall that $X \in \mathcal{L}_R$ is a (co)generator if for every nontrivial $M \in \mathcal{L}_R$ there exists a nontrivial morphism $X \to M$ (respectively, $M \to X$). Clearly, every functorial duality $^\#: \mathcal{L}_R \to \mathcal{L}_R$ takes generators to cogenerators, and vice versa. Furthermore, $^\#$ takes the subcategory $\mathcal{C}_R$ of compact R-modules to the subcategory $\mathcal{D}_R$ of discrete ones, and vice versa (use an argument as in Exercise 13.7.6 tailored for $\mathcal{L}_R$).

Of primary importance is the compact R-module $T = R^\#$, named the *torus* of the functorial duality $^\#: \mathcal{L}_R \to \mathcal{L}_R$ (in particular, $\widehat{R}$ is the torus of the Pontryagin-van Kampen functor). Since R is a discrete projective generator of $\mathcal{L}_R$, T is a compact injective cogenerator of $\mathcal{L}_R$ with $\mathrm{CHom}_R(T, T) \cong R$ canonically. Let us denote by $\mathfrak{T}_R$ the family of isomorphism classes of R-modules T with these properties, i. e., T is a compact injective cogenerator of $\mathcal{L}_R$ with $\mathrm{CHom}_R(T,T) \cong R$. To every functorial duality $^\#: \mathcal{L}_R \to \mathcal{L}_R$ we associated the R-module $T = R^\# \in \mathfrak{T}_R$. For the sake of brevity, we denote by $\mathfrak{D}_R$ the family of all (equivalence classes under natural equivalence of) functorial dualities $^\#$ on $\mathcal{L}_R$. Then the assignment $^\#: R \mapsto R^\#$ defines a correspondence $\Phi_R: \mathfrak{D}_R \to \mathfrak{T}_R$ which need not be one-to-one in general.

Let us see now that Φ_R is surjective. To this end, for $T \in \mathfrak{T}_R$, define the R-module $\Delta_T(X) := \mathrm{CHom}_R(X,T)$ of all continuous R-module homomorphisms equipped with the compact-open topology. The elements of $\Delta_T(X)$ are called *continuous T-characters of X* and it is easily checked that $\Delta_T(X) \in \mathcal{L}_R$. The assignment $\Delta_T: \mathcal{L}_R \to \mathcal{L}_R$, $X \mapsto \Delta_T(X)$, induces a contravariant functor which is a functorial duality. The natural equivalence $\bar{\omega}$ between $1_{\mathcal{L}_R}$ and $\Delta_T \cdot \Delta_T$ is defined by the evaluation homomorphisms $\bar{\omega}_X: X \to \Delta_T(\Delta_T(X))$, i. e., for $x \in X$ and $\chi \in \Delta_T(X)$, $\bar{\omega}_X(x)(\chi) = \chi(x)$ (see [96, Theorem 5.2]). Moreover, $\Delta_T(R) = \mathrm{CHom}_R(R,T) \cong T$, so T is the torus of Δ_T. This shows that each $T \in \mathfrak{T}_R$ is the torus of some functorial duality on $\mathcal{L}_R$. This is why we briefly call *torus* any R-module $T \in \mathfrak{T}_R$.

In order to discuss injectivity of the correspondence $\Phi_R: \mathfrak{D}_R \to \mathfrak{T}_R$, pick a functorial duality $^\#: \mathcal{L}_R \to \mathcal{L}_R$ and its torus T. As we already discussed above, there is already a functorial duality, namely, Δ_T, with torus T. The question is whether Δ_T coincides with (namely, is naturally equivalent to) $^\#$. It is easy to find an algebraic isomorphism $i_X: X^\# \to \Delta_T(X)$ for every $X \in \mathcal{L}_R$, obtained as a composition of the algebraic isomorphisms

$$X^\# \to \mathrm{Hom}_R(R, X^\#) \to \mathrm{CHom}_R(X^{\#\#}, R^\#) \to \mathrm{CHom}_R(X, R^\#) = \Delta_T(X),$$

where the first is given by the assignment $x \mapsto \phi_x$ with $\phi_x(r) = rx$ for $x \in X^\#$ and $r \in R$, the second is induced by the functorial duality, and the third by the isomorphism $e_X: X \to X^{\#\#}$ given by the natural equivalence between $1_{\mathcal{L}_R}$ and $^\# \cdot {}^\#$ (see [96]).

Remark 16.2.6. Denote by X' the R-module $\Delta_T(X)$ equipped with the topology that makes $i_X: X^\# \to X'$ a topological isomorphism. Then $i_X: X^\# \to X'$ defines a natural equivalence between $^\#$ and $'$ (i. e., they coincide in $\mathfrak{D}_R$), so modulo this identification one can assume that the elements of $X^\#$ are the T-characters $X \to T$, but the topology of $X^\#$ need not be the compact-open one (see [96, Proposition 4.2]). This topology will be described later on.

Definition 16.2.7. A functorial duality $^\#: \mathcal{L}_R \to \mathcal{L}_R$ is *continuous* if for each $X \in \mathcal{L}_R$ the isomorphism $i_X: X^\# \to \Delta_T(X)$ is also topological (so, $^\#$ is naturally equivalent to Δ_T); otherwise $^\#$ is *discontinuous*.

The topological isomorphisms $i_X: \widehat{X} \to \Delta_{\widehat{R}}(X)$, $X \in \mathcal{L}_R$, witnesses that:

Fact 16.2.8 ([96]). *The Pontryagin-van Kampen duality functor of $\mathcal{L}_R$ is a continuous functorial duality.*

As stated above, the continuous dualities of $\mathcal{L}_R$ are classified by their tori, hence by means of the discrete Pontryagin-van Kampen duals of those tori, which are the projective finitely generated R-modules V with $\mathrm{End}_R(V) \cong R$. They form the *Picard group* $\mathrm{Pic}(R)$ of R: this is the set of isomorphism classes of finitely generated projective R-modules V satisfying $\mathrm{End}(V) \cong R$ with group operation $[V]+[V'] = [V \otimes_R V']$ (where $[V]$ denotes the isomorphism class of V; the neutral element of the group $\mathrm{Pic}(R)$ is $[R]$ and the opposite of $[V]$ is $\mathrm{Hom}_R(V,R)$). Therefore, roughly speaking, (up to the bijection $\mathfrak{T}_R \to \mathrm{Pic}(R)$, $T \mapsto \widehat{T}$, composed with Φ_R) *the continuous functorial dualities of $\mathcal{L}_R$ form a group, namely,* $\mathrm{Pic}(R)$.

Theorem 16.2.9 ([96, Theorem 5.17], [99, §3.4]). *Let R be a discrete ring. The unique continuous functorial duality on $\mathcal{L}_R$ is the Pontryagin-van Kampen duality if and only if* $\mathrm{Pic}(R) = \{0\}$.

Prodanov proved in [235] (see also [99, §3.4]) that every functorial duality on $\mathcal{L} = \mathcal{L}_{\mathbb{Z}}$ is continuous. In view of $\mathrm{Pic}(\mathbb{Z}) = \{0\}$, this result, combined with Theorem 16.2.9, gives another proof of Roeder theorem 13.4.19.

Continuous dualities were studied in the noncommutative context in [159].

While the Picard group provides an excellent tool to measure the failure of uniqueness for continuous dualities, there is still no completely efficient way to capture it for discontinuous ones. The first example of a discontinuous duality was given in [96, Theorem 11.1] (see Example 16.2.10).

The classification of the discontinuous functorial dualities with a given torus T amounts to the description of the "fiber" $\Phi_R^{-1}(T)$ of $\Phi_R\colon \mathfrak{D}_R \to \mathfrak{T}_R$. According to [96, Theorem 4.4], for every functorial duality $^\#$ with torus T there is a (not necessarily continuous) automorphism $\kappa\colon T \to T$ such that the natural equivalence e between $1_{\mathcal{L}_R}$ and $^{\#.\#}$ satisfies $e_X(x) = \kappa \circ \bar{\omega}_X(x)$ for every $X \in \mathcal{L}_R$ and $x \in X$. Moreover, $\kappa \circ \kappa\colon T \to T$ is a topological isomorphism, hence a multiplication by an invertible element $r \in R$ (see [96, Theorem 4.8]). For this reason, κ is called *the involution of* $^\#$, and it is continuous precisely when κ is a multiplication by an element of R. In these terms, the duality $^\#$ is (dis)continuous precisely when κ is (dis)continuous. The torus T and the involution κ determine the duality $^\#$ up to natural equivalence (see [96, Theorem 9.3]). The question of which involutions of a torus T are involutions of a functorial duality is highly nontrivial.

To summarize, the existence of nonisomorphic tori (i. e., $\mathrm{Pic}(R) \neq \{0\}$) leads to nonuniqueness of (continuous) functorial dualities; nevertheless, the existence of discontinuous functorial dualities does not always lead to nonuniqueness (see Example 16.2.11).

In order to describe very briefly the construction of the duality $^\#$ with assigned torus T and involution κ, we need to recall the following facts (proofs can be found in [96, 99]). For a torus T and $(X,\tau) \in \mathcal{L}_R$, the weak topology of all continuous T-characters $X \to T$ coincides with τ^+. Hence, the topology of $X \in \mathcal{L}_R$ is uniquely determined by its continuous T-characters $\chi: X \to T$ (i. e., if $X, Y \in \mathcal{L}_R$ have the same underlying abstract module and the same continuous T-characters, then they coincide topologically). Using this fact, the following counterpart for $\mathcal{L}_R$ of Exercise 13.7.15 was proved in [96, §6]: if T is a torus and $X, Y \in \mathcal{L}_R$, a homomorphism $f: X \to Y$ of abstract R-modules is continuous if and only if $\chi \circ f: X \to T$ is continuous for every continuous T-character $\chi: Y \to T$.

This permits us to build, for each functorial duality $^{\#}: \mathcal{L}_R \to \mathcal{L}_R$ with torus T and involution κ, a functor (*concrete equivalence*) $\mu: \mathcal{L}_R \to \mathcal{L}_R$ (see [96, §6]). More precisely, μ leaves unchanged the underlying abstract R-modules and their R-module homomorphisms, but the continuous T-characters $\chi: \mu(X) \to T$ are the R-module homomorphisms $\chi: X \to T$ such that $\kappa^{-1} \circ \chi: X \to T$ is continuous. Then $\mu = 1_{\mathcal{L}_R}$ if and only if $\mu(T) = T$ (if and only if $^{\#}$ is continuous). Moreover, μ preserves exact sequences, $\mu(\mathcal{C}_R) = \mathcal{C}_R$ and $\mu(\mathcal{D}_R) = \mathcal{D}_R$. An R-module homomorphism $h: X \to Y$ is continuous if and only if $h: \mu(X) \to \mu(Y)$ is continuous.

On the other hand, starting from T, κ, and μ with the above properties, one obtains the isomorphisms $\eta_X: \Delta_T(\mu(X)) \to \mu(\Delta_T(X))$, letting $\eta_X(\chi) = \kappa^{-1} \circ \chi$ for every continuous T-character $\chi: X \to T$ and $X \in \mathcal{L}_R$. They give a natural equivalence between $\Delta_T \cdot \mu$ and $\mu \cdot \Delta_T$, and the assignment $X \mapsto X^{\#} := \Delta_T(\mu(X))$ induces a functorial duality $^{\#}: \mathcal{L}_R \to \mathcal{L}_R$ out of a torus T with involution κ and a concrete equivalence μ related to T and κ as above (see [96, 6.7]).

Let $\mathcal{L}_R^0$ be the full subcategory of $\mathcal{L}_R$ with objects all R-modules having a compact open submodule. This subcategory contains $\mathcal{C}_R$ and $\mathcal{D}_R$, and each functorial duality $^{\#}: \mathcal{L}_R \to \mathcal{L}_R$ sends $\mathcal{L}_R^0$ to $\mathcal{L}_R^0$. This permits us to consider the restriction of $^{\#}$ to $\mathcal{L}_R^0$ as a functorial duality on $\mathcal{L}_R^0$. On the other hand, we can associate to each functorial duality $^{\#}$ on $\mathcal{L}_R^0$ a torus and an involution as we did for $\mathcal{L}_R$, and speak of (dis)continuity of $^{\#}$. Then a compact module T is a torus in $\mathcal{L}_R^0$ if and only if it is a torus in $\mathcal{L}_R$.

Example 16.2.10. It turns out that $\mathcal{L}_R^0$ has much more functorial dualities than the category $\mathcal{L}_R$. Actually, every pair (T, κ) where T is a torus T and $\kappa: T \to T$ an involution give rise to a functorial duality $^{\#}$ on $\mathcal{L}_R^0$ with torus T and involution κ (see [96, 10.2] – by means of κ, one builds a concrete involutive equivalence $\mu: \mathcal{C}_R \to \mathcal{C}_R$ which extends to $\mathcal{L}_R^0$ and defines $X^{\#} = \Delta_T(\mu(X))$, as above). In particular, $\mathcal{L}_R^0$ *has always discontinuous dualities*. Using this fact, one can produce examples of rings R (e. g., any field R with $\mathrm{char}(R) = 0$ and $|R| > \mathfrak{c}$ – see [96, Theorem 11.1]) with $\mathcal{L}_R = \mathcal{L}_R^0$, so $\mathcal{L}_R$ has discontinuous dualities.

It was conjectured by Prodanov that in case R is an algebraic number ring, all functorial dualities are continuous (so uniqueness is available if R is a principal ideal domain). This conjecture was proved to be true for real algebraic number rings, but it was shown to fail in case $R = \mathbb{Z}[i]$ (see [76] for more detail).

Example 16.2.11. A full subcategory of $\mathcal{L}_{\mathbb{Q}}$ containing $\mathcal{L}_{\mathbb{Q}}^0$ is *finitely closed* if it is closed with respect to isomorphisms, taking quotients (with respect to closed submodules), finite products, and closed submodules. Using the fact that the Adele ring $\mathbf{A}_{\mathbb{Q}}$ has no ring automorphisms beyond the identity (see [117]), it was proved in [94] that all dualities of the finitely closed subcategories of $\mathcal{L}_{\mathbb{Q}}$ are continuous (actually, among all finitely closed subcategories $\mathcal{L}$ of $\mathcal{L}_{\mathbb{Q}}$, exactly those with $\mathbf{A}_{\mathbb{Q}} \in \mathcal{L}$ have this property), whereas $\mathcal{L}_{\mathbb{R}}$ and $\mathcal{L}_{\mathbb{C}}$, where $\mathbb{R}$ and $\mathbb{C}$ are considered with the discrete topology, admit discontinuous dualities. Nevertheless, there is a unique functorial duality both on $\mathcal{L}_{\mathbb{R}}$ and $\mathcal{L}_{\mathbb{C}}$.

16.2.2 Locally linearly compact modules

In the sequel A will be a Hausdorff topological ring.

Definition 16.2.12. A topological A-module M is *linearly topologized* if it is Hausdorff and admits a local base at 0 consisting of open submodules of M.

Discrete modules are obviously linearly topologized.

Given a linearly topologized A-module M, a *linear variety* V of M is a coset $x + N$, where $x \in M$ and N is a submodule of M. A linear variety $x + N$ is *open* (respectively, *closed*) in M if N is open (respectively, closed) in M.

Definition 16.2.13. A linearly topologized A-module M is *linearly compact* if any collection of closed linear varieties of M with the finite intersection property has nonempty intersection. A topological ring A is *linearly compact* if A is a linearly compact A-module.

Inspired by the same notion for vector spaces given by Lefschetz [195] (see §16.2.3), linearly compact modules were introduced by Zelinsky [292]. Linear compactness is largely used for the study of ring and module structure (e. g., see [224]).

We collect basic properties concerning linearly compact modules that are very similar to the properties of compact groups:

Proposition 16.2.14 ([292]). *Let M be a linearly topologized A-module.*

(a) *Then M is linearly compact if and only if any collection of open linear varieties of M with the finite intersection property has a nonempty intersection.*
(b) *If N is a linearly compact submodule of M, then N is closed.*
(c) *If M is linearly compact and N is a closed submodule of M, then N is linearly compact.*
(d) *If M is linearly compact, N a linearly topologized A-module, and $f\colon M \to N$ is a continuous surjective A-module homomorphism, then N is linearly compact.*
(e) *If N is a closed linear submodule of M, then M is linearly compact if and only if N and M/N are linearly compact.*
(f) *The direct product of linearly compact modules is linearly compact.*
(g) *An inverse limit of linearly compact modules is linearly compact.*
(h) *If M is linearly compact, then M is complete.*

Definition 16.2.15. A linearly topologized A-module M is *strictly linearly compact* if every continuous surjective A-module homomorphism $f: M \to N$, onto a linearly topologized A-module N, is open.

There exist linearly compact modules that are not strictly linearly compact. A discrete strictly linearly compact module is Artinian (see Exercise 16.3.12).

Since linearly compact modules are complete, similarly to Proposition of 7.2.11, one can see that a linearly compact module is an inverse limit of discrete linearly compact modules. So, a linearly compact module is *strictly linearly compact* if and only if it is an inverse limit of discrete Artinian modules. In particular, every linearly compact module over an Artinian ring (in particular, a division ring) is necessarily strictly linearly compact.

As in [195] for vector spaces, we introduce the following concept in the obvious way (a slightly weaker notion under the same name was discussed in [2, 204], see Definition 16.2.24).

Definition 16.2.16. A linearly topologized A-module M is *locally linearly compact* if there exists an open submodule of M that is linearly compact.

Thus, an A-module M is locally linearly compact if and only if it admits a local base at 0 consisting of open linearly compact submodules of M. Clearly, linearly compact A-modules and discrete A-modules are locally linearly compact.

Example 16.2.17. Let G be a topological abelian group considered as a topological $\mathbb{Z}$-module. Then G is a linearly compact $\mathbb{Z}$-module if and only if G is linearly compact. If G is locally compact and hereditarily disconnected, then G is linearly topologized by Theorem 8.5.2(a) and G is a locally linearly compact $\mathbb{Z}$-module.

Proposition 16.2.18. *Let M be a linearly topologized A-module.*
(a) *If M is locally linearly compact, then M is complete.*
(b) *If N is a locally linearly compact submodule of M, then N is closed.*
(c) *If N is a closed linear submodule of M, then M is locally linearly compact if and only if N and M/N are locally linearly compact.*

Proof. (a) is a consequence of Proposition 16.2.14(h) and implies (b), since a complete topological subgroup of a Hausdorff group is closed by Proposition 7.1.22.

(c) If M is locally linearly compact, then N and M/N are locally linearly compact as well. Conversely, assume that N and M/N are locally linearly compact and let us show that M has a linearly compact open submodule U. To this end, let W be a linearly compact open submodule of N and O a linearly compact open submodule of M/N. We can assume without loss of generality that there exists an open submodule B of M such that $B \cap N = W$. Moreover, let $\pi: M \to M/N$ be the canonical projection and set $A = \pi^{-1}(O)$, which is an open submodule of M. Therefore, $U = B \cap A$ is an open submodule of M and $U \subseteq A$. Hence, $\pi(U)$, as an open submodule of the linearly compact A-module

0, is linearly compact as well. Since both $U/(U \cap N) \cong \pi(U)$ and $U \cap N = B \cap A \cap N = A \cap W = W$ are linearly compact, we deduce from Proposition 16.2.14(e) that U is linearly compact. □

16.2.3 The Lefschetz–Kaplansky–MacDonald duality

Here we concentrate on the special case proposed by Lefschetz [195], that is, K is a discrete field, and we consider the category LLC_K, with objects being all locally linearly compact vector spaces over K and morphisms all continuous linear transformations. First, we see that locally linearly compact vector spaces have a very simple structure:

Theorem 16.2.19. *Every $V \in \mathrm{LLC}_K$ is topologically isomorphic to $V_c \times V_d$, where V_c and V_d are linear subspaces of V, with V_c linearly compact open and V_d discrete.*

Proof. Let V_c be an open linear subspace of V that is linearly compact. There exists a linear subspace V_d of V such that $V = V_c \oplus V_d$, where V_d is discrete, being topologically isomorphic to V/V_c. It is straightforward to prove that the isomorphism between V and $V_c \times V_d$ is a topological isomorphism. □

The duality theory of LLC_K, for a discrete field K, proposed by Lefschetz [195], was inspired by the Pontryagin-van Kampen duality, with which it shares many common features. Let $\mathrm{CHom}(V,K)$ be the vector space of all continuous functionals $V \to K$. For a linear subspace A of V, the *annihilator* of A in $\mathrm{CHom}(V,K)$ is $A^\perp := \{\chi \in \mathrm{CHom}(V,K) : \chi(A) = \{0\}\}$, while for a linear subspace B of $\mathrm{CHom}(V,K)$, the *annihilator* of B in V is $B^\top := \{v \in V : \chi(v) = 0 \text{ for every } \chi \in B\}$.

Definition 16.2.20. For $V \in \mathrm{LLC}_K$, the *Lefschetz dual* $V^\wedge$ of V is $\mathrm{CHom}(V,K)$ endowed with the topology having as a local base at 0 the family of linear subspaces $\{A^\perp : A \text{ linearly compact linear subspace of } V\}$.

Since, for a linearly compact open linear subspace A of V, Theorem 16.2.19 implies $V = A \times B$ for some discrete subspace B of V, it follows that $A^\perp \cong B^\wedge$ is linearly compact. Therefore, $V^\wedge \in \mathrm{LLC}_K$ and V is discrete if and only if $V^\wedge$ is linearly compact. Let $^\wedge : \mathrm{LLC}_K \to \mathrm{LLC}_K$ be the Lefschetz duality functor, which is defined on the objects by $V \mapsto V^\wedge$ and on the morphisms sending $f : V \to W$ to $f^\wedge : W^\wedge \to V^\wedge$ such that $f(\chi) = \chi \circ f$ for every $\chi \in W^\wedge$. This is a contravariant representable functor. The biduality functor $^{\wedge\wedge} : \mathrm{LLC}_K \to \mathrm{LLC}_K$ is defined by composing $^\wedge$ with itself. Here is a counterpart of the Pontryagin-van Kampen duality theorem.

Theorem 16.2.21 (Lefschetz duality theorem [195, Theorem 29.1]). *The biduality functor $^{\wedge\wedge} : \mathrm{LLC}_K \to \mathrm{LLC}_K$ and $1_{\mathrm{LLC}_K} : \mathrm{LLC}_K \to \mathrm{LLC}_K$ are naturally isomorphic. This induces a duality between the subcategories LC_K of linearly compact vector spaces over K and $\mathbf{Vect}_K$ of discrete vector spaces over K.*

As in Pontryagin-van Kampen duality theorem 13.4.17, one proves that, for every locally linearly compact vector space V, the evaluation map $\omega_V\colon V \to V^{\wedge\wedge}$, $v \mapsto \omega_V(v)$, with $\omega_V(v)(\chi) = \chi(v)$ for every $\chi \in V^{\wedge}$, is a topological isomorphism.

Corollary 16.2.22. *Let V be a linearly compact vector space over a discrete field K. Then V is compact if and only if K is finite. In particular, V is a hereditarily disconnected locally compact abelian group whenever K is finite.*

Proof. By Exercise 16.3.16, $V = \prod_{i\in I} K_i$ with $K_i = K$ for all $i \in I$. If V is compact, then each K_i is compact as well, hence K is finite, being compact and discrete. Conversely, if K is finite, then each K_i is compact, so V is compact. □

Example 16.2.23. The Pontryagin-van Kampen duality functor, when restricted to p-torsion discrete abelian groups, gives a duality between p-torsion discrete abelian groups and abelian pro-p-groups. Both categories consist of $\mathbb{J}_p$-modules; moreover, $\widehat{G} = \mathrm{Hom}(G, \mathbb{T}) = \mathrm{Hom}_{\mathbb{J}_p}(G, \mathbb{Z}(p^\infty))$ for a discrete p-torsion abelian group G, and $\widehat{K} = \mathrm{CHom}(K, \mathbb{T}) = \mathrm{CHom}_{\mathbb{J}_p}(K, \mathbb{Z}(p^\infty))$ for an abelian pro-p-group K, since $\chi(K)$ is a finite cyclic p-group for every $\chi \in \widehat{K}$.

Kaplansky [183] and MacDonald [204] succeeded to cover both Example 16.2.23 and the Lefschetz duality as follows. Assume that R is a commutative local Noetherian ring with maximal ideal $\mathfrak{m}$ such that R is complete in the $\mathfrak{m}$-topology τ. An R-module M is *primary* if for every $x \in M$ the ideal $\mathrm{Ann}(x) = \{a \in R\colon ax = 0\}$ is τ-open in R, i. e., the discrete R-module M is a topological (R, τ)-module.

Definition 16.2.24. Call an R-module M:
(i) *linearly discrete* if every primary quotient of M is discrete;
(ii) *KM-locally linearly compact* if M has a linearly compact submodule N such that M/N is linearly discrete.

Property (ii) is obviously weaker than local linear compactness.

Denote by LLC_R^{KM} the category of KM-locally linearly compact R-modules. Let E be the injective hull of the (unique) simple R-module $R/\mathfrak{m}$ equipped with the discrete topology. For every $M \in \mathrm{LLC}_R^{KM}$, the continuous R-module homomorphisms $M \to E$ (i. e., the continuous E-characters) separate the points of M. Following [204, 8.1], denote by $M^\dagger$ the R-module $\mathrm{CHom}_R(M, E)$ carrying the topology with local base at 0 the family $\{A^\perp\colon A$ linearly compact submodule of $M\}$.

Theorem 16.2.25 ([204, 9.13]). *In the above notations, for every $M \in \mathrm{LLC}_R^{KM}$ also $M^\dagger \in \mathrm{LLC}_R^{KM}$ and the evaluation map $\omega_M\colon M \to M^{\dagger\dagger}$ defined as above is a topological isomorphism. This gives a natural equivalence between the functors $1_{\mathrm{LLC}_R^{KM}}$ and ††, hence † is a duality of LLC_R^{KM}.*

Remark 16.2.26. (a) With $R = K$ a discrete field (so, $\mathfrak{m} = \{0\}$ and $E = K$), one obtains the Lefschetz duality.

(b) According to [204, 9.4], if M is a discrete R-module, then $M^\dagger$ is linearly compact and $\mathfrak{m}$-separated (i. e., $\bigcap_{n\in\mathbb{N}_+} \mathfrak{m}^n M = \{0\}$), while $M^\dagger$ is discrete when M is linearly compact and $\mathfrak{m}$-separated. So, the restriction of the duality $^\dagger\colon \mathrm{LLC}_R^{KM} \to \mathrm{LLC}_R^{KM}$ gives a duality between discrete R-modules and linearly compact $\mathfrak{m}$-separated R-modules.
With $R = \mathbb{J}_p$, the discrete R-modules are the abelian p-groups M, and their duals $M^\dagger$, coinciding algebraically with $\widehat{M}$, are pro-p-finite groups with a possibly finer topology. This is why $M^\dagger$ is $p\mathbb{J}_p$-separated (having a Hausdorff topology coarser than $\varpi^p_{M^\dagger}$, by Corollary 11.6.8). This gives Example 16.2.23.

16.3 Exercises

Exercise 16.3.1. Prove that the product of two Cauchy nets of a topological ring is a Cauchy net.

Hint. Let $\{x_\alpha\}_{\alpha\in A}$ and $\{y_\alpha\}_{\alpha\in A}$ be Cauchy nets in a topological ring. To see that $\{x_\alpha b_\alpha\}_{\alpha\in A}$ is a Cauchy net as well, pick $U \in \mathcal{V}$ and $V \in \mathcal{V}$ with $VV + VV + V + V \subseteq U$. Pick $\alpha_0 \in A$ such that $x_\alpha - x_{\alpha'} \in V$ and $y_\alpha - y_{\alpha'} \in V$ for all $\alpha, \alpha' \geq \alpha_0$ in A. Now choose $W \in \mathcal{V}$ with $x_{\alpha_0} W \subseteq V$, $W y_{\alpha_0} \subseteq V$ and $W \subseteq V$. There exists $\alpha_1 \geq \alpha_0$ in A such that $x_\alpha - x_{\alpha'} \in W$ and $y_\alpha - y_{\alpha'} \in W$ for all $\alpha, \alpha' \geq \alpha_1$ in A. Then $x_\alpha y_\alpha - x_{\alpha'} y_{\alpha'} \in U$ for all $\alpha, \alpha' \geq \alpha_1$ in A, so $\{x_\alpha y_\alpha\}_{\alpha\in A}$ is Cauchy.

Exercise 16.3.2. For a topological ring A prove that:
(a) the family of bounded subsets of A is stable under taking subsets and finite unions;
(b) every compact subset of A is bounded;
(c) if A has a linear topology, then A is bounded;
(d)* if A is a precompact unitary ring, then A is linearly topologized.

Hint. (d) Let $U \in \mathcal{V}(0)$. There exist $\chi_1, \dots, \chi_n \in \widehat{A}$ such that $U' := \bigcap_{i=1}^n \chi_i^{-1}(\Lambda_1) \subseteq U$. By (a) and (b), there exists $V \in \mathcal{V}(0)$ such that $AV \cup VA \subseteq U'$. Then $\chi_i(Av) \cup \chi_i(vA) \subseteq \Lambda_1$ for every $v \in V$. Since Λ_1 contains no nontrivial subgroup of $\mathbb{T}$, this yields $\chi_i(Av) = \chi_i(vA) = 0$. Hence, $\ker\chi_i \supseteq J_i := \langle AV \cup VA\rangle \supseteq V$. Therefore, J_i is an open two-sided ideal and $\bigcap_{i=1}^n J_i \subseteq U' \subseteq U$. So, A is linearly topologized.

Exercise 16.3.3. Deduce from Exercise 16.3.2(d) that compact fields are finite.

Exercise 16.3.4. Prove that a subfield of a locally retrobounded field is locally retrobounded.

Exercise 16.3.5. Prove that the metric topology τ generated by an absolute value on a field K makes (K, τ) a retrobounded topological field.

Exercise 16.3.6. Let A be a commutative unitary ring. Prove that:
(a) if σ, τ are ring topologies on A, then $\sup\{\sigma, \tau\}$ is a ring topology on A;
(b) if A is a field and σ, τ are field topologies on A, then $\sup\{\sigma, \tau\}$ is a field topology on A;

(c) for every prime p, the p-adic topology τ_p on $\mathbb{Q}$ is a locally retrobounded field topology that is not locally compact;
(d) for distinct primes p, q, the topology $\sup\{\tau_p, \tau_q\}$ on $\mathbb{Q}$ is a field topology that is not locally retrobounded.

Hint. (c) Apply Exercise 16.3.5 and note that $\mathbb{Q}$ is countable and nondiscrete, so cannot be locally compact.

(d) According to Theorem 16.1.12, the completion of a locally retrobounded field is a field. It remains to observe that the completion $\mathbb{Q}_p \times \mathbb{Q}_q$ of $(\mathbb{Q}, \sup\{\tau_p, \tau_q\})$ is not a field.

Exercise 16.3.7. Deduce from the above hint that the completion of a topological field need not be a field.

Exercise 16.3.8. Show that $\mathcal{L}_{\mathbb{R}}$, where $\mathbb{R}$ is the ring of reals with the usual topology, consists of all groups topologically isomorphic to $\mathbb{R}^n$ for some $n \in \mathbb{N}$.

Exercise 16.3.9. For a discrete commutative unitary ring R, construct the Bohr compactification in $\mathcal{L}_R$ following the line of the proof of Theorem 8.6.1 (or, in alternative, the proof of Theorem 14.3.19, using the torus $\widehat{R}$ in place of $\mathbb{T}$).

Exercise 16.3.10. Prove that if $M \in \mathcal{L}_R$, then $\widehat{M}$ is a topological R-module.

Hint. Fix a basic neighborhood $W(K, \Lambda_1) \in \mathcal{V}_{\widehat{M}}(0)$, where $U \in \mathcal{V}_M(0)$ and $K = \overline{U}$ is compact.

To check (mt1), fix an open $O \in \mathcal{V}_M(0)$ with $C = \overline{O}$ compact. By Remark 16.2.3, there exists $V \in \mathcal{V}_R(0)$ such that $VK \subseteq C$, then $VW(C, \Lambda_1) \subseteq W(K, \Lambda_1)$ proves (mt1). To check (mt2), pick $\chi \in \widehat{M}$, then $O = \chi^{-1}(\Lambda_1) \in \mathcal{V}_M(0)$ is open. So, again by Remark 16.2.3, there exists $V \in \mathcal{V}_R(0)$ such that $VK \subseteq O$. Now $V\chi \subseteq W(K, \Lambda_1)$, and this proves (mt2). Finally, to check (mt3), pick $r \in R$ and put $C = rK$. Then rC is compact and $rW(rC, \Lambda_1) \subseteq W(K, \Lambda_1)$.

Exercise 16.3.11.* Compute $\mathrm{Pic}(\mathbb{Z})$, $\mathrm{Pic}(\mathbb{Z}[i])$, $\mathrm{Pic}(\mathbb{J}_p)$, $\mathrm{Pic}(\mathbb{Q})$, $\mathrm{Pic}(\mathbb{Q}[x])$, and $\mathrm{Pic}(\mathbb{R})$. Show that $\mathrm{Pic}(R)$ is trivial when R is a principal ideal domain.

Exercise 16.3.12. Prove that a discrete strictly linearly compact module M is Artinian.

Hint. To show that M satisfies the descending chain condition on submodules, assume for a contradiction that $\{N_n : n \in \mathbb{N}\}$ is a proper descending chain of submodules of M and let $N = \bigcap_{n\in\mathbb{N}} N_n$. Then M/N has a nondiscrete linear topology τ with local base $\{N_n/N : n \in \mathbb{N}\}$. Therefore, $M \to (M/N, \tau)$ is not open, a contradiction.

Exercise 16.3.13. Let V be a linearly topologized vector space. Prove that:
(a) if $\dim V < \infty$, then V is linearly topologized if and only if V is discrete;
(b) if V is discrete, then V is linearly compact if and only if it has finite dimension.

Exercise 16.3.14. Let V be a locally linearly compact vector space. Prove that:
(a) $\mathrm{CHom}(V, \mathbb{K})$ separates the points of V;
(b) if $\dim V < \infty$, then V is discrete and $V^\wedge \cong V$ is the algebraic dual of V;
(c) V is discrete if and only if $V^\wedge$ is linearly compact;
(d) V is linearly compact if and only if $V^\wedge$ is discrete;
(e) if B is a linear subspace of $V^\wedge$, then $\varpi_V(B^\top) = B^\perp$.

Exercise 16.3.15. Let $f: V \to W$ be a continuous homomorphism of locally linearly compact vector spaces. Prove that:
(a) if f is injective and open onto its image, then $\widehat{f}$ is surjective;
(b) if f is surjective, then $\widehat{f}$ is injective.

Exercise 16.3.16. Deduce from the Lefschetz duality theorem that every linearly compact vector space is a product of one-dimensional vector spaces.

Exercise 16.3.17. Let V be a locally linearly compact vector space and A, B linear subspaces of V. Prove that:
(a) if $A \subseteq B$, then $B^\perp \subseteq A^\perp$;
(b) $A^\perp = \overline{A}^\perp$ and $A^\perp$ is a closed linear subspace of $V^\wedge$;
(c) if A is a closed, then $(A^\perp)^\top = A$.

Exercise 16.3.18. Let V be a locally linearly compact vector space and U a closed linear subspace of V. Prove that $(V/U)^\wedge \cong U^\perp$ and $U^\wedge \cong V^\wedge/U^\perp$.

Exercise 16.3.19. Prove that the linearly compact abelian groups are strictly linearly compact $\mathbb{Z}$-modules, and the same about $\mathbb{J}_p$-modules.

16.4 Further readings, notes, and comments

Minimal topological rings can be introduced in analogy to the minimal topological groups [72]. As fields equipped with a topology induced by an absolute value are minimal (see [72]), minimal commutative rings need not be precompact (this should be compared with the precompactness of minimal abelian groups, see Theorem 10.5.1).

For the connection of minimal ring topologies to Krull dimension, see [71, 72].

A complete proof of Roeder theorem 13.4.19 on the uniqueness of Pontryagin-van Kampen duality in $\mathcal{L} = \mathcal{L}_\mathbb{Z}$ following the line of §16.2.1 can be found in [99, §3.4]. The backbone of §16.2.1 is the unpublished manuscript [235] of Prodanov. Its results, along with some further development, were published in [96], where the reader can find complete proofs of his results announced in §16.2.1 without proof. For a more general setting of dualities, see [121, 122, 229].

The fact that the Pontryagin-van Kampen functor on $\mathcal{L}_R$ is a functorial duality motivated Prodanov to raise the question of *how many* functorial dualities can carry $\mathcal{L}_R$ (Theorem 16.2.5 answers this question in case R is compact) and to extend this question to other well-known dualities and adjunctions, such as the Stone duality, the spectrum of a commutative rings (see [236]), etc., at his Seminar on dualities (Sofia University, 1979/1983 – see [118] for the part concerning the theory of abstract spectra). His conjecture that the Stone duality (see §C.2.3) is the unique functorial duality between hereditarily disconnected compact Hausdorff spaces and Boolean algebras was proved to be true by Dimov [120] (for more recent results see also [119] and the references therein).

Gregorio [158] extended Theorem 16.2.5 to the general case of not necessarily commutative compact rings R (here left and right R-modules should be distinguished, so that the dualities are no more *endofunctors*). Later Gregorio jointly with Orsatti [160] offered another approach to this phenomenon.

A more general result than Theorem 16.2.25 was obtained by Åhn [2]; we do not formulate it explicitly, while we preferred to explicitly formulate the Kaplansky–MacDonald duality, due to its nicer features: the duality functor, $\mathrm{CHom}(-, E)$, as well as its domain and codomain LLC_R^{KM}, has a rather simple form.

A Background on groups

We start with general notation and terminology, for undefined terms and notation see [134, 138, 139, 174, 177, 184, 203]. The symbols $\mathbb{Z}$, $\mathbb{Q}$, $\mathbb{R}$, $\mathbb{C}$ denote the integers, rationals, reals, and complex numbers, respectively. Let also $\mathbb{R}_{\geq 0} = \{r \in \mathbb{R}: r \geq 0\}$ and $\mathbb{R}_{>0} = \{r \in \mathbb{R}: r > 0\}$. We denote by $\mathbb{P}$, $\mathbb{N}$, and $\mathbb{N}_+$, respectively, the set of positive primes, the set of natural numbers, and the set of positive integers. For $x \in \mathbb{R}$, we denote by $\lfloor x \rfloor$ the unique integer (called *integral part* of x) such that $\lfloor x \rfloor \leq x < \lfloor x \rfloor + 1$. The symbol $\aleph_0 = |\mathbb{N}|$ denotes the smallest infinite cardinal, while $\mathfrak{c} = |\mathbb{R}|$ stands for the cardinality of the continuum. For a set X, we denote by $[X]^{<\omega}$ the family of all nonempty finite subsets of X. For a cardinal κ, let $\log \kappa = \min\{\lambda: 2^\lambda \geq \kappa\}$.

Generally, a group G is written multiplicatively and the neutral element is denoted by e_G, simply e when there is no danger of confusion. For abelian groups, we use the additive notation, consequently 0 denotes the neutral element.

For subsets $A, A_1, A_2, \dots, A_n$ of a group G, we denote

$$A^{-1} = \{a^{-1}: a \in A\} \quad \text{and} \quad A_1 A_2 \cdots A_n = \{a_1 a_2 \cdots a_n: a_i \in A_i, i \in \{1, \dots, n\}\} \tag{A.1}$$

and, if all $A_i = A$ for $A_1 A_2 \cdots A_n$, we write

$$A^n = \underbrace{AA \cdots A}_{n}. \tag{A.2}$$

Clearly, the counterparts of (A.1) and (A.2) for subsets $A, A_1, A_2, \dots, A_n$ of an abelian group G are $-A$, $A_1 + A_2 + \cdots + A_n$, and $nA = \{na: a \in A\}$.

For a family $\{G_i: i \in I\}$ of groups, we denote by $\prod_{i\in I} G_i$ the *direct product* of the groups G_i: the underlying set is the Cartesian product $\prod_{i\in I} G_i$ and the operation is defined coordinatewise. When I is empty, we define $\prod_{i\in I} G_i$ to be the trivial group $\{e\}$.

For $x = (x_i)_{i\in I} \in \prod_{i\in I} G_i$, the *support* of x is $\operatorname{supp}(x) = \{i \in I: x_i \neq e_{G_i}\}$. The *direct sum* $\bigoplus_{i\in I} G_i$ is the subgroup of $\prod_{i\in I} G_i$ consisting of all elements of finite support. If all G_i are isomorphic to the same group G and $|I| = \alpha$, we denote $\prod_{i\in I} G_i = \prod_I G = G^I$ also by $\prod_\alpha G$ or G^α and $\bigoplus_{i\in I} G_i = \bigoplus_I G$ also by $\bigoplus_\alpha G$ or $G^{(\alpha)}$. Moreover, the diagonal subgroup of G^I is $\Delta_G = \{(x)_{i\in I} \in G^I: x \in G\}$.

For every $i \in I$, consider $e_{G_i} \in D_i \subseteq G_i$. The *σ-product* of the family $\{D_i: i \in I\}$ is $\bigoplus_{i\in I} D_i = \{x \in \prod_{i\in I} D_i: |\operatorname{supp}(x)| \text{ is finite}\}$, while the *Σ-product* of $\{D_i: i \in I\}$ is $\Sigma_{i\in I} D_i = \{x \in \prod_{i\in I} D_i: |\operatorname{supp}(x)| \leq \aleph_0\}$; if $G_i = G$ and $D_i = D$ for every $i \in I$, we denote the Σ-product also by ΣD^I.

For groups G, H, we denote by $\operatorname{Hom}(G, H)$ the group of all homomorphisms from G to H where the operation is defined pointwise. Moreover, we denote by $\operatorname{Aut}(G)$ the group of the automorphisms of G; an automorphism f of G is *involutive* (or, an *involution*) if $f \circ f = id_G$.

When the groups G, H are abelian, the group $\operatorname{Hom}(G, H)$ is abelian and written additively. We leave to the reader the verification of the following isomorphisms related

https://doi.org/10.1515/9783110654936-017

to families of abelian groups $\{G_i : i \in I\}$, $\{H_i : i \in I\}$ and abelian groups G, H:

$$\mathrm{Hom}\left(H, \prod_{i\in I} G_i\right) \cong \prod_{i\in I} \mathrm{Hom}(H, G_i) \quad \text{and} \quad \mathrm{Hom}\left(\bigoplus_{i\in I} H_i, G\right) \cong \prod_{i\in I} \mathrm{Hom}(H_i, G). \tag{A.3}$$

For $m \in \mathbb{N}_+$, we use $\mathbb{Z}(m)$ for the finite cyclic group of order m. For an abelian group G and $m \in \mathbb{N}_+$, let $G[m] = \{x \in G : mx = 0\}$. Denoting $\mu_m = m\,id_G$, $x \mapsto mx$, clearly $G[m] = \ker \mu_m$. It is straightforward to verify that

$$\mathrm{Hom}(\mathbb{Z}, G) \cong G \quad \text{and} \quad \mathrm{Hom}(\mathbb{Z}(m), G) \cong G[m]. \tag{A.4}$$

The group $\mathbb{T} = \mathbb{R}/\mathbb{Z}$ and the group $\mathrm{Hom}(G, \mathbb{T})$, for an abelian group G, play a central role in this book, and they are always used in additive notation. The multiplicative form $\mathbb{S} = \{z \in \mathbb{C} : |z| = 1\}$ of $\mathbb{T}$ and $G^* = \mathrm{Hom}(G, \mathbb{S}) \cong \mathrm{Hom}(G, \mathbb{T})$ are used as well, when necessary (e. g., concerning easier computation in $\mathbb{C}$, etc.). The elements of $\mathrm{Hom}(G, \mathbb{T}) \cong \mathrm{Hom}(G, \mathbb{S})$ are called *characters* of G. For $m \in \mathbb{N}_+$, we identify $\mathbb{Z}(m)$ with the unique cyclic subgroup of order m of the circle group $\mathbb{S} \cong \mathbb{T}$.

A standard reference for abelian groups is the monograph [138] (see also its latest version [139]) and for nonabelian groups we refer to [249]. Here we give only those facts or definitions that appear very frequently in the book.

A.1 Torsion abelian groups and torsion-free abelian groups

The torsion elements of an abelian group G form a subgroup of G denoted by $t(G) = \bigcup_{m\in\mathbb{N}_+} G[m]$. For a prime p, $t_p(G) = \bigcup_{n\in\mathbb{N}} G[p^n]$ is a subgroup of G called *p-primary component* of G. It is not hard to check that $t(G) = \bigoplus_{p\in\mathbb{P}} t_p(G)$.

Since the subgroup $G[p]$ is a vector space over the finite field $\mathbb{Z}/p\mathbb{Z}$, the *p-rank* $r_p(G)$ of G is the dimension of $G[p]$ over $\mathbb{Z}/p\mathbb{Z}$. The *socle* of G is the subgroup $\mathrm{Soc}(G) = \bigoplus_{p\in\mathbb{P}} G[p]$. The nonzero elements of $\mathrm{Soc}(G)$ are precisely the elements of square-free order of G.

Let us start with the structure theorem for finitely generated abelian groups, that extends the Frobenius–Stickelberger theorem on the structure of finite abelian groups. We provide a proof of Theorem A.1.1 after Theorem A.3.4.

Theorem A.1.1. *If G is a finitely generated abelian group, then G is isomorphic to a finite direct product of cyclic groups. If $m \in \mathbb{N}_+$ and G has m generators, then every subgroup of G is finitely generated and has at most m generators.*

Definition A.1.2. An abelian group G is:
(a) *torsion* if $t(G) = G$;
(b) a *p-group*, for a prime p, if $t_p(G) = G$;
(c) *torsion-free* if $t(G) = \{0\}$;
(d) *bounded* (or, *bounded torsion*) if $mG = \{0\}$ for some $m \in \mathbb{N}_+$.

In case an abelian group G is bounded and $m = \min\{n \in \mathbb{N}_+ : nG = \{0\}\}$, we say that G *has finite exponent* m, denoted by $m = \exp(G)$.

Example A.1.3. (a) The additive groups $\mathbb{Z}$, $\mathbb{Q}$, $\mathbb{R}$, and $\mathbb{C}$ are torsion-free. The class of torsion-free abelian groups is stable under taking direct products and subgroups.
(b) The groups $\mathbb{Z}(m)$, for $m \in \mathbb{N}_+$, and $\mathbb{Q}/\mathbb{Z}$ are torsion. The class of torsion abelian groups is stable under taking direct sums, subgroups, and quotients.
(c) Let $m_1, \dots, m_k \in \mathbb{N}$ with $m_i > 1$ for $i \in \{1, \dots, k\}$, and let $\alpha_1, \dots, \alpha_k$ be cardinal numbers. Then $G = \bigoplus_{i=1}^k \mathbb{Z}(m_i)^{(\alpha_i)}$ is bounded, as clearly $m_1 \cdots m_k G = \{0\}$. We see next that every bounded abelian group has this form.

Since a direct sum of cyclic groups which has finite exponent has necessarily the form given in Example A.1.3(c), all bounded abelian groups have that form:

Theorem A.1.4 (Prüfer theorem). *Every bounded abelian group G is a direct sum of cyclic groups.*

Proof. Since $G = \bigoplus_{p \in \mathbb{P}} t_p(G)$, we can assume without loss of generality that G is a p-group for some prime p, so $\mathrm{Soc}(G) = G[p]$ and $p^n G = \{0\}$ for some $n \in \mathbb{N}_+$. We proceed by induction on $n \in \mathbb{N}_+$.

If $n = 1$, then $G = G[p]$ is a vector space over the finite field $\mathbb{Z}/p\mathbb{Z}$, so G is a direct sum of cyclic groups all isomorphic to $\mathbb{Z}(p)$.

Assume that $n > 1$ and that the assertion is true for all p-groups of exponent p^{n-1}. The subgroup pG has exponent p^{n-1}, so $pG = \bigoplus_{i \in I} \langle a_i \rangle$ for appropriate elements $a_i \in pG$. Then for all $i \in I$, there exists $x_i \in G$ with $a_i = px_i$. Let $H = \sum_{i \in I} \langle x_i \rangle$; we prove that

$$H = \bigoplus_{i \in I} \langle x_i \rangle. \tag{A.5}$$

Assume, aiming for a contradiction, that for distinct elements $i_1, \dots, i_k \in I$ there exist nonzero elements $y_j \in \langle x_{i_j} \rangle$, with $j \in \{1, \dots, k\}$ such that $\sum_{j=1}^k y_j = 0$. Then $y_j = m_j x_{i_j}$ with $1 \le m_j < p^n$ for all $j \in \{1, \dots, k\}$, and in particular $\sum_{j=1}^k m_j x_{i_j} = 0$. Let us consider two cases.

Case 1. If for all $j \in \{1, \dots, k\}$, $p \mid m_j$, then $m_j = pm_j'$ for some $m_j' \in \mathbb{Z}$, and we obtain $\sum_{j=1}^k m_j' a_{i_j} = 0$ by taking into account the equality $px_{i_j} = a_{i_j}$. Since the sum $\bigoplus_{i \in I} \langle a_i \rangle$ is direct, this yields to the conclusion that, for all $j \in \{1, \dots, k\}$, $m_j' a_{i_j} = 0$, and consequently also $y_j = m_j x_{i_j} = m_j' p x_{i_j} = m_j' a_{i_j} = 0$, a contradiction.

Case 2. Assume that there exists $j \in \{1, \dots, k\}$ such that $p \nmid m_j$. Suppose for simplicity that $j = 1$. From $\sum_{j=1}^k m_j x_{i_j} = 0$, we get $x_{i_1} = \sum_{j=2}^k l_j x_{i_j}$ for appropriate $l_j \in \mathbb{Z}$. Multiplying by p and taking into account the equality $px_i = a_i$ for all $i \in I$, we deduce that $a_{i_1} = \sum_{j=2}^k l_j a_{i_j}$. This contradicts the fact that the sum $pG = \bigoplus_{i \in I} \langle a_i \rangle$ is direct.

This concludes the proof of (A.5). Next we see that $G = H + G[p]$. Indeed, for any $g \in G$, one has $pg \in pG = pH$, so $pg = ph$ for some $h \in H$. Hence, $p(g - h) = 0$, and we conclude that $g - h \in G[p]$. Therefore, $g \in H + G[p]$.

As $G[p]$ is a vector space over $\mathbb{Z}/p\mathbb{Z}$, its $\mathbb{Z}/p\mathbb{Z}$-linear subspace $H \cap G[p]$ splits, namely, there exists a $\mathbb{Z}/p\mathbb{Z}$-linear subspace S_1 of $G[p]$ such that

$$G[p] = (H \cap G[p]) \oplus S_1. \tag{A.6}$$

Then $G = H + G[p] = H + (H \cap G[p]) + S_1 = H + S_1$. Moreover, $H \cap S_1 = H \cap G[p] \cap S_1$, as $S_1 \subseteq G[p]$, and hence $H \cap S_1 = (H \cap G[p]) \cap S_1 = \{0\}$ by (A.6). We proved that $G = H \oplus S_1$. As S_1 is a vector space over $\mathbb{Z}/p\mathbb{Z}$, S_1 is a direct sum of copies of $\mathbb{Z}(p)$. In view of (A.5), G is a direct sum of cyclic subgroups. □

This generalizes the Frobenius–Stickelberger theorem about the structure of the finite abelian groups (see Theorem A.1.1).

A.2 Divisible abelian groups

Divisible abelian groups G are defined by the property that the equations of the form $mx = g$ have solution in G for all $g \in G$ and $m \in \mathbb{N}_+$:

Definition A.2.1. An abelian group G is:
(i) *divisible* if $G = mG$ for every $m \in \mathbb{N}_+$;
(ii) *p-divisible*, for p a prime, if $G = pG$.

As, for G an abelian group, $nG \cap mG = nmG$ whenever $n, m \in \mathbb{N}_+$ are coprime, G is divisible if and only if it is p-divisible for every prime p.

Example A.2.2. (a) The groups $\mathbb{Q}$, $\mathbb{R}$, $\mathbb{C}$, and $\mathbb{T}$ are divisible.
(b) For $p \in \mathbb{P}$, we denote by $\mathbb{Z}(p^\infty)$ the *Prüfer group*, namely, the p-primary component of the torsion abelian group $\mathbb{Q}/\mathbb{Z}$ (so that $\mathbb{Z}(p^\infty)$ has generators $\{c_n = 1/p^n + \mathbb{Z} : n \in \mathbb{N}\}$). The group $\mathbb{Z}(p^\infty)$ is divisible.
(c) A torsion-free divisible abelian group D is a $\mathbb{Q}$-vector space, hence $D \cong \mathbb{Q}^{(\kappa)}$ with $\kappa = \dim_{\mathbb{Q}} D$.

Remark A.2.3. The class of divisible abelian groups is stable under taking direct products, direct sums, and quotients. In particular, every abelian group G has a *maximal divisible subgroup*, denoted by $\operatorname{div}(G)$.

If X is a nonempty set, a set $Y \subseteq Z^X$ of functions from X to a nonempty set Z *separates the points* of X if for every $x, y \in X$ with $x \neq y$, there exists $f \in Y$ such that $f(x) \neq f(y)$. In Corollary A.2.6 of the next result, we see that the characters separate the points of a discrete abelian group.

Theorem A.2.4. *Let G be an abelian group, H a subgroup of G and D a divisible abelian group. Then for every homomorphism $f: H \to D$ there exists a homomorphism $\overline{f}: G \to D$ such that $\overline{f} \restriction_H = f$. If $a \in G \setminus H$ and D contains elements of arbitrary finite order, then $\overline{f}$ can be chosen such that $\overline{f}(a) \neq 0$.*

Proof. Let H' be a subgroup of G such that $H \subseteq H'$ and suppose that $g: H' \to D$ is such that $g \restriction_H = f$. We prove that for every $x \in G$, defining $N = H' + \langle x \rangle$, there exists $\overline{g}: N \to D$ such that $\overline{g} \restriction_{H'} = g$. There are two cases.

If $\langle x \rangle \cap H' = \{0\}$, define $\overline{g}(h + kx) = g(h)$ for every $h \in H'$ and $k \in \mathbb{Z}$. Then $\overline{g}$ is a homomorphism. This definition is correct because every element of N can be represented in a unique way as $h + kx$, where $h \in H'$ and $k \in \mathbb{Z}$.

If $C = \langle x \rangle \cap H'$ is not $\{0\}$, then C is cyclic, being a subgroup of a cyclic group. So, $C = \langle lx \rangle$ for some $l \in \mathbb{Z}$, $l \neq 0$. In particular, $lx \in H'$ and we can consider the element $g(lx) \in D$. Since D is divisible, there exists $y \in D$ such that $g(lx) = ly$. Now define $\overline{g}: N \to D$ by putting $\overline{g}(h+kx) = g(h) + ky$ for every $h + kx \in N$, where $h \in H'$ and $k \in \mathbb{Z}$. To see that this definition is correct, suppose that $h + kx = h' + k'x$ for $h, h' \in H'$ and $k, k' \in \mathbb{Z}$. Then $h - h' = k'x - kx = (k' - k)x \in C$. So $k' - k = sl$ for some $s \in \mathbb{Z}$. Since $g: H' \to D$ is a homomorphism and $lx \in H'$,

$$g(h) - g(h') = g(h - h') = g(s(lx)) = sg(lx) = sly = (k' - k)y = k'y - ky.$$

Thus, from $g(h) - g(h') = k'y - ky$, we conclude that $g(h) + ky = g(h') + k'y$. Therefore, $\overline{g}$ is correctly defined. Moreover, $\overline{g}$ is a homomorphism and extends g.

Let $\mathcal{M}$ be the family of all pairs (H_i, f_i), where H_i is a subgroup of G containing H and $f_i: H_i \to D$ is a homomorphism extending $f: H \to D$. For $(H_i, f_i), (H_j, f_j) \in \mathcal{M}$, let $(H_i, f_i) \leq (H_j, f_j)$ if $H_i \leq H_j$ and f_j extends f_i. In this way $(\mathcal{M}, \leq)$ is partially ordered. Let $\{(H_i, f_i): i \in I\}$ be a totally ordered subset of $(\mathcal{M}, \leq)$. Then $H_0 = \bigcup_{i \in I} H_i$ is a subgroup of G and $f_0: H_0 \to D$, defined by $f_0(x) = f_i(x)$ whenever $x \in H_i$, is a homomorphism that extends f_i for every $i \in I$. This proves that $(\mathcal{M}, \leq)$ is inductive and so we can apply the Zorn lemma to find a maximal element $(H_{\max}, f_{\max})$ of $(\mathcal{M}, \leq)$. Using the first part of the proof, we can conclude that $H_{\max} = G$.

Suppose now that D contains elements of arbitrary finite order. If $a \in G \backslash H$, we can extend f to $\overline{f}$ on $H + \langle a \rangle$ defining it as in the first part of the proof. If $\langle a \rangle \cap H = \{0\}$, then $\overline{f}(h + ka) = f(h) + ky$ for every $k \in \mathbb{Z}$, where $y \in D \backslash \{0\}$. If $\langle a \rangle \cap H \neq \{0\}$, since D contains elements of arbitrary finite order, we can choose $y \in D$ such that $\overline{f}(h + ka) = f(h) + ky$ with $y \neq 0$. In both cases $\overline{f}(a) = y \neq 0$. □

Corollary A.2.5. *Let G be an abelian group and H a subgroup of G. For $a \in G \setminus H$, any $\chi \in \mathrm{Hom}(H, \mathbb{T})$ can be extended to $\overline{\chi} \in \mathrm{Hom}(G, \mathbb{T})$, with $\overline{\chi}(a) \neq 0$.*

Proof. Since $\mathbb{T}$ has elements of arbitrary finite order, Theorem A.2.4 applies. □

Corollary A.2.6. *For an abelian group G, $\mathrm{Hom}(G, \mathbb{T})$ separates the points of G.*

Proof. If $x \neq y$ in G, then $a = x - y \neq 0$, and so, by Corollary A.2.5, there exists $\chi \in \mathrm{Hom}(G, \mathbb{T})$ with $\chi(a) \neq 0$, that is, $\chi(x) \neq \chi(y)$. □

Corollary A.2.7. *If G is an abelian group and D a divisible subgroup of G, then there exists a subgroup B of G such that $G = D \oplus B$. Moreover, if a subgroup H of G satisfies $H \cap D = \{0\}$, then B can be chosen to contain H.*

Proof. Since the first assertion can be obtained from the second one with $H = \{0\}$, let us prove directly the second assertion. As $H \cap D = \{0\}$, we can define a homomorphism $f: D + H \to D$, $(x + h) \mapsto x$. By Theorem A.2.4, we can extend f to a homomorphism $\overline{f}: G \to D$. Then put $B = \ker \overline{f}$ and observe that $H \subseteq B$, $G = D + B$ and $D \cap B = \{0\}$; consequently, $G = D \oplus B$. □

Call a subgroup H of a not necessarily abelian group G *essential* in G if every nontrivial normal subgroup of G nontrivially meets H.

Example A.2.8. Obviously, every group G is an essential subgroup of G itself. On the other hand, a direct summand H of G (in particular, any divisible subgroup of an abelian group G) is essential in G if and only if $H = G$.

Here come some less trivial examples (see also Exercise A.7.8).

(a) Every nontrivial subgroup (in particular, $\mathbb{Z}$) is essential in $\mathbb{Q}$.
(b) Every essential subgroup of an abelian group G contains $\mathrm{Soc}(G)$.
(c) For an abelian group G, $\mathrm{Soc}(G)$ is essential in G if and only if G is torsion.

Lemma A.2.9. *Let G be an abelian group. Then:*
(a) *G has no proper essential subgroups if and only if $G = \mathrm{Soc}(G)$;*
(b) *a subgroup H of G is essential if and only if every homomorphism $f: G \to G_1$ of abelian groups, such that the restriction $f \restriction_H$ is injective, is injective itself.*

Proof. (a) Apply Example A.2.8(b) and Exercise A.7.4.
(b) Apply the definition to the subgroup $N = \ker f$. □

The following property follows from Exercise A.7.4 and Theorem A.1.4.

Lemma A.2.10. *Every abelian group contains an essential subgroup that is a direct sum of cyclic groups.*

Next comes a more general version of Exercise A.7.5.

Lemma A.2.11. *Let $\{G_i : i \in I\}$ be a family of abelian groups and H_i an essential subgroup of G_i for $i \in I$. Then $\bigoplus_{i \in I} H_i$ is an essential subgroup of $\bigoplus_{i \in I} G_i$.*

As $\mathbb{Z}$ is an essential subgroup of $\mathbb{Q}$, while $\mathbb{Z}^{\mathbb{N}}$ is not an essential subgroup of $\mathbb{Q}^{\mathbb{N}}$, so the above property cannot be extended to infinite direct products.

For a proof of the next theorem, see [138].

Theorem A.2.12. *For every abelian group G, there exists a divisible abelian group $D(G)$ containing G as an essential subgroup. If D' is another group with the same properties, there exists an isomorphism $i: D(G) \to D'$ such that $i \restriction_G = id_G$.*

The divisible abelian group $D(G)$ defined above is called *divisible hull* of the abelian group G. When G is torsion-free, $D(G)$ is torsion-free as well, so it is a $\mathbb{Q}$-vector space (it is built explicitly in Lemma A.2.14 below).

Example A.2.13. (a) For p a prime, $D(\mathbb{Z}(p)) = \mathbb{Z}(p^\infty)$ and, for a cardinal α, $D(\mathbb{Z}(p)^{(\alpha)}) = \mathbb{Z}(p^\infty)^{(\alpha)}$.
(b) If G is a torsion divisible abelian group, then $G = D(\mathrm{Soc}(G))$. So, if G is a torsion abelian group, then $D(G) = D(\mathrm{Soc}(G))$.
(c) If G is an abelian p-group for a prime p, then $D(G) \cong \mathbb{Z}(p^\infty)^{(r_p(G))}$.

Item (a) follows from Theorem A.2.12 and Lemma A.2.11, while (b) from Example A.2.8(c). For (c), Theorem A.1.4 gives that $\mathrm{Soc}(G) \cong \mathbb{Z}(p)^{(r_p(G))}$, and so we can use item (a).

A subset X of an abelian group G is *independent* if $\sum_{i=1}^n k_i x_i = 0$, with $k_i \in \mathbb{Z}$ and distinct elements x_i of X for $i \in \{1, \dots, n\}$, imply $k_1 = k_2 = \cdots = k_n = 0$. The maximum size $r_0(G)$ of an independent subset of G is the *free-rank* of G.

The next pair of lemmas takes care of the correctness of the definition of $r_0(G)$.

Lemma A.2.14. *For a torsion-free abelian group G:*
(a) *the divisible hull $D(G)$ is a vector space over the field $\mathbb{Q}$ containing G as a subgroup and such that $D(G)/G$ is torsion; moreover, $\dim_{\mathbb{Q}}(G) = r_0(G)$;*
(b) *for G and $D(G)$ as in (a), a subset X in H is independent (respectively, maximal independent) if and only if it is linearly independent in (respectively, a base of) the $\mathbb{Q}$-vector space $D(G)$;*
(c) *all maximal independent subsets of G have the same size (namely, $\dim_{\mathbb{Q}}(G)$).*

Proof. (a) In this specific case one can build the divisible hull $D(G)$ independently of Theorem A.2.12. To this end, consider the relation $\sim$ in $X = G \times \mathbb{N}_+$ defined by $(g, n) \sim (g', n')$ precisely when $n'g = ng'$. The quotient set $D(G) = X/\sim$ carries a binary operation defined by $[(g, n)] + [(g', n')] = [(n'g + ng'), nn']$. It is easy to show that $D(G)$ is the desired $\mathbb{Q}$-vector space.

Item (b) is immediate and (c) follows from (b). □

Lemma A.2.15. *Let G be an abelian group and consider the canonical projection $q\colon G \to G/t(G)$. Then:*
(a) *if X is a subset of G, then X is independent if and only if $q(X)$ is independent;*
(b) *all maximal independent subsets of G have the same size;*
(c) $r_0(G) = r_0(G/t(G))$.

Proof. The proof of (a) is straightforward, (b) follows from (a) and Lemma A.2.14, and (c) follows from (a) and (b). □

Example A.2.16. (a) The divisible abelian group $\mathbb{R}$ has $r_0(\mathbb{R}) = \mathfrak{c}$, in view of Lemma A.2.14(a).
(b) The divisible abelian group $\mathbb{T}$ has $t(\mathbb{T}) = \mathbb{Q}/\mathbb{Z}$, and so $r_0(\mathbb{T}) = \mathfrak{c}$ by Lemma A.2.15(c). Hence, algebraically $\mathbb{T}$ is isomorphic to $\mathbb{Q}/\mathbb{Z} \oplus \mathbb{Q}^{(\mathfrak{c})}$.
If G is any torsion-free abelian group, then $D(G) \cong \mathbb{Q}^{(r_0(G))}$ by Lemma A.2.14(a), and so G can be algebraically embedded in $\mathbb{T}$ if and only if $r_0(G) \leq \mathfrak{c}$.

The following result describes precisely the structure of divisible abelian groups in terms of their ranks.

Theorem A.2.17. *Every divisible abelian group G has the form* $\mathbb{Q}^{(r_0(G))} \oplus (\bigoplus_{p\in\mathbb{P}} \mathbb{Z}(p^\infty)^{(r_p(G))})$.

Proof. The subgroups $t(G)$ and $t_p(G)$ are divisible (for $t_p(G)$ use Exercise A.7.2), so $G \cong G_1 \oplus \bigoplus_{p\in\mathbb{P}} t_p(G)$, where $G_1 = G/t(G)$ is torsion-free by Exercise A.7.1, with $r_0(G_1) = r_0(G)$ by Lemma A.2.15(c), and divisible by Remark A.2.3. By Example A.2.2(c), $G_1 \cong \mathbb{Q}^{(r_0(G))}$. The divisible abelian group $t_p(G)$ coincides with $D(\mathrm{Soc}(t_p(G))) = D(G[p])$, hence $t_p(G) \cong \mathbb{Z}(p^\infty)^{(r_p(G))}$ by Example A.2.13(c). □

A.3 Free abelian groups

An abelian group G is *free* if G has an independent set of generators X. In such a case, $G \cong \mathbb{Z}^{(X)}$, where the isomorphism is given by $g \mapsto (k_x)_x$ with $g = \sum_{x\in X} k_x x$ (only finitely many $k_x \neq 0$, so effectively $(k_x)_{x\in X} \in \mathbb{Z}^{(X)}$). Since the empty-set is vacuously independent, the trivial group $\{0\}$ is free of rank 0. So, given a cardinal α, we can speak about *the* free abelian group of rank α, that is, $\mathbb{Z}^{(\alpha)}$.

Lemma A.3.1. *An abelian group G is free if and only if G has a set of generators X with the following property:*

(F) *every map $f: X \to H$ to an abelian group H can be extended to a homomorphism $\bar{f}: G \to H$.*

Proof. Without loss of generality, let $F = \mathbb{Z}^{(X)}$ be the free abelian group of rank $|X|$, and let 1_x denote the generator of the xth copy of $\mathbb{Z}$ in F. The set $S = \{1_x : x \in X\}$ generates F. Every map $f: S \to H$ to an abelian group H extends to a homomorphism $\bar{f}: F \to H$ by letting $\bar{f}(\sum_{i=1}^n k_i 1_{x_i}) = \sum_{i=1}^n k_i f(1_{x_i})$.

Now assume that G has a set of generators X with the property (F). To prove that G is free, we show that X is independent. As above, let $F = \mathbb{Z}^{(X)}$ and $S = \{1_x : x \in X\}$, and define $f: X \to F$ by $f(x) = 1_x$ for every $x \in X$. Let $\bar{f}: G \to F$ be the extension of f given by (F). Since $f(X) = S$ is independent in F, we deduce that X is independent in G as well. □

Clearly, a set of generators of a free abelian group G with the property (F) is a maximal independent set of generators. In the sequel we briefly refer to such a set of generators as a *basis* of G.

Next we collect some useful properties of free abelian groups.

Theorem A.3.2. (a) *Every abelian group is (isomorphic to) a quotient group of a free group.*

(b) *If G is an abelian group such that for a subgroup H of G the quotient group G/H is free, then H is a direct summand of G.*

(c) *A subgroup of a free abelian group is free.*

Proof. (a) follows from Lemma A.3.1, while for (c) see [138].

(b) Fix an independent set of generators X of G/H and let $q\colon G \to G/H$ be the canonical projection. For every $x \in X$, pick an element $s(x) \in G$ such that $q(s(x)) = x$. By Lemma A.3.1, there exists an homomorphism $f\colon G/H \to G$ extending $s\colon X \to G$. Then $q \circ f = id_{G/H}$ as $q \circ s = id_X$. This implies that $H \cap f(G/H) = \{0\}$ and $H + f(G/H) = G$, that is, $G = H \oplus f(G/H)$. □

Remark A.3.3. A free abelian group F is divisible if and only if F is trivial. Indeed, if $F = \mathbb{Z}^{(X)}$ with nonempty X, then, for p a prime, $pF = (p\mathbb{Z})^{(X)}$, so $F/pF \cong \mathbb{Z}(p)^{(X)}$ is nontrivial.

Also in order to ensure a proof to Theorem A.1.1 (see below), we provide a proof of Theorem A.3.2(c) for free abelian groups of finite rank:

Theorem A.3.4 (Stacked bases theorem). *Let F_n be a free abelian group of rank $n \in \mathbb{N}_+$. If H is a subgroup of F_n, then H is free. Moreover, if H is nontrivial, there exist bases $\{h_1, h_2, \dots, h_k\}$ and $\{f_1, f_2, \dots, f_n\}$ of H and F_n, respectively, such that $k \le n$, and there exist $m_1, m_2, \dots, m_k \in \mathbb{N}_+$ with $h_i = m_i f_i$ for every $i \in \{1, \dots, k\}$ and $m_i \mid m_{i+1}$ for every $i \in \{1, \dots, k-1\}$.*

Proof. We proceed by induction on $n \in \mathbb{N}_+$. The case $n = 1$ is clear, so assume that $n > 1$ and that the statement holds for $n-1$. Let $\mathcal{B} = \{f_1, b_2, \dots, b_n\}$ be a basis of F_n and let $a \in H \setminus \{0\}$ be such that the pair $(\mathcal{B}, a)$ has the property:

$$m_1 \text{ in } a = m_1 f_1 + \sum_{i=2}^{n} s_i b_i \text{ is positive and minimal.} \tag{A.7}$$

We prove first that $m_1 \mid s_i$ for every $i \in \{2, \dots, n\}$. Indeed, for every $i \in \{2, \dots, n\}$, there exist $q_i, r_i \in \mathbb{Z}$ with $s_i = q_i m_1 + r_i$ and $0 \le r_i < m_1$. Letting $f_1' = f_1 + \sum_{i=2}^{n} q_i b_i$, we have

$$a = m_1 f_1' + \sum_{i=2}^{n} r_i b_i \tag{A.8}$$

with $0 \le r_i < m_1$. If $r_i > 0$ for some $i \in \{2, \dots, n\}$, then

$$\mathcal{B}^* = \{b_i, b_2, \dots, b_{i-1}, f_1', b_{i+1}, \dots, b_n\}$$

is a basis of F_n, such that, due to (A.8), the pair $(\mathcal{B}^*, a)$ gives rise to a positive coefficient r_i relative to b_i smaller than m_1. This contradicts the choice of the pair $(\mathcal{B}, a)$ with (A.7). Hence, $r_i = 0$ for every $i \in \{2, \dots, n\}$; so, $m_1 \mid s_i = q_i m_1$ for every $i \in \{2, \dots, n\}$ and

$$a = m_1 f_1'. \tag{A.9}$$

Let us anticipate that f_1' and a are precisely the first elements of the bases of F_n and H, respectively, we are building. But before exposing the whole bases, we need to establish another property of f_1' and a with respect to the basis $\mathcal{B}' = \{f_1', b_2, \dots, b_n\}$ of F_n. Namely, if $h \in H$ has the presentation $h = mf_1' + \sum_{i=2}^n t_i b_i$, with $m, t_i \in \mathbb{Z}$, with respect to the basis $\mathcal{B}'$, then $m_1 \mid m$. Indeed, let $q, r \in \mathbb{Z}$ be such that $m = qm_1 + r$ with $0 \leq r < m_1$; in view of (A.9),

$$h = qm_1 f_1' + rf_1' + \sum_{i=2}^{n} t_i b_i = qa + rf_1' + \sum_{i=2}^{n} t_i b_i.$$

Therefore, $H \ni h - qa = rf_1' + \sum_{i=2}^n t_i b_i$. By the choice of $(\mathcal{B}, a)$ with (A.7), $r = 0$, hence $m = qm_1$. This shows that letting

$$F_{n-1} = \langle b_2, \dots, b_n \rangle \quad \text{and} \quad K = H \cap F_{n-1} \leq F_{n-1},$$

$h = qa + \sum_{i=2}^n t_i b_i \in \langle a \rangle + K$ for every $h \in H$. Since obviously $\langle a \rangle \cap K = \langle a \rangle \cap F_{n-1} = \{0\}$, this proves that $H = \langle a \rangle \oplus K$, and we can apply the inductive hypothesis to $K \leq F_{n-1}$. Set $a_1 = a$.

By the inductive hypothesis, F_{n-1} and K admit bases $\{f_2, \dots, f_n\}$ and $\{a_2, \dots, a_k\}$, respectively, such that $a_i = m_i f_i$ for every $i \in \{2, \dots, k\}$ and $m_i \mid m_{i+1}$ for every $i \in \{2, \dots, k-1\}$. In order to show that $\mathcal{B}'' = \{f_1', f_2, \dots, f_n\}$ and $\{a_1, a_2, \dots, a_k\}$ are the required bases of F_n and H, respectively, it remains to verify that $m_1 \mid m_2$. To this end, let $q, r \in \mathbb{Z}$ with $m_2 = qm_1 + r$ and $0 \leq r < m_1$. Then $a_2 - a_1 \in H$ and $a_2 - a_1 = m_2 f_2 - m_1 f_1' = m_1(qf_2 - f_1') + rf_2$. With respect to the basis $\{qf_2 - f_1', f_2, f_3, \dots, f_n\}$ of F_n, the coefficient of $a_2 - a_1 \in H$ relative to f_2 is $r < m_1$. By the choice of the pair $(\mathcal{B}, a)$ with (A.7), this implies $r = 0$. Therefore, $m_1 \mid m_2$. □

Proof of Theorem A.1.1. Let G be a finitely generated abelian group. By Theorem A.3.2, G is isomorphic to a quotient $\mathbb{Z}^n/H$ of $\mathbb{Z}^n$ for some $n \in \mathbb{N}_+$ and some subgroup H of $\mathbb{Z}^n$. By Theorem A.3.4, there exist bases $\{h_1, h_2, \dots, h_k\}$ and $\{f_1, f_2, \dots, f_n\}$ of H and $\mathbb{Z}^n$, respectively, such that $k \leq n$, and there exist $m_1, m_2, \dots, m_k \in \mathbb{N}_+$ with $h_i = m_i f_i$ for every $i \in \{1, \dots, k\}$ (and $m_i \mid m_{i+1}$ for every $i \in \{1, \dots, k-1\}$). Then $G \cong \mathbb{Z}^n/H \cong \mathbb{Z}(m_1) \times \cdots \times \mathbb{Z}(m_k) \times \mathbb{Z}^{n-k}$. □

A.4 Reduced abelian groups

Definition A.4.1. An abelian group G is *reduced* if the only divisible subgroup of G is the trivial one, that is, $\operatorname{div}(G) = \{0\}$.

A series of reduced abelian groups can be given right away:

Example A.4.2. According to Remark A.3.3, every free abelian group is reduced. Moreover, every bounded abelian group is reduced as well. Finally, every proper subgroup of $\mathbb{Q}$ is reduced.

Our initial supply of reduced abelian groups can be extended due to the nice properties of this class, which is stable under taking subgroups and direct products (see Exercise A.7.10(a)). Yet this class is not stable under taking quotients (see Exercise A.7.10(b)). Another source of reduced abelian groups is based on Exercise A.7.10(a) and the fact that finite abelian groups are reduced.

Now, as a consequence of Corollary A.2.7, we obtain the following important factorization theorem for arbitrary abelian groups.

Theorem A.4.3. *Every abelian group G can be written as $G = \operatorname{div}(G) \oplus R$, where R is a reduced subgroup of G.*

Proof. By Corollary A.2.7, there exists a subgroup R of G such that $G = \operatorname{div}(G) \oplus R$. To conclude that R is reduced, it suffices to apply the definition of $\operatorname{div}(G)$. □

In particular, this theorem implies that every abelian p-group G can be written as $G \cong \mathbb{Z}(p^\infty)^{(\kappa)} \times R$, where $\kappa = r_p(\operatorname{div}(G))$ and R is a reduced abelian p-group.

Call an abelian group G *almost divisible* if there exists $m \in \mathbb{N}_+$ such that mG is divisible. Obviously, divisible abelian groups, as well as bounded abelian groups, are almost divisible. The class of almost divisible abelian groups is closed under taking finite direct sums and quotients. In this class the factorization from Theorem A.4.3 has more subtle properties:

Proposition A.4.4. *An abelian group G is almost divisible if and only $G/\operatorname{div}(G)$ is bounded iff $G = \operatorname{div}(G) \oplus B$, where B is a bounded subgroup of G. If $f\colon G_1 \to G_2$ is a surjective homomorphism of abelian groups and if G_1 is an almost divisible abelian group, then so is G_2. Moreover, if $G_i = \operatorname{div}(G_i) \oplus B_i$, with B_i bounded for $i = 1, 2$, $f(\operatorname{div}(G_1)) = \operatorname{div}(G_2)$, and $f(B_1) \cong B_2$.*

Proof. Let $G = \operatorname{div}(G) \oplus R$ as in Theorem A.4.3 where R is a reduced group. For every $m \in \mathbb{N}_+$, we have $mG = \operatorname{div}(G) \oplus mR$ and $\operatorname{div}(G)$ is also the maximal divisible subgroup of mG. So, G is almost divisible iff $mR = \{0\}$ for some $m \in \mathbb{N}_+$ iff R is bounded. Since $R \cong G/\operatorname{div}(G)$, the first assertion follows.

If $m \in \mathbb{N}_+$ is such that mG_1 is divisible, then $mG_2 = f(mG_1)$ is divisible as well. So, G_2 is almost divisible.

To see that $f(\operatorname{div}(G_1)) = \operatorname{div}(G_2)$, note first that $f(\operatorname{div}(G_1)) \subseteq \operatorname{div}(G_2)$. The opposite inclusion follows from $\operatorname{div}(G_2) \subseteq mG_2 = f(\operatorname{div}(G_1)) + f(mB_1) = f(\operatorname{div}(G_1))$ for a suitable $m \in \mathbb{N}_+$. Since $f(B_1) \cap f(\operatorname{div}(G_1))$ is a bounded divisible group, it is trivial. This shows that $G_2 = f(\operatorname{div}(G_1)) \oplus f(B_1) = \operatorname{div}(G_2) \oplus f(B_1)$, and hence $f(B_1) \cong B_2$. □

A.4.1 Residually finite groups

Definition A.4.5. A group G is *residually finite* if G is isomorphic to a subgroup of a direct product of finite groups. The *Ulm subgroup* G^1 of an abelian group G is $G^1 = \bigcap_{n\in\mathbb{N}_+} nG$.

We prove item (a) in the following result for not necessarily abelian groups.

Proposition A.4.6. *Let G be an abelian group. Then:*
(a) *(in case G is not necessarily abelian) G is residually finite if and only if the intersection of all normal subgroups of G of finite index is trivial;*
(b) *if G is residually finite, G is reduced;*
(c) *G is residually finite if and only if $G^1 = \{0\}$;*
(d) *G^1 coincides with the intersection of all subgroups of G of finite index;*
(e) *if G is torsion-free, $\operatorname{div}(G) = G^1$; hence, G is reduced if and only if G is residually finite, that is, $G^1 = \{0\}$.*

Proof. (a) Suppose that G is (isomorphic to) a subgroup of the direct product $P = \prod_{i\in I} F_i$ of finite groups F_i. If $p_i: P \to F_i$ is the ith canonical projection, then $\ker p_i$ is a normal subgroup of P of finite index, hence $N_i = G \cap \ker p_i$ is a normal subgroup of G of finite index and obviously $\bigcap_{i\in I} N_i$ is trivial.

Vice versa, if $\{N_i : i \in I\}$ is a family of normal subgroups of G of finite index with $\bigcap_{i\in I} N_i = \{e_G\}$, then the diagonal homomorphism $G \to \prod_{i\in I} G/N_i$ of the family of canonical projections $q_i: G \to G/N_i$ has trivial kernel $\bigcap_{i\in I} N_i$, and so G is isomorphic to a subgroup of the product $\prod_{i\in I} G/N_i$ of finite groups.

(b) follows from Exercise A.7.10(a), since finite groups are reduced.

(c) Assume that $G^1 = \{0\}$. Then the family of canonical projections $q_n: G \to G/nG$, for $n \in \mathbb{N}_+$, gives rise to a diagonal homomorphism $q: G \to \prod_{n\in\mathbb{N}_+} G/nG$ which is injective, since $\ker q = G^1 = \{0\}$. Hence, it suffices to see that each group G/nG is residually finite. Since G/nG is bounded, G is a direct sum $G = \bigoplus_{i\in I} F_i$ of finite cyclic groups F_i by Theorem A.1.4, and hence G is a subgroup of the direct product $\prod_{i\in I} F_i$ of finite cyclic groups.

To prove the converse implication, it suffices to note that the Ulm subgroup has the following remarkable property:

$$\text{if } f: N \to H \text{ is a group homomorphism, then } f(N^1) \subseteq H^1. \tag{A.10}$$

Then for every direct product $H = \prod_{i\in I} H_i$,

$$H^1 = \Big(\prod_{i\in I} H_i\Big)^1 \subseteq \prod_{i\in I} H_i^1, \tag{A.11}$$

by applying (A.10) to the canonical projections of the product.

Now assume that G is a subgroup of a product $P = \prod_{i\in I} F_i$ of finite abelian groups. Then $F_i^1 = \{0\}$ for every $i \in I$, so $P^1 \subseteq \prod_{i\in I} F_i^1 = \{0\}$, by (A.11). Applying (A.10) to the inclusion homomorphism $G \to P$, we deduce that $G^1 = \{0\}$.

(d) For every $m \in \mathbb{N}_+$, each mG is an intersection of finite index subgroups of G, since G/mG has finite exponent and so it is a direct sum of finite cyclic groups by Theorem A.1.4, thus the intersection of the finite index subgroups of G/mG is trivial.

(e) It is enough to prove that $\operatorname{div}(G) = G^1$, where the inclusion $\operatorname{div}(G) \subseteq G^1$ is obvious. To verify the inclusion $G^1 \subseteq \operatorname{div}(G)$, it is enough to see that G^1 is divisible. Fix a prime p and $g \in G^1 \subseteq pG$. Then there exists a unique element $x \in G$ such that $px = g$. To see that $x \in G^1$, fix $m \in \mathbb{N}_+$. As $g \in G^1$, we know that $g \in pmG$, so $g = pmy$ for some $y \in G$. As G is torsion-free, $px = pmy$ gives $x \in mG$, and hence $x \in G^1$. □

Proposition A.4.6(e) allows us to conclude that for torsion-free abelian groups the notions "reduced" and "residually finite" coincide. This fails in the torsion case:

Example A.4.7. Let p be a prime, $G = \bigoplus_{n\in\mathbb{N}_+} \mathbb{Z}(p^n)$, and for every $n \in \mathbb{N}_+$ let b_n denote a generator of $\mathbb{Z}(p^n)$. Let

$$H = \langle b_1 - pb_2, b_1 - p^2 b_3, \dots, b_1 - p^{n-1} b_n, \dots \rangle,$$

$N = G/H$, and $\bar{b}_1 = b_1 + H \in N$. Then $N^1 = \langle \bar{b}_1 \rangle \neq \{\bar{0}\}$, so N is not residually finite. Nevertheless, N is reduced since $\operatorname{div}(N) \subseteq N^1$, consequently, $\operatorname{div}(N) = p\operatorname{div}(N) \subseteq pN^1 = \{\bar{0}\}$.

Clearly, direct sums of cyclic groups are reduced (since cyclic groups are reduced). The notion of a basic subgroup we introduce below allows for a nice "approximation" of reduced abelian p-groups by direct sums of cyclic subgroups. First, we need the following notion that can be introduced for arbitrary abelian groups:

Definition A.4.8. A subgroup H of an abelian group G is *pure* if $mG \cap H = mH$ for every $m \in \mathbb{N}_+$.

Example A.4.9. (a) Direct summands are always pure subgroups. The torsion subgroup $t(G)$ of an abelian group G is pure, as well.

(b) For an abelian group G, one can obviously write the equality in Definition A.4.8 also as $\mu_m(H) = \mu_m(G) \cap H$. Since inverse images commute with intersections, this is equivalent to

$$\mu_m^{-1}(H) = \mu_m^{-1}(\mu_m(G) \cap H) = \mu_m^{-1}(\mu_m(H)) = \ker \mu_m + H = G[m] + H,$$

which means that $\{g \in G : mg \in H\} = \mu_m^{-1}(mH) = G[m] + H$.

If G is an abelian p-group and $m \in \mathbb{N}_+$, $mG = p^n G$ where $p^n \mid m$ and $p^{n+1} \nmid m$. Then a subgroup H of G is pure if and only if $p^n G \cap H = p^n H$ for every $n \in \mathbb{N}$.

Definition A.4.10. A subgroup B of an abelian p-group G is a *basic subgroup* if B is a direct sum of cyclic subgroups, B is pure and G/B is divisible.

Theorem A.4.11 ([138]). *Every abelian p-group admits a basic subgroup.*

The case considered in the next example is the one that we use in this book.

Example A.4.12. Let G be an abelian p-group.

(a) If B is a bounded basic subgroup of G, then B splits off as a direct summand of G, so $G = B \oplus D$ for some subgroup D of G, where $D \cong G/B$ is divisible. Indeed, if $p^n B = \{0\}$, then $p^n G \cap B = \{0\}$. On the other hand, $G = p^n G + B$ (as $p^n(G/B) = (p^n G + B)/B$, so the divisibility of G/B yields $(p^n G + B)/B = G/B$), hence this sum is direct.

(b) If $r_p(G) < \infty$ and G is infinite, then G is almost divisible and $G = \operatorname{div}(G) \oplus B$ where $\operatorname{div}(G) \cong \mathbb{Z}(p^\infty)^k$ for some $k > 0$ and B is a finite subgroup of G. Indeed, fix a basic subgroup B of G which exists by Theorem A.4.11. Then $r_p(B) \leq r_p(G)$ is finite, so B is bounded (actually, finite). Hence, by (a), $G = B \oplus D$ holds with $D \cong \mathbb{Z}(p^\infty)^k$ and $k \leq r_p(G)$. Since B is finite, necessarily $k > 0$, and the group G is almost divisible by Proposition A.4.4.

(c) A reduced abelian p-group of finite rank is finite. Indeed, by Theorem A.4.11, there is a basic subgroup B of G. Since B is a direct sum of cyclic groups and has finite p-rank, B is a bounded p-group, hence finite. By (a), the subgroup B splits, i.e., $G = D \oplus B$ holds for a suitable divisible subgroup D of G. Since G is reduced, D must be trivial. Hence $G = B$ is finite.

Proposition A.4.13. *If G is a torsion abelian group and H is a subgroup of G, then $r_p(G) \geq r_p(G/H)$.*

Proof. Since $G = \bigoplus_{q \in \mathbb{P}} t_q(G)$ and $H = \bigoplus_{q \in \mathbb{P}} t_q(H)$, we may assume that G is a p-group and that G/H has exponent p. Let $f\colon G \to G/H$ be the canonical projection.

Case 1: Let G be a finite p-group. Since pG is contained in the kernel of the projection f, it can be factorized $f\colon G \to G/pG \to G/H$. By Exercise A.7.9, $r_p(G) = r_p(G/pG)$. Since G/pG and G/H are vector spaces over the field $\mathbb{Z}/p\mathbb{Z}$, we obtain $r_p(G/pG) \geq r_p(G/H)$. Combining both results yields the assertion in this case.

Case 2: Assume now that $r_p(G)$ is infinite. Since $G = \bigcup_{n \in \mathbb{N}} G[p^n]$ and $|G[p^n]| = |G[p]| = r_p(G)$ holds for every $n \in \mathbb{N}_+$, we obtain $|G| = r_p(G)$. This gives the desired inequality $r_p(G/H) \leq |G/H| \leq |G| = r_p(G)$.

Case 3: Finally, suppose that $r_p(G)$ is finite and G is infinite. Example A.4.12(b) and Proposition A.4.4 yield that G and H are almost divisible. So, $G = \operatorname{div}(G) \oplus F_G$ and $H = \operatorname{div}(H) \oplus F_H$ for finite subgroups F_G of G and F_H of H.

If G is divisible, by Corollary A.2.7 there exists a divisible subgroup D of G containing F_H such that $G = \operatorname{div}(H) \oplus D$. Hence, $G/H \cong D/F_H \cong D$. So, $r_p(G/H) = r_p(D) \leq r_p(G)$.

In case G itself is not divisible, the quotient G/H is almost divisible hence $G/H = \operatorname{div}(G/H) \oplus F_{G/H}$ for a suitable bounded p-group $F_{G/H} \leq G/H$ by Proposition A.4.4 and the surjective homomorphism $f\colon G \to G/H$ satisfies $f(\operatorname{div}(G)) = \operatorname{div}(G/H)$ and $f(F_G) \cong F_{G/H}$. Since $\operatorname{div}(G)$ is divisible and has finite p-rank, $r_p(\operatorname{div}(G)) \geq r_p(\operatorname{div}(G/H))$, by the above argument. Since f induces a surjective homomorphism $F_G \to F_{G/H}$, the estimate $r_p(F_G) \geq r_p(F_{G/H})$ holds by case 1. So, $r_p(G/H) = r_p(\operatorname{div}(G/H)) + r_p(F_{G/H}) \leq r_p(\operatorname{div}(G)) + r_p(F_G) = r_p(G)$. □

Define the *rank* $r(G)$ of an abelian group G by $r(G) = r_0(G) + \sum_{p\in\mathbb{P}} r_p(G)$. Then one can easily prove the following properties of this rank:

Lemma A.4.14. *Let G be an abelian group. Then:*

(a) $r(G) < \infty$ *if and only if* $G \cong G_0 \oplus F \oplus \bigoplus_{i=1}^m \mathbb{Z}(p_i^\infty)$, *where the primes $p_1, \dots, p_m$ are not necessarily distinct, F is a finite abelian group and G_0 is a subgroup of $\mathbb{Q}^n$ for $n = r_0(G)$;*

(b) *if $r(G)$ is infinite, then $|G| = r(G)$ and G contains a subgroup H such that $H = \bigoplus_{i\in I} C_i$, $|I| = |G|$ and each C_i is cyclic; moreover, H can be chosen to be essential and each group C_i to be either infinite or of finite of prime order.*

A.4.2 The p-adic integers

Here we describe, for $p \in \mathbb{P}$, the ring of endomorphisms of the group $\mathbb{Z}(p^\infty)$, known as the ring of *p-adic integers* and denoted by $\mathbb{J}_p$. We shall see that it is isomorphic to the inverse limit $\varprojlim_{n\in\mathbb{N}} \mathbb{Z}(p^n)$ of the finite rings $\mathbb{Z}(p^n)$.

In order to describe the ring $\mathbb{J}_p = \operatorname{End}(\mathbb{Z}(p^\infty))$, write $\mathbb{Z}(p^\infty) = \bigcup_{n\in\mathbb{N}_+} \mathbb{Z}(p^n)$ and, for $n \in \mathbb{N}_+$, denote by c_n the generator $\frac{1}{p^n} + \mathbb{Z}$ of

$$\mathbb{Z}(p^n) = \left\{0 + \mathbb{Z}, \frac{1}{p^n} + \mathbb{Z}, \dots, \frac{p^n - 1}{p^n} + \mathbb{Z}\right\} \leq \mathbb{Z}(p^\infty) = t_p(\mathbb{T}).$$

Let $\alpha \in \operatorname{End}(\mathbb{Z}(p^\infty))$. Every subgroup $\langle c_n\rangle = \mathbb{Z}(p^n) = \mathbb{Z}(p^\infty)[p^n]$ of $\mathbb{Z}(p^\infty)$ is α-invariant, hence

$$\alpha(c_n) = k_n c_n \tag{A.12}$$

for some $k_n \in \mathbb{Z}$, and the sequence of integers $k = (k_n)_{n\in\mathbb{N}_+} \in \mathbb{Z}^{\mathbb{N}_+}$ satisfies

$$k_{n+1} \equiv k_n (\operatorname{mod} p^n) \tag{A.13}$$

as $pc_{n+1} = c_n$. Vice versa, every $k = (k_n)_{n\in\mathbb{N}_+} \in \mathbb{Z}^{\mathbb{N}_+}$ satisfying (A.13) defines an endomorphism $\alpha_k \in \operatorname{End}(\mathbb{Z}(p^\infty))$ with (A.12).

For $k = (k_n)_{n\in\mathbb{N}_+}$, $k' = (k'_n)_{n\in\mathbb{N}_+} \in \mathbb{Z}^{\mathbb{N}_+}$, put $k \sim k'$ if $k_n \equiv k'_n (\operatorname{mod} p^n)$ for every $n \in \mathbb{N}_+$. Obviously, $\alpha_k = \alpha_{k'}$ if and only if $k \sim k'$. Let

$$B = \{k = (k_n)_{n\in\mathbb{N}_+} \in \mathbb{Z}^{\mathbb{N}_+} : k_{n+1} \equiv k_n (\operatorname{mod} p^n)\ \forall n \in \mathbb{N}\}.$$

Then B is a subring of the ring $\mathbb{Z}^{\mathbb{N}_+}$ and setting $\mu(k) = \alpha_k$ we define a surjective ring homomorphism $\mu: B \to \mathbb{J}_p$. Let J be the ideal $p\mathbb{Z} \times p^2\mathbb{Z} \times \cdots \times p^n\mathbb{Z} \times \cdots$ of the ring $\mathbb{Z}^{\mathbb{N}_+}$

and $I = J \cap B$. Then I is an ideal of B such that $k - k' \in I$ if and only if $k \sim k'$. In other words, $I = \ker \mu$. This gives the isomorphism $\mathbb{J}_p \cong B/I$.

Obviously, one can identify the ring $\mathbb{J}_p$ also with the subring $\bar{B}$ of the quotient $\mathbb{Z}^{\mathbb{N}}/J \cong \mathbb{Z}(p) \times \mathbb{Z}(p^2) \times \cdots \times \mathbb{Z}(p^n) \times \cdots$ defined by the relations

$$\bar{B} = \left\{ x = (x_n)_{n \in \mathbb{N}_+} \in \prod_{n \in \mathbb{N}_+} \mathbb{Z}(p^n) : \varphi_n(x_{n+1}) = x_n \text{ for all } n \in \mathbb{N}_+ \right\},$$

where $\varphi_n : \mathbb{Z}(p^{n+1}) \to \mathbb{Z}(p^n)$ is the canonical homomorphism with kernel the principal ideal (p^n) of the ring $\mathbb{Z}(p^{n+1})$.

Theorem A.4.15. *For p a prime, $\mathbb{J}_p$ is a domain with unique maximal ideal $L = \{\alpha \in \mathbb{J}_p : \alpha(c_1) = 0\} = p\mathbb{J}_p$ and group of units $U(\mathbb{J}_p) = \mathbb{J}_p \setminus p\mathbb{J}_p$.*

Proof. If $\alpha \in \mathbb{J}_p \setminus \{0\}$, then there exists $n \in \mathbb{N}_+$ such that $\alpha(c_n) \neq 0$ (hence, also $\alpha(c_m) \neq 0$ for all $m \geq n$). Let $\beta \in \mathbb{J}_p \setminus \{0\}$. There exists $m \in \mathbb{N}_+$ such that $\alpha(c_m) \neq 0$ and $\beta(c_m) \neq 0$; hence, letting $k = (k_n)_{n \in \mathbb{N}}$ and $k' = (k'_n)_{n \in \mathbb{N}}$ be the sequences in B corresponding to α and β, respectively, $p^m \nmid k_m$ and $p^m \nmid k'_m$. Since

$$k_m \equiv k_{2m} (\operatorname{mod} p^m) \quad \text{and} \quad k'_m \equiv k'_{2m} (\operatorname{mod} p^m),$$

clearly $p^m \nmid k_{2m}$ and $p^m \nmid k'_{2m}$. Then $p^{2m} \nmid k_{2m} k'_{2m}$, and therefore $\alpha(\beta(c_{2m})) \neq 0$. This proves that $\alpha\beta \neq 0$.

Moreover, $\mathbb{J}_p$ is commutative, as $\alpha(\beta(c_m)) = k_m k'_m c_m = k'_m k_m c_m = \beta(\alpha(c_m))$ for every $m \in \mathbb{N}_+$. Since $\mathbb{J}_p$ is a unitary ring, this proves that $\mathbb{J}_p$ is a domain.

Next we recall that $\operatorname{Aut}(\mathbb{Z}(p^\infty)) = U(\mathbb{J}_p)$ is the set of units of $\mathbb{J}_p$. If $\alpha \in U(\mathbb{J}_p)$, then α is injective, and so clearly $\alpha(c_1) \neq 0$ and $\alpha \in \mathbb{J}_p \setminus L$. To prove that $U(\mathbb{J}_p) = \mathbb{J}_p \setminus L$, pick an $\alpha \in \mathbb{J}_p \setminus L$. Then α is injective, as $\ker \alpha$ trivially meets $\operatorname{Soc}(\mathbb{Z}(p^\infty)) = \mathbb{Z}(p)$ which is essential in $\mathbb{Z}(p^\infty)$ by Example A.2.8(c), and so $\ker \alpha = \{0\}$. Hence, $D = \alpha(\mathbb{Z}(p^\infty))$ is a divisible subgroup of $\mathbb{Z}(p^\infty)$ containing $\mathbb{Z}(p)$. As $\mathbb{Z}(p^\infty)$ is divisible, by the uniqueness of the divisible hull, we deduce that $D = \mathbb{Z}(p^\infty)$, so α is also surjective, hence an automorphism. Therefore, $\alpha \in U(\mathbb{J}_p)$.

The last assertion $L = p\mathbb{J}_p$ is deduced from the more general result in Claim A.4.16. □

Claim A.4.16. *Let $\alpha \in L$ satisfy $\alpha(c_1) = \cdots = \alpha(c_n) = 0$ and $\alpha(c_{n+1}) \neq 0$. Then $\alpha = p^n \beta$ for some $\beta \in U(\mathbb{J}_p)$. Consequently, $p^n\mathbb{J}_p$, for $n \in \mathbb{N}$, are exactly all the nonzero ideals of $\mathbb{J}_p$ and $\bigcap_{n \in \mathbb{N}} p^n \mathbb{J}_p = \{0\}$.*

Proof. Let $\alpha = \mu(k)$ with $k = (k_n)_{n \in \mathbb{N}_+} \in B$. Our hypothesis yields $k_1 = \cdots = k_n = 0$. By (A.13), $p^n \mid k_m$ for all $m > n$ and $p^{n+1} \nmid k_{n+1}$, since $\alpha(c_{n+1}) \neq 0$. Let $k'_i = k_{n+i} p^{-n}$. Then $k' = (k'_i)_{i \in \mathbb{N}_+} \in B$, as $k'_n \equiv k'_{n+1} (\operatorname{mod} p^n)$. Moreover, $\beta = \mu(k') \notin L$, as $p \nmid k'_1$ and so $\beta(c_1) \neq 0$. By the above argument, β is an automorphism of $\mathbb{J}_p$, that is, $\beta \in U(\mathbb{J}_p)$. Obviously, $\alpha = p^n \beta$. □

Remark A.4.17. The field of quotients $\mathbb{Q}_p$ of $\mathbb{J}_p$ is called *field of p-adic numbers*. Sometimes we shall consider only the underlying groups of these rings and simply speak of "the group of p-adic integers", or "the group of p-adic numbers". Since all nonzero ideals of $\mathbb{J}_p$ are of the form $p^n\mathbb{J}_p$, for $n \in \mathbb{N}$, one can write $\mathbb{Q}_p = \bigcup_{n\in\mathbb{Z}} p^n\mathbb{J}_p$, i. e., every element of $\mathbb{Q}_p \setminus \mathbb{J}_p$ can be written as $p^{-n}\eta$, with $\eta \in U(\mathbb{J}_p)$ and $n \in \mathbb{N}$.

A.4.3 Indecomposable abelian groups

We have seen that the finitely generated abelian groups, the divisible abelian groups, the bounded abelian groups and the free abelian groups are direct sums of very simple groups, actually indecomposable ones:

Definition A.4.18. An abelian group is *indecomposable* if it is not a direct product of any pair of its proper subgroups.

Example A.4.19. The groups $\mathbb{Z}$, $\mathbb{Q}$, $\mathbb{Z}(p^n)$, for a prime p and $n \in \mathbb{N}_+$, and $\mathbb{Z}(p^\infty)$ for a prime p, are indecomposable.

The reader may have the impression that every abelian group is a direct sum of cyclic groups or copies of $\mathbb{Q}$ or the Prüfer groups $\mathbb{Z}(p^\infty)$. This fails in a spectacular way: there exist arbitrarily large indecomposable abelian groups (hence, nonisomorphic to any of the groups $\mathbb{Z}$, $\mathbb{Q}$, $\mathbb{Z}(p^n)$ and $\mathbb{Z}(p^\infty)$). As the next theorem shows, $\mathbb{J}_p$ is indecomposable (note that $|\mathbb{J}_p| = \mathfrak{c}$, see Exercise A.7.13). There exist also smaller indecomposable groups (e. g., subgroups of $\mathbb{Q}^2$).

Theorem A.4.20. *For p a prime, $\mathbb{J}_p/p\mathbb{J}_p \cong \mathbb{Z}(p)$. Consequently, the group $(\mathbb{J}_p, +)$ is indecomposable.*

Proof. The isomorphism is given by the first theorem of homomorphisms applied to the surjective homomorphism $\rho: \mathbb{J}_p \to \mathbb{Z}(p)$, $\alpha \mapsto \alpha(c_1)$.

Suppose that $\mathbb{J}_p = A \oplus B$ for a pair of subgroups A, B of $\mathbb{J}_p$. Then $p\mathbb{J}_p = pA \oplus pB$, so $\mathbb{J}_p/p\mathbb{J}_p \cong A/pA \oplus B/pB$. As $\mathbb{J}_p/p\mathbb{J}_p \cong \mathbb{Z}(p)$ and both A/pA and B/pB are of exponent p, they must be isomorphic to direct sums of copies of $\mathbb{Z}(p)$. Hence, either $A/pA = \{0\}$ or $B/pB = \{0\}$. Suppose that $A/pA = \{0\}$. We prove that then $A = \{0\}$ as well. In fact, $A/pA = \{0\}$ implies $A = pA$, and so $A = p^nA$ for every $n \in \mathbb{N}$. By Remark A.4.16, this yields $A = \{0\}$. □

A.5 Extensions of abelian groups

Definition A.5.1. Let A, C be abelian groups. An abelian group B is an *extension of A by C* if B has a subgroup $A' \cong A$ such that $B/A' \cong C$.

In such a case, if $i: A \to B$ is the injective homomorphism with $i(A) = A'$ and $B/A' \cong C$, we shall briefly denote this by the diagram

$$0 \longrightarrow A \xrightarrow{i} B \xrightarrow{q} C \longrightarrow 0 \tag{A.14}$$

where q is the composition of the canonical projection $B \to B/i(A)$ and the isomorphism $B/i(A) \cong C$. More generally, we shall refer to (A.14), as well as to any pair of group homomorphisms $i: A \to B$ and $q: B \to C$ with $\ker q = i(A)$, $\ker i = 0$, and $\operatorname{coker} q = 0$, as a *short exact sequence*.

Example A.5.2. A typical extension of a group A by a group C is the direct product $B = A \times C$. We call this extension *trivial extension*.

(a) There exist nontrivial extensions: $\mathbb{Z}$ is a nontrivial extension of $\mathbb{Z}$ and $\mathbb{Z}(2)$.
(b) In some cases only trivial extensions are available of A by C: take $A = \mathbb{Z}(2)$ and $C = \mathbb{Z}(3)$.

For more examples, see Exercise A.7.18.

A property $\mathfrak{G}$ of abelian groups is called *stable under extension* (or, *three space property*) if every abelian group B that is an extension of two abelian groups, both having $\mathfrak{G}$, necessarily has $\mathfrak{G}$.

Lemma A.5.3. *Assume that the two horizontal rows are exact in the following commutative diagram of abelian groups and homomorphisms:*

$$\begin{array}{ccccccccc} 0 & \longrightarrow & G_1 & \xrightarrow{i_1} & G & \xrightarrow{q_1} & G_2 & \longrightarrow & 0 \\ & & \downarrow{\scriptstyle f_1} & & \downarrow{\scriptstyle f} & & \downarrow{\scriptstyle f_2} & & \\ 0 & \longrightarrow & H_1 & \xrightarrow[i_2]{} & H & \xrightarrow[q_2]{} & H_2 & \longrightarrow & 0 \end{array}$$

If the homomorphisms f_1 and f_2 are surjective (respectively, injective), then also f is surjective (respectively, injective).

Proof. Assume that both f_1 and f_2 are surjective. Since the second row is exact, $\ker q_2 = \operatorname{im} i_2$. We prove first that

$$\ker q_2 = \operatorname{im} i_2 \subseteq f(G). \tag{A.15}$$

Indeed, if $x \in H_1$, then $i_2(x) \in i_2(f_1(G_1)) = f(i_1(G_1)) \subseteq f(G)$ by the surjectivity of f_1. Now to check the surjectivity of f, pick $y \in H$. Then $q_2(y) \in f_2(q_1(G)) = q_2(f(G))$, by the surjectivity of f_2 and q_1. Therefore, $q_2(y) = q_2(z)$ for some $z \in f(G)$. This yields $y \in z + \ker q_2 \subseteq f(G) + f(G) = f(G)$, by (A.15).

Now assume that both f_1 and f_2 are injective. To prove that f is injective, assume that $f(x) = 0$ for some $x \in G$. Then $0 = q_2(f(x)) = f_2(q_1(x))$. By the injectivity of f_2, $q_1(x) = 0$. Hence, $x \in i_1(G_1)$. Let $x = i_1(y)$ for some $y \in G_1$. So, $f(x) = f(i_1(y)) = i_2(f_1(y)) = 0$. As both i_2 and f_1 are injective, $y = 0$. Therefore, $x = 0$. □

Let us find a description of an extension B of given abelian groups A and C. Suppose for simplicity that A is a subgroup of B and $C = B/A$. Let $q: B \to C = B/A$ be the canonical projection. Since it is surjective one can fix a *section* $s: C \to B$ of q (namely, a map such that $q(s(c)) = c$ for all $c \in C$) with $s(0) = 0$.

For $b \in B$ the element $r(b) = b - s(q(b))$ belongs to $A = \ker q$. This defines a map $r: B \to A$ such that $r \restriction_A = id_A$. Therefore, every element $b \in B$ is uniquely described by the pair $(q(b), r(b)) \in C \times A$ by $b = s(q(b)) + r(b)$. If s is a homomorphism, the image $s(C)$ is a subgroup of B and $B \cong s(C) \times A$ splits.

From now on we consider the general case. Then, for $c, c' \in C$, the element

$$f_s(c, c') := s(c) + s(c') - s(c + c') \in B \tag{A.16}$$

certainly belongs to A, as q is a homomorphism. This defines a map $f_s: C \times C \to A$ uniquely determined by the extension B and the choice of the section s. When s is a homomorphism, $f_s = 0$.

We leave to the reader the verification of the fact that the commutativity and the associativity of the operation in B yield, for all $c, c', c'' \in C$,

$$f_s(c, c') = f_s(c', c) \quad \text{and} \quad f_s(c, c') + f_s(c + c', c'') = f_s(c, c' + c'') + f_s(c', c''). \tag{A.17}$$

As the section s satisfies $s(0) = 0$, for all $c \in C$,

$$f_s(c, 0) = f_s(0, c) = 0. \tag{A.18}$$

Definition A.5.4. Let C, A be abelian groups. A function $f: C \times C \to A$ satisfying (A.17) and (A.18) is a *factor set* on C to A.

We denote by $\mathrm{Fact}(C, A)$ the set of all factor sets on C to A. One can easily see that $\mathrm{Fact}(C, A)$ is a subgroup of the abelian group $(A^{C \times C}, +)$.

The proof of the next proposition is straightforward.

Proposition A.5.5. *Let A, C be abelian groups. Every $f \in \mathrm{Fact}(C, A)$ gives rise to an extension B_f of A by C defined in the following way. The support of the group is $B_f = C \times A$, with operation*

$$(c, a) + (c', a') = (c + c', a + a' + f(c, c')) \quad \text{for } c, c' \in C, a, a' \in A$$

and the subgroup $A' = \{0\} \times A \cong A$ is such that $B_f / A' \cong C$. For the section $s: C \to B_f$ of the canonical projection $q: B_f \to C$, defined by letting $s(c) = (c, 0)$ for every $c \in C$, we get $f = f_s$.

It is important to note that the subset $C \times \{0\} = s(C)$ need not be a subgroup of B_f; it is a subgroup of B_f precisely when $f = 0$.

Proposition A.5.5 shows that there exists a correspondence

$$\mathrm{Fact}(C, A) \to \{\text{extensions of } A \text{ by } C\}, \quad f \mapsto B_f,$$

and this correspondence is such that, for the section $s: C \to B_f$ with $s(c) = (c, 0)$, we get $f_s = f$.

The trivial extension is determined by the identically zero factor set f, if the section $s(c) = c$ is chosen. More precisely one has:

Example A.5.6. Let A, C be abelian groups and identify C and A with subgroups of the trivial extension $B = C \times A$. Every section $s: C \to B$ of the canonical projection $q: B \to C$, with $s(0) = 0$, is defined by $s(c) = c + h(c)$ for $c \in C$, where $h: C \to A$ is map with $h(0) = 0$. The factor set f_s obtained from (A.16) satisfies, for $c, c' \in C$,

$$f_s(c, c') = h(c) + h(c') - h(c + c'). \tag{A.19}$$

Inspired by this example, for abelian groups C, A, now we consider two sections $s_1, s_2: C \to B$, with $s_1(0) = 0 = s_2(0)$, of the same arbitrary extension B of A by C. For every $c \in C$, let $h: C \to A$ be defined by

$$h(c) = s_1(c) - s_2(c) \in A;$$

in particular, $h(0) = 0$. The factor sets f_{s_1} and f_{s_2} satisfy

$$f_{s_1}(c, c') - f_{s_2}(c, c') = h(c) + h(c') - h(c + c'). \tag{A.20}$$

This motivates the following definition.

Definition A.5.7. Let A, C be abelian groups. Call two factor sets $f_1, f_2: C \times C \to A$ *equivalent* if (A.20) holds for some map $h: C \to A$ with $h(0) = 0$.

Definition A.5.8. Let A, C be abelian groups. Call two extensions B_1, B_2 of A by C *equivalent* if there exists a homomorphism $\xi: B_1 \to B_2$ such that the following diagram, where both horizontal rows describe the respective extension, is commutative:

$$\begin{array}{ccccccccc} 0 & \longrightarrow & A & \xrightarrow{i_1} & B_1 & \xrightarrow{q_1} & C & \longrightarrow & 0 \\ & & {\scriptstyle id_A}\downarrow & & \downarrow{\scriptstyle \xi} & & \downarrow{\scriptstyle id_A} & & \\ 0 & \longrightarrow & A & \xrightarrow[i_2]{} & B_2 & \xrightarrow[q_2]{} & C & \longrightarrow & 0 \end{array} \tag{A.21}$$

It follows from Lemma A.5.3 that the homomorphism ξ in the above definition is necessarily an isomorphism, provided it exists.

Theorem A.5.9. *Let A, C be abelian groups and $f_1, f_2 \in \mathrm{Fact}(C, A)$. The two extensions B_{f_1}, B_{f_2} of A by C are equivalent if and only if the factor sets f_1, f_2 are equivalent. On the other hand, if B is an extension of A by C with section s, then for $f_s \in \mathrm{Fact}(C, A)$ the extension B_{f_s} is equivalent to B.*

Proof. Assume that the factor sets f_1, f_2 are equivalent, namely, there exists a map $h: C \to A$ with $h(0) = 0$ and such that (A.20) holds. For $i = 1, 2$, let and $\xi: B_{f_1} \to B_{f_2}$, $(c, a) \mapsto (c, a + h(c))$. We leave the verification of the fact that ξ is an isomorphism and the extensions B_{f_1}, B_{f_2} are equivalent to the reader.

Now assume that the extensions B_{f_1}, B_{f_2} are equivalent. From the commutativity of (A.21), we deduce that the isomorphism $\xi: B_1 \to B_2$ must be of the form $\xi(c, a) = (c, a + h(c))$, where $h: C \to A$ is a map with $h(0) = 0$. For $c, c' \in C$, one has

$$\begin{aligned}(c + c', h(c + c') + f_1(c, c')) &= \xi(c + c', f_1(c, c')) = \xi((c, 0) +_1 (c', 0)) \\ &= \xi(c, 0) +_2 \xi(c', 0) = (c, h(c)) +_2 (c', h(c')) = (c + c', f_2(c, c') + h(c) + h(c')).\end{aligned}$$

This gives $h(c + c') + f_1(c, c') = f_2(c, c') + h(c) + h(c')$, so f_1, f_2 are equivalent.

To prove the last assertion, assume that (A.14) is an extension of A by C with section s. It is easy to check that the required isomorphism $\xi: B \to B_f$ is defined by $\xi(b) = (r(b), q(b))$, where $b \in B$ and $r = s \circ q$. □

The factor sets f of the form (A.19) obviously form a subgroup $\mathrm{Trans}(C, A)$ of $\mathrm{Fact}(C, A)$. Hence, the above theorem tells us that the set $\mathrm{Ext}(C, A)$ of all equivalence classes of extensions of A by C admits a bijection with the set of all equivalence classes of factor sets. But since these equivalence classes of factor sets form precisely the quotient group $\mathrm{Fact}(C, A)/\mathrm{Trans}(C, A)$, we see that $\mathrm{Ext}(C, A)$ carries a structure of abelian group transported from the group $\mathrm{Fact}(C, A)/\mathrm{Trans}(C, A)$. In particular, using the fact that $\mathrm{Ext}(C, A)$ is a group, we write $\mathrm{Ext}(C, A) = \{0\}$ to say that there are only trivial extensions of A by C.

See [202] for a different equivalent definition of the group structure of $\mathrm{Ext}(C, A)$.

A.6 Nonabelian groups

Recall that a *simple* group is a nontrivial group whose only normal subgroups are the trivial group and the group itself.

In a group G, the *centralizer* of $a \in G$ is the subgroup $c_G(a) = \{x \in G: xa = ax\}$ of G, and more generally for $X \subseteq G$, $c_G(X) = \{x \in G: xg = gx \text{ for every } g \in X\}$. The *center* of G is $Z(G) = c_G(G)$; it is a normal subgroup.

For $x, y \in G$, denote by $[x, y] = x^{-1}y^{-1}xy$ the *commutator* of x and y in G. Moreover, for subgroups H, K of G, the *commutator* of H and K in G is the subgroup of G generated by $\{[x, y]: x \in H, y \in K\}$. The *commutator* of G is $G' = [G, G]$, which is a characteristic (so, normal) subgroup of G.

The *lower central series* of G is defined inductively by $\gamma_1(G) = G$ and, for every $n \in \mathbb{N}_+$, by $\gamma_{n+1}(G) = [\gamma_n(G), G]$. Each $\gamma_n(G)$ is a characteristic subgroup of G.

The *upper central series* of G is defined inductively by $Z_0(G) = \{e_G\}$, $Z_1(G) = Z(G)$ and, for every $n \in \mathbb{N}$, by $Z_{n+1}(G) = \{x \in G: \forall y \in G, [x, y] \in Z_n(G)\}$ (equivalently, $Z_{n+1}(G)/Z_n(G) = Z(G/Z_n(G))$).

Definition A.6.1. A group G is *nilpotent* if $\gamma_{c+1}(G) = \{e_g\}$ for some $c \in \mathbb{N}$. We say that G has *nilpotency class c* if $c \in \mathbb{N}$ is the minimum such that $\gamma_{c+1}(G) = \{e_G\}$ (equivalently, $Z_{c+1}(G) = G$).

It is easy to verify that subgroups and quotients of nilpotent groups are nilpotent.

For a group G, the *derived series* is defined inductively by $G^{(0)} = G$ and, for every $n \in \mathbb{N}$, $G^{(n+1)} = [G^{(n)}, G^{(n)}]$. Each $G^{(n)}$ is a characteristic subgroup of G.

Definition A.6.2. A group G is *solvable* if $G^{(d)} = \{e_G\}$ for some $d \in \mathbb{N}$. We say that G has *derived length d* if $d \in \mathbb{N}$ is the minimum such that $G^{(d)} = \{e_G\}$.

A group G is *metabelian* if G is solvable of derived length at most 2.

Subgroups and quotients of solvable groups are solvable. Clearly, a group G is metabelian in case G' is abelian.

Nilpotent groups are solvable, but, for example, S_3 is metabelian but not nilpotent. Moreover, we recall that a torsion finitely generated solvable group is necessarily finite.

For a unitary ring R the *Heisenberg group over R* is the subgroup $\mathcal{H}(R)$ of $\mathrm{GL}_3(R)$ of all upper unitriangular 3×3 matrices) over the ring R. The group $\mathcal{H}(R)$ is always nilpotent of class 2.

The *Frattini subgroup* $\mathrm{Fratt}(G)$ of a group G is the intersection of all maximal subgroups of G. Note that $\mathrm{Fratt}(G) = G$ in case G has no maximal subgroups.

Let us recall that for groups K, H and a group homomorphism $\theta\colon K \to \mathrm{Aut}(H)$, one defines the semidirect product $G = H \rtimes_\theta K$, as the group with support the Cartesian product $G = H \times K$ and group operation defined by $(h,k) \cdot (h_1, k_1) = (h\theta(k)(h_1), kk_1)$. Identifying H and K with the subgroups $H \times \{e_K\}$ and $\{e_H\} \times K$, respectively, of G, the conjugation in G by an element k of K restricted to H is precisely the automorphism $\theta(k)$ of H. For a subgroup $K \leq \mathrm{Aut}(H)$ and the inclusion $\theta\colon K \to \mathrm{Aut}(H)$, we simply write $H \rtimes K$ in place of $H \rtimes_\theta K$ (in particular, for an automorphism $\phi \in \mathrm{Aut}(H)$, we write $H \rtimes \langle \phi \rangle$).

A.7 Exercises

Exercise A.7.1. Verify that for every abelian group G, the quotient $G/t(G)$ is torsion-free.

Exercise A.7.2. For a prime p and G an abelian p-group, prove that G is divisible if and only if G is p-divisible.

Exercise A.7.3. Let G be a torsion-free abelian group.
(a) Let M be an independent subset of G. Prove that M is maximal if and only if the subgroup $\langle M \rangle$ is essential.
(b) Prove that a subgroup H of G is essential if and only if G/H is torsion.

Exercise A.7.4. Prove that a subgroup H of an abelian group G is essential in G if and only if H contains $\mathrm{Soc}(G)$ and a maximal independent subset of G.

Exercise A.7.5. Prove that if G, H are abelian groups and N, K are essential subgroups of G, H, respectively, then $N \times K$ is an essential subgroup of $G \times H$.

Exercise A.7.6. Let G be an abelian group and H a subgroup of G. Prove that $r_0(G) = r_0(H) + r_0(G/H)$. If G is a p-group for a prime p, prove that $r_p(G) = r_p(H) + r_p(G/H)$.

Exercise A.7.7. For an abelian group G, prove that:
(a) $r_0(G^{\mathbb{N}}) > 0$ if and only if $r_0(G^{\mathbb{N}}) \geq \mathfrak{c}$, if and only if G is not bounded;
(b) if G is not bounded and $|G| \leq \mathfrak{c}$, then $r_0(G^\rho) = 2^\rho$ for any infinite cardinal ρ.

Exercise A.7.8. Prove that every subgroup of $\mathbb{Q}^2$ of rank 2 is essential in $\mathbb{Q}^2$.

Exercise A.7.9. For a prime p and a bounded abelian p-group B, prove that $r_p(B) = r_p(B/pB)$.

Hint. Use the fact that $B \cong \mathbb{Z}(p)^{(\alpha_1)} \oplus \cdots \oplus \mathbb{Z}(p^n)^{(\alpha_n)}$ for some $n \in \mathbb{N}_+$ and cardinals $\alpha_1, \ldots, \alpha_n$, and $r_p(\mathbb{Z}(p^k)^{(\kappa)}) = \kappa$ for every $k \in \mathbb{N}_+$ and a cardinal κ.

Exercise A.7.10. Prove that:
(a) subgroups and direct products of reduced abelian groups are reduced;
(b) every abelian group is a quotient of a reduced group.

Hint. Use Theorem A.3.2(a) and Example A.4.2.

Exercise A.7.11. Prove that G/G^1 is residually finite for every abelian group G.

Exercise A.7.12. Prove that, for an odd prime p, if $\xi^p = 1$ for $\xi \in \mathbb{J}_p$ then $\xi = 1$. For $\xi \in \mathbb{J}_2$, if $\xi^2 = 1$ then $\xi = \pm 1$.

Hint. Use Claim A.4.16.

Exercise A.7.13. Prove that $|\mathbb{J}_p| = \mathfrak{c}$.

Hint. Show that to every sequence of integers $k = \{k_n\}_{n\in\mathbb{N}}$ satisfying (A.13) and $0 \leq k_n < p^n$, one can assign a sequence $\{a_n\}_{n\in\mathbb{N}} \in \{0, 1, \ldots, p-1\}^{\mathbb{N}}$ such that

$$k_n = \sum_{j=0}^{n-1} a_j p^j \tag{A.22}$$

for every $n \in \mathbb{N}_+$. Vice versa, every sequence $\{a_n\}_{n\in\mathbb{N}} \in \{0, 1, \ldots, p-1\}^{\mathbb{N}}$ defines, via (A.22), a sequence $k = \{k_n\}_{n\in\mathbb{N}} \in \mathbb{Z}^{\mathbb{N}}$ satisfying (A.13) and $0 \leq k_n < p^n$.

Exercise A.7.14. Prove that a finite abelian group G is indecomposable if and only if $G \cong \mathbb{Z}(p^n)$ for some prime p and $n \in \mathbb{N}$.

Exercise A.7.15. Prove that an abelian group is indecomposable if its endomorphism ring has no nontrivial idempotents.

Exercise A.7.16. An abelian group G is called *rigid* if its endomorphism ring consists only of the endomorphisms $m \cdot id_G$ with $m \in \mathbb{Z}$. Prove that an infinite rigid abelian group is indecomposable.

Exercise A.7.17. Prove that the following properties of the abelian groups are stable under extension: torsion, torsion-free, divisible, reduced, p-torsion, having no non-trivial p-torsion elements, having finite exponent. Prove that residual finiteness is not stable under extension.

Hint. In Example A.4.7, N is not residually finite, N^1 is finite, and N/N^1 is residually finite.

Exercise A.7.18. Let A, C be abelian groups. Prove that $\mathrm{Ext}(C, A) = \{0\}$ in the following cases:

(a) A is divisible;

(b) C is free;

(c) both A and C are torsion and for every prime p either $r_p(A) = 0$ or $r_p(C) = 0$;

(d) * (Prüfer theorem) C is torsion-free and A has finite exponent. (So, in every abelian group G such that $t(G)$ is bounded, the torsion subgroup splits as $G = t(G) \oplus G/t(G)$.)

Hint. For (a) use Corollary A.2.7, for (b) use Lemma A.3.1. For (c), deduce first that every extension B of A by C is torsion and then argue using the hypothesis on $r_p(A)$ and $r_p(C)$. A proof of (d) can be found in [138].

B Background on topological spaces

For the sake of completeness, we recall some frequently used notions and notations from general topology. We omit most of the proofs, since these are classical results in general topology. See, for example, the monograph [134].

B.1 Basic definitions

B.1.1 Filters

We start with the definition of filter, filter base, and topology.

Definition B.1.1. Let X be a set. A family $\mathcal{F} \subseteq \mathcal{P}(X) \setminus \{\emptyset\}$ is:
(i) a *filter prebase* on X if it has the *finite intersection property*, that is, $A_1 \cap \cdots \cap A_n \neq \emptyset$ for every $A_1, \ldots, A_n \in \mathcal{F}$;
(ii) a *filter base* on X if for $A, B \in \mathcal{F}$ there exists $C \in \mathcal{F}$ such that $C \subseteq A \cap B$;
(iii) *up-closed* if $F_1 \in \mathcal{F}$, $F_2 \in \mathcal{P}(X)$ and $F_1 \subseteq F_2$ imply $F_2 \in \mathcal{F}$;
(iv) a *filter* on X if $\mathcal{F}$ is an up-closed filter base;
(v) an *ultrafilter* of X if $\mathcal{F} \subseteq \mathcal{F}'$ for some filter $\mathcal{F}'$ of X yields $\mathcal{F}' = \mathcal{F}$.

Clearly, every filter is a filter base, and every filter base is a filter prebase.

Every family $\mathcal{B} \subseteq \mathcal{P}(X) \setminus \{\emptyset\}$ generates a smallest up-closed family $\mathcal{B}^*$ containing $\mathcal{B}$, namely, $\mathcal{B}^* = \{F \in \mathcal{P}(X) : \exists G \in \mathcal{B},\ G \subseteq F\}$. Then $\mathcal{F} = \mathcal{B}^*$ is a filter if and only if $\mathcal{B}$ is a filter base. In such a case we say that the filter base $\mathcal{B}$ *generates* the filter $\mathcal{F}$ and we call $\mathcal{B}$ a *base* of $\mathcal{F}$.

Example B.1.2. If A is a nonempty subset of a set X, let $[A]$ denote the family of all subsets of X containing A. If $A = \{a\}$, we briefly write $[a]$ in place of $[\{a\}]$.
(a) It is easy to see that $[A]$ is a filter; moreover, if $A \subseteq B \subseteq X$, then $[B] \subseteq [A]$.
(b) One can deduce from (a), that $[a]$ is an ultrafilter for every $a \in X$.

A filter of the form $[A]$ is called *principal* or *fixed*, and we say that it is *generated* by the set A. Clearly, a filter $\mathcal{F}$ is fixed if and only if $\bigcap \mathcal{F} \neq \emptyset$ belongs to $\mathcal{F}$.

All filters on a finite set are fixed.

Example B.1.3. An example of a nonfixed filter on an infinite set X is the *Fréchet filter* defined as the family of all the cofinite subsets of X, where $A \subseteq X$ is called *cofinite* if the complement $X \setminus A$ is finite.

More generally, if X is an infinite set and α is a cardinal with $\alpha \leq |X|$, the family $\{A \subseteq X : |X \setminus A| < \alpha\}$ is a filter on X.

https://doi.org/10.1515/9783110654936-018

B.1.2 Topologies, bases, prebases, and neighborhoods

Definition B.1.4. Let X be a set. A family $\tau \subseteq \mathcal{P}(X)$ is a *topology* on X if:
(a) $\emptyset, X \in \tau$;
(b) if $U_1, \dots, U_n \in \tau$ for $n \in \mathbb{N}_+$, then $U_1 \cap \dots \cap U_n \in \tau$;
(c) if $\{U_i : i \in I\} \subseteq \tau$, then $\bigcup_{i \in I} U_i \in \tau$.

The pair (X, τ) is called a *topological space*; when there is no possibility of confusion, we simply say that X is a topological space omitting its topology τ.

The members of τ are called *open*, or τ*-open*, the complement of an open set is called *closed*, or τ*-closed*, and a set is *clopen* if it is both closed and open.

In the sequel, all topological spaces are assumed to be nonempty.

Example B.1.5. Let X be a set.
(a) The *discrete* topology δ_X on X has as open sets all subsets of X, namely, $\delta_X = \mathcal{P}(X)$. The *indiscrete* topology ι_X on X has as open sets only the sets X and $\emptyset$, namely, $\iota_X = \{\emptyset, X\}$.
(b) If X is infinite, the *cofinite topology* on X is $\gamma_X = \{\emptyset\} \cup \{A \in \mathcal{P}(X) : X \setminus A \text{ finite}\}$. So, a proper subset $Y \subseteq X$ is closed in γ_X precisely when Y is finite.

Recall that a point x of a topological space X is *isolated* if $\{x\}$ is open in X.

Definition B.1.6. For a topological space (X, τ), a family $\mathcal{B} \subseteq \tau$ is:
(i) a *base* of (X, τ) if for every $x \in X$ and for every $x \in U \in \tau$ there exists $B \in \mathcal{B}$ such that $x \in B \subseteq U$;
(ii) a *prebase* of (X, τ) if the family $\mathcal{B}^*$ of all nonempty finite intersections of members of $\mathcal{B}$ is a base of (X, τ).

A base of a topological space need not be stable for finite intersections. Clearly, for a set X, $\{\{x\} : x \in X\}$ is a base of δ_X and $\{X\}$ is a base of ι_X.

Example B.1.7. Consider $\mathbb{R}$ with its usual Euclidean topology τ induced by the usual metric d of $\mathbb{R}$ (see §B.3.2). Then $\mathcal{B} = \{(a, b) : a, b \in \mathbb{Q}, a < b\}$ is a (countable) base of τ while $\mathcal{P} = \{(a, +\infty) : a \in \mathbb{Q}\} \cup \{(-\infty, b) : b \in \mathbb{Q}\}$ is a prebase of τ.

We omit the immediate proof of the next lemma.

Lemma B.1.8. *Let X be a set and $\mathcal{B} \subseteq \mathcal{P}(X) \setminus \{\emptyset\}$. Then:*
(a) *$\mathcal{B}$ is base of some topology τ on X if and only if $\bigcup_{B \in \mathcal{B}} B = X$ and for every pair $B, B' \in \mathcal{B}$ and $x \in X$ with $x \in B \cap B'$ there exists $B'' \in \mathcal{B}$ such that $x \in B'' \subseteq B \cap B'$;*
(b) *$\mathcal{B}$ is prebase of some topology τ on X if and only if $\bigcup_{B \in \mathcal{B}} B = X$.*

The following is another fundamental concept when dealing with topological spaces.

Definition B.1.9. Let (X, τ) be a topological space and $x \in X$. A *neighborhood U* of x in τ is any subset V of X such that there exists $U \in \tau$ with $x \in U \subseteq V$.

The neighborhoods of a point x in a topological space (X, τ) form a filter $\mathcal{V}_{(X,\tau)}(x)$ of X. When there is no possibility of confusion, we denote it simply by $\mathcal{V}_\tau(x)$, $\mathcal{V}_X(x)$, or $\mathcal{V}(x)$.

Definition B.1.10. Let (X, τ) be a topological space and $x \in X$. A base $\mathcal{B}$ of the filter $\mathcal{V}(x)$ is called *base of the neighborhoods* of x (or briefly, *local base at* x) in (X, τ); the elements of $\mathcal{B}$ are called *basic neighborhoods* of x in (X, τ).

A prebase $\mathcal{B}$ of the filter $\mathcal{V}(x)$ is called *prebase of the neighborhoods* of x in (X, τ); the elements of $\mathcal{B}$ are called *prebasic neighborhoods* of x in (X, τ).

Remark B.1.11. If (X, τ) is a topological space and, for $x \in X$, $\mathcal{B}_x$ is a local base at x in (X, τ) consisting of open neighborhoods, then $\bigcup_{x \in X} \mathcal{B}_x$ is a base of τ.

We recall also the following known fact.

Lemma B.1.12. *Let* (X, τ) *be a topological space and* $V \subseteq X$. *Then* $V \in \tau$ *if and only* $V \in \mathcal{V}_\tau(x)$ *for every* $x \in V$.

Moreover, for a topological space (X, τ) and $x \in X$, the family $\{U \in \mathcal{V}_\tau(x) \colon U \in \tau\}$ of all open neighborhoods of x in τ is a base of $\mathcal{V}_\tau(x)$.

B.1.3 Ordering topologies, closure, and interior

The set $\mathcal{T}(X)$ of all topologies on a given set X is ordered by inclusion.

Definition B.1.13. Let X be a set and $\tau_1, \tau_2 \in \mathcal{T}(X)$. We say that τ_1 is *coarser* than τ_2, or that τ_2 is *finer* than τ_1, if $\tau_1 \subseteq \tau_2$. We denote this also by $\tau_1 \leq \tau_2$.

Remark B.1.14. For a set X, $\mathcal{T}(X)$ becomes a complete lattice with top element the discrete topology and bottom element the indiscrete topology. If $\{\tau_i \colon i \in I\}$ is a family of topologies of X, then the intersection $\tau = \bigcap_{i \in I} \tau_i$ is a topology on X and so $\tau = \inf_{i \in I} \tau_i$, that is, τ is the *infimum* of the family $\{\tau_i \colon i \in I\}$. In other words, τ is the finest topology on X coarser than τ_i for every $i \in I$.

This is enough to claim that $\mathcal{T}(X)$ is a complete lattice, but it is convenient to get an explicit description of the *supremum* $\tau' = \sup_{i \in I} \tau_i$ of $\{\tau_i \colon i \in I\}$ in $\mathcal{T}(X)$. This is the topology τ' on X with prebase $\bigcup_{i \in I} \tau_i$. In other words, for every $x \in X$ a local base at x in (X, τ') is the family $\{U_1 \cap \cdots \cap U_n \colon U_k \in \mathcal{V}_{\tau_{i_k}}(x), k \in \{1, \ldots, n\}, n \in \mathbb{N}_+\}$. So, τ' is the coarsest topology on X that is finer than τ_i for every $i \in I$.

We conclude this section with the following basic definition.

Definition B.1.15. Let (X,τ) be a topological space and $M \subseteq X$.

(a) The *closure* $\overline{M}$ (denoted also by $\overline{M}^{\tau}$, when it is necessary to enhance the topology) of M in X is the smallest closed set of X containing M. This is the intersection of all closed sets containing M. Clearly, it coincides also with the set of all $x \in X$ such that every $U \in \mathcal{V}_\tau(x)$ meets M.

(b) Dually, the *interior* $\mathrm{Int}(M)$ of M in X is the largest open set of X contained in M. This is the union of all open sets contained in M, that is, the set of all $x \in M$ such that there exists $U \in \tau$ with $x \in U \subseteq M$.

Obviously, M is closed if and only if $\overline{M} = M$, while M is open if and only if $\mathrm{Int}(M) = M$. An open set U is a *regular open* set if $U = \mathrm{Int}(\overline{U})$.

Example B.1.16. A topology τ on a set X is an *Alexandrov topology*, and (X,τ) is an *Alexandrov topological space*, if arbitrary intersections of τ-open sets are τ-open. So, in this case $\bigcap \mathcal{V}(x) \in \tau$ for every $x \in X$, i. e., every point $x \in X$ has a smallest neighborhood $\bigcap \mathcal{V}(x)$ (obviously, it coincides with $\{x\}$ for all $x \in X$, for an Alexandrov space X, precisely when τ is discrete).

The Alexandrov topological spaces are exactly the topological spaces (X,τ) such that arbitrary unions of τ-closed sets are τ-closed. Hence, for every $M \subseteq X$ there is a smallest open set containing M (namely, the intersection of all open sets containing M) and a largest closed set contained in M (namely, the union of all closed sets contained in M).

The reader is maybe tempted to compare the set $\bigcap \mathcal{V}(x)$, that prominently appeared in the above example, with the point-closure $\overline{\{x\}}$, and even expect that they coincide. Let us anticipate here that the equality $\bigcap \mathcal{V}(x) = \overline{\{x\}}$ occurs very rarely (actually, only when the space is indiscrete), as we discuss in the extended Exercise B.7.9.

Definition B.1.17. A subset D of a topological space (X,τ) is *dense* in (X,τ) if $\overline{D}^{\tau} = X$ (we say also that D is τ*-dense* in X). A topological space X is *separable* if X has a dense countable subset.

Example B.1.18. (a) For a nonempty set X, a subset D of X is dense in (X,δ_X) precisely when $D = X$ and every nonempty subset of X is dense in (X,ι_X).

(b) As $\mathbb{Q}$ is dense in $\mathbb{R}$ with its usual Euclidean topology, $\mathbb{R}$ is separable.

Lemma B.1.19. *Let X be a topological space and U an open set of X. If D is a dense subset of X, then $U \cap D$ is dense in U.*

Proof. Let V be a nonempty open set of U. Then V is open in X, and so $V \cap (U \cap D) = V \cap D \neq \emptyset$. □

B.2 Convergent nets and filters

B.2.1 Convergent sequences

Definition B.2.1. A sequence $\{x_n\}_{n\in\mathbb{N}}$ in a topological space (X,τ) is said to *converge* to $x \in X$, and we write $x_n \to x$, if every neighborhood U of x in τ contains all but finitely many members of the sequence.

A convergent sequence $\{x_n\}_{n\in\mathbb{N}}$ is *trivial* if all but finitely many members x_n coincide with the limit point x, otherwise we call it *nontrivial.*

Unlike in the case of metric spaces, where the topology can be completely determined by convergence of sequences (see §B.3.2), topological spaces may have few convergent sequences, or at least not sufficiently many to provide a description of the topology (or the closure) in terms of convergent sequences.

Definition B.2.2. A subset A of a topological space X is *sequentially closed* if A contains the limits of all convergent sequences $\{a_n\}_{n\in\mathbb{N}}$ entirely contained in A. A topological space X is *sequential* if every sequentially closed set of X is closed.

Clearly, every closed set is also sequentially closed, but the inverse implication may fail (for example, in a nondiscrete space without nontrivial convergent sequences). Moreover, sequential spaces are the most general class of spaces for which sequences suffice to determine the topology.

There is a more subtle (stronger) connection between the topology and convergent sequences, given by the following notion.

Definition B.2.3. A topological space X is *Fréchet–Urysohn* if for every subset A of X and every $x \in \overline{A}$, there exists a sequence $\{a_n\}_{n\in\mathbb{N}}$ in A such that $a_n \to x$.

The Fréchet–Urysohn spaces X are obviously sequential, as sequentially closed sets in a Fréchet–Urysohn space are obviously closed.

B.2.2 Convergent nets

An alternative to sequences are nets, invented by Moore and Smith.

A partially ordered set $(A,\leq)$ is *directed* if for every $\alpha,\beta \in A$ there exists $\gamma \in A$ such that $\alpha \leq \gamma$ and $\beta \leq \gamma$. A subset B of $(A,\leq)$ is *cofinal* in A if for every $\alpha \in A$ there exists $\beta \in B$ with $\alpha \leq \beta$.

A *net* S in a topological space X is a map from a directed set $(A,\leq)$ to X. We write x_α for the image of $\alpha \in A$ in X, so that the net S is usually written in the form $S = \{x_\alpha\}_{\alpha\in A}$ in order to keep the similarity with the leading example, that is, with the sequences, obtained with $A = (\mathbb{N},\leq)$.

Definition B.2.4. Let X be a topological space. A net $S = \{x_\alpha\}_{\alpha\in A}$ in X *converges* to $x \in X$ if for every $U \in \mathcal{V}(x)$ there exists $\beta \in A$ such that $\alpha \in A$ and $\beta \leq \alpha$ imply $x_\alpha \in U$. In such a case, we call x *limit point* of the net S.

A net $S = \{x_\alpha\}_{\alpha\in A}$ in a topological space X can have more than one limit point. We denote by $\lim_{\alpha\in A} x_\alpha$ or by $\lim S$ the set of all limit points of S, and we write $x \in \lim_{\alpha\in A} x_\alpha$, or $x_\alpha \to x$, or $S \to x$, to denote that the net $S = \{x_\alpha\}_{\alpha\in A}$ converges to x. We write as usual $x = \lim_{\alpha\in A} x_\alpha$ in case the limit is unique.

Here comes the leading example of a (convergent) net.

Example B.2.5. Let X be a topological space and $x \in X$. Consider the set $\mathcal{A}$ of neighborhoods of x with the reversed order, that is, $U \leq V$ for $U, V \in \mathcal{A}$ if $U \supseteq V$. Then $(\mathcal{A}, \leq)$ is directed and, if for every $U \in \mathcal{A}$ one picks a point $x_U \in U$, then the net $\{x_U\}_{U\in\mathcal{A}}$ converges to x.

From the above example we deduce:

Lemma B.2.6. *Let X be a topological space. Then:*
(a) *for $Z \subseteq X$, $x \in \overline{Z}$ if and only if there exists a convergent net $\{x_\alpha\}_{\alpha\in A}$ with $x_\alpha \to x$ and $x_\alpha \in Z$ for every $\alpha \in A$;*
(b) *$Z \subseteq X$ is closed if and only if for every convergent net $\{x_\alpha\}_{\alpha\in A}$, with $x_\alpha \in Z$ for every $\alpha \in A$, one has $\lim_{\alpha\in A} x_\alpha \subseteq Z$.*

The concept of subnet is a generalization of that of subsequence. While every subsequence of a sequence is a subnet, a sequence may admit subnets that are not subsequences.

Definition B.2.7. Let X be a topological space. A *subnet* of the net $S = \{x_\alpha\}_{\alpha\in A}$ in X is a net $\{y_\beta\}_{\beta\in B}$ in X, where B is a directed set such that there exists a function $h\colon B \to A$ with the properties:
(i) for every $\alpha \in A$ there exists $\beta_0 \in B$ such that $h(\beta) \geq \alpha$ for all $\beta \geq \beta_0$ in B;
(ii) $y_\beta = x_{h(\beta)}$ for all $\beta \in B$.

Definition B.2.8. Let X be a topological space, $S = \{x_\alpha\}_{\alpha\in A}$ a net in X, and $U \subseteq X$. The net S is *frequently in U* if for every $\alpha \in A$ there exists $\beta \in A$, $\beta \geq \alpha$ such that $x_\beta \in U$. A point $x \in X$ is an *accumulation point* (or *cluster point*) of the net S if for every $U \in \mathcal{V}(x)$ the net S is frequently in U.

Clearly, a point $x \in X$ is an accumulation point of a net S precisely when some subnet of S converges to x. Now we see other similarities between the behavior of subnets of nets and subsequences of sequences.

Proposition B.2.9. *Let $S = \{x_\alpha\}_{\alpha\in A}$ be a net in a topological space X.*
(a) *If $S \to x \in X$, then $S' \to x$ for every subnet S' of S.*

(b) *If $x \in X$ is an accumulation point of a subnet of S, then x is an accumulation point of S.*
(c) *If $x \in X$ is an accumulation point of S, then x is a limit point of some subnet of S.*

Proof. (a) and (b) are obvious.

(c) Assume that $x \in X$ is an accumulation point of S. The set $I = \mathcal{V}(x) \times A$ has a partial order $\alpha' \leq \beta'$, defined for $\alpha' = (U, \alpha) \in I$ and $\beta' = (V, \beta) \in I$, by $\alpha \leq \beta$ and $U \supseteq V$. Now consider its subset

$$B = \{(U, \alpha) \in I : x_\alpha \in U\}$$

with the induced order. Let us see that this makes the poset $(B, \leq)$ directed. Indeed, fix $\alpha' = (U, \alpha), \beta' = (V, \beta) \in B$. Since $U \cap V \in \mathcal{V}(x)$ and $x \in X$ is an accumulation point of S, the subset $\{\gamma \in A : x_\gamma \in U \cap V\}$ of A is cofinal. Hence, there exists $x_\gamma \in U \cap V$ with $\gamma \geq \alpha$ and $\gamma \geq \beta$. Now $\gamma' = (U \cap V, \gamma) \in B$ and $\gamma' \geq \alpha', \gamma' \geq \beta'$.

Define $h: B \to A$ simply as the restriction to B of the second projection $\mathcal{V}(x) \times A \to A$. Then h is monotone and $h(B)$ is cofinal in A. Therefore, by setting $y_{\alpha'} = x_\alpha$ for $\alpha' = (U, \alpha) \in B$, we obtain a subnet $S' = \{y_{\alpha'}\}_{\alpha' \in B}$ of S. To check that $S' \to x$, fix $U \in \mathcal{V}(x)$. Since x is an accumulation point of S, there exists $\alpha \in A$ such that $x_\alpha \in U$. Let $\alpha' = (U, \alpha) \in B$. If $\beta' = (V, \beta) \in B$ is such that $\beta' \geq \alpha'$, then $V \subseteq U$ and $y_{\beta'} = x_\beta \in V \subseteq U$. Therefore, $S' \to x$. □

It follows from the above proposition that a net S in a topological space X converges to $x \in X$ if and only if every subnet of S converges to x.

B.2.3 Convergent filters

Now we introduce convergence of filters and filter bases.

Definition B.2.10. Let X be a topological space. A filter $\mathcal{F}$ on X *converges* to $x \in X$ when $\mathcal{V}(x) \subseteq \mathcal{F}$. A point $x \in X$ is a *limit point* of $\mathcal{F}$ if $\mathcal{F}$ converges to x. Moreover, x is an *adherent point* of $\mathcal{F}$, if $x \in \operatorname{ad} \mathcal{F} = \bigcap\{\overline{F} : F \in \mathcal{F}\}$.

Similarly, let $\mathcal{B}$ be a filter base on X and denote by $\mathcal{B}^*$ the filter generated by $\mathcal{B}$. We say that $\mathcal{B}$ *converges* to $x \in X$ when $\mathcal{V}(x) \subseteq \mathcal{B}^*$, and x is an *adherent point* of $\mathcal{B}$ if $x \in \operatorname{ad} \mathcal{B} = \bigcap\{\overline{B} : B \in \mathcal{B}\}$. Since $\operatorname{ad} \mathcal{B}^* = \operatorname{ad} \mathcal{B}$, x is an adherent point of $\mathcal{B}^*$ if and only if x is an adherent point of $\mathcal{B}$. Obviously, $\mathcal{B}$ converges to x if and only if $\mathcal{B}^*$ converges to x.

The following lemma reveals the close connection between convergence of filters and convergence of nets.

Lemma B.2.11. *Let X be a topological space and $x \in X$.*
(a) *Assume that $S = \{x_\alpha\}_{\alpha \in A}$ is a net in X. Then $\mathcal{B} = \{x_\beta : \exists \alpha \in A, \beta \geq \alpha\}$ is a filter base on X. Moreover:*

(a_1) $\mathcal{B}$ *converges to* $x \in X$ *if and only if the net* $\{x_\alpha\}_{\alpha\in A}$ *converges to* x;

(a_2) x *is an accumulation point of the net* S *if and only if* x *is an adherence point of* $\mathcal{B}$.

(b) *For a filter* $\mathcal{F} = \{F_a : a \in A\}$ *on* X *and* $\alpha, \beta \in A$, *let* $\alpha \leq \beta$ *if* $F_\alpha \supseteq F_\beta$. *Then:*

(b_1) *the partially ordered set* $(A, \leq)$ *is directed;*

(b_2) *the filter* $\mathcal{F}$ *converges to* x *if and only if for every choice of a point* $x_a \in F_a$ *the net* $\{x_\alpha\}_{\alpha\in A}$ *converges to* x;

(b_3) x *is an adherence point of the filter* $\mathcal{F}$ *if and only if for every* $F_\alpha \in \mathcal{F}$ *one can choose a point* $x_\alpha \in F_\alpha$ *such that* x *is an accumulation point of the net* $\{x_\alpha\}_{\alpha\in A}$.

B.3 Continuous maps and cardinal invariants of topological spaces

B.3.1 Continuous maps and their properties

Here we introduce continuity and discuss properties of continuous maps.

Definition B.3.1. For a map $f: (X, \tau) \to (Y, \tau')$ between topological spaces and a point $x \in X$ we say that:

(i) f is *continuous* at x if for every $U \in \mathcal{V}_Y(f(x))$ there exists $V \in \mathcal{V}_X(x)$ with $f(V) \subseteq U$;

(ii) f is *open* at $x \in X$ if for every $V \in \mathcal{V}_X(x)$ there exists $U \in \mathcal{V}_Y(f(x))$ with $f(V) \supseteq U$.

Moreover (global properties):

(iii) f is *continuous* (respectively, *open*) if f is continuous (respectively, open) at every point $x \in X$;

(iv) f is *closed* if the subset $f(A)$ of Y is closed for every closed set A of X;

(v) f is a *homeomorphism* if f is continuous, open, and bijective;

(vi) f is a *local homeomorphism* if for every $x \in X$ there exists an open set U of X containing x such that $f(U)$ is open in Y and $f\restriction_U : U \to f(U)$ is a homeomorphism;

(vii) f is a *(topological) embedding* if $f: X \to f(X)$ is a homeomorphism.

In item (i) and (ii), one can limit the test to only basic neighborhoods. Clearly, a homeomorphism is a local homeomorphism, while the converse is not true (for example, the map $f: \mathbb{R} \to \mathbb{S}$, $x \mapsto \cos(x) + i\sin(x)$, is a local homeomorphism which is not a homeomorphism).

The next lemma describes continuity at a point in terms of filters and nets similarly to the description of continuity in metric spaces in terms of sequences.

Lemma B.3.2. *Let* X, Y *be topological spaces,* $f: X \to Y$ *a map, and* $x \in X$. *Then the following conditions are equivalent:*

(a) f *is continuous at* x;

(b) $f(x_\alpha) \to f(x)$ *in* Y *for every net* $\{x_\alpha\}_{\alpha\in A}$ *in* X *with* $x_\alpha \to x$;

(c) $f(\mathcal{F}) \to f(x)$ *in* Y *for every filter* $\mathcal{F}$ *on* X *with* $\mathcal{F} \to x$.

For item (c) of the above lemma, see Exercise B.7.1(a).

The next lemma describes global continuity (in all points of the space).

Lemma B.3.3. *Let X, Y be topological spaces and $f: X \to Y$ a map. Then the following conditions are equivalent:*
(a) *f is continuous;*
(b) *the inverse image under f of open sets of Y are open in X;*
(c) *the inverse image under f of prebasic open sets of Y are open in X;*
(d) *the inverse image under f of closed sets of Y are closed in X.*

Continuity can be conveniently described by means of the closure:

Theorem B.3.4. *A map $f: X \to Y$ between topological spaces is continuous if and only if*

$$f(\overline{M}) \subseteq \overline{f(M)} \text{ for every subset } M \text{ of } X. \tag{B.1}$$

Proof. Assume that $x \in \overline{M}$. By Lemma B.2.6, there exists a net $\{x_\alpha\}_{\alpha \in A}$ contained in M such that $x_\alpha \to x$. Since f is continuous at x, $f(x_\alpha) \to f(x)$ in Y by Lemma B.3.2, so $f(x) \in \overline{f(M)}$ by Lemma B.2.6(a).

Now assume that (B.1) holds for every subset M of X. We intend to use Lemma B.3.3(d) to check the continuity of f. Let F be a closed set of Y and let $M = f^{-1}(F)$. According to (B.1), $f(\overline{M}) \subseteq \overline{f(M)} \subseteq \overline{F} = F$. This proves that $\overline{M} \subseteq M$, that is, M is closed. □

A topological space X is *homogeneous* if for every pair of points $x, y \in X$ there exists a homeomorphism $f: X \to X$ such that $f(x) = y$. Clearly, a nondiscrete homogeneous space cannot have isolated points.

B.3.2 Metric spaces and the open ball topology

We recall that a *pseudometric* on a set X is a function $d: X \times X \to \mathbb{R}_{\geq 0}$ such that:
(i) $d(x, x) = 0$ for every $x \in X$;
(ii) $d(x, y) = d(y, x)$ for every $x, y \in X$;
(iii) $d(x, z) \leq d(x, y) + d(y, z)$ for every $x, y, z \in X$.

In case $d(x, y) = 0$ always implies $x = y$, the function d is called a *metric*. A metric d on X is an *ultrametric* if (iii) is replaced by the stronger axiom
(iii*) $d(x, z) \leq \max\{d(x, y), d(y, z)\}$ for all $x, y, z \in X$.

A pair (X, d), where X is a set provided with a metric (respectively, a pseudometric, an ultrametric) d, is called *metric space* (respectively, *pseudometric space*, *ultrametric space*).

The *diameter* of a nonempty subset Y of (X, d) is defined as $\operatorname{diam}(Y) = \sup\{d(y_1, y_2) : y_1, y_2 \in Y\}$.

For (X, d) a pseudometric space, $x \in X$, and $\varepsilon > 0$,

$$B_\varepsilon^d(x) = B_\varepsilon(x) = \{y \in X : d(y, x) < \varepsilon\}$$

is the *open disk* (or *open ball*) in X with center x and radius ε.

Example B.3.5. Let (X, d) be a (pseudo)metric space. The family $\mathcal{B} = \{B_\varepsilon(x) : x \in X, \varepsilon > 0\}$ is a base of a topology τ_d on X called the *metric topology* (or *open ball topology*) of (X, d). For every $x \in X$, the family $\{B_\varepsilon(x) : \varepsilon > 0\}$ is a base of $\mathcal{V}_{\tau_d}(x)$. Also $\{B_{1/n}(x) : n \in \mathbb{N}_+\}$ is a (countable) base of $\mathcal{V}_{\tau_d}(x)$.

Clearly, $\mathbb{R}$ with its metric topology induced by the Euclidean metric is a homogeneous space.

Example B.3.6. On $\mathbb{R}^n$:
(a) the *Euclidean distance* is defined by $d(x, y) = \sqrt{\sum_{i=1}^n (x_i - y_i)^2}$ for every $x, y \in \mathbb{R}^n$;
(b) the *Chebyshev distance* (or *chessboard distance*) is defined by $d(x, y) = \sup_{i=1,\dots,n} |x_i - y_i|$ for every $x, y \in \mathbb{R}^n$;
(c) the *Manhattan distance* is defined by $d(x, y) = \sum_{i=1}^n |x_i - y_i|$ for every $x, y \in \mathbb{R}^n$; roughly speaking, for $n = 2$ this is how we compute distances when "walking in a city whose streets follow a grid pattern", so this metric carries also the name *taxi driver metric*.

Definition B.3.7. A topological space (X, τ) is *(pseudo)metrizable* if there exists a (pseudo)metric d on X such that $\tau = \tau_d$.

Definition B.3.8. Let (X, d) and (Y, d') be metric spaces. A map $f : X \to Y$ is:
(i) *continuous* if for every $x \in X$ and every $\varepsilon > 0$ there exists $\delta_x > 0$ such that $d'(f(x), f(y)) < \varepsilon$ whenever $d(x, y) < \delta_x$.
(ii) *uniformly continuous* if for every $\varepsilon > 0$ there exists $\delta > 0$ such that $d'(f(x), f(y)) < \varepsilon$ whenever $d(x, y) < \delta$.

Obviously, uniformly continuous maps between metric spaces are continuous. Moreover, a map $f : (X, d) \to (Y, d')$ between metric spaces is continuous if and only if $f : (X, \tau_d) \to (Y, \tau_{d'})$ is continuous.

Definition B.3.9. A sequence $\{x_n\}_{n \in \mathbb{N}}$ in a metric space (X, d) is a *Cauchy sequence* if for every $\varepsilon > 0$ there exists $n_0 \in \mathbb{N}$ such that $d(x_n, x_m) < \varepsilon$ for every $m, n \geq n_0$.

Remark B.3.10. Every convergent sequence is a Cauchy sequence. On the other hand, some metric spaces have nonconvergent Cauchy sequences, while others do not (e. g., $\mathbb{R}^n$ as well as $\mathbb{Z}^n$, for every $n \in \mathbb{N}$, have none).

This justifies the next fundamental notion.

Definition B.3.11. A metric space X is *complete* if every Cauchy sequence in X is convergent.

Every metric space X has a *completion*, namely, there exists a complete metric space $\widetilde{X}$ containing X as a dense subspace, having the property that if Y is a dense subspace of a metric space Y' and $f: Y \to X$ is a uniformly continuous map, then there exists a unique uniformly continuous extension $\widetilde{f}: Y' \to \widetilde{X}$ of f. This implies that the completion $\widetilde{X}$ is unique up to isometry leaving the points of X fixed.

Example B.3.12. For a prime p, the *p-adic norm* $| - |_p: \mathbb{Q} \to \mathbb{R}_{\geq 0}$ is defined as follows. Let $|0|_p = 0$. For $r \in \mathbb{Q} \setminus \{0\}$ define $v_p(r)$ as the unique $k \in \mathbb{Z}$ such that $r = p^k \frac{a}{b}$, with $a, b \in \mathbb{Z} \setminus \{0\}$ coprime and $p \nmid a$, $p \nmid b$. Now let $|r|_p = p^{-v_p(r)}$. This norm induces the *p-adic metric d_p* on $\mathbb{Q}$, which is an ultrametric.

The completion of $(\mathbb{Q}, d_p)$ is the field $\mathbb{Q}_p$ of p-adic numbers. One can easily show that for every choice of the integer coefficients $(a_n)_{n \geq n_0}$ and $n_0 \in \mathbb{Z}$, the series $\sum_{n=n_0}^{\infty} a_n p^n$ converges in $\mathbb{Q}_p$. The closure of $\mathbb{Z}$ in $\mathbb{Q}_p$ is the ring $\mathbb{J}_p$ of p-adic integers.

B.3.3 Cardinal invariants

A *cardinal invariant* of topological spaces is a cardinal number $i(X)$ attached to every topological space X in such a way that if X and Y are homeomorphic spaces, then $i(X) = i(Y)$. Here we recall several cardinal invariants (see Exercise B.7.5).

The *weight* of a topological space X is the cardinal number

$$w(X) = \min\{|\mathcal{B}| : \mathcal{B} \text{ is a base of } X\}.$$

The topological spaces with countable base are called *second countable*.

Clearly, $\mathbb{R}$ endowed with its metric topology induced by the Euclidean metric is second countable, in view of Example B.1.7.

Example B.3.13. (a) Every base of a discrete space $X = (X, \delta_X)$ contains the family of all singletons of X, and hence $w(X) = |X|$.
(b) If $X = (X, \iota_X)$ is a nonempty indiscrete space, $\{X\}$ is a base and $w(X) = 1$.

For a topological space X and $x \in X$, the cardinal number

$$\chi(X, x) = \min\{|\mathcal{B}| : \mathcal{B} \text{ base of } \mathcal{V}(x)\}$$

is the *character of X at x*. We use the symbol $\chi(X)$ for denoting the *character* of X defined by $\chi(X) = \sup\{\chi(X, x) : x \in X\}$. The topological spaces with countable character are called *first countable*. Clearly, a second countable space is also first countable. Metric spaces are obviously first countable (see §B.3.2).

First countable spaces are obviously Fréchet–Urysohn, as sequences are sufficient to describe the closure of a subset.

If $\bigcap \mathcal{V}(x) = \{x\}$ for a point x of a topological space X, the cardinal number

$$\psi(X,x) = \min\left\{|\mathcal{B}| : \mathcal{B} \subseteq \mathcal{V}(x),\ \bigcap_{B\in\mathcal{B}} B = \{x\}\right\}$$

is called *pseudocharacter of X at x*. If $\bigcap \mathcal{V}(x) = \{x\}$ for all points $x \in X$, we let $\psi(X) = \sup_{x\in X} \psi(X,x)$, and we call it *pseudocharacter* of X.

For a topological space X, the *density character* of X is

$$d(X) = \min\{|D| : D \text{ is dense in } X\}.$$

Clearly, X is separable precisely when $d(X)$ is countable.

For a discrete space X one has $d(X) = w(X) = |X|$, while $\chi(X) = 1$. For a nonempty indiscrete space X, one has $d(X) = w(X) = \chi(X) = 1$.

Lemma B.3.14. *Let (X,τ) be a topological space. Then:*
(a) $\chi(X) \le w(X) \le |\tau| \le 2^{w(X)}$ *and* $|\tau| \le 2^{|X|}$*;*
(b) $d(X) \le w(X)$ *and* $d(X) \le |X|$.

Proof. Fix a base $\mathcal{B}$ of τ of size $w(X)$.

(a) The inequalities $|\tau| \le 2^{|X|}$ and $\chi(X) \le w(X) \le |\tau| \le 2^{|X|}$ are obvious. Moreover, every $U \in \tau$ is a union of a subfamily $\mathcal{B}_1$ of $\mathcal{B}$. Since the powerset of $\mathcal{B}$ has size $2^{w(X)}$, we conclude that $|\tau| \le |\mathcal{P}(\mathcal{B})| = 2^{w(X)}$.

(b) Fix $x_B \in B$ for every $B \in \mathcal{B}$. Since the set $D = \{x_B : B \in \mathcal{B}\}$ is dense in X and $|D| \le w(X)$, this proves the first inequality. The second one is trivial. □

Theorem B.3.15 (Hewitt–Marczewski–Pondiczery theorem). *If $\{X_i : i \in I\}$ is a family of separable topological spaces and $|I| \le 2^\kappa$ for an infinite cardinal κ, then $d(\prod_{i\in I} X_i) \le \kappa$ holds. In particular, $\prod_{i\in I} X_i$ is separable whenever $|I| \le \mathfrak{c}$.*

B.3.4 Borel sets, zero-sets, and Baire sets

For a set X, a subfamily $\mathfrak{B}$ of $\mathcal{P}(X)$ is called a *σ-algebra* on X if $X \in \mathfrak{B}$ and $\mathfrak{B}$ is closed under taking complements and countable unions. For a topological space (X,τ), denote by $\mathcal{B}(X)$ the smallest σ-algebra containing τ. The members of $\mathcal{B}(X)$ are called *Borel sets*. Some of the Borel sets of X have special names:

Definition B.3.16. Let X be a topological space. A subset A of X is:
(i) a *G_δ-set* if A is the intersection of a countable family of open sets;
(ii) an *F_σ-set* if A is the union of a countable family of closed sets;
(iii) a *$G_{\delta\sigma}$-set* if A the intersection of a countable family of F_σ-sets;
(iv) an *$F_{\sigma\delta}$-set* if A is the union of a countable family of G_δ-sets.

Obviously, A is a G_δ-set precisely when $X \setminus A$ is an F_σ-set. Similarly, for $F_{\sigma\delta}$-sets and $G_{\delta\sigma}$-sets. Moreover, every open set is also a G_δ-set, while every closed set is also an F_σ-set.

Definition B.3.17. A subset A of a topological space X is G_δ-*dense* in X if every nonempty G_δ-set of X nontrivially meets A.

Example B.3.18. If X is a metric space, then:
(a) all closed sets (in particular, all singletons) are also G_δ-sets;
(b) by (a), a subset A of X is G_δ-dense if and only if $A = X$.

If X is a topological space and $f: X \to \mathbb{R}$ is continuous, then $f^{-1}(0)$ is named a *zero-set* of X. A zero-set is obviously closed, but it is also a G_δ-set. So, the σ-algebra generated by the zero-sets is contained in the σ-algebra $\mathcal{B}(X)$ of all Borel sets of X, and its members are named *Baire sets*. These two σ-algebras need not coincide.

B.4 Subspace, quotient, product, and coproduct topologies

A subset Y of a topological space (X, τ) becomes a topological space when endowed with the *topology induced* by X, namely, $\tau\restriction_Y = \{Y \cap U : U \in \tau\}$. This is also called *subspace topology* and $(Y, \tau\restriction_Y)$ a *subspace* of (X, τ).

Let $\{X_i : i \in I\}$ be a family of topological spaces. Consider the Cartesian product $X = \prod_{i\in I} X_i$ with its canonical projections $p_i: X \to X_i$, for $i \in I$. Then X usually carries the *product topology* (or *Tichonov topology*), having as a prebase $\mathcal{B}'$ the family $\{p_i^{-1}(U_i) : U_i \text{ open in } X_i\}$. Hence, a base of the product topology is the family

$$\mathcal{B} = \{W_J(\{U_j\}_{j\in J}) : J \subseteq I,\ J \text{ finite}\},$$

where $U_i \subseteq X_i$ is open for all $i \in J$ and $W_J(\{U_j\}_{j\in J}) = \bigcap_{i\in J} p_i^{-1}(U_i)$. For every finite subset $J \subseteq I$, let $X_J = \prod_{i\in J} X_i$ and let $p_J: X \to \prod_{i\in J} X_i$ be the projection.

Theorem B.4.1. *Let $\{X_i : i \in I\}$ be a family of topological spaces and let $X = \prod_{i\in I} X_i$ be equipped with the product topology. Then:*
(a) *the projections $p_i: X \to X_i$ are both open and continuous;*
(b) *for every topological space Y, a map $f: Y \to X$ is continuous if and only if for every $i \in I$, the composition $p_i \circ f: Y \to X_i$ is continuous;*
(c) *if $x \in X$ and F is a closed set of X with $x \notin F$, there exists a finite subset J of I such that $p_J(x) \notin p_J(F)$;*
(d) *a sequence $\{x_n\}_{n\in\mathbb{N}}$ in X is convergent if and only if for every $i \in I$ the sequence $\{p_i(x_n)\}_{n\in\mathbb{N}}$ is convergent in X_i;*
(e) *a net $\{x_d\}_{d\in D}$ in X is convergent if and only if the net $\{p_i(x_d)\}_{d\in D}$ is convergent in X_i for every $i \in I$.*

If I is countable and all X_i are metric spaces, then the product topology of $X = \prod_{i\in I} X_i$ is induced by a metric appropriately defined on X. If I is uncountable and uncountably many X_i are nontrivial (i. e., nonsingletons), then the product topology is not first countable, so cannot be induced by any metric defined on the product $X = \prod_{i\in I} X_i$.

Using Theorem B.4.1(c), one can get the following useful rule for computing the closure in infinite products by reduction to the case of finite subproducts, that is, for every $M \subseteq \prod_{i\in I} X_i$,

$$\overline{M} = \bigcap_{J\subseteq I \text{ finite}} p_J^{-1}(\overline{p_J(M)}).$$

In particular, for every $M \subseteq \prod_{i\in I} X_i$,

$$\overline{M} \subseteq \prod_{i\in I} \overline{p_i(M)}.$$

This becomes an equality when $M = \prod_{i\in I} M_i$, where $M_i = p_i(M)$ for every $i \in I$.

Example B.4.2. For a family $\{X_i : i \in I\}$ of topological spaces, on the Cartesian product $X = \prod_{i\in I} X_i$ the *box topology* on X has as a base the family $\{\prod_{i\in I} U_i : U_i \text{ open in } X_i\}$. The box topology is finer than the product topology.

Let X be a topological space, let $\sim$ be an equivalence relation on X, and let $q: X \to X/_\sim$ be the canonical projection. This set carries the so-called *quotient topology* τ_q defined as follows: a set $U \subseteq X/_\sim$ is open in τ_q if and only if $q^{-1}(U)$ is open in X. Obviously, this defines a topology τ_q on $X/_\sim$. Moreover, the quotient topology τ_q is the finest among all topologies on $X/_\sim$ such that q is continuous. This determines the following important property of the quotient topology.

Lemma B.4.3. *Let X be a topological space, let $\sim$ be an equivalence relation on X, let $X/_\sim$ be equipped with the quotient topology and let $q: X \to X/_\sim$ be the canonical projection. Then a map $g: X/_\sim \to Z$ is continuous if and only if the composition $g \circ q: X \to Z$ is continuous.*

For a family $\{(X_i, \tau_i) : i \in I\}$ of topological spaces, the *coproduct* is the disjoint union $X = \bigsqcup_{i\in I} X_i$ equipped with the topology having as a base the union of all topologies τ_i (here and in the sequel we consider X_i as a subset of X). In other words, the *coproduct topology* of $\bigsqcup_{i\in I} X_i$ is defined in such a way that every X_i is open in $\bigsqcup_{i\in I} X_i$. So, a subset A of $\bigsqcup_{i\in I} X_i$ is open in the coproduct topology if and only if every intersection $A \cap X_i$ is open.

Example B.4.4. Let X be the space $\mathbb{N} \times [0,1]$ equipped with the product topology, where $\mathbb{N}$ is discrete and $[0,1]$ has the Euclidean topology. This topology coincides also with the topology of the coproduct of $\mathbb{N}$-many copies of the interval $[0,1]$. Now define $(n,x) \sim (m,y)$ if and only if either $n = m$ and $x = y$, or $x = y = 0$. The quotient space $V = X/_\sim$ is called a *fan*.

B.5 Separation axioms and compactness-like properties

B.5.1 Separation axioms

Now we recall the so-called separation axioms for topological spaces.

Definition B.5.1. A topological space X is:

(i) a T_0-*space* (or, a *Kolmogorov space*) if for every pair of distinct points $x, y \in X$ there exists an open set U of X such that either $x \in U \not\ni y$, or $y \in U \not\ni x$;

(ii) a T_1-*space* if for every pair of distinct points $x, y \in X$ there exist open sets U and V of X such that $x \in U \not\ni y$ and $y \in V \not\ni x$ (or, equivalently, every singleton of X is closed);

(iii) a T_2-*space* (or, a *Hausdorff space*) if for every pair of distinct points $x, y \in X$ there exist disjoint open sets U and V of X such that $x \in U$ and $y \in V$;

(iv) a *regular space* if for every $x \in X$ and every $U \in \mathcal{V}(x)$ there exists $V \in \mathcal{V}(x)$ with $\overline{V} \subseteq U$; a T_3-*space* if it is a regular T_1-space;

(v) a *completely regular space* if for every $x \in X$ and every $U \in \mathcal{V}(x)$ there exists a continuous function $f: X \to [0, 1]$ with $f(x) = 1$ and vanishing on $X \setminus U$; a $T_{3.5}$-*space* (or, a *Tichonov space*) if X is a completely regular T_1-space;

(vi) a *normal space* if for every pair of closed disjoint subsets F and G of X there exists a pair of open disjoint subsets U and V of X such that $F \subseteq U$ and $G \subseteq V$; a T_4-*space* if X is a normal T_1-space.

The following implications hold true among these properties:

$$T_0 \Leftarrow T_1 \Leftarrow T_2 \Leftarrow T_3 \Leftarrow T_{3.5} \Leftarrow T_4$$

While the first four implications are more or less easy to verify, the last implication $T_4 \Rightarrow T_{3.5}$ requires Theorem B.5.2. All these properties (beyond T_4) are preserved by taking subspaces.

Theorem B.5.2 (Urysohn lemma). *Let X be a normal space. Then for every pair of closed nonempty disjoint sets F, G of X there exists a continuous function $f: X \to [0, 1]$ such that $f(F) = \{1\}$ and $f(G) = \{0\}$.*

Theorem B.5.3 (Tietze theorem). *Let X be a T_4-space and Y a closed set of X. For every continuous function $f: Y \to \mathbb{R}$, there exists a continuous extension $g: X \to \mathbb{R}$ of f.*

It is easy to see that a topological space X is Hausdorff if and only if every net in X converges to at most one point in X.

Theorem B.5.4 (Tichonov embedding theorem). *Let X be a Tichonov space. Then X is homeomorphic to a subspace of the product $[0, 1]^{w(X)}$.*

For a cardinal γ, the space $[0, 1]^\gamma$ is called the *Tichonov cube of weight γ*, the Tichonov cube of countable weight is called *Hilbert cube.*

B.5.2 Compactness-like properties

For the sake of completeness, we recall here some frequently used properties of topological spaces related to compactness.

For a topological space X, a family $\mathcal{U} = \{U_i : i \in I\}$ of nonempty open sets of X is an *open cover* of X if $X = \bigcup_{i \in I} U_i$. A subfamily $\mathcal{V} = \{U_i : i \in J\}$ of $\mathcal{U}$, where $J \subseteq I$, is a *subcover* of $\mathcal{U}$ if $X = \bigcup_{i \in J} U_i$.

Definition B.5.5. A topological space X is:
(i) *compact* if every open cover of X admits a finite subcover;
(ii) *countably compact* if every countable open cover of X admits a finite subcover;
(iii) *Lindelöff* if every open cover of X admits a countable subcover;
(iv) *pseudocompact* if every continuous function $X \to \mathbb{R}$ is bounded;
(v) *locally compact* if every point of X has a compact neighborhood in X;
(vi) *σ-compact* if X is the union of countably many compact subsets;
(vii) *hemicompact* if X is σ-compact and has a countable family of compact subsets such that every compact set of X is contained in one of them;
(viii) a *k-space* if a subspace A of X is closed in X if and only if $A \cap K$ is closed in K for all compact subsets K of X.

Example B.5.6. Clearly, $\mathbb{R}^n$ with the Euclidean topology is locally compact, not compact, but hemicompact.

Let B be a subset of $\mathbb{R}^n$ equipped with the usual metric topology. Then B is compact if and only if B is closed and bounded (i. e., B has finite diameter).

Obviously, $\mathbb{C}^n$ has the same property, being topologically isomorphic to $\mathbb{R}^{2n}$.

The following is a criterion for (countable) compactness in terms of nets and filters.

Lemma B.5.7. *Let X be a topological space. Then:*
(a) *X is (countably) compact if and only if every (countable) family of closed sets of X with the finite intersection property has nonempty intersection;*
(b) *X is compact if and only if every ultrafilter on X is convergent;*
(c) *X is compact if and only if every net in X has a convergent subnet.*

Proof. (a) Every family $\mathcal{F}$ of closed sets of X with the finite intersection property and having empty intersection corresponds to an open cover of X without finite subcovers (simply take the complement of the members of $\mathcal{F}$).

(b) follows from (a) and Exercise B.7.3, while (c) can be deduced from (b) and Lemma B.2.11(b_3). □

Proposition B.5.8. *A T_1 topological space is countably compact if and only if every sequence in X has an accumulation point.*

Lemma B.5.9. *Let X be a topological space and Y a compact subset of X. If $\{F_i : i \in I\}$ is a family of closed sets of X such that $Y \cap \bigcap_{i \in I} F_i = \emptyset$, then there exists a finite subset J of I such that $Y \cap \bigcap_{i \in J} F_i = \emptyset$.*

Proof. Suppose for a contradiction that $F_J = Y \cap \bigcap_{i \in J} F_i$ is nonempty for every finite subset J of I. Then $\{F_J : J \subseteq I, J \text{ finite}\}$ gives rise to a filter base of closed sets of Y with $\bigcap_J F_J = Y \cap \bigcap_{i \in I} F_i = \emptyset$, against the compactness of Y by Lemma B.5.7(a). □

Compactness-like properties "improve" separation properties in the following sense.

Theorem B.5.10. *Let X be a topological space.*
(a) *If X is Hausdorff and compact, then X is normal, and if $\psi(X)$ is countable, then X is first countable.*
(b) *If X is regular and Lindelöff, then X is normal.*
(c) *If X is Hausdorff and locally compact, then X is Tichonov.*

It follows from Theorem B.5.10(a) that every subspace of a compact Hausdorff space is necessarily a Tichonov space. According to Theorem B.5.4, every Tichonov space X is a subspace of a compact space K, so taking the closure Y of X in K one obtains also a compact space Y containing X as a dense subspace, namely, a *compactification* of X.

Theorem B.5.11 (Arhangel'skij). *If a compact Hausdorff space X is countable, then X is metrizable. In particular, a countably infinite compact Hausdorff space has a nontrivial convergent sequence.*

Proof. By Theorem B.5.10(c), X is Tichonov. Since X has countable pseudocharacter, X is first countable (being compact) by Theorem B.5.10(a). Since X countable and first countable, it is second countable, as well. By Theorem B.5.4, X is metrizable. □

By βX we denote the *Čech–Stone compactification* of a Tichonov space X, that is, the compact space βX together with the dense embedding $i: X \to \beta X$ such that for every continuous function $f: X \to [0,1]$ there exists a continuous function $f^\beta: \beta X \to [0,1]$ which extends f (this is equivalent to asking that every continuous map from X to a compact space Y be extendable to βX).

For a topological space X, the *one-point compactification* (or, *Alexandrov compactification*) αX of X is obtained as $\alpha X = X \cup \{\infty\}$, where the open sets of αX are the open sets of X together with the sets of the form $U \cup \{\infty\}$, where U is an open set of X such that $X \setminus U$ is compact. The topological space αX is Hausdorff if and only if X is Hausdorff and locally compact.

Now we discuss several properties of the notions from Definition B.5.5.

Lemma B.5.12. *Let X, Y be topological spaces and $f: X \to Y$ a continuous surjective map. Then Y is compact (respectively, Lindelöff, countably compact, σ-compact) whenever X has the same property.*

Most of the above properties are preserved by taking closed subspaces:

Lemma B.5.13. *If X is a closed subspace of a topological space Y, then X is compact (respectively, Lindelöff, countably compact, σ-compact, locally compact) whenever Y has the same property.*

Now we discuss the preservation of properties under unions.

Lemma B.5.14. *Let X be a topological space and assume $X = \bigcup_{i \in I} X_i$, where X_i are subspaces of X.*
(a) *If I is finite and each X_i is (countably) compact, X is (countably) compact.*
(b) *If I is countable and each X_i is σ-compact (respectively, Lindelöff), then X has the same property.*

The next theorem shows that many of the properties of topological spaces are preserved under taking products. As far as compactness is concerned, this is known as *Tichonov theorem*:

Theorem B.5.15. *Let $\{X_i : i \in I\}$ be a family of topological spaces and let $X = \prod_{i \in I} X_i$ be endowed with the product topology. Then:*
(a) *X is compact (respectively, T_0, T_1, T_2, T_3, $T_{3.5}$) if and only if every X_i has the same property;*
(b) *if I is finite, the same holds for local compactness and σ-compactness.*

Let us mention here that countable compactness, as well as the Lindelöff property and T_4, are not stable even under taking finite products.

We add a few more compactness-like properties (see the next section for their connection to the above ones).

Definition B.5.16. A topological space X is:
(i) a *Baire space* if any countable intersection of dense open sets of X is dense in X;
(ii) *of first category* if there exist a family $\{A_n : n \in \mathbb{N}\}$ of closed sets of X with empty interior such that $X = \bigcup_{n \in \mathbb{N}} A_n$;
(iii) *of second category* if X is not of first category.

Equivalently, a topological space X is a Baire space if any countable intersection of dense G_δ-sets of X is still a dense G_δ-set of X. For example, every countable T_1-space without isolated points is of first category (e. g., $\mathbb{Q}$); indeed, the singletons are closed sets with empty interior.

Theorem B.5.17 (Baire category theorem). *Every complete metric space is a Baire space.*

In the sequel we prove that locally compact, as well as countably compact spaces, are Baire spaces.

B.5.3 Relations among compactness-like properties

Here we show the relations among the properties introduced in the previous section (see the diagram at the end of this subsection). Obviously, a topological space is compact if and only if it is both Lindelöff and countably compact. Compact spaces are locally compact and σ-compact.

Lemma B.5.18. *If X is a σ-compact space, then X is a Lindelöff space.*

Proof. Let $\{U_i : i \in I\}$ be an open cover of X. Since X is σ-compact, $X = \bigcup_{n\in\mathbb{N}_+} K_n$ where each K_n is a compact subset of X. For every $n \in \mathbb{N}_+$, there exists a finite subset F_n of I such that $K_n \subseteq \bigcup_{i\in F_n} U_i$. Now $J = \bigcup_{n\in\mathbb{N}_+} F_n$ is a countable subset of I, and $K_n \subseteq \bigcup_{i\in J} U_i$ for every $n \in \mathbb{N}_+$. Therefore, $X = \bigcup_{i\in J} U_i$. □

Lemma B.5.19. *A Baire space X is of second category.*

Proof. Assume that $X = \bigcup_{n\in\mathbb{N}_+} A_n$ where each A_n is a closed set of X with empty interior. For every $n \in \mathbb{N}_+$, the set $D_n = X \setminus A_n$ is open and dense in X. Then $\bigcap_{n\in\mathbb{N}_+} D_n$ is dense in X, in particular nonempty, so $X \neq \bigcup_{n\in\mathbb{N}_+} A_n$, a contradiction. □

According to Theorem B.5.17, complete metric spaces are Baire. Now we prove that also locally compact spaces are Baire.

Theorem B.5.20. *A Hausdorff locally compact space X is a Baire space.*

Proof. For every $n \in \mathbb{N}_+$, let D_n be a dense open set of X. To show that $\bigcap_{n\in\mathbb{N}_+} D_n$ is dense in X, fix an open set $V \neq \emptyset$ of X. According to Theorem B.5.10, X is a regular space, so there exists an open set $U_0 \neq \emptyset$ of X with $\overline{U}_0$ compact and $\overline{U}_0 \subseteq V$. Since D_1 is dense in X, $U_0 \cap D_1 \neq \emptyset$. Pick $x_1 \in U_0 \cap D_1$ and let U_1 be an open set of X with $x_1 \in U_1$, $\overline{U}_1$ compact and $\overline{U}_1 \subseteq U_0 \cap D_1$. Proceeding in this way, for every $n \in \mathbb{N}_+$ we can find an open set $U_{n+1} \neq \emptyset$ of X with $\overline{U}_{n+1}$ compact and $\overline{U}_{n+1} \subseteq U_n \cap D_{n+1}$. Since every $\overline{U}_n$ is compact, there exists $x \in \bigcap_{n\in\mathbb{N}_+} U_n$ by Lemma B.5.7(a). Obviously, $x \in V \cap \bigcap_{n\in\mathbb{N}_+} D_n$. □

The above proof works also for complete metric spaces, that is, in Theorem B.5.17, but the neighborhoods U_n must be chosen each time with $\mathrm{diam}(U_n) \leq 1/n$. Then the Cantor theorem (for complete metric spaces) guarantees $\bigcap_{n\in\mathbb{N}_+} \overline{U}_n \neq \emptyset$.

Next we collect all implications between the properties we have discussed so far.

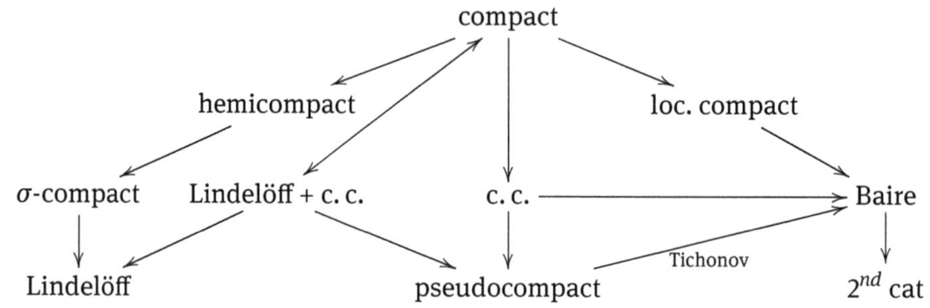

B.5.4 The Stone–Weierstraß theorem

Throughout this book, for a nonempty topological space X we denote by:

- $C(X)$ the $\mathbb{C}$-algebra of all continuous complex-valued functions on X,
- $C^*(X)$ the $\mathbb{C}$-algebra of all bounded continuous complex-valued functions on X,
- $C_0(X)$ the $\mathbb{C}$-algebra of all continuous complex-valued functions on X with compact support (i. e., functions vanishing outside of a compact subset of X).

Note that $C_0(X) \subseteq C^*(X)$. Sometimes we adopt this notation also for a nonempty set X, assuming silently that X *carries the discrete topology*, so that $C(X) = \mathbb{C}^X$, $C^*(X)$ is the family of all bounded complex-valued functions on X, and $C_0(X)$ the family of all complex-valued functions on X with finite support.

On $C^*(X)$ we consider the sup-norm, defined by letting, for every $f \in C^*(X)$, $\|f\| = \sup\{|f(x)|: x \in X\}$. In particular, if X is a compact space and $f \in C(X)$, then $f \in C^*(X)$, and so $\|f\|$ is well-defined.

Theorem B.5.21 (Stone–Weierstraß theorem). *Let X be a compact space. A $\mathbb{C}$-subalgebra $\mathcal{A}$ of $C(X)$, containing all constants and closed under complex conjugation, is dense in $(C(X), \|-\|)$ if and only if $\mathcal{A}$ separates the points of X.*

We need the following local form of the Stone–Weierstraß theorem.

Corollary B.5.22. *For a compact space X, $f \in C(X)$ can be uniformly approximated by a $\mathbb{C}$-subalgebra $\mathcal{A}$ of $C(X)$ containing all constants and closed under complex conjugation if and only if $\mathcal{A}$ separates the points of X separated by $f \in C(X)$.*

Proof. The necessity of the condition is obvious, so we only check the sufficiency.

Denote by $G: X \to \mathbb{C}^{\mathcal{A}}$ the diagonal map of the maps in $\mathcal{A}$. Then $Y = G(X)$ is a compact subspace of $\mathbb{C}^{\mathcal{A}}$ and, by the compactness of X, its subspace topology coincides with the quotient topology of the map $G: X \to Y$. The equivalence relation $\sim$ in X determined by this quotient is as follows: $x \sim y$ for $x, y \in X$ if and only if $G(x) = G(y)$ (if and only if $g(x) = g(y)$ for every $g \in \mathcal{A}$).

Clearly, every continuous function $h: X \to \mathbb{C}$ such that $h(x) = h(y)$ for every pair $x, y \in X$ with $x \sim y$ can be factorized as $h = h^* \circ G$, where $h^* \in C(Y)$. In particular, this holds true for all $g \in \mathcal{A}$ and for f (for the latter case this follows from our hypothesis).

Let $\mathcal{A}^*$ be the $\mathbb{C}$-subalgebra $\{h^*: h \in \mathcal{A}\}$ of $C(Y)$. Then $\mathcal{A}^*$ is closed under complex conjugation and contains all constants; moreover, it separates the points of Y: if $y \neq y'$ in Y with $y = G(x)$, $y' = G(x')$, $x, x' \in X$, then $x \not\sim x'$; so, there exists $h \in \mathcal{A}$ with $h^*(y) = h(x) \neq h(x') = h^*(y')$. Now we can apply Theorem B.5.21 to Y and $\mathcal{A}^*$ to deduce that we can uniformly approximate f^* by functions of $\mathcal{A}^*$. This produces a uniform approximation of f by functions of $\mathcal{A}$. □

B.6 Connected and hereditarily disconnected spaces

Connected sets were introduced by Lennes in 1911 and by Hausdorff in his book *Zusammenhängende Mengen* in 1914.

Definition B.6.1. A topological space X is *connected* if every proper clopen set of X is empty.

Remark B.6.2. It is easy to see that a topological space X is connected if and only if every continuous function $X \to \{0, 1\}$, where $\{0, 1\}$ is equipped with the discrete topology, is constant.

Example B.6.3. The space $\mathbb{R}$ is connected. Moreover, a subset X of $\mathbb{R}$ is connected if and only if it is an interval. The same occurs in $\mathbb{T}$.

Lemma B.6.4. (a) *A continuous image of a connected space is connected.*
(b) *The closure of a connected set is connected.*

Proof. (a) follows easily from the definition.

(b) Let D be a dense connected subset of a topological space X. If $f: X \to \{0, 1\}$ is continuous, then f is constant on D, and consequently also on X by Theorem B.3.4. So, X is connected by Remark B.6.2. □

Lemma B.6.5. *Let $X = \bigcup_{i \in I} X_i$ be a topological space, where each X_i is a subspace of X. If $\bigcap_{i \in I} X_i \neq \emptyset$ and each X_i is connected, then X is connected.*

Proof. Let $f: X \to \{0, 1\}$ be a continuous function. By Remark B.6.2, for every $i \in I$, $f \restriction_{X_i}$ is constant. Since there exists a common point to all X_i, this constant is common for all the subsets X_i. Consequently, f is constant as well, hence X is connected by Remark B.6.2. □

The same argument proves that if $\{C_i : i \in I\}$ is a family of connected subspaces of a topological space X having a common point, then the set $\bigcup_{i\in I} C_i$ is connected.

Lemma B.6.5 and the above comment imply that, for a topological space X, for every $x \in X$ there is a largest connected subset C_x of X with $x \in C_x$, called the *connected component of* x in X. In view of Lemma B.6.4, each C_x is a closed set of X, and $\{C_x : x \in X\}$ is a partition of X.

Definition B.6.6. A topological space X is:
(i) *hereditarily disconnected* if every connected component of X is a singleton;
(ii) *zero-dimensional* if X has a base of clopen sets; we denote this by $\dim X = 0$.

Example B.6.7. In $\mathbb{R}$, clearly, $\mathbb{Z}$ is zero-dimensional since it is discrete. Moreover, $\mathbb{Q}$ is zero-dimensional since $\{(a,b) \cap \mathbb{Q} : a, b \in \mathbb{R} \setminus \mathbb{Q}\}$ is a base of clopen sets of the topology induced on $\mathbb{Q}$ by $\mathbb{R}$.

Zero-dimensional T_0-spaces are $T_{3.5}$ and hereditarily disconnected (as every point is an intersection of clopen sets). Hereditary disconnectedness and zero-dimensionality are preserved under taking subspaces and products:

Theorem B.6.8. *Let $\{X_i : i \in I\}$ be a family of topological space and let $X = \prod_{i\in I} X_i$ be endowed with the product topology. Then X is hereditarily disconnected (respectively, connected, zero-dimensional) if and only if every X_i is hereditarily disconnected (respectively, connected, zero-dimensional) for every $i \in I$.*

In a topological space X, for a point $x \in X$, the *quasicomponent* Q_x of x is the intersection of all clopen sets of X containing x, namely,

$$Q_x = \bigcap \{O \subseteq X : x \in O,\ O \text{ clopen}\}.$$

Since, for every $x \in X$, $x \in C_x \subseteq O$ whenever $O \subseteq X$ is clopen, $C_x \subseteq Q_x$.

Definition B.6.9. A topological space X is *totally disconnected* if all quasicomponents are trivial.

In particular, for T_0 topological spaces, the following implications hold:

$$\text{zero-dimensional} \implies \text{totally disconnected} \implies \text{hereditarily disconnected.}$$

These implications become equivalences for locally compact spaces:

Theorem B.6.10 (Vedenissov theorem). *Every hereditarily disconnected locally compact space is zero-dimensional.*

Actually, one can say something more precise for compact spaces:

Lemma B.6.11 (Shura–Bura lemma). *In a compact space, the quasicomponents and the connected components coincide.*

B.7 Exercises

Exercise B.7.1. Let X, Y be nonempty sets and $f: X \to Y$ a map. Prove that:
(a) if $\mathcal{F}$ is a filter on X, then $f(\mathcal{F}) = \{f(F): F \in \mathcal{F}\}$ is a filter base on Y;
(b) if $\mathcal{F}$ is a filter on Y and f is surjective (or, more generally, $\mathcal{F} \cup \{f(X)\}$ has the finite intersection property), then $f^{-1}(\mathcal{F}) = \{f^{-1}(F): F \in \mathcal{F}\}$ is a filter base on X.

Exercise B.7.2. Let X be a nonempty set. Prove that every filter $\mathcal{F}$ on X is contained in some ultrafilter.

Hint. Apply the Zorn lemma to the ordered by inclusion set of all filters of X containing $\mathcal{F}$.

Exercise B.7.3. Prove that if x is an adherent point of an ultrafilter $\mathcal{U}$ of a topological space X, then x is also a limit point of $\mathcal{U}$.

Exercise B.7.4. Let X, Y be topological spaces, M a subset of X, and $f: X \to Y$ a continuous map. Prove that:
(a) $f\restriction_M: M \to Y$ is continuous;
(b) if $f: X \to Y$ is a homeomorphism, then also M and $f(M)$ are homeomorphic (so, being homeomorphic is preserved by taking restrictions to subspaces);
(c) if $f: X \to Y$ is injective and open (respectively, closed, an embedding), then also $f\restriction_M: M \to f(M)$ is open (respectively, closed, an embedding) (in other words, also the properties "open", "closed", and "embedding" of an injective map are preserved by restrictions);
(d) the composition of continuous maps is a continuous map;
(e) provide an example of topological spaces X, Y, an open map $f: X \to Y$ and $M \subseteq X$, such that $f\restriction_M: M \to f(M)$ is not open.

Exercise B.7.5. Prove that $|-|, d(-), w(-), \chi(-), \psi(-)$ are cardinal invariants of topological spaces.

Exercise B.7.6. Let X be a set, $\{(Y_i, \tau_i): i \in I\}$ a family of topological spaces, and for every $i \in I$, let $f_i: X \to Y_i$ be a map. The topology τ on X having as a prebase $\{f_i^{-1}(U): U \subseteq Y_i \text{ open}, i \in I\}$ is called *initial topology* of the family $\{f_i: i \in I\}$. Prove that:
(a) every map $f_i: (X, \tau) \to (Y_i, \tau_i)$ is continuous;
(b) τ is the coarsest topology on X with the property in (a);
(c) the product topology of $X = \prod_{i\in I}(X_i, \tau_i)$ coincides with the initial topology of the family $\{p_i: i \in I\}$ of all projections $p_i: X \to X_i$ for $i \in I$;

(d) if $I = \{0\}$ and $f_0 \colon X \to Y_0$ is an injective map, then the topology on X induced by the subspace topology of $f_0(X)$ in Y_0 coincides with the initial topology of the family $\{f_0\}$.

Exercise B.7.7. Let X be a set, let $\{(Y_i, \tau_i) \colon i \in I\}$ be a family of topological spaces and, for every $i \in I$, let $f_i \colon Y_i \to X$ be a map. The topology τ on X having as open sets all the sets $U \subseteq X$ such that $f_i^{-1}(U)$ is open in Y_i for every $i \in I$, is called *final topology* of the family $\{f_i \colon i \in I\}$. Prove that:

(a) every map $f_i \colon (Y_i, \tau_i) \to (X, \tau)$ is continuous;
(b) τ is the finest topology on X with the property in (a);
(c) the coproduct topology of $X = \bigsqcup_{i \in I}(X_i, \tau_i)$ coincides with the final topology of the family $\{\iota_i \colon i \in I\}$ of all inclusions $\iota_i \colon X_i \hookrightarrow X$ for $i \in I$;
(d) if $I = \{0\}$ and $f_0 \colon Y_0 \to X$ is a surjective map, then the quotient topology on X induced by f_0 coincides with the final topology of the family $\{f_0\}$.

Exercise B.7.8. Prove that the only convergent sequences with limit point the common point 0 of the fan V (see Example B.4.4) are those that are eventually contained in the sets $q(\{0, \dots, n\} \times [0, 1])$, where $q \colon X = \mathbb{N} \times [0, 1] \to V = X/_{\sim}$ is the canonical projection.

Exercise B.7.9. For a topological space X and $x \in X$, let $V_x = \bigcap \mathcal{V}(x)$. Prove that:

(a) X is T_0 if and only if $V_x \cap \overline{\{x\}} = \{x\}$ for every $x \in X$; conclude that X is indiscrete if and only if $V_x = \overline{\{x\}}$ for every $x \in X$;
(b) if X is an Alexandrov T_0-space and $x, y \in X$, $y \in \overline{\{x\}}$ if and only if $x \in V_y$;
(c) for a T_0-space (X, τ) and $x, y \in X$, putting $x \preceq_\tau y$ whenever $y \in \overline{\{x\}}$, $\preceq_\tau$ is a partial order on X; this is called *specialization order* (some authors prefer to use this name for the dual order).

Exercise B.7.10. For a partially ordered set $(X, \leq)$, let τ_{AT} be the topology having as a base the family of all downward closed sets B of X (i. e., if $b \in B$ and $a \leq b$, then $a \in B$ as well). Show that τ_{AT} is a topology on X which is Alexandrov and T_0; the topology τ_{AT} is usually named *Alexandroff–Tucker topology*.

Moreover, the specialization order of the topological space (X, τ_{AT}) coincides with $\leq$, and τ_{AT} is the finest topology on $(X, \leq)$ with this property.

Exercise B.7.11. Let X be a topological space. Show that for every subset Y of X the set $\mathrm{Int}(\overline{Y})$ is a regular open set and deduce from this that every regular space has a base (and local bases at each point) consisting of regular open sets.

Exercise B.7.12. Let X, Y be topological spaces such that Y is Hausdorff. Prove that if $f, g \colon X \to Y$ are continuous maps, then $\{x \in X \colon f(x) = g(x)\}$ is closed in X.

Hint. Let $N = \{x \in X \colon f(x) \neq g(x)\}$. Let $z \in N$ and let U, V be disjoint open sets of Y with $f(z) \in U$ and $g(z) \in V$. Then $f^{-1}(U) \cap g^{-1}(V)$ is an open neighborhood of z contained in N. Thus, N is a union of open sets, so open.

Exercise B.7.13. Show that a locally compact σ-compact space X is hemicompact.

Hint. Use the definitions. If X is Hausdorff, then X is Tichonov by Theorem B.5.10(c), so one can consider the one-point compactification of X.

C Background on categories and functors

C.1 Categories

Definition C.1.1. A *category* $\mathcal{X}$ consists of:
(i) a class $\mathcal{O}b(\mathcal{X})$ whose members X are called *objects* of the category;
(ii) a class $\mathrm{Hom}(\mathcal{X})$ of *morphisms* f, so that each f has a domain X_1 and codomain X_2 belonging to $\mathcal{O}b(\mathcal{X})$, and the morphisms with domain X_1 and codomain X_2 form a set $\mathrm{Hom}_{\mathcal{X}}(X_1, X_2)$ for every ordered pair (X_1, X_2) of objects of $\mathcal{X}$; a morphism $f \in \mathrm{Hom}_{\mathcal{X}}(X_1, X_2)$ is usually written as $f: X_1 \to X_2$ (or shortly as f);
(iii) an associative composition law $\circ: \mathrm{Hom}_{\mathcal{X}}(X_2, X_3) \times \mathrm{Hom}_{\mathcal{X}}(X_1, X_2) \to \mathrm{Hom}_{\mathcal{X}}(X_1, X_3)$, for every ordered triple (X_1, X_2, X_3) of objects of $\mathcal{X}$, that associates to every pair of morphisms $(f, g) \in \mathrm{Hom}_{\mathcal{X}}(X_2, X_3) \times \mathrm{Hom}_{\mathcal{X}}(X_1, X_2)$, a morphism $f \circ g \in \mathrm{Hom}_{\mathcal{X}}(X_1, X_3)$ called *composition* of f and g.

The following conditions must be satisfied:
(1) the sets $\mathrm{Hom}_{\mathcal{X}}(X, X')$ and $\mathrm{Hom}_{\mathcal{X}}(Y, Y')$ are disjoint if the pairs of objects (X, X') and (Y, Y') of $\mathcal{X}$ do not coincide;
(2) for every object X of $\mathcal{X}$ there exists a morphism $id_X \in \mathrm{Hom}_{\mathcal{X}}(X, X)$ such that $id_X \circ f = f$ and $g \circ id_X = g$ for every $f \in \mathrm{Hom}_{\mathcal{X}}(X', X)$ and $g \in \mathrm{Hom}_{\mathcal{X}}(X, X')$.

Example C.1.2. The following categories are frequently used in this book:
(a) **Set** – sets and maps;
(b) $\mathbf{Vect}_K$ – vector spaces over a field K and linear transformations;
(c) **Grp** – groups and group homomorphisms;
(d) **AbGrp** – abelian groups and group homomorphisms;
(e) **Rng** – rings and ring homomorphisms;
(f) $\mathbf{Rng}_1$ – unitary rings and homomorphisms of unitary rings;
(g) **Top** – topological spaces and continuous maps.

Definition C.1.3. A morphism $f: X \to Y$ in a category $\mathcal{X}$ is:
(i) an *isomorphism* if there exists a morphism $g: Y \to X$ in $\mathcal{X}$ such that $g \circ f = id_X$ and $f \circ g = id_Y$;
(ii) an *epimorphism* (or, *right cancellable*) if for every pair of morphisms $g, h: Y \to Z$ in $\mathcal{X}$ with $g \circ f = h \circ f$ one has $g = h$;
(iii) a *monomorphism* (or, *left cancellable*) if for every pair of morphisms $g, h: Z \to X$ in $\mathcal{X}$ with $f \circ g = f \circ h$ one has $g = h$;
(iv) a *bimorphism* if it is simultaneously an epimorphism and a monomorphism.

If M, X are objects of a category $\mathcal{X}$ and $m: M \to X$ is a monomorphism, one often refers to M (and the monomorphism m) as a *subobject* of X.

https://doi.org/10.1515/9783110654936-019

Nevertheless, in some cases, as typically in **Top**, a smaller class of monomorphisms is more adapted to be treated as subobject:

Example C.1.4. In **Set**, **Vect**$_K$, **Grp**, **AbGrp**, and **Top**, the epimorphisms are precisely the surjective morphisms, while the monomorphisms are precisely the injective morphisms. Usually, as subobjects in **Top** one takes only those monomorphisms that are topological embeddings.

We give below further examples to show that an epimorphism need not be surjective and that a monomorphism need not be injective (whenever "surjective" or "injective" still may make sense).

For a nonempty family of objects $\mathcal{F} = \{X_i : i \in I\}$ in a category $\mathcal{X}$, an object X of $\mathcal{X}$ is a *product* of this family of objects if there exists a family of morphisms $\{p_i : X \to X_i : i \in I\}$ (often referred to as *projections*) such that for every family of morphisms $\{f_i : Y \to X_i : i \in I\}$ in $\mathcal{X}$ there exists a unique morphism $f : Y \to X$ in $\mathcal{X}$ such that $f_i = p_i \circ f$ for every $i \in I$. The product X is often denoted by $\prod_{i \in I} X_i$.

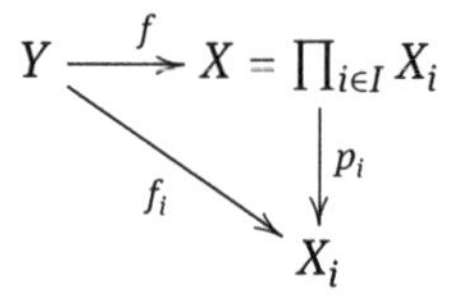

Dually, the *coproduct* of the family $\mathcal{F}$ is the object X of $\mathcal{X}$ provided with a family of morphisms $\{\nu_i : X_i \to X : i \in I\}$ such that for every family of morphisms $\{f_i : X_i \to Y : i \in I\}$ in $\mathcal{X}$ there exists a unique morphism $f : X \to Y$ such that $f_i = f \circ \nu_i$ for every $i \in I$. The coproduct X is usually denoted by $\coprod_{i \in I} X_i$.

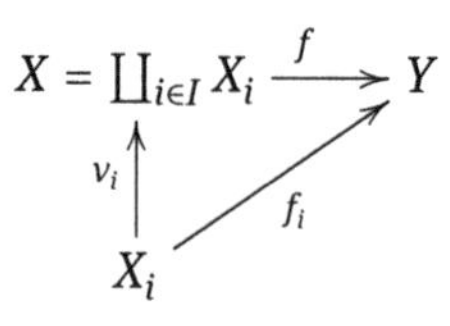

Both products and coproducts are uniquely determined up to isomorphism.

When the family $\mathcal{F}$ is empty, its product T has the simple property that $|\mathrm{Hom}(X, T)| = 1$ for every object X in $\mathcal{X}$. The uniquely determined object T with this property is called a *terminal object*. Dually, the coproduct I of the empty family has the dual property $|\mathrm{Hom}(I, X)| = 1$ for every object X in $\mathcal{X}$. The uniquely determined object I with this property is called an *initial object*.

Clearly, products and coproducts (in particular, terminal and initial objects) need not exist in general.

Example C.1.5. (a) In **Set** the product is the Cartesian product equipped with its canonical projections. In **Vect**$_K$, **Grp**, and **AbGrp**, the product is the direct product, while in **Top** the product is the topological product. In all these cases the canonical projections accompany the product.

(b) In **Set** the coproduct of a nonempty family of sets $\{X_i : i \in I\}$ is the disjoint union $\bigsqcup_{i\in I} X_i$, in **Top** it is the coproduct $\coprod_{i\in I} X_i$ with the coproduct topology. In both cases the morphisms $\nu_i : X_i \to \coprod_{i\in I} X_i$ are the canonical inclusions.
(c) The coproduct of a family of objects $\{X_i : i \in I\}$ in $\mathbf{Vect}_K$ and **AbGrp** is simply the direct sum $\bigoplus_{i\in I} X_i$ equipped with the canonical inclusions $\nu_i : X_i \to \bigoplus_{i\in I} X_i$.

We say that a category $\mathcal{Y}$ is a *subcategory* of a category $\mathcal{X}$ if $\mathcal{O}b(\mathcal{Y}) \subseteq \mathcal{O}b(\mathcal{X})$ and $\mathrm{Hom}_{\mathcal{Y}}(Y, Y') \subseteq \mathrm{Hom}_{\mathcal{X}}(Y, Y')$ for any pair $Y, Y' \in \mathcal{O}b(\mathcal{Y})$. Moreover, the subcategory $\mathcal{Y}$ of $\mathcal{X}$ is said to be a *full* subcategory of $\mathcal{X}$ if $\mathrm{Hom}_{\mathcal{Y}}(Y, Y') = \mathrm{Hom}_{\mathcal{X}}(Y, Y')$ for any pair $Y, Y' \in \mathcal{O}b(\mathcal{Y})$.

Compare the following example with Remark C.2.18.

Example C.1.6. (a) **AbGrp** is a full subcategory of **Grp**.
(b) The class $\mathcal{T}_{\mathbf{t}}$ (respectively, $\mathcal{F}_{\mathbf{t}}$) of all torsion (respectively, torsion-free) abelian groups and group homomorphisms is a full subcategory of **AbGrp**.
(c) Similarly, the class $\mathcal{T}_{\mathbf{div}}$ (respectively, $\mathcal{F}_{\mathbf{div}}$) of all divisible (respectively, reduced) abelian groups and group homomorphisms is a full subcategory of **AbGrp**.
(d) For every $m \in \mathbb{N}_+$, the class $\mathbf{AbGrp_m}$ of all abelian groups G with the property $G = G[m]$ and group homomorphisms is a full subcategory of $\mathcal{T}_{\mathbf{t}}$.

Some of these subcategories (as **AbGrp**, $\mathcal{F}_{\mathbf{t}}$, $\mathcal{F}_{\mathbf{div}}$, and $\mathbf{AbGrp_m}$) are stable under taking direct products and subgroups, while others (as **AbGrp**, $\mathcal{T}_{\mathbf{t}}$, $\mathcal{T}_{\mathbf{div}}$, and $\mathbf{AbGrp_m}$) are stable under taking direct sums and quotients.

There is a series of full subcategories of **Top** determined by separation axioms, the most prominent among them is $\mathbf{Top_2}$, having as objects all Hausdorff spaces. Similarly, one defines $\mathbf{Top_i}$ for $i \in \{0, 1, 3, 3.5, 4\}$. All $\mathbf{Top_i}$, for $i \neq 4$, are stable under taking direct products and subspaces. We see below that this stability property determines a relevant global property of the subcategory in question.

Example C.1.7. Here come examples where the epimorphisms and monomorphisms are not precisely what one may expect.
(a) In $\mathcal{F}_{\mathbf{t}}$ the epimorphisms are precisely the group homomorphisms $f : G \to H$ such that $f(G)$ is an essential subgroup of H (e. g., the nonsurjective inclusion map $\mathbb{Z} \to \mathbb{Q}$ is an epimorphism in $\mathcal{F}_{\mathbf{t}}$).
(b) In the category $\mathcal{T}_{\mathbf{div}}$ the noninjective quotient homomorphism $\mathbb{Q} \to \mathbb{Q}/\mathbb{Z}$ is a monomorphism.
(c) For $i \in \{2, 3, 3.5\}$, in the full subcategory $\mathbf{Top_i}$ of **Top** the nonsurjective embedding $\mathbb{Q} \to \mathbb{R}$ is an epimorphism. More generally, a continuous map $f : X \to Y$ in $\mathbf{Top_i}$ is an epimorphism if and only if $f(X)$ is dense in Y.

Definition C.1.8. A full subcategory $\mathfrak{V}$ of **Grp** is called a *variety (of groups)* if $\mathfrak{V}$ is stable under taking subgroups, direct products, and quotients. In such a case, $\mathfrak{V} \cap \mathbf{AbGrp}$ is called a *variety of abelian groups*.

In particular, **AbGrp** is a variety of groups.

The notion of a variety can be introduced also by using identities in a group:

Remark C.1.9. Let F be the free group with countably many free generators, namely, $\{x_n : n \in \mathbb{N}\}$, and denote by 1 its neutral element. An element $w = x_{i_1}^{\epsilon_1} \cdots x_{i_n}^{\epsilon_n} \in F \setminus \{1\}$, with $\epsilon_k \in \{1, -1\}$ for $k \in \{1, \dots, n\}$ and $\epsilon_k \neq -\epsilon_{k+1}$ whenever $x_k = x_{k+1}$ for $k \in \{1, \dots, n-1\}$, is usually referred to as a *word* (in the *alphabet* $\{x_n : n \in \mathbb{N}_+\}$).

We say that the identity $w = 1$ holds in a group G if for every homomorphism $f : F \to G$ one has $f(w) = e_G$. For example, the identity $[x_1, x_2] = 1$ holds in every abelian group G.

It is easy to see that, for every identity $w = 1$, the class $\mathfrak{V}_w$ of groups G where it holds forms a variety. More generally, if W is a set of words in F, then the class $\mathfrak{V}_W$ of groups G, where the identity $w = 1$ holds for every $w \in W$, is a variety which obviously coincides with $\bigcap_{w \in W} \mathfrak{V}_w$.

According to a theorem of Birkhoff, every variety of groups $\mathfrak{V}$ can be obtained in this way, i. e., there exists a set of words W in F such that $\mathfrak{V} = \mathfrak{V}_W$.

C.2 Functors

Definition C.2.1. Consider two categories $\mathcal{X}$ and $\mathcal{Y}$. A *covariant* (respectively, *contravariant*) functor $\mathsf{F} : \mathcal{X} \to \mathcal{Y}$ assigns to each object X of $\mathcal{X}$ an object $\mathsf{F}X$ of $\mathcal{Y}$ and to each morphism $f : X \to X'$ in $\mathcal{X}$ a morphism $\mathsf{F}f : \mathsf{F}X \to \mathsf{F}X'$ (respectively, $\mathsf{F}f : \mathsf{F}X' \to \mathsf{F}X$) such that

$$\mathsf{F}id_X = id_{\mathsf{F}X} \quad \text{and} \quad \mathsf{F}(g \circ f) = \mathsf{F}g \circ \mathsf{F}f \quad (\text{respectively, } \mathsf{F}(g \circ f) = \mathsf{F}f \circ \mathsf{F}g)$$

for every pair of morphisms $f : X \to X'$ and $g : X' \to X''$ in $\mathcal{X}$. A functor $\mathsf{F} : \mathcal{X} \to \mathcal{Y}$ with $\mathcal{X} = \mathcal{Y}$ is named *endofunctor* (of $\mathcal{X}$).

For a category $\mathcal{X}$, denote by $1_{\mathcal{X}}$ the functor $1_{\mathcal{X}} : \mathcal{X} \to \mathcal{X}$ such that

$$1_{\mathcal{X}}(X) = X \quad \text{for every } X \in \mathcal{O}b(\mathcal{X}) \quad \text{and} \quad 1_{\mathcal{X}}(f) = f \quad \text{for every } f \in \operatorname{Hom}(\mathcal{X}).$$

If $\mathsf{F} : \mathcal{X} \to \mathcal{Y}$ and $\mathsf{G} : \mathcal{Y} \to \mathcal{Z}$ are functors, let $\mathsf{G} \cdot \mathsf{F} : \mathcal{X} \to \mathcal{Z}$ be the functor defined by letting $(\mathsf{G} \cdot \mathsf{F})X = \mathsf{G}(\mathsf{F}X)$ for every object X in $\mathcal{X}$ and $(\mathsf{G} \cdot \mathsf{F})f = \mathsf{G}(\mathsf{F}f)$ for every morphism f in $\mathcal{X}$. It is easy to see that the functor $\mathsf{G} \cdot \mathsf{F}$ is covariant whenever both functors are simultaneously covariant or contravariant, otherwise the functor $\mathsf{G} \cdot \mathsf{F}$ is contravariant.

A functor $\mathsf{F} : \mathcal{X} \to \mathcal{Y}$ defines a map $\operatorname{Hom}_{\mathcal{X}}(X, X') \to \operatorname{Hom}_{\mathcal{Y}}(\mathsf{F}X, \mathsf{F}X')$ for every pair (X, X') of objects of $\mathcal{X}$. We say that F is *faithful* (respectively, *full*) if for every pair (X, X') of objects of $\mathcal{X}$ the above map is injective (respectively, surjective).

If $\mathcal{Y}$ is a subcategory of a category $\mathcal{X}$, then the inclusion $\mathsf{I} : \mathcal{Y} \hookrightarrow \mathcal{X}$ is a covariant functor. It is full precisely when the subcategory is full.

Definition C.2.2. A category $\mathcal{X}$ is *concrete* if it admits a faithful functor $\mathsf{U}: \mathcal{X} \to \mathbf{Set}$. In such a case the functor U is called *forgetful*.

All examples above are concrete categories.

Example C.2.3. (a) For every abelian group G, the group $G/t(G)$ is torsion-free (see Exercise A.7.1), and for every morphism $f: G \to H$ in **AbGrp**, one has an induced homomorphism $\bar{f}: G/t(G) \to H/t(H)$ between torsion-free abelian groups. Therefore, the assignments $G \mapsto G/t(G)$ and $f \mapsto \bar{f}$ define a covariant functor $\mathbf{AbGrp} \to \mathcal{F}_{\mathbf{t}}$. Similarly, the assignment $G \mapsto t(G)$ defines a covariant functor $\mathbf{AbGrp} \to \mathcal{T}_{\mathbf{t}}$, by assigning to every morphism $f: G \to H$ in **AbGrp** its restriction $f \restriction_{t(G)}: t(G) \to t(H)$.

(b) For every abelian group G, the group $G/\mathrm{div}(G)$ is reduced (see Theorem A.4.3) and the assignment $G \mapsto G/\mathrm{div}(G)$ defines, as in (a), a covariant functor $\mathbf{AbGrp} \to \mathcal{F}_{\mathbf{div}}$.
On the other hand, as in (a), the assignment $G \mapsto \mathrm{div}(G)$ induces a covariant functor $\mathbf{AbGrp} \to \mathcal{T}_{\mathbf{div}}$.

(c) For every abelian group G, the quotient G/G^1 is residually finite (see Exercise A.7.11), and for every morphism $f: G \to H$ in **AbGrp** one has an induced homomorphism $\bar{f}: G/G^1 \to H/H^1$ between residually finite abelian groups. Hence, the assignment $G \mapsto G/G^1$ induces a covariant functor from **AbGrp** to its full subcategory **ResFinGrp** of residually finite abelian groups.

The next important notions connect two "parallel" functors.

Definition C.2.4. Let $\mathcal{X}, \mathcal{Y}$ be categories and $\mathsf{F}, \mathsf{F}': \mathcal{X} \to \mathcal{Y}$ be covariant functors. A *natural transformation* γ from F to F′ assigns to each object X of $\mathcal{X}$ a morphism $\gamma_X: \mathsf{F}X \to \mathsf{F}'X$ such that for every morphism $f: X \to X_1$ in $\mathcal{X}$ the following diagram is commutative:

$$\begin{array}{ccc} \mathsf{F}X & \xrightarrow{\mathsf{F}f} & \mathsf{F}X_1 \\ {\scriptstyle \gamma_X}\downarrow & & \downarrow{\scriptstyle \gamma_{X_1}} \\ \mathsf{F}'X & \xrightarrow[\mathsf{F}'f]{} & \mathsf{F}'X_1 \end{array}$$

A *natural equivalence* is a natural transformation γ such that each γ_X is an isomorphism.

These notions allow us to define *equivalence* (respectively, *duality*) *of categories* as follows. This is a pair of covariant (respectively, contravariant) functors $\mathsf{F}: \mathcal{X} \to \mathcal{Y}$ and $\mathsf{G}: \mathcal{Y} \to \mathcal{X}$ such that there exists a pair of natural equivalences $\eta: 1_{\mathcal{X}} \to \mathsf{G} \cdot \mathsf{F}$ and $\varepsilon: \mathsf{F} \cdot \mathsf{G} \to 1_{\mathcal{Y}}$. In case $\mathcal{X} = \mathcal{Y}$ and the contravariant functors $\mathsf{F} = \mathsf{G}$ coincide, we say that F is an *involutive duality*.

C.2.1 Reflectors and coreflectors

Beyond the inclusion functor $\mathsf{I}: \mathcal{Y} \to \mathcal{X}$, where $\mathcal{Y}$ is a subcategory of the category $\mathcal{X}$, other frequently used functors are the reflectors and the coreflectors defined below. For the sake of simplicity, in the sequel we do not use I, so we write Y (respectively, f) in place of $\mathsf{I}Y$ (respectively, $\mathsf{I}f$) for an object Y (respectively, morphism f) of $\mathcal{Y}$.

Definition C.2.5. For a subcategory $\mathcal{Y}$ of a category $\mathcal{X}$, we say that:
(a) a functor $\mathsf{R}: \mathcal{X} \to \mathcal{Y}$ is a *reflector* if there exists a natural transformation $\gamma: 1_{\mathcal{X}} \to \mathsf{R}$ such that R coincides with $1_{\mathcal{Y}}$ on $\mathcal{Y}$ (i. e., $\mathsf{R}Y = Y$ for every object Y of $\mathcal{Y}$ and $\mathsf{R}f = f$ for every morphism f of $\mathcal{Y}$);
(b) a functor $\mathsf{C}: \mathcal{X} \to \mathcal{Y}$ is a *coreflector* if there exists a natural transformation $\eta: \mathsf{C} \to 1_{\mathcal{X}}$ such that C coincides with $1_{\mathcal{Y}}$ on $\mathcal{Y}$.

In particular we have the following commutative diagrams, where $f: X \to Y$ is a morphism in $\mathcal{X}$ with codomain Y an object of $\mathcal{Y}$, and $g: Y \to X$ is a morphism in $\mathcal{X}$ with domain Y in $\mathcal{Y}$:

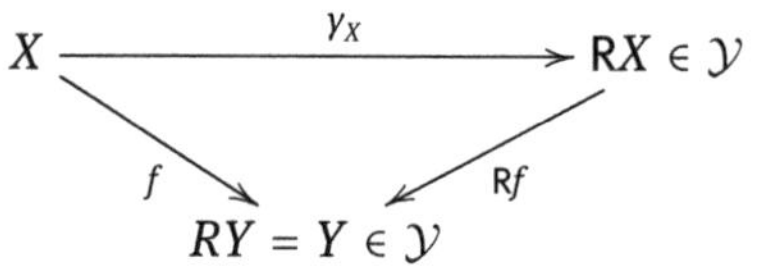

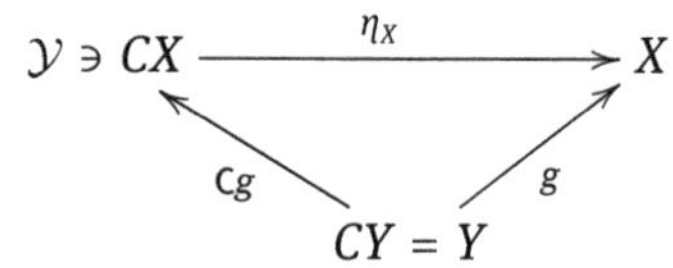

Very often, the reflectors are actually *epireflectors* (i. e., all γ_Y are epimorphisms), while the coreflectors are most often *monocoreflectors* (i. e., all η_X are monomorphisms).

A full subcategory $\mathcal{Y}$ of a category $\mathcal{X}$ is called *(epi)reflective* if there is an (epi)reflector $\mathsf{R}: \mathcal{X} \to \mathcal{Y}$. This means that for each object X of $\mathcal{X}$ there exist an object $\mathsf{R}X$ of $\mathcal{Y}$ and a morphism $\gamma_X: X \to \mathsf{R}X$ in $\mathcal{X}$ such that for each morphism $f: X \to Y$ in $\mathcal{X}$ to an object Y of $\mathcal{Y}$ there exists a unique morphism $\mathsf{R}f: \mathsf{R}X \to Y$ in $\mathcal{Y}$ with $\mathsf{R}f \circ \gamma_X = f$. The morphism γ_X (and sometimes, only $\mathsf{R}X$) is referred to as being the *$\mathcal{Y}$-reflection* of X.

Remark C.2.6. The $\mathcal{Y}$-reflection $\gamma_X: X \to \mathsf{R}X$ is unique (up to isomorphism) with the above property, in the following sense. If for every $X \in \mathcal{X}$ a morphism $h_X: X \to K_X$ in $\mathcal{X}$ with $K_X \in \mathcal{Y}$ is assigned such that for every morphism $f: X \to C$ where $C \in \mathcal{Y}$, there exists a unique morphism $f': K_X \to C$ with $f' \circ h_X = f$, then there exists an isomorphism $\mathsf{s}: \mathsf{R}X \to K_X$ such that $\mathsf{s} \circ \gamma_X = h_X$ (see Exercise C.3.16).

Similarly is defined a *(mono)coreflective* subcategory. In many cases some natural stability properties of a subcategory turn out to be equivalent to epireflectivity or monocoreflectivity (see Exercises C.3.18 and C.3.19).

Example C.2.7. (a) The covariant functor **AbGrp** $\to \mathcal{F}_{\mathsf{t}}$ defined by the assignment $G \mapsto G/t(G)$ in Example C.2.3(a) is an epireflector with natural transformation

γ defined by letting $\gamma_G: G \to G/t(G)$ be the canonical projection. The covariant functor **AbGrp** $\to \mathcal{T}_t$ defined there by $G \mapsto t(G)$ is a monocoreflector, with natural transformation η defined by letting each $\eta_G: t(G) \to G$ be the inclusion of $t(G)$ in G.

(b) The covariant functor **AbGrp** $\to \mathcal{F}_{\mathbf{div}}$ defined by the assignment $G \mapsto G/\mathrm{div}(G)$ in Example C.2.3(b) is an epireflector with natural transformation γ defined by letting each $\gamma_G: G \to G/\mathrm{div}(G)$ be the canonical projection; on the other hand, the covariant functor **AbGrp** $\to \mathcal{T}_{\mathbf{div}}$ defined by the assignment $G \mapsto \mathrm{div}(G)$ is a monocoreflector with natural transformation defined by letting each $\eta_G: \mathrm{div}(G) \to G$ be the inclusion of $\mathrm{div}(G)$ in G.

(c) The covariant functor **AbGrp** $\to$ **ResFinGrp** defined in Example C.2.3(c) by the assignment $G \mapsto G/G^1$ is an epireflector with natural transformation γ defined by letting each $\gamma_G: G \to G/G^1$ be the canonical projection.

Unlike items (a) and (b), in (c) we did not mention anything about the assignment $G \mapsto G^1$. Indeed, it need not be a coreflector, as the subgroup G^1 need not satisfy $(G^1)^1 = G^1$. Actually, an abelian group G satisfies $G = G^1$ if and only if $G = \mathrm{div}(G)$, that is, G is divisible. We further discuss this issue in §C.2.2.

A very natural and useful epireflection in **Top** is obtained as follows:

Example C.2.8. For a topological space X, set $x \sim y$ for $x, y \in X$ when $x \in \overline{\{y\}}$ and $y \in \overline{\{x\}}$. This binary relation is an equivalence relation on X. The quotient $\mathsf{T}_0 X = X/_\sim$ is a T_0-space. Moreover, for every continuous map $f: X \to Z$, where Z is a T_0-space, there exists a unique continuous map $f': \mathsf{T}_0 X \to Z$ with $f = f' \circ \gamma_X$, where $\gamma_X: X \to \mathsf{T}_0 X$ is the canonical projection. Therefore, the assignments $X \mapsto \mathsf{T}_0 X$ and $f \mapsto \mathsf{T}_0 f$, where $\mathsf{T}_0 f = (\gamma_X \circ f)'$, define an epireflector $\mathsf{T}_0: \mathbf{Top} \to \mathbf{Top}_0$ with natural transformation γ.

The (co)reflectors in all cases above were built *explicitly*. Now we give a simple construction of (epi)reflectors in **Top** that is not explicit. The same construction can be carried out also in **Grp** and **AbGrp** for a subcategory stable under taking subgroups and direct products.

Theorem C.2.9. *Let* **A** *be a subcategory of* **Top** *stable under taking subspaces and products. Then there exists an epireflector* $\mathsf{R}: \mathbf{Top} \to \mathbf{A}$.

Proof. We show that every topological space X admits a continuous surjective map $r_X: X \to \mathsf{R}X$, where $\mathsf{R}X \in \mathbf{A}$ and for every continuous map $f: X \to Z$ with $Z \in \mathbf{A}$, there exists a unique continuous map $f': \mathsf{R}X \to Z$ with $f = f' \circ r_X$; in other words, $f' = \mathsf{R}f$.

Fix a set $\{f_i: X \to A_i : i \in I\}$ of representatives of all continuous surjective maps $X \to A$ with $A \in \mathbf{A}$ (i. e., for every continuous surjective map $f: X \to A$ with $A \in \mathbf{A}$ there exist $i \in I$ and a homeomorphism $\xi: A_i \to A$ such that $f = \xi \circ f_i$). Let $g: X \to \prod_{i \in I} A_i$ be the diagonal map of the family $\{f_i : i \in I\}$, and let $\mathsf{R}X$ be the image $g(X)$ equipped with the topology induced by the product. Then $r_X: X \to \mathsf{R}X$ has the desired proprieties. Indeed, if $f: X \to Z$ is a continuous surjective map and $Z \in \mathbf{A}$, then there exists $i \in I$ such that f coincides with f_i up to homeomorphism, so $Z = A_i$ up to homeomorphism.

Hence, we can take as $f'\colon \mathrm{R}X \to Z = A_i$ the restriction of the projection $p_i\colon \prod_{i\in I} A_i \to A_i$ to $\mathrm{R}X$. In case $f\colon X \to Z$ is not surjective, apply the above argument to the surjective map $f\colon X \to f(X)$. □

In particular, for $\mathbf{A} = \mathbf{Top_i}$ with $i \in \{0, 1, 2, 3, 3.5\}$ one obtains an epireflection $\mathrm{T}_i\colon \mathbf{Top} \to \mathbf{Top_i}$ relative to the subcategory $\mathbf{Top_i}$ of **Top**. For a topological space X, we call *T_i-reflection* of X the space $\mathrm{T}_i X$; the T_2-reflection (respectively, $T_{3.5}$-reflection) of X is called also *Hausdorff reflection* (respectively, *Tichonov reflection*) of X.

We omit the proof of the next theorem which is similar to that of Theorem C.2.9.

Theorem C.2.10. *Let* **A** *be a subcategory of* **Top** *stable under taking closed subspaces and products. Then every topological space admits a continuous map with dense image* $\gamma_X\colon X \to \mathrm{R}X$, *where* $\mathrm{R}X \in \mathbf{A}$ *and for every continuous map* $f\colon X \to Z$, *with* $Z \in \mathbf{A}$, *there exists a unique continuous map* $f'\colon \mathrm{R}X \to Z$ *with* $f = f' \circ \gamma_X$. *In other words,* $\mathrm{R}\colon \mathbf{Top} \to \mathbf{A}$ *is a reflector.*

Remark C.2.11. In case **A** is the class of compact Hausdorff spaces, the theorem gives the Čech–Stone compactification $\mathrm{R}X = \beta Y$ of the Tichonov reflection $Y = \mathrm{T}_{3.5}X$ of the topological space X. In this case R is a reflector, but not an epireflector, since the maps $\gamma_X\colon X \to \mathrm{R}X$ need not be surjective.

C.2.2 Reflectors and coreflectors vs (pre)radicals in AbGrp

Definition C.2.12. A *preradical* $\mathbf{r}$ in **AbGrp** is an assignment of a subgroup $\mathbf{r}G$ to every abelian group G, in such a way that if $f\colon G \to H$ is a homomorphism in **AbGrp**, then $f(\mathbf{r}G) \subseteq \mathbf{r}H$. Call a preradical $\mathbf{r}$:

(i) *idempotent* if $\mathbf{r}(\mathbf{r}G) = \mathbf{r}G$ for every abelian group G;
(ii) *radical* if $\mathbf{r}(G/\mathbf{r}G) = \{0\}$ for every abelian group G;
(iii) *hereditary* if $\mathbf{r}(H) = H \cap \mathbf{r}G$ for every abelian group G and $H \leq G$.

Example C.2.13. The property (A.10) in the proof of Proposition A.4.6 shows that letting $\mathbf{r}G = G^1$ for every abelian group G defines a preradical, which is not idempotent (see Example C.2.7(c)).

Similarly, one checks that the subgroups of an abelian group G listed below give rise to preradicals with specific properties:

(a) $t(G)$ induces a hereditary radical;
(b) $\mathrm{div}(G)$ induces an idempotent radical that is not hereditary;
(c) $\mathrm{Soc}(G)$, as well as $G[p]$ for p a prime, induces an hereditary preradical that is not a radical;
(d) mG, for every $m \in \mathbb{N}_+$, induces a nonidempotent radical.

Definition C.2.14. For a preradical $\mathbf{r}$ in **AbGrp**, call an abelian group G $\mathbf{r}$*-torsion-free* (respectively, $\mathbf{r}$*-torsion*) if $\mathbf{r}G = 0$ (respectively, if $G = \mathbf{r}G$). Denote by $\mathcal{F}_{\mathbf{r}}$ (respectively, $\mathcal{T}_{\mathbf{r}}$) the full subcategory of **AbGrp** with objects the $\mathbf{r}$-torsion-free abelian groups (respectively, the $\mathbf{r}$-torsion abelian groups).

Exactly as in the proof of Proposition A.4.6, one can show that for every direct product $H = \prod_{i \in I} H_i$ of a family $\{H_i : i \in I\}$ of abelian groups,

$$\mathbf{r}H = \mathbf{r}\left(\prod_{i \in I} H_i\right) \subseteq \prod_{i \in I} \mathbf{r}H_i, \tag{C.1}$$

by applying the definition of preradical to the canonical projections of the product.

Remark C.2.15. For a preradical $\mathbf{r}$ in **AbGrp**, clearly $\mathcal{F}_{\mathbf{r}}$ is stable under taking subgroups. From (C.1) one can deduce that $\mathcal{F}_{\mathbf{r}}$ is stable also under taking products. On the other hand, one can see that $\mathcal{T}_{\mathbf{r}}$ is stable under taking direct sums and quotients (but need not be stable also under taking subgroups or products).

Remark C.2.16. (a) If $\mathbf{r}$ is a preradical in **AbGrp**, the assignment $G \mapsto \mathbf{r}G$ induces a covariant functor by letting also, for a morphism $f: G \to H$ in **AbGrp**,

$$\mathbf{r}f = f\restriction_{\mathbf{r}G} : \mathbf{r}G \to \mathbf{r}H.$$

This functor has a special property, namely, there is a natural transformation $\gamma: \mathbf{r} \to 1_{\mathbf{AbGrp}}$ such that $\gamma_G: \mathbf{r}G \to G$ is a subgroup embedding for every abelian group G. Nevertheless, $\mathbf{r}$ is not a coreflection in general; indeed, it is a coreflection if and only if $\mathbf{r}$ is idempotent.
Vice versa, if $\mathsf{C}: \mathbf{AbGrp} \to \mathcal{Y} \subseteq \mathbf{AbGrp}$ is a monocoreflector, then putting $\mathbf{r}G = \mathsf{C}G$ for every abelian group G (i. e., assuming that $\mathsf{C}G$ is simply a subgroup of G) one obtains an idempotent preradical $\mathbf{r}$ and $\mathcal{Y} = \mathcal{T}_{\mathbf{r}}$.

(b) Similarly, a preradical $\mathbf{r}$ in **AbGrp** gives rise to another covariant functor defined by the assignment $G \mapsto G/\mathbf{r}G$ for every abelian group G; in fact, every morphism $f: G \to H$ in **AbGrp** gives rise to a morphism $\bar{f}: G/\mathbf{r}G \to H/\mathbf{r}H$ in **AbGrp**. Since $\mathbf{r}(G/\mathbf{r}G) = 0$ for all abelian groups G precisely when $\mathbf{r}$ is a radical, we deduce that the functor $G \mapsto G/\mathbf{r}G$ is a reflector precisely when $\mathbf{r}$ is a radical.
On the other hand, if $\mathsf{R}: \mathbf{AbGrp} \to \mathcal{Y} \subseteq \mathbf{AbGrp}$ is an epireflector with natural transformation γ, then putting $\mathbf{r}G = \ker \gamma_G$ for every abelian group G, where $\gamma_G: G \to \mathsf{R}G$, one gets a radical of **AbGrp**, such that $G/\mathbf{r}G = \mathsf{R}G \in \mathcal{Y}$, i. e., $\mathcal{Y} = \mathcal{F}_{\mathbf{r}}$.

We resume our observations from Remark C.2.16 in the following:

Theorem C.2.17. *Every epireflective subcategory of* **AbGrp** *has the form* $\mathcal{F}_{\mathbf{r}}$ *for some radical* $\mathbf{r}$*, while every monocoreflective subcategory of* **AbGrp** *has the form* $\mathcal{T}_{\mathbf{r}}$ *for some idempotent preradical* $\mathbf{r}$.

Consequently, the epireflective subcategories of **AbGrp** *are stable under taking direct products and subgroups, while the monocoreflective subcategories of* **AbGrp** *are stable under taking direct sums and quotients.*

We conclude by noting that if $\mathbf{r}$ is an idempotent radical in **AbGrp**, then the pair of subcategories $(\mathcal{T}_{\mathbf{r}}, \mathcal{F}_{\mathbf{r}})$ has the following properties:
(a) $\mathcal{T}_{\mathbf{r}} \cap \mathcal{F}_{\mathbf{r}} = \{\{0\}\}$;
(b) every morphism $\mathcal{T}_{\mathbf{r}} \ni T \to F \in \mathcal{F}_{\mathbf{r}}$ is zero;
(c) every abelian group G has a subgroup $T \in \mathcal{T}_{\mathbf{r}}$ such that $G/T \in \mathcal{F}_{\mathbf{r}}$;
(d) the class $\mathcal{T}_{\mathbf{r}}$ is stable under taking extensions, direct sums, and quotients;
(e) the class $\mathcal{F}_{\mathbf{r}}$ is stable under taking extensions, direct products, and subgroups.

Remark C.2.18. A pair $(\mathcal{T}, \mathcal{F})$ of full subcategories of **AbGrp** with the properties (a)–(e) above is called a *torsion theory*. Leading examples to this effect are the pairs $(\mathcal{T}_{\mathbf{t}}, \mathcal{F}_{\mathbf{t}})$ and $(\mathcal{T}_{\mathbf{div}}, \mathcal{F}_{\mathbf{div}})$, namely, the radicals $\mathbf{t} = t(-)$ and $\mathbf{div} = \mathrm{div}(-)$.

Every torsion theory $(\mathcal{T}, \mathcal{F})$ has this form for an appropriate idempotent radical $\mathbf{r}$, that is, $\mathcal{T} = \mathcal{T}_{\mathbf{r}}$ and $\mathcal{F} = \mathcal{F}_{\mathbf{r}}$. (Just note that the subgroup T of a group G, as of item (c), is uniquely determined, and put $\mathbf{r}G = T$. The remaining properties imply that $\mathbf{r}$ is an idempotent radical with $\mathcal{T}_{\mathbf{r}} = \mathcal{T}$ and $\mathcal{F}_{\mathbf{r}} = \mathcal{F}$.)

C.2.3 Contravariant Hom-functors

The functors in the previous subsection were all covariant. Here we give a series of examples of contravariant functors.

Example C.2.19. The simplest example is the contravariant functor $\mathsf{P}: \mathbf{Set} \to \mathbf{Set}$, which takes each set X to its power set $\mathcal{P}(X)$ and each function $f: X \to Y$ to its inverse image map $\mathsf{P}(f): \mathsf{P}(Y) \to \mathsf{P}(X)$, $B \mapsto f^{-1}(B)$.

In this example one can obtain $\mathcal{P}(X)$ as a "function space", identifying the elements of $\mathcal{P}(X)$ with functions $X \to D = \{0, 1\}$. More precisely, $\mathcal{P}(X) \ni A \mapsto \chi_A \in D^X$, where χ_A is the characteristic function of $A \subseteq X$. Moreover, under this identification, for a map $f: X \to Y$ and $B \in \mathsf{P}(Y)$, $\mathsf{P}(f)(\chi_B) = \chi_{f^{-1}(B)} = \chi_B \circ f$.

In the language of category theory, D^X is $\mathrm{Hom}(X, D)$ in **Set**. Hence, in the above example we obtain $\mathsf{P}(X)$ as the Hom-set $\mathrm{Hom}(X, D)$. This is the leading idea behind the following fundamental notion.

Definition C.2.20. A *representable functor* $F: \mathcal{X} \to \mathbf{Set}$ is defined by $\mathsf{F}(X) = \mathrm{Hom}(X, D)$ for some fixed object D of the category $\mathcal{X}$ and $\mathsf{F}(f): \mathsf{F}(Y) \to \mathsf{F}(X)$ for a morphism $f: X \to Y$ in $\mathcal{X}$, defined by $\mathsf{F}(f)(g) = g \circ f$ for every $g \in \mathrm{Hom}(Y, D)$.

In all examples that follow the Hom-set $\mathrm{Hom}(X, D)$ has also some additional structure that allows one to consider it as an object of some concrete category $\mathcal{Y}$, obtaining in this way a functor $\overline{\mathsf{F}}: \mathcal{X} \to \mathcal{Y}$ such that the composition with the forgetful functor $\mathsf{U}: \mathcal{Y} \to \mathbf{Set}$ gives $\mathsf{F} = \mathsf{U} \cdot \overline{\mathsf{F}}$.

Example C.2.21. Let K be a field and $\mathcal{X} = \mathbf{Vect}_K$. Now for every object V in $\mathcal{X}$, the set $\mathrm{Hom}(V, K)$ carries an obvious structure of K-vector space, usually called the *dual* vector space of V. Put $V^* = \mathrm{Hom}(V, K)$ equipped with this structure.

We get a contravariant functor $^*: \mathcal{X} \to \mathcal{X}$ by assigning V^* to every V in $\mathcal{X}$, while to every $f: V \to W$ we assign the morphism $f^*: W^* \to V^*$ defined as above by letting $f^*(\varphi) = \varphi \circ f$ for every $\varphi \in W^*$.

There is a natural transformation $\alpha: 1_{\mathcal{X}} \to {}^{**}$, where we denote by ** the composition of * with itself: α is given by the evaluation map $\alpha_V: V \mapsto V^{**}$ defined by $\alpha_V(v)(\chi) = \chi(v)$ for every $v \in V$ and every $\chi \in V^*$.

It is well known that α_V is an isomorphism precisely when $\dim_K V < \infty$. More precisely, if $\mathcal{Y}$ denotes the full subcategory of $\mathcal{X}$ of all finite dimensional vector spaces, then * gives an involutive duality $\mathcal{Y} \to \mathcal{Y}$.

Example C.2.22. Let $D = \{0, 1\}$ be the discrete doubleton. For every topological space X, the set $\mathrm{Hom}(X, D)$ of all continuous functions is also a Boolean ring (recall that a unitary ring B is *Boolean* if $x^2 = x$ for every $x \in B$): in fact, D can be provided with the ring structure inherited by the identification $D \equiv \mathbb{Z}/2\mathbb{Z}$. An equivalent way to get a Boolean ring structure on $\mathrm{Hom}(X, D)$ is to observe that $\mathrm{Hom}(X, D)$ is (identifying each $f \in \mathrm{Hom}(X, D)$ with its clopen support) isomorphic to the Boolean ring $\mathsf{B}(X)$ of all clopen sets of X with the operations of symmetric difference and intersection. Therefore, one finds a contravariant functor

$$\mathsf{B}: \mathbf{Top} \to \mathbf{BRng},$$

where $\mathbf{BRng}$ is the full subcategory of $\mathbf{Rng_1}$ with objects all Boolean rings, by assigning also to each morphism $f: X \to Y$ in $\mathbf{Top}$ its inverse image map $\mathsf{B}(f): \mathsf{B}(Y) \to \mathsf{B}(X)$, $A \mapsto f^{-1}(A)$ for $A \in \mathsf{B}(Y)$.

On the other hand, for every $B \in \mathbf{BRng}$ the set $\mathrm{Hom}(B, \mathbb{Z}/2\mathbb{Z})$ coincides, up to natural bijection (identifying each $f \in \mathrm{Hom}(B, \mathbb{Z}/2\mathbb{Z})$ with $\ker f$), with the set $\mathsf{Spec}\, B$ of all prime ideals of B (note that in a Boolean ring a prime ideal is also maximal). It is well known that on $\mathsf{Spec}\, B$ one can consider its *Zariski topology*: a base of this topology is given by the sets $O_b = \{\mathfrak{p} \in \mathsf{Spec}\, B: b \notin \mathfrak{p}\}$, with $b \in B$; in particular, each O_b is clopen. Then $\mathsf{Spec}\, B$ with its Zariski topology is a zero-dimensional compact Hausdorff space. Therefore, also in this case the assignment $B \mapsto \mathrm{Hom}(B, \mathbb{Z}/2\mathbb{Z})$ gives rise to a contravariant functor $\mathbf{BRing} \to \mathbf{Top}$. This coincides with the restriction of the contravariant functor Spec from the category of all commutative unitary rings to $\mathbf{Top}$, so we still keep on denoting it by

$$\mathsf{Spec}: \mathbf{BRing} \to \mathbf{Top}.$$

Actually, the range of the restriction $\mathsf{Spec}\colon \mathbf{BRng} \to \mathbf{Top}$ is the full subcategory **StoneSp** of **Top** consisting of *Stone spaces* (i. e., zero-dimensional compact Hausdorff spaces). It can be verified that $\mathsf{B}(\mathsf{Spec}(A)) \cong A$ for every $A \in \mathbf{BRng}$, and that X is homeomorphic to $\mathsf{Spec}(\mathsf{B}(X))$ for every $X \in \mathbf{StoneSp}$. Both isomorphisms induce natural transformations, respectively

$$1_{\mathbf{BRng}} \to \mathsf{B} \cdot \mathsf{Spec} \quad \text{and} \quad 1_{\mathbf{StoneSp}} \to \mathsf{Spec} \cdot \mathsf{B}.$$

What we have described above can shortly be summarized by saying that the pair of functors Spec and B gives a duality between the categories **BRing** and **StoneSp**, known as *Stone duality*.

This is triggered by Hom-sets with target the very special double-faced object $D \equiv \mathbb{Z}/2\mathbb{Z}$ carrying both a structure of a compact Hausdorff space D and also a structure of a two-point Boolean ring $\mathbb{Z}/2\mathbb{Z}$. This allows one to enrich the structure of the corresponding Hom-sets $\mathrm{Hom}(X, D)$ and $\mathrm{Hom}(B, \mathbb{Z}/2\mathbb{Z})$, respectively, from a set to a Boolean ring or Stone space, respectively: making use of the identification $D \equiv \mathbb{Z}/2\mathbb{Z}$, by the obvious inclusions $\mathrm{Hom}(X, D) \subseteq \mathbb{Z}/2\mathbb{Z}^X$ and $\mathrm{Hom}(B, \mathbb{Z}/2\mathbb{Z}) \subseteq D^B$, we get that $\mathrm{Hom}(X, D)$ is a Boolean ring (as products and subrings of Boolean rings are Boolean) and that $\mathrm{Hom}(B, \mathbb{Z}/2\mathbb{Z})$ is a Stone space (since it results to be a closed subspace of the zero-dimensional compact Hausdorff space D^B). Due to this double-faced appearance of this object $D \equiv \mathbb{Z}/2\mathbb{Z}$, it was very appropriately named *schizophrenic object* by Peter Johnstone.

C.3 Exercises

Exercise C.3.1. Determine the terminal and initial objects in the categories **Set**, $\mathbf{Vect}_K$, **Grp**, **AbGrp**, and **Top**.

Hint. In $\mathbf{Vect}_K$, **Grp**, and **AbGrp** the initial and terminal object coincide with the trivial group (vector space). In **Set** and **Top** the initial object is the empty-set (space), while the terminal object, denoted by T in the sequel, is the singleton set (in **Top** equipped with the unique topology on T).

Exercise C.3.2. Build forgetful functors $\mathbf{Vect}_K \to \mathbf{AbGrp}$, $\mathbf{Rng} \to \mathbf{AbGrp}$.

Exercise C.3.3. Show that:
(a) **AbGrp** and $\mathbf{AbGrp_m}$, for $m \in \mathbb{N}_+$, are all varieties of abelian groups;
(b) for every $m \in \mathbb{N}_+$, the nilpotent groups of class $\leq m$ form a variety.

Exercise C.3.4. For a natural number $m > 1$, the variety of groups $\mathcal{B}_m$ satisfying the identity $x^m = 1$ is called a *Burnside variety*. Show that $\mathcal{B}_m$ is an epireflective subcategory of **Grp** and build an explicit epireflection $\mathsf{B}_m\colon \mathbf{Grp} \to \mathcal{B}_m$.

Exercise C.3.5. Show that the nilpotent groups of class 2 form a variety of groups $\mathcal{N}_2$ and build an explicit epireflection $\mathbf{Grp} \to \mathcal{N}_2$.

Exercise C.3.6. Show that the metabelian groups form a variety $\mathcal{R}_2$ and build an explicit epireflection R: **Grp** $\to \mathcal{R}_2$.

Is the group S_4 metabelian? Compute RS_4. What about S_5, or more generally, S_n for $n > 4$?

Exercise C.3.7. (a) Is the subcategory $\mathcal{N}$ of all nilpotent groups reflexive in **Grp**? Is it a variety?
(b) Is the subcategory $\mathcal{R}$ of all solvable groups reflexive in **Grp**? Is it a variety?
(c) Is the subcategory $\mathcal{R}_{\mathrm{fin}}$ of all residually finite groups a reflective subcategory in **Grp**? If yes, then build an explicit reflection. Is $\mathcal{R}_{\mathrm{fin}}$ a variety?

Exercise C.3.8. Let $\mathbf{r}$ be a preradical in **AbGrp**. Show that for every $G \in$ **AbGrp** the subgroup $\mathbf{r}G$ is fully invariant (i. e., $f(\mathbf{r}G) \subseteq \mathbf{r}G$ for every endomorphism $f: G \to G$).

Exercise C.3.9. Introduce a partial order in the class PR(**AbGrp**) of all preradicals in **AbGrp** by letting $\mathbf{r} \le \mathbf{s}$ for $\mathbf{r}, \mathbf{s} \in$ PR(**AbGrp**) whenever $\mathbf{r}G \le \mathbf{s}G$ for every abelian group G.
(a) Show that PR(**AbGrp**) is a large complete lattice where, for a family $\{\mathbf{r}_i : i \in I\}$ of preradicals in **ABGrp**, $\bigvee_{i\in I} \mathbf{r}_i$ and $\bigwedge_{i\in I} \mathbf{r}_i$ are defined by $(\bigvee_{i\in I} \mathbf{r}_i)G = \sum_{i\in I} \mathbf{r}_i(G)$ and $(\bigwedge_{i\in I} \mathbf{r}_i)G = \bigcap_{i\in I} \mathbf{r}_i G$ for every abelian group G.
(b) Describe the top and bottom elements **1** and **0** of PR(**AbGrp**).
(c) Show that the subclass RAD(**AbGrp**) of PR(**AbGrp**) consisting of all radicals is stable under meet and deduce that every preradical $\mathbf{r}$ admits a smallest radical $\mathbf{r}_\infty$ larger than $\mathbf{r}$ (called, *radical hull* of $\mathbf{r}$).
(d) Show that the subclass IPR(**AbGrp**) of *PR*(**AbGrp**) consisting of all idempotent preradicals is stable under join and deduce that every preradical $\mathbf{r}$ admits a largest preradical $\mathbf{r}^\infty$ smaller than $\mathbf{r}$ (called, *idempotent core* of $\mathbf{r}$).

Exercise C.3.10. For preradicals $\mathbf{r}, \mathbf{s}$ in **AbGrp**, define the composite $(\mathbf{r} : \mathbf{s})$ by letting, for every abelian group G, $(\mathbf{r}: \mathbf{s})G = q^{-1}(\mathbf{r}(G/\mathbf{s}G))$, where $q: G \to G/\mathbf{s}G$ is the canonical projection.

Furthermore, for every ordinal α define $\mathbf{r}_\alpha$ by $(\mathbf{r}: \mathbf{r}_\beta)$, provided $\alpha = \beta + 1$ is a successor, otherwise let $\mathbf{r}_\alpha G = \bigcup_{\beta<\alpha} \mathbf{r}_\beta G$. Let $\hat{\mathbf{r}} = \bigvee_\alpha \mathbf{r}_\alpha$.
(a) Show that $(\mathbf{r}: \mathbf{s})$ and $\mathbf{r}_\alpha$, for every ordinal α, are preradicals.
(b) Show that $\hat{\mathbf{r}}$ is a radical and $\hat{\mathbf{r}} = \mathbf{r}_\infty$.
(c) Show that $\mathbf{r}$ is a radical if and only if $(\mathbf{r}: \mathbf{r}) = \mathbf{r}$ if and only if $\mathbf{r} = \hat{\mathbf{r}}$.
(d) Let $\mathbf{r}$ be one of the preradicals in **AbGrp** given by $\mathbf{r}G = \mathrm{Soc}(G)$ or $\mathbf{r}G = G[p]$, p a prime, for every abelian group G. Compute $\hat{\mathbf{r}}$.

Exercise C.3.11. For preradicals $\mathbf{r}, \mathbf{s}$ in **AbGrp**, define the composition $\mathbf{r} \cdot \mathbf{s}$ by $(\mathbf{r} \cdot \mathbf{s})G = \mathbf{r}(\mathbf{s}G)$ for $G \in$ **AbGrp**. Furthermore, for every ordinal α define $\mathbf{r}^\alpha$ by $\mathbf{r}\mathbf{r}^\beta$, if $\alpha = \beta + 1$ is a successor, otherwise let $\mathbf{r}^\alpha G = \bigcap_{\beta<\alpha} \mathbf{r}^\beta G$. Let $\check{\mathbf{r}} = \bigwedge_\alpha \mathbf{r}^\alpha$.

(a) Show that $\check{\mathbf{r}}$ is an idempotent preradical and $\check{\mathbf{r}} = \mathbf{r}^\infty$, so $\mathbf{r}$ is idempotent if and only if $\mathbf{r} = \check{\mathbf{r}}$ if and only if $\mathbf{r}^2 = \mathbf{r}$.
(b) Compute $\check{\mathbf{r}}$ for the preradicals $\mathbf{r}$ given by $\mathbf{r}G = G^1$ or $\mathbf{r}G = pG$, p a prime, for every abelian group G.
(c) For an abelian p-group G define by transfinite induction $p^\alpha G$ by letting $p^0 G = G$, $p^\alpha G = p(p^\beta G)$, when $\alpha = \beta + 1$ and $p^\alpha G = \bigcap_{\beta<\alpha} p^\beta G$ if α is a limit ordinal. Show that $p^\omega G = G^1$. Moreover, $\bigcap_\alpha p^\alpha G$ coincides with $\check{\mathbf{r}}G$, where the preradical $\mathbf{r}$ is defined by $\mathbf{r}H = pH$ for every abelian group H, as well as with $\check{\mathbf{s}}G$, where $\mathbf{s}H = H^1$ for every abelian group H.
(d) Show that the abelian group G in Example A.4.7 satisfies $p^\omega G \neq \{0\}$, but $p^{\omega+1} G = \{0\}$.

Exercise C.3.12. Show that for an abelian group G one has the equality $\mathrm{Fratt}(G) = \bigcap_{p\in\mathbb{P}} pG$ and deduce that Fratt is a radical. Is this radical idempotent? Compute its idempotent core.

Exercise C.3.13. Let A be an abelian group.
(a) For G in **AbGrp**, let $\mathbf{r}_A(G) = \bigcap\{\ker f : f \in \mathrm{Hom}(G, A)\}$.
  (a_1) Show that $\mathbf{r}_A$ is a radical.
  (a_2) If $\mathcal{A} = \{A_i : i \in I\}$ is a *strongly rigid system* in **AbGrp** (i. e., $\mathrm{Hom}(A_i, A_j) = \{0\}$ whenever $i \neq j$ in I and $\mathrm{End}(A_i) \cong \mathbb{Z}$ for all $i \in I$), then all radicals $\mathbf{r}_{A_i}$ are pairwise distinct.
(b) For G in **AbGrp**, let $\mathbf{r}^A(G) = \mathrm{Tr}_A(G) = \sum\{\mathrm{im} f : f \in \mathrm{Hom}(A, G)\}$.
  (b_1) Show that $\mathbf{r}^A$ is an idempotent preradical.
  (b_2) If $\mathcal{A} = \{A_i : i \in I\}$ is a strongly rigid system in **AbGrp**, then all preradicals $\mathbf{r}^{A_i}$ are pairwise distinct.

Deduce that both RAD(**AbGrp**) and IPR(**AbGrp**) are proper classes.

Hint. For (a_2), note that when $i \neq j$ in I, then $\mathbf{r}_{A_i}(A_j) = A_j$, while $\mathbf{r}_{A_j}(A_j) = \{0\}$. Hence, the group A_j witnesses the inequality $\mathbf{r}_{A_i} \neq \mathbf{r}_{A_j}$. Argue similarly for ($b_2$).

For the final assertion, use the fact that Shelah proved that there exist strongly rigid systems $\mathcal{A} = \{A_i : i \in I\}$ of torsion-free abelian groups of arbitrarily large size $|I|$. This answers the last question taking into account both (a_2) and (b_2).

Exercise C.3.14. For a topological space (X, τ), consider the collection of all sequentially closed sets of X and show that they form the family of all closed sets of a topology τ_s on X finer than τ.

Show that if (Y, τ') is a topological space and $f : (X, \tau) \to (Y, \tau')$ is a continuous map, then also $f : (X, \tau_s) \to (Y, \tau'_s)$ is continuous. Conclude that the assignment $(X, \tau) \mapsto (X, \tau_s)$ defines a coreflection $\mathrm{C} : \mathbf{Top} \to \mathcal{Y}$, where $\mathcal{Y}$ is the full subcategory of **Top** consisting of sequential spaces with natural transformation y given by the identity map $y_{(X,\tau)} = id_X : (X, \tau_s) \to (X, \tau)$.

Exercise C.3.15. Show that the full subcategory **Alex** of Alexandroff spaces in **Top** is coreflective and exhibit a coreflector.

Exercise C.3.16. Prove the uniqueness claimed in Remark C.2.6.

Hint. In the notation of Remark C.2.6, applied to $f = \gamma_X : X \to RX$ (so, $C = RX$), there exists a unique morphism $t : K_X \to RX$ with $t \circ h_X = \gamma_X$. Moreover, there exists a unique morphism $s : RX \to K_X$ such that $s \circ \gamma_X = h_X$, by the properties of the $\mathcal{Y}$-reflection. Then $id_{RX} \circ \gamma_X = \gamma_X = t \circ h_X = (t \circ s) \circ \gamma_X$. So, by the uniqueness property applied to the morphism $\gamma_X : X \to RX$ composed with the morphisms id_{RX} and $t \circ s$, we deduce that $id_{RX} = t \circ s$. Similarly, $id_{K_X} = s \circ t$. So, s is the desired isomorphism.

Exercise C.3.17. Show that the subcategory of **Top** consisting of only the empty space is coreflective in **Top**.

Assume that $\mathcal{Y}$ is a coreflective subcategory of **Top**. Show that:

(a) if $\mathcal{Y}$ contains a nonempty space, then $T \in \mathcal{Y}$;
(b) $\mathcal{Y}$ is stable under taking coproducts;
(c) $\mathcal{Y}$ contains all discrete spaces;
(d) $\mathcal{Y}$ is stable under taking quotients;
(e) $\mathcal{Y}$ is bicorefletive, i. e., for every $X \in$ **Top** the coreflection map $\gamma_X : CX \to X$ is bijective.

Hint. (b) If $C : \mathbf{Top} \to \mathcal{Y}$ is the coreflection, and $\{Y_i : i \in I\}$ is a family in $\mathcal{Y}$, consider the coproduct $Y = \coprod_{i \in I} Y_i$ and let $Z = C(\coprod_{i \in I} Y_i)$ with the family of morphisms $\{Y_i \to Z : i \in I\}$ obtained from the coreflector. Show that it has the property of the coproduct and conclude, by the uniqueness of the coproduct, that they coincide, i. e., $\gamma_Y : Z = CY \to Y$ is an isomorphism. Therefore, $Y \in \mathcal{Y}$.

(c) follows from (a) and (b).

(d) Argue similarly with the quotient $q : Y \to Z$ of some $Y \in \mathcal{Y}$, to show that $\gamma_Z : CZ \to Z$ is an isomorphism.

Exercise C.3.18. Prove that a subcategory $\mathcal{Y}$ of **Top** is coreflective if and only if $\mathcal{Y}$ is stable under taking coproducts and quotients.

Hint. For the necessity, use Exercise C.3.17. Assume now that $\mathcal{Y}$ is stable under taking coproducts and quotients. For $X \in$ **Top**, consider a set of continuous injective maps $n_i : Y_i \to X$, $i \in I$, with $Y_i \in \mathcal{Y}$, such that for every continuous injective map $n : Y \to X$ with $Y \in \mathcal{Y}$, there exists a homeomorphism $\xi : Y \to Y_i$ such that $n = n_i \circ \xi$. Let $S = \coprod_{i \in I} Y_i$, let $v_i : Y_i \to S$ be the related embeddings, and let $n : S \to X$ be the map ensured by the coproduct property. Since the space X is covered by the images $n_i(Y_i)$, the map n is surjective. Put on X the quotient topology τ_q of the map $n : S \to X$. Since n was continuous, this topology is finer that the original topology of X. Denote by CX the space (X, τ_q) so obtained, by $l : S \to CX$ the quotient map, and by $\gamma_X : CX \to X$ the identity map. Since CX is a quotient of $S \in \mathcal{Y}$, we deduce that $CX \in \mathcal{Y}$. Let $Y \in \mathcal{Y}$ and $f : Y \to X$ be a continuous map. Factorize f through the quotient $q : Y \to Y'$, where $Y' = f(Y)$, but which carries the quotient topology. Then $Y' \in \mathcal{Y}$ and for the inclusion $j : Y' \to X$ there exists $i \in I$ and a homeomorphism $\eta : Y' \to Y_i$ such that $j = n_i \circ \eta$. Now $q' = l \circ v_i \circ \eta \circ q : Y \to CX$ witnesses the coreflection property.

Exercise C.3.19. For $\mathcal{X}$ being any of the categories **Top**, **Grp**, or **AbGrp**, show that a subcategory $\mathcal{Y}$ of $\mathcal{X}$ is epireflective if and only if $\mathcal{Y}$ is stable under taking products and subobjects (in the case of **Top** consider as subobjects the subspaces).

Hint. For **Top** apply an argument dual to that of the hint to Exercise C.3.18 and apply Theorem C.2.9 for the sufficiency. For **Grp** and **AbGrp** argue similarly or use Theorem C.2.17.

Exercise C.3.20. Denote by **Met** the subcategory of **Top** having as objects the topological spaces with a metrizable topology and as morphisms all uniformly continuous maps. Show that **Met** is a nonfull subcategory of **Top**.

Hint. Use a continuous not uniformly continuous map between metric spaces.

Exercise C.3.21. (a) Show that taking the class of all posets $(X, \leq)$ as objects and the class of all monotone (i. e., order preserving) maps $(X, \leq) \to (X, \leq)$ as morphism one obtains a category that we shall denote by **PoSet**.

(b) For a T_0-space (X, τ), consider the specialization order $\preceq_\tau$ on X and show that if $f: (X, \tau) \to (Y, \tau')$ is continuous, then the map $f: (X, \preceq_\tau) \to (Y, \preceq_{\tau'})$ is order preserving. Deduce that this gives a covariant functor $\mathsf{O}: \mathbf{Top_0} \to \mathbf{PoSet}$.

(c) Show that assigning to every $(X, \leq) \in \mathbf{PoSet}$ the Alexandroff–Tucker topological space (X, τ_{AT}) we obtain a functor $\mathsf{AT}: \mathbf{PoSet} \to \mathbf{Alex_0} = \mathbf{Top_0} \cap \mathbf{Alex}$.

(d) Let O' be the restriction of O to the subcategory **Alex**. Show that the pair of functors $\mathsf{AT}: \mathbf{PoSet} \to \mathbf{Alex_0}$ and $\mathsf{O}': \mathbf{Alex_0} \to \mathbf{PoSet}$ give an equivalence between the categories $\mathbf{Alex_0}$ and **PoSet**.

Bibliography

[1] J. Aaronson, M. Nadkarni, L_∞ *eigenvalues and* L_2 *spectra of no-singular transformations*, Proc. London Math. Soc. 55 (3) (1987), 538–570.

[2] P. N. Áhn, *Duality of modules over topological rings*, J. Algebra 75 (1982), no. 2, 395–425.

[3] M. Ajtai, I. Havas, J. Komlós, *Every group admits a bad topology*, Studies in Pure Mathematics, 21–34, Birkhäuser, Basel, 1983.

[4] H. Anzai, S. Kakutani, *Bohr compactifications of a locally compact Abelian group. I*, Proc. Imp. Acad. Tokyo 19 (1943), 476–480.

[5] S. Ardanza Trevijano, M. J. Chasco, X. Domínguez, M. Tkachenko, *Precompact noncompact reflexive Abelian groups*, Forum Math. 24 (2012), no. 2, 289–302.

[6] A. Arhangel′skiĭ, *Mappings connected with topological groups* (Russian), Dokl. Akad. Nauk SSSR 181 (1968), 1303–1306.

[7] A. Arhangel′skiĭ, M. Tkachenko, *Topological groups and related structures*, Atlantis Studies in Mathematics, 1. Atlantis Press/World Scientific Publishing Co. Pte. Ltd., Paris/Hackensack, NJ, 2008.

[8] D. L. Armacost, *The structure of locally compact abelian groups*, Monographs and Textbooks in Pure and Applied Mathematics, 68. Marcel Dekker, Inc., New York, 1981.

[9] D. L. Armacost, W. L. Armacost, *Uniqueness in structure theorems for LCA groups*, Canad. J. Math. 30 (1978), 593–599.

[10] G. Artico, V. Malykhin, U. Marconi, *Some large and small sets in topological groups*, Math. Pannon. 12 (2001), no. 2, 157–165.

[11] L. Außenhofer, *Contributions to the duality theory of abelian topological groups and to the theory of nuclear groups*, Dissertationes Math. 384 (1999), 113 pp.

[12] L. Außenhofer, *On the non-existence of the Mackey topology for locally quasi-convex groups*, Forum Math. 30 (2018), no. 5, 1119–1127.

[13] L. Außenhofer, *The group* $\mathbb{Z}^{(\mathbb{N})}$ *does not have a Mackey topology*, Topology Appl. 281 (2020), 107188.

[14] L. Außenhofer, D. Dikranjan, *Mackey groups and Mackey topologies*, Dissertationes Math., to appear.

[15] T. Banakh, *A quantitative generalization of Prodanov–Stoyanov theorem on minimal Abelian topological groups*, Topology Appl. 271 (2020), 106983.

[16] T. Banakh, I. Guran, I. Protasov, *Algebraically determined topologies on permutation groups*, Topology Appl. 159 (2012), no. 9, 2258–2268.

[17] T. Banakh, A. Ravsky, *Each regular paratopological group is completely regular*, Proc. Amer. Math. Soc. 145 (2017), no. 1, 1373–1382.

[18] W. Banaszczyk, *Additive subgroups of topological vector spaces*, Lecture Notes in Mathematics, 1466. Springer-Verlag, Berlin, 1991.

[19] W. Banaszczyk, E. Martín Peinador, *The Glicksberg theorem on weakly compact sets for nuclear groups*, Ann. New York Acad. Sci. 788 (1996), 34–39.

[20] G. Barbieri, D. Dikranjan, A. Giordano Bruno, H. Weber, *Dirichlet sets vs characterized subgroups*, Topology Appl. 231 (2017), 50–76.

[21] G. Barbieri, D. Dikranjan, C. Milan, H. Weber, *Answer to Raczkowski's questions on convergent sequences of integers*, Topology Appl. 132 (2003), no. 1, 89–101.

[22] G. Barbieri, D. Dikranjan, C. Milan, H. Weber, *Convergent sequences in precompact group topologies*, Appl. Gen. Topol. 6 (2005), no. 2, 149–169.

[23] G. Barbieri, D. Dikranjan, C. Milan, H. Weber, *Topological torsion related to some recursive sequences of integers*, Math. Nachr. 281 (2008), no. 7, 930–950.

[24] G. Barbieri, A. Giordano Bruno, H. Weber, *Inclusions of characterized subgroups*, Topology Appl. 221 (2017), 534–555.

https://doi.org/10.1515/9783110654936-020

[25] H. Bauer, G. Nöbeling, *Allgemeine Approximationskriterien mit Anwendungen* (German), Jber. Deutsch. Math.-Verein. 58 (1955), Abt. 1, 54–72.
[26] G. Baumslag, A. Myasnikov, V. Remeslennikov, *Algebraic geometry over groups. I. Algebraic sets and ideal theory*, J. Algebra 219 (1999), no. 1, 16–79.
[27] B. Bazigaran, D. Dikranjan, H. Khass, *Atomic topological groups*, Adv. Group Theory Appl. 6 (2018), 23–53.
[28] M. Beiglböck, C. Steineder, R. Winkler, *Sequences and filters of characters characterizing subgroups of compact abelian groups*, Topology Appl. 153 (2006), no. 11, 1682–1695.
[29] A. Bella, V. Malykhin, *On certain subsets of a group*, Questions Answers Gen. Topology 17 (1999), no. 2, 183–197.
[30] G. Birkhoff, *Lattice Theory*, American Math. Society Colloquium Publications, Vol. XXV. American Math. Society, Providence, R. I., 1967.
[31] A. Biró, *Characterizations of groups generated by Kronecker sets*, J. Théor. Nombres Bordeaux 19 (2007), no. 3, 567–582.
[32] A. Bíró, J.-M. Deshouillers, V. Sós, *Good approximation and characterization of subgroups of* $\mathbb{R}/\mathbb{Z}$, Studia Sci. Math. Hungar. 38 (2001), 97–113.
[33] S. Bochner, *Beiträge zur Theorie der fastperiodischen Funktionen* (German), Math. Ann. 96 (1927), 119–147.
[34] P. Bohl, *Über die Darstellung von Funktionen einer Variablen durch trigonometrische Reihen mit mehreren einer Variablen proportionalen Argumenten* (German), Dorpat, 1893.
[35] H. Bohr, *Sur les fonctions presque périodiques* (French), Acad. Sci. Paris 177 (1923), 737–739.
[36] J. P. Borel, *Sur certains sous-groupes de* $\mathbb{R}$ *liés à la suite des factorielles* (French), Colloq. Math. 62 (1991), no. 1, 21–30.
[37] K. Bose, P. Das, W. He, *Generating subgroups of the circle using statistical convergence of order* α, Acta Math. Hungar. 162 (2020), no. 2, 633–646.
[38] N. Bourbaki, *Elements of Mathematics. General topology. Part 1.* Hermann/Addison-Wesley Publishing Co., Paris/Reading, Mass.–London–Don Mills, Ont., 1966.
[39] A. Bouziad, *Every Čech-analytic Baire semitopological group is a topological group*, Proc. Amer. Math. Soc. 124 (1996), no. 3, 953–959.
[40] D. Boyer, A. Mader, *Functorial topologies on abelian groups*, Rocky Mountain J. Math. 10 (1980), no. 4, 695–708.
[41] J. Braconnier, *Sur les groupes topologiques localement compacts* (French), Math. Pures Appl. (9) 27 (1948), 1–85.
[42] R. Bryant, *The verbal topology of a group*, J. Algebra 48 (1977), 340–346.
[43] B. Charles, *Méthodes topologiques en théorie des groupes abéliens* (French), In: 1964 Proc. Colloq. Abelian Groups (Tihany, 1963), 29–42. Akadémiai Kiadó, Budapest.
[44] M. J. Chasco, *Pontryagin duality for metrizable groups*, Arch. Math. (Basel) 70 (1998), no. 1, 22–28.
[45] M. J. Chasco, D. Dikranjan, E. Martín Peinador, *A survey on reflexivity of abelian topological groups*, Topology Appl. 159 (2012), no. 9, 2290–2309.
[46] M. J. Chasco, E. Martín Peinador, V. Tarieladze, *On Mackey topology for groups*, Stud. Math. 132 (1999), no. 3, 257–284.
[47] H. Chu, *Compactification and duality of topological groups*, Trans. Amer. Math. Soc. 123 (1966), 310–324.
[48] B. Clark, V. Schneider, *The extending topologies*, Internat. J. Math. Math. Sci. 7 (1984), no. 3, 621–623.
[49] B. Clark, V. Schneider, *The normal extensions of subgroup topologies*, Proc. Amer. Math. Soc. 97 (1986), no. 1, 163–166.
[50] B. Clark, V. Schneider, *Atoms and antiatoms in the lattice of group topologies*, Arch. Math. (Basel) 52 (1989), no. 2, 173–174.

[51] J. Cleary, S. Morris, *Topologies on locally compact groups*, Bull. Austral. Math. Soc. 38 (1988), 105–111.

[52] M. Clementino, W. Tholen, *Tychonoff's theorem in a category*, Proc. Amer. Math. Soc. 124 (1996), no. 11, 3311–3314.

[53] W. Comfort, S. Hernández, F. Javier Trigos-Arrieta, *Cross sections and homeomorphism classes of abelian groups equipped with the Bohr topology*, Topology Appl. 115 (2001), no. 2, 215–233.

[54] W. Comfort, *Problems on topological groups and other homogeneous spaces*, In: Open Problems in Topology, J. van Mill and G. M. Reed, editors, 313–347. North-Holland, Amsterdam, 1990.

[55] W. Comfort, F. Gould, *Some classes of minimally almost periodic topological groups*, Appl. Gen. Topol. 16 (2015), 141–165.

[56] W. Comfort, S. Hernández, D. Remus, F. Javier Trigos-Arrieta, *Some open questions on topological groups*, In: Nuclear Groups and Lie Groups (Madrid, 1999), 57–76, Res. Exp. Math., 24. Heldermann, Lemgo, 2001.

[57] W. Comfort, F. Javier Trigos-Arrieta, Ta Sun Wu, *The Bohr compactification, modulo a metrizable subgroup*, Fund. Math. 143 (1993), 119–136.

[58] W. W. Comfort, K. A. Robertson, *Proper pseudocompact extensions of compact abelian group topologies*, Proc. Amer. Math. Soc. 86 (1982), 173–178.

[59] W. W. Comfort, K. A. Robertson, *Extremal phenomena in certain classes of totally bounded groups*, Dissertationes Math. 272 (1988), 48 pp.

[60] W. W. Comfort, K. A. Ross, *Topologies induced by groups of characters*, Fund. Math. 55 (1964), 283–291.

[61] W. W. Comfort, K. A. Ross, *Pseudocompactness and uniform continuity in topological groups*, Pacific J. Math. 16 (1966), 483–496.

[62] W. W. Comfort, J. van Mill, *Extremal pseudocompact abelian groups are compact metrizable* Proc. Amer. Math. Soc. 135 (2007), no. 12, 4039–4044.

[63] M. Cotlar, R. Ricabarra, *On the existence of characters in topological groups*, Amer. J. Math. 76 (1954), 375–388.

[64] M. Day, *The spaces L_p with* $0 < p < 1$, Bull. Amer. Math. Soc. 46 (1940), 816–823.

[65] L. De Leo, D. Dikranjan, *Straightening theorem for bounded abelian groups*, Topology Appl. 155 (2008), no. 17–18, 2158–2176.

[66] R. Di Santo, D. Dikranjan, *Answer to Armacost's quest on topologically torsion elements of the circle group*, Comm. Algebra 32 (2004), 133–146.

[67] R. Di Santo, D. Dikranjan, *A characterization of the circle group via uniqueness of roots*, Topology Appl. 158 (2011), no. 2, 159–166.

[68] R. Di Santo, D. Dikranjan, A. Giordano Bruno, *Characterized subgroup of the circle group*, Ric. Mat. 67 (2018), no. 2, 625–655.

[69] S. Dierolf, U. Schwanengel, *Un exemple d'un groupe topologique $\mathcal{Q}$-minimal mais non précompact* (French), Bull. Sci. Math. (2) 101 (1977), no. 3, 265–269.

[70] S. Dierolf, S. Warken, *Some examples in connection with Pontryagin's duality theorem*, Arch. Math. 30 (1978), 599–605.

[71] D. Dikranjan, *Minimal ring topologies and Krull dimension*, Colloq. Math. Soc. Janos Bolyai, 23, 357–366. Topology, Budapest (Hungary) (1978).

[72] D. Dikranjan, *Minimal topological rings*, Serdica 8 (1982), 139–165.

[73] D. Dikranjan, *Sur la minimalité des produits de groupes topologiques abéliens* (French), C. R. Acad. Sci. Paris Sér. I Math. 299 (1984), no. 8, 303–306.

[74] D. Dikranjan, *Disconnectedness and dimension in pseudocompact groups*, Comp. Rend. Acad. Sci. Paris, Série I 316 (1993), 309–314.

[75] D. Dikranjan, *Zero-dimensionality of some pseudocompact groups*, Proc. Amer. Math. Soc. 120 (1994), no. 4, 1299–1308.

[76] D. Dikranjan, *Uniqueness of dualities*, In: Abelian Groups and Modules (Padova, 1994), 123–133, Math. Appl., 343. Kluwer Acad. Publ., Dordrecht, 1995.

[77] D. Dikranjan, *Recent advances in minimal topological groups*, Topology Appl. 85 (1998), no. 1–3, 53–91.

[78] D. Dikranjan, *Compactness and connectedness in topological groups*, Topology Appl. 84 (1998), no. 1–3, 227–252.

[79] D. Dikranjan, *Quotients of zero-dimensional precompact abelian groups, Special issue on topological groups.* Topology Appl. 86 (1998), no. 1, 47–62.

[80] D. Dikranjan, *Countably compact groups satisfying the open mapping theorem*, Topology Appl. 98 (1999), no. 1–3, 81–129.

[81] D. Dikranjan, *Topologically torsion elements of topological groups*, Topology Proc. 26 (2001/02), no. 2, 505–532.

[82] D. Dikranjan, *Closure operators in topological groups related to von Neumann's kernel*, Topology Appl. 153 (2006), no. 11, 1930–1955.

[83] D. Dikranjan, P. Das, K. Bose, *Statistically characterized subgroups of the circle*, Fund. Math. 249 (2020), no. 2, 185–209.

[84] D. Dikranjan, M. V. Ferrer, S. Hernández, *Dualities in topological groups*, Sci. Math. Jpn. 72 (2010), no. 2, 197–235.

[85] D. Dikranjan, A. Giordano Bruno, *w-Divisible groups*, Topology Appl. 155 (2008), no. 4, 252–272.

[86] D. Dikranjan, A. Giordano Bruno, *Functorial topologies*, Topology Appl. 158 (2011), no. 17, 2391–2407.

[87] D. Dikranjan, A. Giordano Bruno, D. Impieri, *Characterized subgroups of topological abelian groups*, Axioms 4 (2015), 459–491.

[88] D. Dikranjan, A. Giordano Bruno, C. Milan, *Weakly metrizable pseudocompact groups*, Appl. Gen. Topol. 7 (2006), 1–39.

[89] D. Dikranjan, D. Impieri, *Topologically torsion elements of the circle group*, Comm. Algebra 42 (2014), 600–614.

[90] D. Dikranjan, D. Impieri, *On the Borel complexity of characterized subgroups*, Topology Appl. 201 (2016), 372–387.

[91] D. Dikranjan, K. Kunen, *Characterizing countable subgroups of compact abelian groups*, J. Pure Appl. Algebra 208 (2007), no. 1, 285–291.

[92] D. Dikranjan, M. Megrelishvili, *Relative minimality and co-minimality of subgroups of topological groups*, Topology Appl. 157 (2010), no. 1, 62–76.

[93] D. Dikranjan, M. Megrelishvili, *Minimality conditions in topological groups*, In: Recent Progress in General Topology III, 229–327. Atlantis Press, Paris, 2014.

[94] D. Dikranjan, C. Milan, *Dualities of locally compact modules over the rationals*, J. Algebra 256 (2002), no. 2, 433–466.

[95] D. Dikranjan, C. Milan, A. Tonolo, *A characterization of the maximally almost periodic abelian groups*, J. Pure Appl. Algebra 197 (2005), no. 1–3, 23–41.

[96] D. Dikranjan, A. Orsatti, *On an unpublished manuscript of Ivan Prodanov concerning locally compact modules and their dualities*, Comm. Algebra 17 (1989), no. 11, 2739–2771.

[97] D. Dikranjan, Iv. Prodanov, *Totally minimal topological groups*, Annuaire Univ. Sofia Fac. Math. Méc. 69 (1974/75), 5–11 (1979).

[98] D. Dikranjan, Iv. Prodanov, *A class of compact abelian groups*, Annuaire Univ. Sofia, Fac. Math. Méc. 70 (1975/76), 191–206 (1981).

[99] D. Dikranjan, Iv. Prodanov, L. Stoyanov, *Topological groups. Characters, dualities and minimal group topologies*, Monographs and Textbooks in Pure and Applied Mathematics, 130. Marcel Dekker, Inc., New York, 1990.

[100] D. Dikranjan, I. Protasov, *Every infinite group can be generated by a P-small subset*, Appl. Gen. Topol. 7 (2006), no. 2, 265–268.

[101] D. Dikranjan, D. Shakhmatov, *Algebraic structure of pseudocompact groups*, Mem. Amer. Math. Soc., 133, no. 633, 1998.

[102] D. Dikranjan, D. Shakhmatov, *Selected topics from the structure theory of topological groups*, In: Open problems in Topology II, 389–406. E. Pearl, editor. Elsevier 2007.

[103] D. Dikranjan, D. Shakhmatov, *Reflection principle characterizing groups in which unconditionally closed sets are algebraic*, J. Group Theory 11 (2008), no. 3, 421–442.

[104] D. Dikranjan, D. Shakhmatov, *The Markov–Zariski topology of an abelian group*, J. Algebra 324 (2010), no. 6, 1125–1158.

[105] D. Dikranjan, D. Shakhmatov, *A Kronecker–Weyl theorem for subsets of abelian groups*, Adv. Math. 226 (2011), no. 6, 4776–4795.

[106] D. Dikranjan, D. Shakhmatov, *Final solution of Protasov–Comfort's problem on minimally almost periodic group topologies*, arXiv:1410.3313, 2014.

[107] D. Dikranjan, D. Shakhmatov, *A complete solution of Markov's problem on connected group topologies*, Adv. Math. 286 (2016), 286–307.

[108] D. Dikranjan, D. Shakhmatov, *Topological groups with many small subgroups*, Topology Appl. 200 (2016), 101–132.

[109] D. Dikranjan, L. Stoyanov, *An elementary approach to Haar integration and Pontryagin duality in locally compact abelian groups*, Topology Appl. 158 (2011), no. 15, 1942–1961.

[110] D. Dikranjan, M. Tkachenko, *Weak completeness of free topological groups*, Topology Appl. 112 (2001), no. 3, 259–287.

[111] D. Dikranjan, M. Tkachenko, *Sequentially complete groups: dimension and minimality*, J. Pure Appl. Algebra 157 (2001), 215–239.

[112] D. Dikranjan, M. Tkachenko, I. Yaschenko, *On transversal group topologies*, Topology Appl. 153 (2005), 786–817.

[113] D. Dikranjan, D. Toller, *The Markov and Zariski topologies of some linear groups*, Topology Appl. 159 (2012), no. 13, 2951–2972.

[114] D. Dikranjan, D. Toller, *Zariski topology and Markov topology on groups*, Topology Appl. 241 (2018), 115–144.

[115] D. Dikranjan, V. Uspenskij, *Categorically compact topological groups*, J. Pure Appl. Algebra 126 (1998), no. 1–3, 149–168.

[116] D. Dikranjan, S. Watson, *A solution to van Douwen's problem on Bohr topologies*, J. Pure Appl. Algebra 163 (2001), no. 2, 147–158.

[117] D. Dikranjan, U. Zannier, *On a property of the adele ring of the rationals*, Rings, modules, algebras, and abelian groups, 133–137, Lecture Notes in Pure and Appl. Math., 236. Dekker, New York, 2004.

[118] G. Dimov, D. Vakarelov, *On the investigations of Ivan Prodanov in the theory of abstract spectra*, God. Sofiĭ. Univ. "Sv. Kliment Okhridski" Fac. Mat. Inform. 102 (2015), 31–70.

[119] G. Dimov, D. Vakarelov, *Topological representation of precontact algebras and a connected version of the Stone duality theorem – I*, Topology Appl. 227 (2017), 64–101.

[120] G. D. Dimov, *An axiomatic characterization of Stone duality*, Serdica 10 (1984), no. 2, 165–173.

[121] G. D. Dimov, W. Tholen, *A characterization of representable dualities*, In: Categorical Topology and Its Relation to Analysis, Algebra and Combinatorics (Prague, 1988), 336–357. World Sci. Publ., Teaneck, NJ, 1989.

[122] G. D. Dimov, W. Tholen, *Group of dualities*, Trans. Amer. Math. Soc. 336 (1993), no. 2, 901–913.

[123] J. Dixmier, *Quelques propriétés des groupes abéliens localement compacts* (French), Bull. Sci. Math. 81 (1957), 38–48.

[124] J. Dixmier, *C^*-algebra*, North-Holland Publ. Company, New York, 1977.

[125] J. Dixon, P. Neumann, S. Thomas, *Subgroups of small index in infinite symmetric groups*, Bull. London Math. Soc. 18 (1986), no. 6, 580–586.

[126] D. Doïtchinov, *Produits de groupes topologiques minimaux* (French), Bull. Sci. Math. (2) 96 (1972), 59–64.

[127] H. G. Eggleston, *Sets of fractional dimensions which occur in some problems of number theory*, Proc. London Math. Soc. (2) 54 (1952), 42–93.

[128] P. Eklof, A. Mekler, *Almost free modules: Set theoretic methods*, North Holland, Amsterdam, 1990.

[129] P. Eliaš, *On inclusions between Arbault sets*, Acta Univ. Carol. Math. Phys. 44 (2003), 65–72.

[130] P. Eliaš, *Dirichlet sets, Erdös–Kunen–Mauldin theorem, and analytic subgroups of the reals*, Proc. Amer. Math. Soc. 139, (2011), no. 6, 2093–2104.

[131] R. Ellis, *Locally compact transformation groups*, Duke Math. J. 24 (1957), 119–125.

[132] R. Ellis, *A note on the continuity of the inverse*, Proc. Amer. Math. Soc. 8 (1957), 372–373.

[133] R. Ellis, H. B. Keynes, *Bohr compactifications and a result of Følner*, Israel J. Math. 12 (1972), 314–330.

[134] R. Engelking, *General topology*, Second edition. Sigma Series in Pure Mathematics, 6. Heldermann Verlag, Berlin, 1989.

[135] P. Erdös, K. Kunen, R. D. Mauldin, *Some additive properties of sets of real numbers*, Fund. Math. 113 (1981), 187–199.

[136] E. Følner, *Generalization of a theorem of Bogolioùboff to topological abelian groups. With an appendix on Banach mean values in non-abelian groups*, Math. Scand. 2 (1954), 5–18.

[137] L. Fuchs, *Note on linearly compact abelian groups*, J. Austral. Math. Soc. 9 (1969), 433–440.

[138] L. Fuchs, *Infinite abelian groups. Vol. I*, Pure and Applied Mathematics, 36. Academic Press, New York-London, 1970.

[139] L. Fuchs, *Abelian groups*, Springer Monographs in Mathematics. Springer, Cham, 2015.

[140] L. Fuchs, K.-H. Hofmann, *Extensions of compact abelian groups by discrete ones and their duality theory. I*, J. Algebra 196 (1997), no. 2, 578–594.

[141] L. Fuchs, K.-H. Hofmann, *Extensions of compact abelian groups by discrete ones and their duality theory. II*, Abelian groups, module theory, and topology (Padova, 1997), 205–225, Lecture Notes in Pure and Appl. Math., 201. Dekker, New York, 1998.

[142] H. Fujita, D. Shakhmatov, *Topological groups with dense compactly generated subgroups*, Appl. Gen. Topol. 3 (2002), no. 1, 85–89.

[143] H. Fujita, D. Shakhmatov, *A characterization of compactly generated metric groups*, Proc. Amer. Math. Soc. 131 (2003), no. 3, 953–961.

[144] H. Furstenberg, *On the infinitude of primes*, Amer. Math. Monthly 62 (1955), 353.

[145] S. Gabriyelyan, *Group of quasi-invariance and the Pontryagin duality*, Topology Appl. 157 (2010), no. 18, 2786–2802.

[146] S. Gabriyelyan, *On T-sequences and characterized subgroups*, Topology Appl. 157 (2010), no. 18, 2834–2843.

[147] S. Gabriyelyan, *Characterizable groups: some results and open questions*, Topology Appl. 159 (2012), no. 9, 2378–2391.

[148] S. Gabriyelyan, *Finitely generated subgroups as von Neumann radicals of an Abelian group*, Mat. Stud. 38 (2012), no. 2, 124–138.

[149] S. Gabriyelyan, *Minimally almost periodic group topologies on countably infinite Abelian groups*, Proc. Amer. Math. Soc. 143 (2015), no. 4, 1823–1829.

[150] S. Gabriyelyan, *A locally quasi-convex abelian group without a Mackey group topology*, Proc. Amer. Math. Soc. 146 (2018), no. 8, 3627–3632.

[151] J. Galindo, S. Hernández, *Pontryagin-van Kampen reflexivity for free abelian topological groups*, Forum Math. 11 (1999), no. 4, 399–415.

[152] J. Galindo, S. Hernández, *Interpolation sets and the Bohr topology of locally compact groups*, Adv. Math. 188 (2004), no. 1, 51–68.
[153] J. Galindo, S. Hernández, T. S. Wu, *Recent results and open questions relating Chu duality and Bohr compactifications of locally compact groups*, In: Open problems in Topology II, 439–450, E. Pearl, editor. Elsevier, 2007.
[154] J. Galindo, L. Recoder-Núñez, M. Tkachenko, *Nondiscrete P-groups can be reflexive*, Topology Appl. 158 (2011), no. 2, 194–203.
[155] A. Giordano Bruno, *Extremal α-pseudocompact abelian groups*, Forum Math. 21 (2009), no. 4, 639–659.
[156] B. N. Givens, K. Kunen, *Chromatic numbers and Bohr topologies*, Topology Appl. 131 (2003), no. 2, 189–202.
[157] E. Glasner, *On minimal actions of Polish groups*, Topology Appl. 85 (1998), no. 1–3, 119–125. 8th Prague Topological Symposium on General Topology and Its Relations to Modern Analysis and Algebra (1996).
[158] E. Gregorio, *Dualities over compact rings*, Rend. Sem. Mat. Univ. Padova 80 (1988), 151–174, (1989).
[159] E. Gregorio, *Tori and continuous dualities*, Rend. Accad. Naz. Sci. XL Mem. Mat. (5) 13 (1989), no. 1, 211–221.
[160] E. Gregorio, A. Orsatti, *Uniqueness and existence of dualities over compact rings*, Tsukuba J. Math. 18 (1994), no. 1, 39–61.
[161] M. Gromov, *Groups of polynomial growth and expanding maps*, Publ. Math. IHES 53 (1981), 53–73.
[162] I. I. Guran, *Topological groups similar to Lindelöf groups* (Russian), Dokl. Akad. Nauk SSSR 256 (1981), no. 6, 1305–1307.
[163] P. R. Halmos, *Comment on the real line*, Bull. Amer. Math. Soc. 50 (1944), 877–878.
[164] P. R. Halmos, *Measure theory*, D. Van Nostrand Co., Inc., New York, 1950.
[165] J. E. Hart, K. Kunen, *Bohr compactifications of discrete structures*, Fund. Math. 160 (1999), no. 2, 101–151.
[166] J. E. Hart, K. Kunen, *Limits in function spaces*, Topology Appl. 151 (2005), no. 1–3, 157–168.
[167] J. E. Hart, K. Kunen, *Limits in compact abelian groups*, Topology Appl. 153 (2006), no. 7, 991–1002.
[168] S. Hartman, J. Mycielski, *On the imbedding of topological groups into connected topological groups*, Colloq. Math. 5 (1958), 167–169.
[169] W. Herfort, K.-H. Hofmann, F. G. Russo, *Periodic locally compact groups. A study of a class of totally disconnected topological groups*, De Gruyter Studies in Mathematics, 71. De Gruyter, Berlin, 2019.
[170] S. Hernández, *Pontryagin duality for topological abelian groups*, Math. Z. 238 (2001), no. 3, 493–503.
[171] S. Hernández, *Extension of continuous functions on product spaces, Bohr compactification and almost periodic functions*, Topology Proc. 28 (2004), no. 2, 527–540. Proceedings of the 18th Summer Conference on Topology and its Applications.
[172] S. Hernández, V. Uspenskij, *Pontryagin duality for spaces of continuous functions*, J. Math. Anal. Appl. 242 (2000), no. 2, 135–144.
[173] G. Hesse, *Zur Topologisierbarkeit von Gruppen*, Dissertation, Univ. Hannover, 1979.
[174] E. Hewitt, K. Ross, *Abstract harmonic analysis. Vol. I. Structure of topological groups, integration theory, group representations*, Grundlehren der Mathematischen Wissenschaften, 115. Springer-Verlag, Berlin–New York, 1979.
[175] H. Heyer, *Dualität lokalkompakter Gruppen* (German), Lecture Notes in Mathematics, 150. Springer-Verlag, Berlin–New York, 1970.

[176] H. Heyer, *Groups with Chu duality*, Probability and information theory, II, 181–215. Lecture Notes in Math., 296. Springer, Berlin, 1973.

[177] K.-H. Hofmann, S. Morris, *The structure of compact groups. A primer for the student – a handbook for the expert*, De Gruyter Studies in Mathematics, 25. Walter de Gruyter & Co., Berlin, 1998.

[178] P. Holm, *On the Bohr compactification*, Math. Ann. 156 (1964), 34–46.

[179] M. Hrušák, J. van Mill, U. Ramos-García, S. Shelah, *Countably compact groups without non-trivial convergent sequences*, Trans. Amer. Math. Soc. 374 (2021), no. 2, 1277–1296.

[180] N. Jacobson, *Totally disconnected locally compact rings*, Amer. J. Math. 58 (1936), 433–449.

[181] S. Kaplan, *A zero-dimensional topological group with a one-dimensional factor group*, Bull. Amer. Math. Soc. 54 (1948), 964–968.

[182] S. Kaplan, *Extensions of the Pontryagin duality, I: Infinite products*, Duke Math. J. 15 (1948), 649–658.

[183] I. Kaplansky, *Dual modules over a valuation ring. I*, Proc. Amer. Math. Soc. 4 (1953), 213–219.

[184] A. Kechris, *Classical descriptive set theory*, Graduate Texts in Mathematics, 156. Springer-Verlag, New York, 1995.

[185] A. Klyachko, A. Yu. Olshanskii, D. Osin, *On topologizable and non-topologizable groups*, Topology Appl. 160 (2013), no. 16, 2104–2120.

[186] T. Kōmura, Y. Kōmura, *Über die Einbettung der nuklearen Räume in* $(s)^A$ (German), Math. Ann. 162 (1965/66), 284–288.

[187] H.-J. Kowalsky, *Zur topologischen Kennzeichnung von Körpern* (German), Math. Nachr. 9 (1953), 261–268.

[188] C. Kraaikamp, P. Liardet, *Good approximations and continued fractions*, Proc. Amer. Math. Soc. 112 (1991), no. 2, 303–309.

[189] L. Kuipers, H. Niederreiter, *Uniform distribution of sequences*, Pure and Applied Mathematics. Wiley-Interscience, New York–London–Sydney, 1974.

[190] K. Kunen, *Bohr topology and partition theorems for vector spaces*, Topology Appl. 90 (1998), no. 1–3, 97–107.

[191] K. Kunen, W. Rudin, *Lacunarity and the Bohr topology*, Math. Proc. Cambridge Philos. Soc. 126 (1999), no. 1, 117–137.

[192] V. Kuz'minov, *Alexandrov's hypothesis in the theory of topological groups* (Russian), Dokl. Akad. Nauk SSSR 125 (1959), 727–729.

[193] S.-H. Kye, *Pontryagin duality in real linear topological spaces*, Chinese J. Math. 12 (1984), no. 2, 129–136.

[194] G. Larcher, *A convergence problem connected with continued fractions*, Proc. Amer. Math. Soc. 103 (1988), no. 3, 718–722.

[195] S. Lefschetz, *Algebraic topology*, American Mathematical Society Colloquium Publications, 27. American Mathematical Society, New York, 1942.

[196] H. Leptin, *Zur Dualitätstheorie projektiver Limites abelscher Gruppen* (German), Abh. Math. Sem. Univ. Hamburg 19 (1955), 264–268.

[197] L. H. Loomis, *An introduction to abstract harmonic analysis*, D. Van Nostrand Company, Inc., New York, 1953.

[198] P. Loth, *Classifications of abelian groups and Pontrjagin duality*, Algebra, Logic and Applications, 10. Gordon and Breach Science Publishers, Amsterdam, 1998.

[199] G. Lukács, *Precompact abelian groups and topological annihilators*, J. Pure Appl. Algebra 208 (2007), no. 3, 1159–1168.

[200] G. Lukács, *Compact-like topological groups*, Research and Exposition in Mathematics, 31. Heldermann Verlag, Lemgo, 2009.

[201] G. Lukács, *Report on the open problems session*, Topology Appl. 159 (2012), no. 9, 2476–2482.

[202] S. Mac Lane, *Homology*, Die Grundlehren der mathematischen Wissenschaften, 114. Academic Press, Inc., Publishers/Springer-Verlag, New York/Berlin–Göttingen–Heidelberg, 1963.
[203] S. Mac Lane, *Categories for the working mathematician*, second edition. Graduate Texts in Mathematics, 5. Springer-Verlag, New York, 1998.
[204] I. G. MacDonald, *Duality over complete local rings*, Topology 1 (1962), 213–235.
[205] G. W. Mackey, *Harmonic analysis as the exploitation of symmetry – a historical survey*, Bull. Amer. Math. Soc. (N. S.) 3 (1980), no. 1, part 1, 543–698.
[206] A. Mader, *Basic concepts of functorial topologies*, Abelian group theory (Oberwolfach, 1981), 251–271, Lecture Notes in Math. 874, Springer, Berlin–New York, 1981.
[207] A. Mader, R. Mines, *Functorial topologies with totally ordered neighborhood bases*, Arch. Math. (Basel) 34 (1980), no. 3, 272–275.
[208] A. Mann, *How groups grow*, London Mathematical Society Lecture Notes Series, 395. Cambridge University Press, 2012.
[209] A. A. Markov, *On unconditionally closed sets*, C. R. (Doklady) Acad. Sci. URSS (N. S.) 44 (1944), 180–181.
[210] A. A. Markov, *On the existence of periodic connected topological groups* (Russian), Bull. Acad. Sci. URSS. Sér. Math. [Izvestia Akad. Nauk SSSR] 8, (1944), 225–232.
[211] A. A. Markov, *On free topological groups*, Topology and Topological Algebra, Transl. Ser. I, 8. Amer. Math. Soc., 1962, 195–272; Russian original in: Izv. Akad. Nauk SSSR 9 (1945), 3–64.
[212] J. Milnor, *Problem 5603*, Amer. Math. Monthly 75 (1968), 685–686.
[213] C. C. Moore, *Groups with finite dimensional irreducible representations*, Trans. Amer. Math. Soc. 166 (1972), 401–410.
[214] S. A. Morris, *Pontryagin duality and the structure of locally compact abelian groups*, London Mathematical Society Lecture Note Series, 29. Cambridge University Press, Cambridge–New York–Melbourne, 1977.
[215] M. Mutylin, *Completely simple commutative topological rings* (Russian), Mat. Zametki 5 (1969), 161–171.
[216] L. Nachbin, *On strictly minimal topological division rings*, Bull. Amer. Math. Soc. 55 (1949), 1128–1136.
[217] K. Nagami, *Dimension-theoretical structure of locally compact groups*, J. Math. Soc. Japan 14 (1962), 379–396.
[218] I. Namioka, *Right topological groups, distal flows, and a fixed-point theorem*, Math. Systems Theory 6 (1972), 193–209.
[219] G. Negro, *Polish LCA groups are strongly characterizable*, Topology Appl. 162 (2014), 66–75.
[220] P. Nickolas, *Reflexivity of topological groups*, Proc. Amer. Math. Soc. 151 (1970), 551–561.
[221] J. W. Nienhuys, *A solenoidal and monothetic minimally almost periodic group*, Fund. Math. 73 (1971/72), no. 2, 167–169.
[222] A. Yu. Ol′shanskii, *A remark on a countable nontopologized group* (Russian), Vestnik Moskov. Univ. Ser. I, Mat. Mekh. (1980), no. 3, 103.
[223] A. Orsatti, *Introduzione ai gruppi astratti e topologici* (Italian), Quaderni dell'Unione Matematica Italiana. Pitagora Ed., Bologna, 1979.
[224] A. Orsatti, *Una introduzione alla teoria dei moduli* (Italian), Aracne Ed., Roma, 1995.
[225] B. A. Pasynkov, *On coincidence of different definitions of dimensions for factor spaces of locally compact groups* (Russian), Uspehi Mat. Nauk 17 (1962), no. 5 (107), 129–135.
[226] V. Pestov, *Free abelian topological groups and the Pontryagin-van Kampen duality*, Bull. Austral. Math. Soc. 52 (1995), 297–311.
[227] V. Pestov, *Forty-plus annotated questions about large topological groups*, In: Open problems in Topology II, 439–450, E. Pearl, editor. Elsevier, 2007.

[228] L. S. Pontryagin, *Selected works. Vol. 2. Topological groups*, Edited and with a preface by R. V. Gamkrelidze. Translated from the Russian and with a preface by Arlen Brown. With additional material translated by P. S. V. Naidu. Third edition. Classics of Soviet Mathematics. Gordon & Breach Science Publishers, 1986.

[229] H.-E. Porst, W. Tholen, *Concrete dualities*, Category theory at work (Bremen, 1990), 111–136, Res. Exp. Math., 18. Heldermann, Berlin, 1991.

[230] I. Prodanov, *Elementary example of a group without characters*, In: Mathematics and Mathematical Education, 79–81. Proc. Ninth Spring Conf. (April, 1980) of the Bulgarian Math. Soc. at Slunchev brjag (Bulgarian Acad. Sci. Sophia, Bulgaria).

[231] Iv. Prodanov, *Precompact minimal group topologies and p-adic numbers*, Annuaire Univ. Sofia Fac. Math. 66 (1971/72), 249–266.

[232] Iv. Prodanov, *Minimal topologies on countable abelian groups*, Annuaire Univ. Sofia Fac. Math. Méc. 70 (1975/76), 107–118 (1981).

[233] Iv. Prodanov, *Some minimal group topologies are precompact*, Math. Ann. 227 (1977), no. 2, 117–125.

[234] Iv. Prodanov, *Maximal and minimal topologies on abelian groups*, Colloq. Math. Soc. János Bolyai, 23, 985–997. Topology, Budapest (Hungary), 1978.

[235] Iv. Prodanov, *An axiomatic characterization of the Pontryagin duality*, preprint, 1979.

[236] Iv. Prodanov, *An abstract approach to the algebraic notion of spectrum* (Russian). Topology, Moscow, 1979. Trudy Mat. Inst. Steklov. 154 (1983), 200–208.

[237] Iv. Prodanov, *Elementary proof of the Peter–Weyl theorem* (Russian), C. R. Acad. Bulgare Sci. 34 (1981), no. 3, 315–318.

[238] Iv. Prodanov, *Compact representations of continuous algebraic structures* (Russian), In: 1967 General Topology and its Relations to Modern Analysis and Algebra, II (Proc. Second Prague Topological Sympos., 1966), 290–294. Academia, Prague.

[239] Iv. Prodanov, L. N. Stojanov, *Every minimal abelian group is precompact*, C. R. Acad. Bulgare Sci. 37 (1984), no. 1, 23–26.

[240] I. Protasov, E. Zelenyuk, *Topologies on groups determined by sequences*, Mathematical Studies Monograph Series, 4. VNTL Publishers, L'viv, 1999.

[241] I. V. Protasov, E. G. Zelenyuk, *Topologies on abelian groups* (Russian), Izv. Akad. Nauk SSSR Ser. Mat. 54 (1990), no. 5, 1090–1107; translation in Math. USSR-Izv. 37 (1991), no. 2, 445–460.

[242] S. Raczkowski, *Totally bounded topological group topologies on the integers*, Topology Appl. 121 (2002), no. 1–2, 63–74. Proceedings of the First Joint Japan-Mexico Meeting in Topology (Morelia, 1999).

[243] D. Remus, *On the structure of the lattice of group topologies*, Results Math. 6 (1983), no. 2, 251–252.

[244] D. Remus, *Topological groups without non-trivial characters*, In: General Topology and Its Relations to Modern Analysis and Algebra VI, Z. Frolík, editor, pages 477–484. Heldermann Verlag, Berlin, Proc, Sixth 1986 Prague Topological Symposium.

[245] D. Remus, L. Stojanov, *Complete minimal and totally minimal groups*, Topology Appl. 42 (1991), no. 1, 57–69.

[246] D. Remus, F. J. Trigos-Arrieta, *The Bohr topology of Moore groups*, Topology Appl. 97 (1999), no. 1–2, 85–98. Special issue in honor of W. W. Comfort (Curaçao, 1996).

[247] N. Rickert, *Arcs in locally compact groups*, Math. Ann. 172 (1967), 222–228.

[248] L. C. Robertson, *A note on the structure of Moore groups*, Bull. Amer. Math. Soc. 75 (1969), 594–599.

[249] D. J. S. Robinson, *A course in the theory of groups*, Second edition. Graduate Texts in Mathematics, 80. Springer-Verlag, New York, 1996.

[250] D. W. Roeder, *Functorial characterizations of Pontryagin duality*, Trans. Amer. Math. Soc. 154 (1971), 151–175.

[251] K. Ross, K. Stromberg, *Baire sets and Baire measures*, Ark. Mat. 6 (1965), 151–160.

[252] W. Rudin, *Real and complex analysis*, McGraw-Hill International Editions, New York, 1987.

[253] U. Schwanengel, *An example of a $\mathcal{Q}$-minimal precompact topological group containing a non-minimal closed normal subgroup*, Manuscripta Math. 27 (1979), no. 3, 323–327.

[254] D. Shakhmatov, *A problem of coincidence of dimensions in topological groups*, Topology Appl. 33 (1989), no. 1, 105–113.

[255] D. Shakhmatov, *Imbeddings into topological groups preserving dimensions*, Topology Appl. 36 (1990), no. 2, 181–204.

[256] D. Shakhmatov, *A direct proof that every infinite compact group G contains* $\{0,1\}^{w(G)}$, Papers on general topology and applications (Flushing, NY, 1992), 276–283, Ann. New York Acad. Sci., 728. New York Acad. Sci., 1994.

[257] D. Shakhmatov, V. Yañez, *SSGP topologies on abelian groups of positive finite divisible rank*, Fund. Math. 244 (2019), no. 2, 125–145.

[258] D. Shakhmatov, V. Yañez, *SSGP topologies on free groups of infinite rank*, Topology Appl. 259 (2019), 384–410.

[259] S. Shelah, *Infinite abelian groups – Whitehead's problem and some constructions*, Israel J. Math. 18 (1974), 243–256.

[260] S. Shelah, *On a problem of Kurosh, Jónsson groups, and applications*, In: Word problems, II (Conf. on Decision Problems in Algebra, Oxford, 1976), 373–394, Stud. Logic Foundations Math., 95. North-Holland, Amsterdam–New York, 1980.

[261] O. V. Sipacheva, *Consistent solution of Markov's problem about algebraic sets*, arXiv:math.GR/0605558v2 (2007).

[262] M. F. Smith, *The Pontrjagin duality theorem in linear spaces*, Ann. of Math. (2) 56 (1952), 248–253.

[263] S. Solecki, *Polish group topologies*, In: Sets and Proofs, London Math. Society Lecture Notes Series 258. Cambridge University Press (1999), 339–364.

[264] R. M. Stephenson Jr., *Minimal topological groups*, Math. Ann. 192 (1971), 193–195.

[265] L. Stojanov, *Weak periodicity and minimality of topological groups*, Annuaire Univ. Sofia Fac. Math. Méc. 73 (1979), 155–167.

[266] L. Stoyanov, *Precompactness of minimal metrizable periodic abelian groups*, Annuaire Univ. Sofia Fat. Math. Méc. 71, part II (1976/77), 45–52.

[267] L. Stoyanov, *Some minimal abelian groups are precompact*, Pliska Stud. Math. Bulgar. 6 (1983), 75–89.

[268] L. Stoyanov, *Dualities over compact commutative rings*, Rend. Accad. Naz. Sci. XL Mem. Mat. (5) 7 (1983), 155–176.

[269] L. Stoyanov, *Total minimality of the unitary groups*, Math. Z. 187 (1984), 273–283.

[270] M. Stroppel, *Locally compact groups*, EMS Textbooks in Mathematics. European Mathematical Society (EMS), Zürich, 2006.

[271] A. D. Taĭmanov, *Topologizable groups. II* (Russian), Sibirsk. Mat. Zh. 19 (1978), no. 5, 1201–1203, 1216.

[272] M. Tkachenko, *Factorization theorems for topological groups and their applications*, Topol. Appl. 38 (1991), 21–37.

[273] M. Tkachenko, *Subgroups, quotient groups and products of $\mathbb{R}$-factorizable groups*, Topology Proc. 16 (1991), 201–231.

[274] M. Tkachenko, *Paratopological and semitopological groups versus topological groups*, In: Recent Progress in General Topology III, 825–882. Atlantis Press, Paris, 2014.

[275] F. J. Trigos-Arrieta, *Every uncountable abelian group admits a nonnormal group topology*, Proc. Amer. Math. Soc. 122 (1994), no. 3, 907–909.

[276] D. van Dantzig, *Studien over topologische algebra* (German), Dissertation, Amsterdam, 1931.
[277] J. van Mill, *Open problems in van Douwen's papers*, In: Eric K. van Douwen Collected papers Vol. I, II, North Holland Publishing Co., Amsterdam, 1994.
[278] N. Varopoulos, *Studies in Harmonic analysis*, Proc. Cambridge Philos. Soc. 60 (1964), 465–516.
[279] W. A. Veech, *The equicontinuous structure relation for minimal Abelian transformation groups*, Amer. J. Math. 90 (1968), 723–732.
[280] R. Venkataraman, *A characterization of Pontryagin duality*, Math. Z. 149 (1976), no. 2, 109–119.
[281] N. Vilenkin, *A contribution to the theory of direct decompositions of topological groups*, C. R. (Doklady) Acad. Sci. URSS (N. S.) 47 (1945), 611–613.
[282] N. Ya. Vilenkin, *The theory of characters of topological Abelian groups with boundedness given*, Izvestiya Akad. Nauk SSSR. Ser. Mat. 15 (1951), 439–462.
[283] A. Vitolo, U. Zannier, *On small sets in a group*, Atti Accad. Naz. Lincei Rend. Cl. Sci. Fis. Mat. Nat. (8) 82 (1988), no. 3, 413–418 (1990).
[284] J. von Neumann, *Almost periodic functions in a group. I*, Trans. Amer. Math. Soc. 36 (1934), no. 3, 445–492.
[285] J. von Neumann, E. Wigner, *Minimally almost periodic groups*, Ann. of Math. (2) 41 (1940), 746–750.
[286] Yuan Wang (editor), *The Goldbach conjecture*, Second edition, Series in Pure Mathematics, 4. World Scientific Publishing Co., Inc., River Edge, NJ, 2002.
[287] S. Warner, *Topological rings*, North-Holland Mathematics Studies, 178. North-Holland Publishing Co., Amsterdam, 1993.
[288] A. Weil, *L'intégration dans les groupes topologiques et ses applications* (French), Actual. Sci. Ind., 869. Hermann et Cie, Paris, 1940.
[289] A. Weil, *Basic number theory*, Third edition, Die Grundlehren der Mathematischen Wissenschaften, 144. Springer-Verlag, New York–Berlin, 1974.
[290] R. Winkler, *Ergodic group rotations, Hartman sets and Kronecker sequences*, Monatsh. Math. 135 (2002), no. 4, 333–343.
[291] Y. Zelenyuk, *On topologizing groups*, J. Group Theory 10 (2007), no. 2, 235–244.
[292] D. Zelinsky, *Linearly compact modules and rings*, Amer. J. Math. 75 (1953), 79–90.

Index of symbols

Index

De Gruyter Studies in Mathematics

Volume 82
Friedrich Wagemann
Crossed Modules, 2021
ISBN 978-3-11-075076-8, e-ISBN 978-3-11-075095-9,
e-ISBN (ePUB) 978-3-11-075099-7

Volume 81
Ľubica Holá, Dušan Holý, Warren Moors
USCO and Quasicontinuous Mappings, 2021
ISBN 978-3-11-075015-7, e-ISBN 978-3-11-075018-8,
e-ISBN (ePUB) 978-3-11-075022-5

Volume 76
Steven Lord, Fedor Sukochev, Dmitriy Zanin
Singular Traces. Volume 1: Theory, 2021
ISBN 978-3-11-037779-8, e-ISBN 978-3-11-037805-4,
e-ISBN (ePUB) 978-3-11-039231-9

Volume 80
Xiao-Xiong Gan
Formal Analysis. An Introduction, 2021
ISBN 978-3-11-059785-1, e-ISBN 978-3-11-059945-9,
e-ISBN (ePUB) 978-3-11-059860-5

Volume 79
Rajendra V. Gurjar, Kayo Masuda, Masayoshi Miyanishi
Affine Space Fibrations, 2021
ISBN 978-3-11-057736-5, e-ISBN 978-3-11-057756-3,
e-ISBN (ePUB) 978-3-11-057742-6

Volume 78
Kai Liu, Ilpo Laine, Lianzhong Yang
Complex Delay-Differential Equations, 2021
ISBN 978-3-11-056016-9, e-ISBN 978-3-11-056056-5,
e-ISBN (ePUB) 978-3-11-056040-4

Volume 77
Laura Costa, Rosa Maria Miró-Roig, Joan Pons-Llopis
Ulrich Bundles. From Commutative Algebra to Algebraic Geometry, 2021
ISBN 978-3-11-064540-8, e-ISBN 978-3-11-064768-6,
e-ISBN (ePUB) 978-3-11-064580-4

www.degruyter.com

www.ingramcontent.com/pod-product-compliance
Ingram Content Group UK Ltd.
Pitfield, Milton Keynes, MK11 3LW, UK
UKHW050954290926
15470UKWH00006B/1812

9 783110 653496